AF362054

Developing "Smartness" in Emerging Environments and Applications with Focus on the Internet of Things (IoT)

Developing "Smartness" in Emerging Environments and Applications with Focus on the Internet of Things (IoT)

Editors

Rashid Mehmood
Juan M. Corchado
Tan Yigitcanlar

MDPI • Basel • Beijing • Wuhan • Barcelona • Belgrade • Manchester • Tokyo • Cluj • Tianjin

Editors

Rashid Mehmood
King Abdul Aziz University
Saudi Arabia

Juan M. Corchado
University of Salamanca
Spain

Tan Yigitcanlar
Queensland University of
Technology
Australia

Editorial Office
MDPI
St. Alban-Anlage 66
4052 Basel, Switzerland

This is a reprint of articles from the Special Issue published online in the open access journal *Sensors* (ISSN 1424-8220) (available at: https://www.mdpi.com/journal/sensors/special_issues/smartness_IoT).

For citation purposes, cite each article independently as indicated on the article page online and as indicated below:

LastName, A.A.; LastName, B.B.; LastName, C.C. Article Title. *Journal Name* **Year**, *Volume Number*, Page Range.

ISBN 978-3-0365-6183-7 (Hbk)
ISBN 978-3-0365-6184-4 (PDF)

Cover image courtesy of Denys Nevozhai, Dennis Kummer and Dhruvansh Soni.

Contents

About the Editors

Rashid Mehmood

Rashid Mehmood is a Research Professor of Big Data Systems and the Director of Research, Training, and Consultancy at the High-Performance Computing Center, King Abdulaziz University, Saudi Arabia. His qualifications and work experience stem from universities in the UK, including Cambridge and Oxford Universities. Rashid has 27 years of experience in computational modelling, simulations, and design using computational intelligence, big data, high-performance computing, and distributed systems. Broadly, his research aim is the development of multi-disciplinary science and technology to enable a better quality of life and smart economies with a focus on real-time intelligence and autonomous system management. He has published over 200 research papers including six edited books. He has organized and chaired international conferences and workshops including EuropeComm 2009, Nets4Cars 2010–2013, SCE 2017–2019, SCITA 2017, HPC Saudi 2018–2020, and DCAI 2022. He has led and contributed to academia–industry collaborative projects funded by the EPSRC, the EU, UK regional funds, and the Technology Strategy Board UK with a value of over GBP 50 million. He is a founding member of the Future Cities and Community Resilience (FCCR) Network, a member of ACM and OSA, a Senior Member of IEEE, and the former Vice Chairman of IET Wales SW Network.

Juan M. Corchado

Juan M. Corchado (born 15 May 1971, Salamanca, Spain) is a Professor at the University of Salamanca. He was the Vice Rector of Research from 2013 to 2017 and is the Director of the Science Park of the University of Salamanca. Elected as the Dean of the Faculty of Science twice, he holds a PhD in Computer Science from the University of Salamanca and a PhD in Artificial Intelligence from the University of the West of Scotland. He currently leads the renowned BISITE (Bioinformatics, Intelligent Systems, and Educational Technology) Research Group, which was created in 2000.

He is the Director of the IoT Digital Innovation Hub, president of the AIR Institute, and project leader of the DIGIS3 consortium. J. M. Corchado has also been a Visiting Professor at the Osaka Institute of Technology since January 2015, a Visiting Professor at the Universiti Malaysia Kelantan, and a Member of the Advisory Group on Online Terrorist Propaganda of the European Counter Terrorism Centre (EUROPOL). He has recently been included in the board of trustees of the AstraZeneca Foundation, along with other health professionals and researchers recognized for bringing scientific knowledge closer to society.

J. M. Corchado has been the president of IEEE Systems and the Man and Cybernetics Society, the academic coordinator of the University Institute for Research in Art and Animation Technology at the University of Salamanca, and a researcher at the Universities of Paisley (UK), Vigo (Spain), and the Plymouth Marine Laboratory (UK).

Tan Yigitcanlar

Tan Yigitcanlar is an internationally recognized Australian researcher who has had a profound impact on the field of urban studies and planning. He is a Professor of Urban Studies and Planning at the School of Architecture and Built Environment, Queensland University of Technology, Brisbane, Australia. Along with this post, he holds the following positions: Honorary Professor at the School of Technology, Federal University of Santa Catarina, Florianopolis, Brazil; Director of the Australia–Brazil Smart City Research and Practice Network; Lead of the QUT Smart City Research Group; and Co-Director of the QUT City 4.0 Lab. His research is clustered around the following three interdisciplinary themes: smart technologies, communities, cities, and urbanism; sustainable and resilient cities, communities, and urban ecosystems; and the knowledge-based development of cities and innovation districts. His research findings have been disseminated in over 275 articles published in high-impact journals and 22 key reference books published by esteemed international publishing houses. His research outputs have been widely cited and have influenced urban policy, practice, and research internationally. As of December 2022, his research was cited over 17,000 times, resulting in an h-index of 76 (via Google Scholar). He was recognized as an 'Australian Research Superstar' in the Social Sciences Category in The Australian's 2020 Research Special Report. According to the 2022 Science-Wide Author Databases of Standardized Citation Indicators, amongst the urban- and regional-planning scholars, he is ranked #1 in Australia and #8 worldwide.

Preface to "Developing "Smartness" in Emerging Environments and Applications with Focus on the Internet of Things (IoT)"

This book is being released at a critical juncture in the history of smart cities and societies. Globally, numerous new smart districts and cities are being created, while many existing cities, such as New York, Singapore, Helsinki, and Seoul, are evolving or transforming into smart cities. Among these cities is the USD one trillion-dollar smart city NEOM, which is being created in Saudi Arabia. This Special Issue will help influence and contribute to these smart advancements in Saudi Arabia, Spain, Australia, and throughout the world.

The guest editors thank the authors of the articles included in this book for their excellent work and the reviewers for their efforts in reviewing the articles and recommending changes that improved the articles. The guest editors also thank the staff of the MDPI journal *Sensors* including Fancy Chai and Dr. Constanze Schelhorn for their assistance in publishing this book.

Rashid Mehmood, Juan M. Corchado, and Tan Yigitcanlar
Editors

Editorial

Developing Smartness in Emerging Environments and Applications with a Focus on the Internet of Things

Rashid Mehmood [1,*], Juan M. Corchado [2,3,4] and Tan Yigitcanlar [5]

1   High Performance Computing Center, King Abdulaziz University, Jeddah 21589, Saudi Arabia
2   Bisite Research Group, University of Salamanca, 37007 Salamanca, Spain
3   Air Institute, IoT Digital Innovation Hub, 37188 Salamanca, Spain
4   Department of Electronics, Information and Communication, Faculty of Engineering, Osaka Institute of Technology, Osaka 535-8585, Japan
5   City 4.0 Lab, School of Architecture and Built Environment, Queensland University of Technology, 2 George Street, Brisbane, QLD 4000, Australia
*   Correspondence: rmehmood@kau.edu.sa

Citation: Mehmood, R.; Corchado, J.M.; Yigitcanlar, T. Developing Smartness in Emerging Environments and Applications with a Focus on the Internet of Things. *Sensors* **2022**, *22*, 8939. https://doi.org/10.3390/s22228939

Received: 26 October 2022
Accepted: 11 November 2022
Published: 18 November 2022

Publisher's Note: MDPI stays neutral with regard to jurisdictional claims in published maps and institutional affiliations.

1. Introduction

The smartness that underpins smart cities and societies is defined by our ability to engage with our environments, analyze them, and make decisions, all in a timely manner [1]. We are witnessing a rapid evolution, or rather, a transformation of our societies. Novel solutions are being developed and adopted in work and life, benefiting from the growing ability to monitor and analyze our environments in near real time. A range of devices and technologies are being used for monitoring purposes, including the Internet of Things (IoT), the Global Positioning System (GPS), sensors, cameras, Radio Frequency Identification (RFID) devices, smartphones, smartwatches, other smart wearables, and social media platforms. These devices produce diverse data that are analyzed using Artificial Intelligence (AI) and other computational intelligence methods and are used for decision-making purposes. While significant advances have been made in developing smart applications and technologies, a systematic effort to define and develop "smartness" is missing. An investigation into the theoretical and technological foundations of this "smartness" can help systemize and mass-produce technologies for autonomous production and for the operation of smart environments.

This Special Issue's focus is on the IoT, and it is concerned with bringing "smartness" to the IoT and other system layers using technologies such as Cloud, Fog, and Edge Computing; High-Performance Computing (HPC); Big Data; Blockchain; and/or AI. In addition to this Editorial piece, a collection of 13 articles is featured in this Special Issue, covering a range of topics, including mobility, healthcare, image analysis, permeable pavements, solid waste management, sensor node and gateway architectures, air quality monitoring, thermal anomalies and smart helmets in industrial environments, smart airports, smart districts, smart travel choices, sensor cities, artificially intelligent cities, and platform urbanism. Figure 1 provides a word cloud which represents the themes explored by this Special Issue.

Smartness is a multidisciplinary topic and can be defined from different perspectives. We see through the articles included in this Special Issue that smartness can be seen to have four dimensions (however, this is not the only way to look at it). These dimensions are: (i) Sensors, IoT, and Data Generation; (ii) Data and Information Processing; (iii) Actuation; and (iv) Digital Systems and Infrastructure. To elaborate, we can see smartness in the way sensing is embedded in a system, the way data and information are processed, how a system interacts internally and with its environment, and whether a system is ubiquitous or limited by space (cloud-based or edge-enabled). What follows is a brief review of the articles included in this Special Issue, which highlights their contributions with respect to these four dimensions. They are grouped according to their application areas: mobility and transportation, healthcare, industrial environments, and other urban infrastructures.

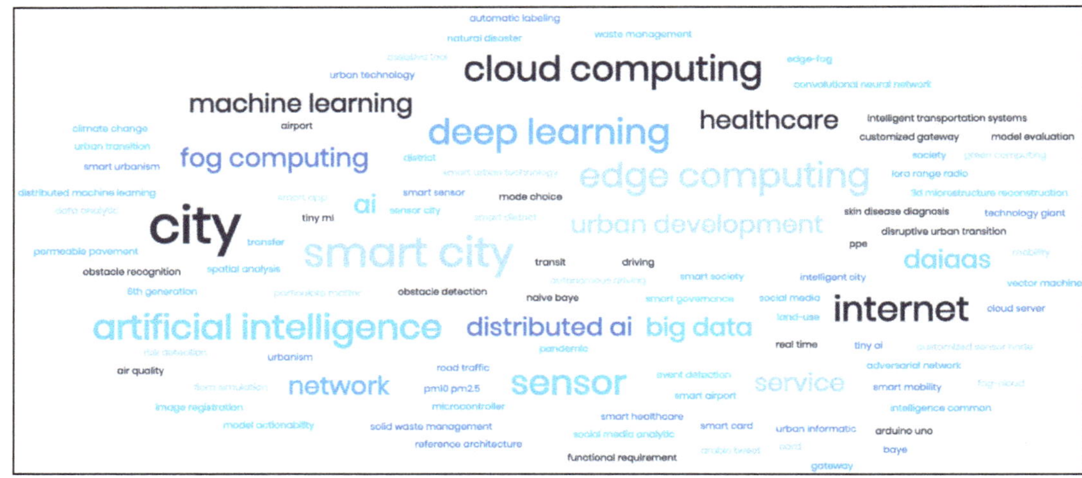

Figure 1. A Word Cloud of the Research Topics in this Special Issue.

2. Mobility and Transportation

Transportation is the backbone of modern economies, albeit at massive human, environmental, and economic costs [2,3]. Lana et al. [4] assess that advances in data science have permeated into every field of transportation science and engineering, resulting in data-driven transportation developments. The authors describe how data from various intelligent transport system (ITS) sources can be used to learn and adapt data-driven models for the efficient operation of ITS assets, systems, and processes and how data-based models can become fully actionable. Furthermore, the authors define the characteristics, engineering requirements, and challenges inherent to the three compounding stages, namely, data fusion, adaptive learning, and model evaluation, based on the described data modelling pipeline for ITS. This work's theoretical framework contributes to the first three dimensions of smartness, namely, sensors, data processing, and actuation.

Chia et al. [5] use smart card data to investigate the relationship between the spatial distribution of relative transfer locations and the attractiveness of the transit service. Transfers are an important part of transit trips because they enable people to reach more destinations; however, they are also the main factor that deters people from using public transportation. The authors' findings imply that smart transit users may value travel direction in addition to travel time, which influences their mode choice. Depending on their relative location, travelers may prefer even adjacent transfer locations. The authors' findings will help improve our understanding of transit user behavior and the impact of transfer smartness, as well as smart transportation planning and the design of new transit routes and services to improve transfer performance. This work contributes to the second dimension (data processing) of smartness.

Alomari et al. [6] introduce their tool, Iktishaf+, which combines Big Data, Distributed Machine Learning, social media analytics, and an automatic data labeling method to detect road traffic-related events. It uses a range of technologies, including Apache Spark, Parquet, and MongoDB. The tool can detect and validate several real-world events in Saudi Arabia, including a fire in Jeddah, rain in Makkah, and an accident in Riyadh, without prior knowledge. The findings demonstrate the effectiveness of Twitter media in detecting important events when no prior knowledge of them is available. This work contributed to the first two dimensions of smartness, sensing, and data processing.

Motivated by the fact that over a billion people are disabled worldwide, with 253 million of them being visually impaired or blind, Busaeed et al. [7] propose an approach for collecting data and predicting objects for environment perception, mobility, and navigation using a LiDAR with a servo motor and an ultrasonic sensor. The authors use this approach with a pair of smart glasses called LidSonic V2.0 to help the visually impaired identify obstacles. The LidSonic system consists of a smart glasses-integrated Arduino Uno edge computing device and a smartphone app that transmits data via Bluetooth. Arduino collects data, controls the smart glasses' sensors, detects obstacles with simple data processing, and provides buzzer feedback to visually impaired users. This work is related to indoor and outdoor mobility, and hence we group this with mobility and transportation research. This work contributes to the first two dimensions of smartness, namely, sensing, and data processing, and moreover, it touches upon the fourth dimension as it discusses its potential to be extended to Cloud and Edge Computing.

3. Industry 4.0

The detection of anomalies in harsh industrial environments is a challenging task. To address this, Ghazal et al. [8] propose an Edge–Fog–Cloud architecture, based on mobile IoT edge nodes carried by autonomous robots, for detecting thermal anomalies in aluminum plants. The authors use companion drones as fog nodes and a cloud back-end to analyze thermal anomalies. Moreover, the authors propose a self-driving, deep learning architecture and a thermal anomaly detection and visualization algorithm. Their results show that the proposed robot surveyors are less expensive, have a shorter response time, and detect anomalies more accurately than human surveyors or fixed IoT nodes monitoring the same industrial area. This work contributes to the first, second, and fourth dimensions of smartness, i.e., sensing, data processing, and computing infrastructure.

Campero-Jurado et al. [9] discuss the role of information and communication technologies (ICTs) in advancing occupational health and safety and increasing worker security. Personal Protective Equipment (PPE) based on ICTs reduces the risk of workplace accidents due to the equipment's ability to make decisions based on environmental factors. Paradigms such as the Industrial Internet of Things (IIoT) and Artificial Intelligence (AI) enable the generation of PPE models and the development of devices with more advanced capabilities such as monitoring, sensing the environment, and risk detection, among others. These models continuously monitor the working environment and notify employees and supervisors of any anomalies or threats. With this context, they propose a smart helmet prototype that monitors the working environment and performs a near real-time risk assessment. The sensor data are sent to an AI-powered platform for analysis. A comparative study of supervised learning models is carried out as part of this research. Furthermore, the use of a Deep Convolutional Neural Network (ConvNet/CNN) is proposed for the detection of potential occupational risks. This work contributes to the first and second dimensions of smartness, namely, sensing and data processing.

4. Healthcare

Janbi et al. [10] propose, implement, and evaluate Imtidad, a reference architecture that provides Distributed Artificial Intelligence (AI) as a Service (DAIaaS) over the cloud, fog, and edge. For this purpose, the authors develop a service catalog case study containing 22 AI skin disease diagnosis services. The services are divided into four service classes based on software platforms and are run on a variety of hardware platforms as well as four network types. Two standard Deep Neural Networks (DNNs) and two Tiny AI deep models were trained and tested using real-life dermatoscopic images to create the AI models for diagnosis. Several benchmarks were used to evaluate the services, including model service value, response time, energy consumption, and network transfer time. The services are intended to enable a variety of use cases, such as home patient diagnosis or sending diagnosis requests to traveling medical professionals via a fog device or cloud. This work contributed to the first, second, and fourth dimensions of smartness, sensing, data processing, and computing infrastructure.

5. Urban Infrastructure

Yigitcanlar et al. [11] investigate whether artificially intelligent cities can protect humanity from natural disasters, pandemics, and other disasters. By charting the evolution of AI and the potential impacts of systematic AI adoption on cities and societies, the authors generate insights and identify prospective research questions. This viewpoint provides theoretical contributions to all four dimensions of smartness and puts forward directions for future research, as well as listing large number of critical research questions that need to be answered. In another work, Yigitcanlar et al. [12] derive insights from sensor city best practices by scrutinizing some well-known projects implemented by Huawei, Cisco, Google, Ericsson, Microsoft, and Alibaba. The authors highlight that platform urbanism is becoming a critical tool to support smart urban governance in an era of digitalization of urban services and processes. On the basis of the lessons learned from the best practices of leading innovation and technology companies, the study advocates the need for further research on the conceptualization and practice of the sensor city notion.

Feri et al. [13] propose a three-dimensional microstructure reconstruction framework based on a 3D improved Wasserstein Generative Adversarial Network (3D-IWGAN) with an enhanced gradient penalty. It is a computational system based on images for analyzing clogging in the permeable pavement. The physical property values extracted from their model are comparable to those obtained from real pavement samples. The authors are motivated by the fact that there is an increasing demand for research into how to improve the functionality of permeable pavement. Their proposed system starts with a two-dimensional image as input and extracts latent features from it. It generates a 3D microstructure image using their model's generative adversarial network. This work contributes to the data processing dimension of smartness.

Akram et al. [14] propose an architecture for designing and developing a customized sensor node and gateway based on LoRa (Long-Range radio) technology for solid waste management, specifically, to achieve the filling level of the bins while using the least amount of energy. The authors also include distinct evaluation metrics for the sensor node's long-range data rate, time on-air (ToA), LoRa sensitivity, link budget, and battery life. LoRa is a popular communication protocol that provides long-range transmission and low data rates while consuming little power. Only a small amount of data needs to be sent to the remote server in the context of solid waste management, hence the use of LoRa. This work contributes to the sensing dimension of smartness.

Jo et al. [15] examined the changes in particulate matter concentrations due to land use over time, as well as the spatial characteristics of the distribution of particulate matter concentrations in Daejeon, Korea, as measured by Private Air Quality Monitoring Smart Sensors (PAQMSSs). According to the primary land use around the 650 m sensor radius, land uses were classified into residential, commercial, industrial, and green groups. The results show that particulate matter concentrations in Daejeon decreased in the order of industrial, housing, commercial, and green groups overall; however, the concentrations of the commercial group were higher than those of the residential group between 21:00 and 23:00, reflecting the commercial group's vital night-time lifestyle in Korea. The study contributes to the data processing dimension of smartness. Janbi et al. [16] propose a framework for Distributed AI as a Service (DAIaaS) provisioning for the Internet of Everything and 6G environments with the aim to help standardize the mass production of technologies for smarter environments. To investigate the design choices and performance bottlenecks of DAIaaS, multiple DAIaaS provisioning configurations for distributed training and inference are proposed, including three case studies (a smart airport, a smart district, and distributed AI provisioning) with eight scenarios, nine applications and AI delivery models, and 50 distinct sensor and software modules. This work contributes to all four smartness dimensions.

6. Concluding Remarks

The smartness that underpins smart cities and societies is defined by our ability to engage with our environments, analyze them, and make decisions, all in a timely manner. The IoT has been the focus of this Special Issue, and its concern has been to bring "smartness" to the IoT and other system layers using emerging technologies. The articles included in this issue cover a wide range of applications, including image analysis, permeable pavements, solid waste management, air quality monitoring, thermal anomalies and smart helmets in industrial environments, smart airports, smart districts, and smart travel choices.

The field of smartness is exciting, and while a lot has been achieved, the future possibilities with technologies such as Deep Learning, Edge Computing, Virtual Reality, and more are endless. There are many works that are complementary to the research presented in this Special Issue, such as deep journalism [17], smartization [18], smart families and homes [19], data-driven smart governance [20,21], responsible innovation [22], and green AI [23]. This is an exciting time for disruptive technologies, and this Special Issue is expected to clarify the concept of smartness, helping more researchers to contribute to this area and lead to the development of truly smart environments.

Funding: The authors would like to thank for and acknowledge the technical and financial support from the Deanship of Scientific Research (DSR) at the King Abdulaziz University (KAU), Jeddah, Saudi Arabia, under Grant No. RG-11-611-38.

Acknowledgments: The work carried out in this paper is supported by the HPC Center at the King Abdulaziz University.

Conflicts of Interest: The authors declare no conflict of interest.

References

1. Alotaibi, S.; Mehmood, R.; Katib, I.; Rana, O.; Albeshri, A. Sehaa: A Big Data Analytics Tool for Healthcare Symptoms and Diseases Detection Using Twitter, Apache Spark, and Machine Learning. *Appl. Sci.* **2020**, *10*, 1398. [CrossRef]
2. AlOmari, E.; Katib, I.; Mehmood, R. Iktishaf: A Big Data Road-Traffic Event Detection Tool Using Twitter and Spark Machine Learning. *Mob. Networks Appl.* **2020**. [CrossRef]
3. Aqib, M.; Mehmood, R.; Alzahrani, A.; Katib, I.; Albeshri, A.; Altowaijri, S.M. Smarter Traffic Prediction Using Big Data, In-Memory Computing, Deep Learning and GPUs. *Sensors* **2019**, *19*, 2206. [CrossRef] [PubMed]
4. Laña, I.; Sanchez-Medina, J.; Vlahogianni, E.; Del Ser, J. From Data to Actions in Intelligent Transportation Systems: A Prescription of Functional Requirements for Model Actionability. *Sensors* **2021**, *21*, 1121. [CrossRef]
5. Chia, J.; Lee, J.; Han, H. How Does the Location of Transfer Affect Travellers and Their Choice of Travel Mode?—A Smart Spatial Analysis Approach. *Sensors* **2020**, *20*, 4418. [CrossRef]
6. Alomari, E.; Katib, I.; Albeshri, A.; Yigitcanlar, T.; Mehmood, R. Iktishaf+: A Big Data Tool with Automatic Labeling for Road Traffic Social Sensing and Event Detection Using Distributed Machine Learning. *Sensors* **2021**, *21*, 2993. [CrossRef]
7. Busaeed, S.; Katib, I.; Albeshri, A.; Corchado, J.M.; Yigitcanlar, T.; Mehmood, R. LidSonic V2.0: A LiDAR and Deep-Learning-Based Green Assistive Edge Device to Enhance Mobility for the Visually Impaired. *Sensors* **2022**, *22*, 7435. [CrossRef]
8. Ghazal, M.; Basmaji, T.; Yaghi, M.; Alkhedher, M.; Mahmoud, M.; El-Baz, A. Cloud-Based Monitoring of Thermal Anomalies in Industrial Environments Using AI and the Internet of Robotic Things. *Sensors* **2020**, *20*, 6348. [CrossRef]
9. Campero-Jurado, I.; Márquez-Sánchez, S.; Quintanar-Gómez, J.; Rodríguez, S.; Corchado, J.M. Smart Helmet 5.0 for Industrial Internet of Things Using Artificial Intelligence. *Sensors* **2020**, *20*, 6241. [CrossRef]
10. Janbi, N.; Mehmood, R.; Katib, I.; Albeshri, A.; Corchado, J.M.; Yigitcanlar, T. Imtidad: A Reference Architecture and a Case Study on Developing Distributed AI Services for Skin Disease Diagnosis over Cloud, Fog and Edge. *Sensors* **2022**, *22*, 1854. [CrossRef]
11. Yigitcanlar, T.; Butler, L.; Windle, E.; DeSouza, K.C.; Mehmood, R.; Corchado, J.M. Can Building "Artificially Intelligent Cities" Safeguard Humanity from Natural Disasters, Pandemics, and Other Catastrophes? An Urban Scholar's Perspective. *Sensors* **2020**, *20*, 2988. [CrossRef] [PubMed]
12. D'Amico, G.; L'Abbate, P.; Liao, W.; Yigitcanlar, T.; Ioppolo, G. Understanding Sensor Cities: Insights from Technology Giant Company Driven Smart Urbanism Practices. *Sensors* **2020**, *20*, 4391. [CrossRef] [PubMed]
13. Feri, L.; Ahn, J.; Lutfillohonov, S.; Kwon, J. A Three-Dimensional Microstructure Reconstruction Framework for Permeable Pavement Analysis Based on 3D-IWGAN with Enhanced Gradient Penalty. *Sensors* **2021**, *21*, 3603. [CrossRef] [PubMed]
14. Akram, S.V.; Singh, R.; AlZain, M.A.; Gehlot, A.; Rashid, M.; Faragallah, O.S.; El-Shafai, W.; Prashar, D. Performance Analysis of IoT and Long-Range Radio-Based Sensor Node and Gateway Architecture for Solid Waste Management. *Sensors* **2021**, *21*, 2774. [CrossRef]

15. Jo, S.S.; Lee, S.; Leem, Y. Temporal Changes in Air Quality According to Land-Use Using Real Time Big Data from Smart Sensors in Korea. *Sensors* **2020**, *20*, 6374. [CrossRef]
16. Janbi, N.; Katib, I.; Albeshri, A.; Mehmood, R. Distributed Artificial Intelligence-as-a-Service (DAIaaS) for Smarter IoE and 6G Environments. *Sensors* **2020**, *20*, 5796. [CrossRef]
17. Ahmad, I.; Alqurashi, F.; Abozinadah, E.; Mehmood, R. Deep Journalism and DeepJournal V1.0: A Data-Driven Deep Learning Approach to Discover Parameters for Transportation. *Sustainability* **2022**, *14*, 5711. [CrossRef]
18. Alahmari, N.; Alswedani, S.; Alzahrani, A.; Katib, I.; Albeshri, A.; Mehmood, R. Musawah: A Data-Driven AI Approach and Tool to Co-Create Healthcare Services with a Case Study on Cancer Disease in Saudi Arabia. *Sustainability* **2022**, *14*, 3313. [CrossRef]
19. Alqahtani, E.; Janbi, N.; Sharaf, S.; Mehmood, R. Smart Homes and Families to Enable Sustainable Societies: A Data-Driven Approach for Multi-Perspective Parameter Discovery Using BERT Modelling. *Sustainability* **2022**, *14*, 13534. [CrossRef]
20. Alswedani, S.; Mehmood, R.; Katib, I. Sustainable Participatory Governance: Data-Driven Discovery of Parameters for Planning Online and In-Class Education in Saudi Arabia During COVID-19. *Front. Sustain. Cities* **2022**, *4*, 97. [CrossRef]
21. Alomari, E.; Katib, I.; Albeshri, A.; Mehmood, R. Covid-19: Detecting government pandemic measures and public concerns from twitter arabic data using distributed machine learning. *Int. J. Environ. Res. Public Health* **2021**, *18*, 1–36. [CrossRef] [PubMed]
22. Yigitcanlar, T.; Corchado, J.M.; Mehmood, R.; Li, R.Y.M.; Mossberger, K.; Desouza, K. Responsible Urban Innovation with Local Government Artificial Intelligence (AI): A Conceptual Framework and Research Agenda. *J. Open Innov. Technol. Mark. Complex.* **2021**, *7*, 71. [CrossRef]
23. Yigitcanlar, T.; Mehmood, R.; Corchado, J.M. Green Artificial Intelligence: Towards an Efficient, Sustainable and Equitable Technology for Smart Cities and Futures. *Sustainability* **2021**, *13*, 8952. [CrossRef]

sensors

MDPI

Article

From Data to Actions in Intelligent Transportation Systems: A Prescription of Functional Requirements for Model Actionability

Ibai Laña [1,*], Javier J. Sanchez-Medina [2], Eleni I. Vlahogianni [3] and Javier Del Ser [1,4]

[1] TECNALIA, Basque Research & Technology Alliance (BRTA), P. Tecnologico Bizkaia, Ed. 700, 48160 Derio, Spain; javier.delser@tecnalia.com or javier.delser@ehu.eus
[2] CICEI, Department of Computer Science, University of Las Palmas de Gran Canaria, 35001 Las Palmas, Spain; javier.sanchez@uplgc.es
[3] Department of Transportation Planning and Engineering, National Technical University of Athens, 15780 Zografou, Greece; elenivl@mail.ntua.gr
[4] Department of Communications Engineering, University of the Basque Country UPV/EHU, Alameda Urquijo S/N, 48013 Bilbao, Spain
[*] Correspondence: ibai.lana@tecnalia.com

Abstract: Advances in Data Science permeate every field of Transportation Science and Engineering, resulting in developments in the transportation sector that are data-driven. Nowadays, Intelligent Transportation Systems (ITS) could be arguably approached as a "story" intensively producing and consuming large amounts of data. A diversity of sensing devices densely spread over the infrastructure, vehicles or the travelers' personal devices act as sources of data flows that are eventually fed into software running on automatic devices, actuators or control systems producing, in turn, complex information flows among users, traffic managers, data analysts, traffic modeling scientists, etc. These information flows provide enormous opportunities to improve model development and decision-making. This work aims to describe how data, coming from diverse ITS sources, can be used to learn and adapt data-driven models for efficiently operating ITS assets, systems and processes; in other words, for data-based models to fully become *actionable*. Grounded in this described data modeling pipeline for ITS, we define the characteristics, engineering requisites and challenges intrinsic to its three compounding stages, namely, data fusion, adaptive learning and model evaluation. We deliberately generalize model learning to be adaptive, since, in the core of our paper is the firm conviction that most learners will have to adapt to the ever-changing phenomenon scenario underlying the majority of ITS applications. Finally, we provide a prospect of current research lines within Data Science that can bring notable advances to data-based ITS modeling, which will eventually bridge the gap towards the practicality and actionability of such models.

Keywords: Intelligent Transportation Systems; functional requirements; machine learning; model actionability; model evaluation

Citation: Laña, I.; Sanchez-Medina, J.J.; Vlahogianni, E.I.; Del Ser, J. From Data to Actions in Intelligent Transportation Systems: A Prescription of Functional Requirements for Model Actionability. *Sensors* **2021**, *21*, 1121. https://doi.org/10.3390/s21041121

Academic Editor: Rashid Mehmood
Received: 7 January 2021
Accepted: 2 February 2021
Published: 5 February 2021

1. Introduction

In the last years Intelligent Transportation Systems (ITS) have experienced an unparalleled expansion for many reasons. The availability of cost-effective sensor networks, pervasive computation in assorted flavors (distributed/edge/fog computing) and the so-called Internet of Things are all accelerating the evolution of ITS [1]. On top of them, Smart Cities cannot be understood anyhow without Smart Mobility and ITS as technological pillars sustaining their operation [2]. Smartness springs from connectivity and intelligence, which implies that massive flows of information are acquired, processed, modeled and used to enable faster and informed decisions.

For the last couple of decades, ITS have grown enough to cross pollinate with previously distant areas such as Machine Learning and its superset in the Artificial Intelligence taxonomy: Data Science. These days Data Science is placed at the methodological core of works ranging from traffic and safety analysis, modeling and simulation, to transit network optimization, autonomous and connected driving and shared mobility. Since the early 90's most ITS systems exclusively relied on traditional statistics, econometric methods, Kalman filters, Bayesian regression, auto-regressive models for time series and Neural Networks, to mention a few [3,4]. What has changed dramatically over the years is the abundance of available data in ITS application scenarios as a result of new forms of sensing (e.g., crowd sensing) with unprecedented levels of heterogeneity and velocity. Zhang et al. [3] have defined this new form of data-driven ITS as the systems that have vision, multisource, and learning algorithms driven to optimize its performance and augment its privacy-aware people-centric character.

The exploitation of this upsurge of data has been enabled by advances in computational structures for data storage, retrieval and analysis, which have rendered it feasible to train and maintain extremely complex data-based models. These baseline technologies have laid a solid substrate for the proliferation of studies dealing with powerful modeling approaches such as Deep Learning or bio-inspired computation [5], which currently protrude in the literature as the *de facto* modeling choice for a myriad of data-intensive applications.

However, significant consideration must be placed to the systematic and myopic selection of complex data-based solutions over well-established modeling choices. The current research mainstream seems to be misleadingly focusing on performance-biased studies, in a fast-paced race towards incorporating sophisticated data-based models to manifold research area, leaving aside or completely disregarding the operational aspects for the applicability of such models in ITS environment. The scope of this work is to review existing literature on data-driven modeling and ITS, and identify the functional elements and specific requirements of engineering solutions, which are the ultimate enablers for data-based models to lead towards efficient means to operate ITS assets, systems and processes; in other words, for data-based models to fully become *actionable*. Bearing the above rationale in mind, this work underscores the need for formulating the requirements to be met by forthcoming research contributions around data-based modeling in ITS systems. To this end, we focus mainly on system-level on-line operations that hinge on data-based pipelines. However, ITS is a wide research field, encompassing operations held at longer time scales (e.g., long-term and mid-term planning) that may not demand some of the functional requirements discussed throughout our work. Furthermore, our discussions target system-level operations rather than user-level or vehicle-level applications, since in the latter the information flow from and to the system is scarce. Nevertheless, some of the described functional requirements for system-level real-time decisions can be extrapolated to other levels and time scales seamlessly. From this perspective, our ultimate goal is to prescribe – or at least, set forth – the main guidelines for the design of models that rely heavily on the collection, analysis and exploitation of data. To this end, we delve into a series of contributions that are summarized below:

- In the first place, we identify the gap between the data-driven research reported so far, and the practical requirements that ITS experts demand in operation. We capitalize on this gap to define what we herein refer to as *actionable data-based modeling workflow*, which comprises all data processing stages that should be considered by any actionable data-based ITS model. Although diverse data-based modeling workflows can be found in literature with different purposes, most of them count on recognized stages, that are presented in this work from an actionability perspective, i.e., what to take into account from the operational point of view when designing the workflow, how to capture and preprocess data, how to develop a model and how to prescribe its output. These guidelines are proposed and argued within an ITS application context. However, they can be useful for any other discipline in which data-based modeling is performed.
- Next, functional requirements to be satisfied by the aforementioned workflow are described and framed in the context of ITS systems and processes, with examples exposing their relevance and consequences if they are not fulfilled.The contributions

of this section are twofold: on the one hand, we identify and define the holistically actionable ITS model along with its main features; on the other hand, we enumerate requirements for each feature to be considered actionable, as well as a review of the latest literature dealing with these features and requisites.

- Finally, on a prospective note we elaborate on current research areas of Data Science that should progressively enter the ITS arena to bridge the identified gap to actionability. Once the challenges of modeling and ITS requirements have been stated, we review emerging research areas in Artificial Intelligence and Data Science that can contribute to the fulfilment of such requirements. We expect that our reflexive analysis serves as a guiding material for the community to steer efforts towards modeling aspects of more impact for the field than the performance of the model itself.

As a summary, the contributions of this work consist of identifying the main actionability gaps in the data-based modeling workflow, gathering and describing the fundamental requirements for a system to be actionable, and considering both the requirements and the usual data-based processing workflow, proposing solutions through the most recent technologies. These contributions are organized throughout the rest of the paper as follows: Section 2 delves into the *actionable data-based modeling workflow*, i.e., the canonical data processing pipeline that should be considered by a fully actionable ITS system with data-based models in use. Section 3 follows by elaborating on the functional features that an ITS system should comply with so as to be regarded as *actionable*. Once these requirements are listed and argued in detail, Section 4 analyzes research paths under vibrant activity in areas related to Data Science that could bring profitable insights in regards to the actionability of data-based models for the ITS field, such as explainable AI, the inference of causality from data, online learning and adaptation to non-stationary data flows. Finally, Section 5 concludes the paper with summarizing remarks drawn from our prospect.

2. From Data to Actions: An Actionable Data-Based Modeling Workflow

ITS applications with data driven modeling problems underneath range from the characterization of driving behavioral patterns, the inference of typical routes or traffic flow forecasting, among others. Data driven modeling can be considered to include the family of problems where a computational model or system must be characterized or learned from a set of inputs and their expected outputs [6]. In the context of this definition, actionability complements the data-driven model by prescribing the actions (in the form of rules, optimized variable values or any other representation alike) that build upon the output knowledge enabled by the model.

In general, a design workflow for data-based modeling consists of 4 sequential stages: (1) data acquisition (*sensing*), which usually considers different sources; (2) data preprocessing, which aims at building consistent, complete, statistically robust datasets; (3) data modeling, where a model is learned for different purposes; and (4) model exploitation, which includes the definition of actions to be taken with respect to the insights provided by models in real life application scenarios. These 4 stages can be regarded as the core of off-line data-driven modeling; however, when time dimension joins the game, a fifth stage—adaptation—must be considered as an iterative stage of this data pipeline, aimed at maintaining learned models updated and adapted to eventual changes in the data distribution. This adaptation is crucial for real-life scenarios, where changes can happen in all stages, from variations of the input data sources, to interpretation adjustments and other sources of non-stationarity imprinting the so-called *concept drift* in the underlying phenomenon to be modelled [7]. We now delve into these five data processing stages in the context of their implementation in ITS applications, following the diagram in Figure 1.

The stages provided in Figure 1 can be considered as a standard workflow in any data-based work; however, although these steps are easily recognisable, they are not always regarded, and it is common to observe that practitioners put the focus only on a subset of them, disregarding their interactions or omitting some of them. For instance, the prescription stage is not frequently considered, while it is an essential link between the modeling outcome and the final decision/action derived from the modeling result. Besides, each step can have implications for the final actionability of the model, reason for which all of them are analyzed below.

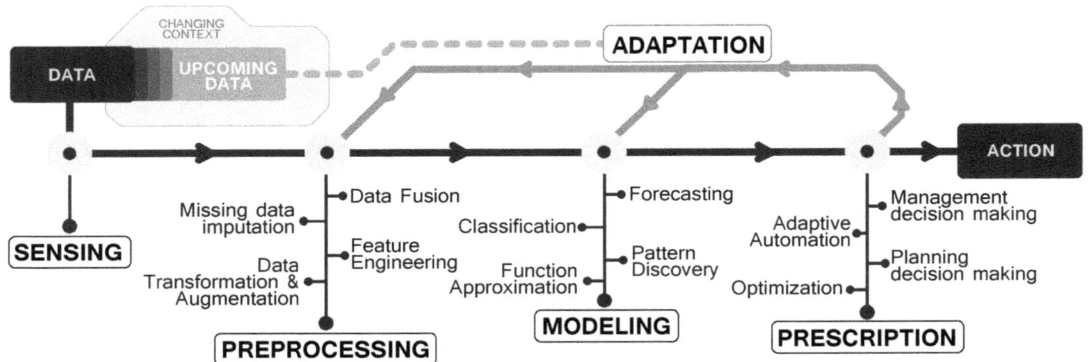

Figure 1. Data-based modeling workflow showing its main processing stages and their principal technology areas.

2.1. Data Acquisition (Sensing)

The path towards concrete data-based actions departs from the capture of available ITS information, which in this specific sector is plentiful and highly diverse. The advent of data science for ITS has come along with the unfolding of copious data sources related to transportation. Indeed, ITS are pouring volumes of sensed data, from the environment perception layer of intelligent and connected vehicles, to human behaviour detection/estimation (drivers, passengers, pedestrians) and the multiple technologies deployed to sense traffic flow and behaviour. Concurrently, many other non-traditional sources that were useful to infer behavioral needs and expectations of people that use transportation, such as social media, have started to become increasingly available and exploited augmenting the more conventional sensing sources towards more efficient mobility solutions. Some of these data sources are currently used in almost any domain of ITS, from operational perspectives such as the estimation of future transportation demands, adaptive signaling or the discovery of mobility patterns, to the provision and of practical solutions, such as the development of autonomous vehicles, although not all sources are suitable for all applications. The model actionability is dependant on this early stage too, reason for which the data selection (when possible) should not be neglected. For instance, a model that consumes speed data will probably require some other measurements (maximum speed of a road segment) to provide in the end something meaningful, while a model that consumes travel-time data will be more straight-forward.

Five main categories can be established to describe the spectrum of ITS data sources:

1. Roadside sensing, which brings together tools and mechanisms that directly capture and convey data measurements from the road, obtaining valuable metrics such as speed, occupation, flow or even which vehicles are traversing a given road segment. These are the most commonly used sensors in ITS, most frequently based on computer vision and radar, as they directly provide traffic information close to the point where it originates. This kind of sensed metrics are useful for traffic flow or speed modeling, allowing practitioners to identify mobility patterns and to model them, so future behavior in sensorized locations can be estimated. Counting vehicles or detecting their speed at a certain point of the road also allows to obtain network wide mobility patterns that can be compared to those provided by a simulation engine. This can help traffic managers and city planners take long-term decisions, such as which road should be extended or how a road cut could affect other segments. However, this information is tethered to the exact points where the sensors are placed, thus the

actionability of a system built upon these data is subject to the geographical area where such sensing devices are deployed and their range.

2. In-vehicle sensing, which includes a broad range of transponder devices that are part of the on-board equipment of certain fleets. Commercial vehicles on land, air and water usually have location devices that record and emit the position and other metrics of the vehicles at all times. This opens up a wide range of ITS applications, such as fleet management [8], route optimization [9], delay analysis and detection [10], airport/port management [11] or, when the vehicles are a part of traffic, a detailed analysis of their behavior along a complete route (not only in certain sensorized locations) [12]. This technology is highly extended in commercial fleets and its multiple analytic applications are nowadays remarkably actionable, due to industry standards requirements. However, machine learning modeling based approaches are starting to emerge, and should consider actionability as a core concern.

3. Cooperative sensing, which denotes the general family of data collection strategies that regards the information provided by different users of the ITS ecosystem as a whole, thus being grouped and jointly processed forward. This inner perspective of traffic and transportation can be obtained through many mechanisms, and, although it is more specific and scarce, it is also more complete than the one obtained from road-side sensing. These data open the door to mobility profiling and anomaly detection, enriching the outlook of a transportation model by means of the fusion of different data-based *views* of an ITS scenario. This includes all forms of mobile sensing data, from call detail record data that can be used to obtain users trajectories [13], to GPS data [14]. These sources are the foundation of abundant research [15,16], but in most cases the data fusion part is obviated. Crowdsourced and Social Media sensing can be analogously considered in this category. These data sources can also contribute to data-based ITS models by means of sentiment analysis and geolocation. The use of crowd-sourced data is well established among technology-based companies (Google, Uber etc), yet not very often available to research community and private and public authorities in the transport operations management. The limited information that becomes available is deprived from the necessary statistical representativeness and truthfulness in order to be easily integrated to legacy management systems.

4. External data sources, which include all data that are not directly related to traffic of demand, but have an impact on it, such as weather, calendar, or planned events, social and economic indicators, demographic characteristics etc. These data are usually easy to obtain, and their incorporation to ITS models augments in general the quality of their produced insights and ultimately, the actionability of the actions yielded therefrom. It is also true that this data source is typically unstructured, which can pose a challenge regarding its automatic integration.

5. Structured/static data, which refers to data sources that provide information of elements that have a direct impact on transportation, such as public transportation lines and timetables, or municipal bike rental services. Due to their inherently structured nature, data provided by these sources are often arranged in a fixed format, making it easier to incorporate to subsequent data-based modeling stages. Any of the previous data and applications can be enriched with these kind of data; a model that is able to represent the mobility of a city would probably enhance its capabilities if it considered these data. For instance, a bus timetable can help understand traffic in the street segments that are traversed by the bus service or where its stops are located. These information sources must be considered for an intelligent transportation system to be actionable, being a particularly essential piece of urban and interurban mobility.

2.2. Data Preprocessing

The variety of the above mentioned sensing sources comes with promises and perils. These data is produced in various forms and formats, various time resolutions, synchronously or asynchronously and different rates of accumulation. To leverage the full spectrum of knowledge these data can bring to the sake of informed decision making, the more the sensing opportunities the larger the needs for powerful preprocessing and skills are before reaching the stage of modeling.

A principled data-driven modeling workflow requires more than just applying off-the-shelf tools. In this regard, preprocessing raw data is undoubtedly an elementary step of the modeling process [17], but still persists nowadays as a step frequently overlooked by researchers in the ITS field [18].

To begin with, when a model is to be built on real ITS data, an important fact to be taken into account is the proneness of real environments to present missing or corrupted data due to many uncertain events that can affect the whole collection, transformation, transmission and storage process [19]. This issue needs to be assessed, controlled and suitably tackled before proceeding further with next stages of the processing pipeline. Otherwise, missing and/or corrupted instances within the captured data may distort severely the outcome of data-based models, hindering their practical utility [20]. A wide extent of missing data imputation strategies can be found in literature [21,22], as well as methods to identify, correct or discriminate abnormal data inputs [23]. However, they are often loosely coupled to the rest of the modeling pipeline [24]. An actionable data preprocessing should focus not only on improving the quality of the captured data in terms of completeness and regularity, but also on providing valuable insights about the underlying phenomena yielding missing, corrupted and/or outlying data, along with their implications on modeling [25].

Next, the cleansed dataset can be engineered further to lie an enriched data substrate for the subsequent modeling [26,27]. A number of operations can be applied to improve the way in which data are further processed along the chain. For instance, data transformation methods can be applied for different purposes related to the representation and distribution of data (e.g., dimensionality reduction, standardization, normalization, discretization or binarization). Although these transformations are not mandatory in all cases, a deep knowledge of what input data represent and how they contribute to modeling is a key aspect to be considered in this preprocessing stage.

Furthermore, data enrichment can be held from two different perspectives that can be adopted depending on the characteristics of the dataset at this point. As such, feature selection/engineering refers to the implementation of methods to either discard irrelevant features for the modeling problem at hand, or to produce more valuable data descriptors by combining the original ones through different operations. Likewise, instance selection/generation implies a transformation of the original data in terms of the examples. Removing instances can be a straight solution for corrupted data and/or outliers, whereas the addition of synthetic instances can help train and validate models for which scarce real data instances are available. Besides, these approaches are among the most predominant techniques to cope with class imbalance [28], a very frequent problem in predictive modeling with real data. Whether each of these operations is required or not depends entirely on the input data, their quality, abundance and the relations among them. This entails a deep understanding of both data and domain, which is not always a common ground among the ITS field practitioners [29].

Finally, data fusion embodies one of the most promising research fields for data-driven ITS [3,30], yet remains marginally studied with respect to other modeling stages despite its potential to boost the actionability of the overall data-based model. Indeed, an ITS model can hardly be actionable if it does not exploit interactions among different data sources. Upon their availability, ITS models can be enriched by fusing diverse data sources. A recent review on different operational aspects of data-driven ITS developments states that these models rarely count on more than one source of data [16]. This fact clearly unveils a niche of research when taking into account the increasing availability of data provided by the growing amount of sensors, devices and other data capturing mechanisms that are deployed in transportation networks, in all sorts of vehicles, or even in personal devices held by the infrastructure users. Despite the relative scarcity of contributions dealing with this part of the data-based modeling workflow, the combination of multiple sources of information has been proven to enrich the model output along different axis, from accuracy to interpretability [31–34].

2.3. Modeling

Once data are obtained, fused, preprocessed and curated, the modeling phase implies the extraction of knowledge by constructing a model to characterize the distribution of such data or their evolution in time. The distillation of such knowledge can be performed for different purposes: to represent unsupervised data in a more valuable manner (as in e.g., clustering or manifold learning), for instance, to insight patterns relating the input data to a set of supervised outputs (correspondingly, classification/regression) aiming to automatically label unseen data observations, to predict future values based on the previous values (time series forecasting), or to inspect the output produced by a model when processing input data (simulation). To do so, in data-based modeling machine learning algorithms are often put to use, which allow automating the modeling process itself.

The above purposes can serve as a discrimination criterion for different algorithmic approaches for data-based modeling. However, when the goal is to model data interactions within complex systems such as transportation networks, it is often the case that the modeling choice resorts to ensembles of different learner types. For instance, when applying regression models for road traffic forecasting, a first clustering stage is often advisable to unveil typicalities in the historical traffic profiles and to feed them as priors for the subsequent predictive modeling [35–37]. However, when it comes to model action-ability, a key feature of this stage is the *generalization* of the developed model to unseen data. This characteristic implies making a model useful beyond the data on which it is trained, which implies that the model design efforts should not only be put on making the model achieve a marginally superior performance, but also to be useful in other spatial or temporal circumstances. Achieving good generalization properties for the developed can be tackled by diverse means, which often depend on the modeling purpose at hand (e.g., cross-validation, regularization, or the use of ensembles in predictive modeling). Essentially, the design goal is to find the trade-off between performance (through representing much of the intrinsic variance of data) and generalization (staying away from an overfitted model to a particular training set). This aspect becomes utterly relevant when data modeling is done on time-varying data produced by dynamic phenomena. ITS are, in point of fact, complex scenarios subject to strong sources of non-stationarity, thereby calling for an utmost focus on this aspect.

The complexity met in traffic and transportation operations is usually treated with heterogeneous modeling approaches that aim to complement each other to improve accuracy [38–40]. This can be done either by comparing different models and selecting the most appropriate one every time, or by combining different models to produce the final outcome. Additionally, in some fields of ITS, such as traffic modeling, physical (namely, theory- or simulation-based) models have been available for decades. Their integration into data-based modeling workflows, considering the knowledge they can provide, can become crucial for a manifold of purposes, e.g., to enforce traffic theory awareness in models learned from ITS data. Indeed, the hybridization of physical and data-based models has a yet to be developed potential that has only been timidly explored in some recent works [41–43].

Interestingly, complex data driven modeling solutions to transportation phenomena have been numerous and resourceful ranging from modular structures, to model combinations, surrogate modeling [44] and so on. Regardless of the approach, literature emphasizes on the critical issue of model hyperparameter optimization using for example nature inspired algorithms, namely Evolutionary Computation or Swarm Intelligence [39,45]. Assuming that there is a feasible and acceptable solution to the problem of selecting the proposed parameters for a data drive model, when dealing with complex modeling structure this task should be conducted automatically by optimizing the hyperparameter space usually based on the models' predictive error. It is to note that, the greater the number of models involved the more difficult the optimization task becomes. Moreover, relying on nature inspired stochastic approaches, full determinism in the solution and convergence stability can not be formally guaranteed [5].

2.4. Prescription

Once the modeling phase itself has been completed, the resulting model faces its application to a real ITS environment. It is at this stage when actions deriving from the data insights are defined/learned/decided, and when the actionability of the model can be best assessed. Yet, this stage is frequently overlooked in most ITS research, where most works conclude at presenting the good performance of a model; it is uncommon to find evaluations of a given model in terms of its final application in a certain environment. Are the actions that can be taken as a result of the outcome of a data-based model aimed at a strategic, tactical or operational decision making? Is the output of the data-based model able to support decisions made by transportation networks managers? Can the output be consumed directly without any need for further modeling, or exploited by means of a secondary modeling process aimed at optimizing the decision making process? This latter case can be exemplified, for instance, by the formulation of the decision making process as an optimization problem, in which actions are represented by the variables compounding a solution to the problem, and the output of the previous data-based modeling phase can be used to quantitatively estimate the quality or fitness of the solution. One of the most prominent examples of this prescription mode deals with routing problems, since they often use simulation tools or predictive models to assess the travel time, pollutant emissions or any other optimization objective characterizing the fitness of the tested routes [46,47]. Other examples of prescription based on data emerge in tactical and strategic planning, such as the modification of public transportation lines [48], the establishment of special lanes (e.g., taxi, bike) [49], the improvement of road features [50], the adaptive control of traffic signaling [51], the identification of optimal delivery (or pickup) routes for different kinds of transportation services [52], the incident detection and management [53], learning for automated driving [54], or the design of sustainable urban mobility plans based on the current and future demand or the drivers' behavior [55,56].

In any of the above presented ITS cases, a data-based model should be equipped with a certain set of features that guarantee its actionability. For instance, if a traffic manager is not able to interpret a model or understand its outcome in terms of confidence, it can be hardly applied for practical decision making. When the model is used for adaptive control purposes (as in automated traffic light scheduling), the adaptability of the model to contextual changes is a key requirement for prescribed actions to be matched to the current traffic status [57]. Interestingly, some control techniques with a long history in the field (e.g., Stochastic Model Predictive Control, SMPC, [58]) serve as a good example of the triple-play between application requirements, decision making and data-based models. When dealing with the design of control methods in ITS, SMPC has been proven to perform efficiently in highly-complex systems subject to the probabilistic occurrence of uncertainties [59]. Specifically, SMPC leverages at its core data-based prediction modeling and low-complexity chance-constrained optimization to deal with control problems that impose that the method to be used must operate in real time. In this case, and in most actionable data-based workflows where decision making is formulated as an optimization problem, we note a clear entanglement between application requirements (e.g., real-time processing), decision making (low-complexity, dynamic optimization techniques) and data-based models (predictive modeling for system dynamics forecasting).

2.5. Adaptation

Finally, the proposed actionable data processing workflow considers model adaptation as a processing layer that can be applied over different modeling stages along the pipeline. When models are based on data, they are subject to many kinds of uncertainties and non-stationarities that can affect all stages of the process. Streaming data initially used to build the model can experience long-term drifts (for instance, an increase of the average number of vehicles), sudden changes (a newly available road), or unexpected events (for example, a public transportation strike) [60–62]. A closed lane, a new tram line, the opening of a tunnel or simply the opening of a new commercial center, may change completely the way in which network users behave, and thus, affect the data-based models that are intended to reflect such a mobility. Therefore, data-based modeling cannot be conceived as a static design process. This critical adaptation should be considered in all parts of the workflow, and constantly updated with new data:

1. In the preprocessing stage, adaptation could be understood from many perspectives: the incorporation of new sources of data, the partial or total failure of data capturing sensors, which lead to an increased need for data fusion, imputation, engineering or augmentation.
2. In the modeling stage, adaptations could range from model retraining, adaptation to new data or alternative model switching, to the change of the learning algorithm due to a change in the requested system requirements (for instance, in terms of processing latency any other performance indicator).
3. In the prescription stage, adaptation is intended to dynamically support decisions accounting for changes in data that propagate to the output of preceding modeling stages. Data-based modeling can deal with such changes and adapt their output accordingly, yet they are effective to a point. For instance, online learning strategies devised to overcome from concept drift in data streams can speed up the learning process after the drift occurs (by e.g., diversity induction in ensembles or active learning mechanisms). Unfortunately, even when model adaptation is considered the performance of the adapted model degrades at different levels after the drift. Extending adaptation to the prescription stage provides an additional capacity of the overall workflow to adapt to changes, leveraging techniques from prescriptive analysis such as dynamic or stochastic optimization.

Adaptations within the above stages can be observed from two perspectives: automatic adaptations that the system is prepared to do when certain circumstances occur, or adaptations that are derived from changes that are introduced by the user. Thus, the adaptation layer is strongly linked to actionability: an ITS model will be more actionable if adaptations, either needed or imposed, are accessible to its final users. For instance, a system could be required to introduce a new set of data, and its impact on all the stages should be controlled by the transportation network manager, or if a drift is detected, the system should consider if it is relevant to inform the user.

3. Functional Requirements for Model Actionability

Any data-based modeling process should embrace actionability as its most desirable feature for the engineered model to yield insights of practical value, so that field stakeholders can harness them in their decision making processes. This is certainly the case of ITS, in which managers, transportation users and policy makers rely on models and research results to make better and more informed decisions. Thus, once the main stages of data-driven modeling have been outlined, this section places the spotlight on the main functional features that should be mandatory to produce fully-actionable ITS data-based models. These functional requirements, which are shown in Figure 2, should not be understood as a compulsory list of features, but rather as an enumeration of possibilities to make a model actionable. Not all ITS scenarios requiring actionable data-based models should impose all these requirements, nor can actionability be thought to be a Boolean property. Different loosely defined degrees of actionability may hold depending on the practicality of decisions stemming from the model.

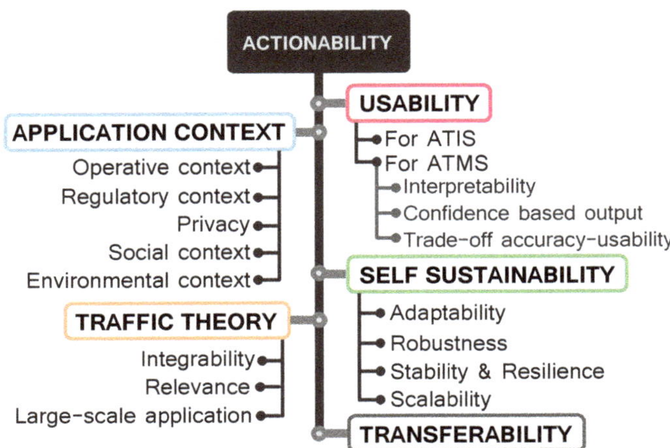

Figure 2. Functional requirements for actionable data-based models in Intelligent Transportation Systems (ITS). ATIS: Advanced Traveler Information Systems; ATMS: Advanced Transportation Management Systems.

3.1. Usability

The way in which humans interact with information systems has been thoroughly studied in last decades and formalized under the general *usability* term [63]. Although usability is a feature that can be associated to any system in which there is some kind of interaction with the user, most of its definitions to date gravitate around the design of software systems [64–66], which is not necessarily the case of ITS research. Usable designs imply defining a clear purpose for a system, and helping users making use of it to reach their objectives [67]. Within ITS, there are domains where this definitions apply directly [68], such as vehicle user interfaces [69–71], the development of navigation systems [72], road signalization [73], or even the way in which public transportation systems information is shown to users [74,75].

The aforementioned domains of application, and mostly any system lying at the core of Advanced Traveler Information Systems (ATIS), have an explicit interaction component. On the other hand, models developed for Advanced Transportation Management Systems (ATMS) are less related to user interaction (beyond the interface design of decision making tools), hence this canonical definition of usability seems to be less applicable. However, the general concept of usability can also accommodate the notion of *utility* as the quality of a system of being useful for its purpose, or the concept of *effectiveness*, in regards to how effective is the information provided by them [76]. Since ITS are systems developed as tools designed to help the different stakeholders that take part in transportation activities, the actionability of data-based models used for this regard depends stringently in this general idea of usability [77]. Models' usability is a feature largely disregarded in literature. A clear example of this situation is traffic forecasting, a preeminent subfield of ATMS, in which the link between the high end deep learning models with the requirements by the road operators in forecasts to support the decision making is very weak [78].

Usability may relate to the person that is going to operate the model, and to the type and complexity of the model, which relate to specific skills. Achieving usable ITS models does not entail the same efforts for all ITS subdomains. Thus, while for research contributions related to ATIS there is a clear interest in this matter [79], for ATMS developments some extra considerations need to be made. Usability in ITS has, therefore, a facet oriented towards user interface, where interfaces reflect at least one of the outputs of an ITS data-based model, and another facet towards creating models that are more aware of the way their outputs are going to be consumed afterwards by the decision maker.

3.1.1. User Interface

For the first of these facets, Spyridakis et al. [77] propose general software usability measuring tools and scales such as System Usability Scale (SUS) [65], ethnographic field studies, or even questionnaires. These basic techniques are also proposed in [80] in order to evaluate navigation systems interfaces. There are also many other evaluation measures that are more specific to the field, such as [81], or those defined by public authorities [82]. Some of the main techniques to appraise ITS interface usability are:

- Usability techniques: if the output of the developed model is consumed through the use of an interface, common techniques like asking directly the users about their experience can be adopted [80]. Among them SUS surveys are the standard to provide interpretable metrics that can be used for the evaluation of passenger information systems [83] or any other kind of automated traveler information system [79].
- Quality of the provided information: in [76], another perspective is proposed, based on estimating the quality of the information provided by the model. Characteristics such as the means to access the information, the reliability of the information provider, or the awareness of the information availability can be measured for assessing the model's usability.
- Transportation-aware strategies: an alternative way to measure usability is to take into account the transportation context and how the use of the model impacts the system. As many of these systems are used during the course of transportation, the environment must be considered in order to provide an adequate and pertinent output [82]. This particular aspect is regarded below in Section 3.4.
- Public transportation guidelines: when ITS developments are intended for the public domain, inclusion of disadvantaged collectives in the usability evaluation is a must [81]. The extent in which these concerns are addressed by the ITS solution should not be disregarded.

3.1.2. Consumption of the Model's Output

For this second usability facet, there are no scales or measurements in literature that provide an objective (or even subjective) usability assessments, but we propose some angles that should be considered when designing this kind of models:

- *Confidence-based outputs:* data-driven models are often subject to stochasticity as a result of their learning procedure or the uncertainty/randomness of their input data (as specially occurs in crowdsourced and Social Media data). This randomness imprints a certain degree of uncertainty in their outputs, which can be estimated values, predicted categories, solutions to an optimization problem or any other alike. Such outcomes are often assessed in terms of their similarity to a ground truth in order to quantitatively assess the performance of the data-based model. Thus, a practitioner aiming to make decisions based on the model's output is informed with a nominal performance score (which has been computed over test data), and the predicted output for a given input. However, when one of such data-based models is intended to work in a real environment, there is no ground truth to evaluate the quality of the result they are providing towards making a decision.For instance, a predictive model could score high on average as per the errors made during the testing phase. However, predictions produced by the model could be less reliable during peak hours than during the night, being less trustworthy in the first case as per the variability of the data from which it was learned, and/or the model's learning algorithm itself. For this reason, the estimation of the confidence of outputs from a data-based model must be analyzed for the sake of its usability. For example, a public transportation model that provides outlooks of future demand could be more usable if, besides the estimation itself, some kind of confidence metric was provided. Elaborating on this aspect is not very frequent in academic research, mainly due to the fact that confidence is not always that easy to obtain and the estimation procedure is, in most cases, model-specific, requiring a previous statistical analysis of input data to properly understand their variability and characteristics. Unfortunately, such a confidence analysis is usually left out of the scope of research contributions, which rather focus on finding the best scoring model for a particular problem. Exceptions to this scarcity of related works are [84], in which the uncertainty inherent to artificial neural networks is analyzed in a real ITS context; [85], in which a committee of different models provides intervals of confidence to predictions;or the more recent contribution in [86], which departs from previous findings in [87,88] to estimate the uncertainty of traffic demand. This uncertainty estimation is then used as an input to assess the confidence of traffic demand predictions. These few references exemplify good practices that should be universally considered in contributions to appear.

- *Interpretability:* a stream of work has been lately concentrated around the noted need for *explaining* how complex models process input data and produce decisions/actions therefrom. Under the so-called XAI (eXplainable Artificial Intelligence) term, a torrent of techniques have been reported lately to explain the rationale behind traditional black-box models, mainly built for prediction purposes [89,90]. Nowadays, Deep Learning is arguably the family of data-driven models mostly targeted by XAI-related studies [91,92].

  The interest of transport researchers to interpretable data-driven models is not new; intuitively, any decision in transportation and traffic operations should be based on a solid understanding of the mechanism by which different factors interact and influence transportation phenomena [93]. In the transportation context explainability is closely related to integrability, when it comes to traffic managers, as ensuring that data-based models can be understood by non-AI expert can make them appropriately trust and favor the inclusion of data-based models in their decisional processes. When framed within ITS systems and processes, the need for explainable data-based models can help decision makers understand how information is processed along the data modeling pipeline, including the quantification of insightful managerial aspects such as the relationship and sensibility of a predicted output with respect to their inputs.

- *Trade-off between accuracy and usability:* when ITS data-based models aim at superior performances, they often work in ideal scenarios where the real context of application is disregarded; should that context apply in practice, the claimed suitability of the developed model for its particular purpose could be compromised. For instance, the goodness of an ITS model devised to detect users' typical trajectories can be measured with regard to the exactitude of the detected trajectories. If the pursuit of a superb performance relies on a constant stream of data (hence, eventually depleting the user's phone battery), it could be a pointless achievement when put to practice. This particular example has been already considered by plenty of researchers [94,95]. However, there is a long way to go in this aspect, as most ITS research developments consider only ideal circumstances without regarding the implications that an accurate design could have on its final usability.

3.2. Self-Sustainability

In general, self-sustainability of a model refers to its ability to survive—hence, to continue to be useful—in a dynamic environment. ITS models and developments are usually intended to operate during long periods of time. However, it is widely accepted that traffic and transportation phenomena are strongly dynamic in nature, meaning that these phenomena exhibit long term trends, evolve in space and time, but also, at the occurrence of an unexpected event, they are susceptible to abrupt changes and exhibit long term memory effects. For instance, a trip information system based on traffic forecasts on a certain part of the network trained with historical data coming from recurrent traffic conditions may not be easily transferable to other road networks or not efficient in case of a severe disruption in traffic operation (accident). What is more, if the specific system does not undergo constant training with new data over time, eventually it will fail to correctly operate even for the network location it was originally designed to operate due to contextually induced non-stationarities. Thus, an intelligent transportation system developed based on data-based approaches should at least follows a set of minimum self-sustainability requirements during the design workflow.

To better understand the importance of self-sustainability as a significant aspect of model's actionability, one should bring to mind the case of cooperative ITS systems (e.g., advanced vehicle control systems) and the automated driving. To this end, a self-sustainable data-based model should bridge the gap between the development of a model prototype and its deployment in a real, potentially non-stationary environment.

When an ITS system or model is deployed to operate in changing conditions, self-sustainability involves dealing with the effects of such changes in the learned knowledge. To this end, different strategies and design approaches could be required depending on the nature of the change and its effects on the model. We next delve into several attributes that can be desirable to deploy data-driven systems or models in changing environments, rendering them actionable:

1. *Adaptable*: Data-driven models for ITS applications created in controlled conditions, with static, self-contained datasets, can provide great performance metrics, but could also fail if data evolve along time [96]. Adaptation is the reaction of a system, model or process to new circumstances intended to reduce its performance deterioration in comparison to the one expected before the change in the environment happened. If data change over time, their evolution is not detected by the model and it does not adapt to it whatsoever, then the developed model will eventually provide an obsolete output. When these contextual variations occur over data streams and models are learned on-line therefrom (for e.g., on-line clustering or classification), such variations can imprint changes in the statistical distribution of input and/or output data, making it necessary to update such models to reflect this change in their learned knowledge. This phenomenon is known as *concept drift* [97], and has been identified as an active research challenge for most of fields connected to machine learning in non-stationary environment [98]. Many of those fields are already studying this topic, from spam detection [99,100] to medicine [101].
   There are two main lines related to concept drift: how to detect drift, and how to adapt to it. Both lines should be scheduled in the research agenda of data-driven ITS, as they have obvious implications when analyzing traffic [102]. Situations like road works can modify completely traffic profiles over a certain area during a period of time, after which the situation goes back to normal. A similar casuistry happens with road design changes (i.e., new lanes, transformation of types of lanes, new accesses, roundabouts, etc), although in those cases there is a new stable traffic profile largely after the change. Even without man-crafted changes, traffic profiles may change for social-economical reasons [103]. Besides, analysis of drift can be used to detect anomalies in the normal operation of roads [104], or to analyze patterns in maritime traffic flow data [105]. However, the adaptability of ITS models to evolving data is scarcely found in literature, and certainly, in many cases concept drift management is the scope of the work, and not a circumstance that is considered to achieve a greater goal [104,106]. There are though some online approaches to typical ITS problems that consider the effects of drift in data [36,107,108], and we consider this kind of initiatives should lead the way for an actionable ITS research.

2. *Robust*: When an ITS system is deployed in a real-life environment, diverse kinds of setbacks can affect its normal operation, from power failures that preclude its functioning to the interruption of the input data flow. Robustness is a self-sustainability trait that prevents a system to stop working when external disruptions occur. Although in most research-level designs this is not a relevant feature, it is essential for actionable, self-sustainable designs. Robustness, defined as the ability to recover from failures, would have, however, different requirements depending on the criticality of the ITS system. Thus, in a traffic flow forecasting system robustness could only imply that the system does not crack when input data fail [109], and it continues to operate; on the

other hand, for critical systems such as air traffic management, robustness would require additional measurements to contain damage [110,111]. All in all, robust data-based workflows should be able to accommodate unseen operational circumstances, such as data distribution shifts or unprecedented levels of information uncertainty, which particularly prevail in crowdsourced and Social Media data [112,113].

3. *Stable and resilient*: Actionable systems require a certain output stability in order to be understandable by their users. This notion is apparently opposed to adaptability, but while the latter is the ability to adapt the output to environment or data changes, stability pursues maintaining the output statistically bounded even when contextual changes occur, through e.g., model adaptation techniques. When adaptation is not perfect and the model violates a given level of statistical stability, stability requires another kind of adaptation, namely *resilience*, to make the model return to its normal operation and thus, minimize the impact of external changes on the quality of its output [114]. This entails, in essence, going one step further in the knowledge of the environment and taking into account those circumstances that can affect the system, and it could be linked to transferable models, which would be addressed below. For instance, a traffic volume characterization model would be adaptable if it considers the changes inherent to traffic volume (an increase over time due to economical factors), and it would be stable if a change in the weather conditions does not deteriorate its performance, or in other words, it has considered this essential circumstance. These kind of considerations are almost nonexistent in literature [78], but however crucial for a model to be self-sustainable.

4. *Scalable*: In the research environment, tests are run in a delimited scale, constrained to the size of data, and useful for the experiments, in contrast with large, multi-variate real environments. Scaling up is not, of course, a matter of ITS research, but an engineering problem. However, models should be designed to be scalable since their conception.

   Leaving aside calibration and training phases, classic transportation theories tend in general to be computationally more affordable than data-driven models. However, the unprecedented amount of computing power available nowadays discards any real pragmatic limitation due to the computational complexity of learning algorithms in data-based modeling. An exception occurs with models falling within the Deep Learning category which, depending on their architecture and size of training data, may require specialized computing hardware such as GPU or multi-core equipment. Nevertheless, the rising trend in terms of scalability is to make data-based models incremental and adaptable [3], which finds its rationale not only in the environmental sustainability of data centers (lower energy consumption and thereby, carbon foot-print), but also in the deployment of scalable model architectures on edge devices, usually with significantly less computing resources than data centers.

   Although some ITS problems are easier to scale and this feature would not be trouble-some, there are some fields that can be very sensitive to scalability. For instance, route planners frequently consist of shortest-path problem and travel-salesman problem implementations that increase in complexity when the number of nodes grow [115]. This is a good example where artificial intelligence and optimization tools provide solutions that are actionable in terms of scalability, and where cases are found effortlessly [116,117]. Caring about aspects like the easiness to introduce new variables when needed, the complexity of tuning if applies, or the execution time, would make a model more actionable, by increasing its self-sustainability. This need for scalability is not just a matter related to the computational complexity of modeling elements along the pipeline, but also links to the feasibility of migrating the designed models from a lab setup to a, e.g., Big Data computing architecture. Unfortunately, scarce publications reflect nowadays on whether their proposed data-based workflows can be deployed and run on legacy ITS systems, thereby avoiding costly upgrade investments in computing equipment.

3.3. Traffic Theory Awareness

Theoretical representations of traffic attempt to construct (mostly simple) models with causal aspects. These models are usually of a closed form and are frequently dictated by simplifying assumptions, which leads to limited performance when modeling complex spatio-temporal dynamics in the microscopic analysis context. In these models, data are instrumental to estimate how well they fit real world conditions. On the other hand, and since their upsurge in the 80s, data driven models rely exclusively on the data to extract the dynamics that govern the phenomena. This, at least theoretically, makes them more adaptable and more efficient in complex conditions when compared to theory based models. But, they can hardly claim applicability in large scale scenarios (city level traffic management) due to significant computational resources requirements. Such data-driven traffic models have been systematically implemented as proof of concepts and are now dominant in Traffic Engineering literature [16], incorporating most well-known advanced techniques, and, in many cases, ignoring the elementary knowledge of traffic and focusing blindly on performance.

Owing to the above, researchers in traffic modeling have diversified the way in which their models are developed and evaluated, fitting them to the technology that is introduced, as opposed to fitting the model to the well established knowledge described in well established theories of traffic flow. This results in models that are hardly actionable for traffic engineers, in terms of integration to legacy traffic control and management systems and relevance to the decision making process of road network operators. Besides, there is a lack of standards in what regards to data and scenarios used to assess the performance, usually due to the availability of real data for each researcher. This was already identified in [78], where test-beds were proposed, either generating them or using some of the existent as standards. This would help compare models, understand them better, as they can be evaluated in a known environment, and obtain their insights concerning traffic theory. Besides, as we anticipated in Section 2.1 there is a industrial trend towards the consideration of different data sources when modeling traffic dynamics. In many cases, these data sources do not have any straightforward relationship to traffic itself. The integration of these sources of data, the models learned from them and theoretical representations of transportation scenarios remains an open challenge that has started to be addressed in literature [118–120].

In this line of reasoning, linking data-driven to theory based models in transportation may resort to efficient and physically consistent representations of transportation phenomena. In fields like traffic modeling and forecasting, this hybrid approach permits to consider theoretical aspects of traffic, such as the relationship among speed, flow and density, the three phases of traffic [121], or the Breakdown Minimization Principle [122] when modeling bottlenecks. The consideration of these theoretical concepts takes effect mainly in the preprocessing, modeling and prescription phases of the modeling workflow. In preprocessing, domain knowledge can be crucial for feature engineering, by describing how available features are related to each other, estimating collinearities in advance, deleting irrelevant predictors, or obtaining feature combinations with improved modeling power [78]. Applying traffic theories and principles can also be useful for data augmentation and missing data imputation, by simulating or generating data that are more akin to what the context can provide [36]. In the modeling phase, previously defined mathematical frameworks can help define the constraints, operation ranges and correct the output of data-based models, which do not take into account the compliance of their output with respect to well-established theories. Lastly, in the prescription phase, model outputs can be linked to traffic theory knowledge to improve the way in which they are applied: a predicted flow value can be more useful if the travel time or the bottleneck probability can be computed afterwards. Furthermore, in the case of predictive models, they can reach a point in which the provided predictions ultimately affect the future behavior of the models themselves, if they are trained only with observed past data. For instance, a model that assists traffic management decisions, like closing a lane, might lead to a situation that has not been observed by the model before, thus making the knowledge captured by the traffic model obsolete and useless until the data captured from the environment is exploited for retraining. Physical models can be highly useful to anticipate scenarios and complement data-based models, providing additional information of what theories or simulations determine that the behavior of the scenario should be.

This emergent modeling paradigm is known as Theory Guided Data Science, and aims to enhance data driven models by integrating scientific knowledge [123]. The main objective of this approach is to enable an insightful learning using theoretic foundations of a specific discipline to tackle the problem of data representativeness, spurious patterns found in datasets, as well as providing physically inconsistent solutions. From the algorithmic point of view, this induction of domain knowledge can be done in assorted means, such as the use of specially devised regularization terms in predictive models (e.g., in the loss function of Deep Learning models), data cleansing strategies that account for known data correlations, or memetic solvers that incorporate local search methods embedding problem-specific heuristics. In transportation, there has been several example of theory enhanced models departing from traffic conditions identification and characterization [124,125], to data driven and agent based traffic simulation models for control and management [3,42,126,127], or cooperative intelligent driving services [128].

Awareness with domain-specific knowledge can be also enforced at the end of the workflow. When decision making is formulated as an optimization problem, the family of optimization strategies known as Memetic Computing [129,130] has been used for years to incorporate local search strategies compounded by global search techniques and low-level local search heuristics. These heuristics can be driven by intuition when tackling the optimization problem at hand or, more suitably for actionability purposes, by a priori knowledge about the decision making process gained as a result of human experience or prevailing theories. For instance, traffic management under incidences in the road network can largely benefit from the human knowledge acquired for years by the manager in charge, since this knowledge may embed features of the traffic dynamics that are not easily observable from historical data. This knowledge can be inserted in an optimization algorithm devised to decide e.g., which lanes should be rerouted in an accident.

3.4. Application Context Awareness

Transportation is exceptionally diverse around the world, with notable differences in modes, preferences and availability due to social, economic and cultural disparity. Moreover, Intelligent Transportation Systems with different purposes have also characteristic requirements that can also be very divergent with respect to space and time. To address this landscape of complex and some times conflicting goals, policies and decision making should span from few seconds (traffic management and control) to years (planning and designing of new systems). It is strongly argued that data driven framework are able to cope with context aware datasets, due to their inherent capabilities of learning patterns hindering in resourceful data and reconstruct-in a sense-the context of the application. Typical examples of such context aware systems are the extraction of Origin-Destination matrices from cellphone based data [131], the mobility applications that aim to improve the the mobility footprint of users [132], as well as the smartphone based driving insurance systems [133]. Although these approaches seem to be appropriate to complement the user or system's experience on a problem, significant uncertainty lies in their transferability and accuracy, owing to the lack of context-aware knowledge.

A certain degree of awareness of the context should be a matter of concern when developing ITS models that intend to be actionable. Context aware information is usually introduced in the modeling, for example accounting fro the demographic characteristics of the application area, the type of the road or network, the mode, the travel purpose etc. However, what is usually disregarded is a much broader consideration of the operational and system's characteristics, such as how models can be introduced to the operations at hand, what the privacy concerns are with respect to data and information flows, what is the regulatory framework and policy level restrictions and goals to be reached.

First, within the operation, the deployment context where a developed model is intended to be implemented can enforce a series of operative constraints. Creating and proposing an ITS model without observing these requirements is an exercise of futility, for its lack of actionability. From this operation perspective, the context covers from deployment and operation costs—is the system cost-efficient considering its potential service?—to functioning modes—has the model the expected response times? can it operate in reduced computational power environments? As an illustration, a system designed to detect and identify pedestrians can be very effective in terms of performance, but if it does not operate at an appropriate speed, or it needs more demanding computations that cannot be embarked in a vehicle, it is useless for an autonomous driving context [134]. A similar reasoning holds if by *operation cost* one thinks about the energy consumption of the model at hand. Questions such as whether the energy consumed by the model compliant with the system should be kept in mind at design time, but also from the academic perspective, where efforts should be directed to the development of models that are consequent with the actual operative circumscription.

Second, regulations constitute a hard and highly contextual constraint in the implementation of ITS. Besides the wide regulatory differences that can be found across regions, there are transport frameworks where regulations are specially rigid. A typical example is the case of airports [135], and where there is a broad field for specialized ITS. Another example is the constantly rising use of drone systems to monitor traffic [136]. Models that fail to relate to the application's regulatory environment are not actionable.

Third, data privacy and sovereignty constitute a growing concern in a connected world where, after a decade of handing over data with complacency, an awareness about personal information sharing is springing. A recent example is the introduction of the EU General Data Protection Regulation (GDPR) framework, that severely disputed the manner data were introduced to models, as well as data availability. ITS models that are based on personal data are common nowadays, for instance in floating car data based developments [137]. However, there are fields where this aspect is becoming crucial (autonomous driving connectivity [138], security in public transport environments [139]), and research is steering to privacy-preserving approaches [140], spheres where technologies such as Blockchain can have a major dominance [141,142].

Fourth, social aspects of the application play a major role in modeling. Social transportation is the subfield in ITS where the "social" information coming from mobile devices, wearable devices and social media is used for a number of ITS management related applications [143]. The outcomes from social transportation may be, to name a few, traffic analysis and forecasting [144,145], transportation based social media [146], transportation knowledge automation in the form of recommending systems and decision support systems [147], and services for the collection of further signal to be used later for the already mentioned purposes or others. However, cultural differences can have a relevant impact in how these systems operate, as social data are most commonly strongly linked to geographical information. This is a key aspect for their actionability.

Fifth, transportation is currently a large source of greenhouse-gas emissions [148]. These concerns are gaining momentum in a wide range of ITS applications, such as the discovery of parking spots [149], multimodality applications that grant travelers the chance of using collective transportation systems efficiently and conveniently [150], the improvement of logistics operations [151], shared mobility applications, which help reducing the number of one-passenger vehicles in the road network [152], or driving analytics to improve safety and ecological footprint [153–155].

Of course, research goes beyond the application context and does not need to be always connected to a certain application scenario. A prototype can be far from the practical requirements of its eventual deployment; still, knowing the essential application common grounds is key to converge to actionable models. Unfortunately, this is a matter frequently disregarded in ITS research.

3.5. Transferability

Within the research context, it is common to employ test data to assess the models. Regardless if these data are obtained from real sources or synthetically generated, the resulting models have been built around them, and can be heavily linked to that experimentation context. Would these models work in other context or with other input data? Transferability could be defined as the quality of a data-driven model to be applied in other environment with other data, and it is directly linked with actionability: the application of a model should be generalizable to different datasets and transportation settings. This definition stems from the more general concept of *Transfer Learning* [156], which can entail that models trained in a certain domain are applied to other domains, so that the previous knowledge obtained from the first makes them perform better in the latter than models without it.

Depending on the subcategory of ITS, this requirement can be easily met or arduous to achieve, as some subcategories are more oriented towards the application and rely less on the environment than others; the key is defining what is *environment*. For example, a travel time forecasting model developed with data of a certain location could be transferable to another location without great complications, if it is built considering this feature [157]. In fact, many ITS models that are spatial-sensitive are developed using real data, but within the experimentation context, they are evaluated only in certain locations. Transferability for these scenarios would imply that the obtained results are reproducible (with certain degree of tolerance) in other locations.This could entail from plainly extrapolating the model to other locations [158], to implementing of techniques such as *soft-sensing*, aimed at modeling situations where no sensor is available [159], and the environment information is enough to obtain these models. A similar case in terms of spatial contexts, but with more parameter complexity, requires plenty of information about the environment. As an illustration, the case of crash risk estimation implies a higher calibration and adjustment needs due to the higher number of parameters that take part in this type of estimations. In these circumstances, works such as [160] or [161] work with posterior probability models and give more relevance to models that behave with a certain performance in many contexts than to models that perform better in a particular location. On the other extreme, for cases like autonomous driving, the change of environment is connatural to the domain (a moving vehicle constantly changing its location), and the parameters of these models are abundant and highly variable. Thus, these applications need transferable solutions, transferability that is specifically sought by researchers, for instance in LIDAR based localization [162] or pedestrian motion estimation [163]. In any case, and regardless the domain, ITS research is in an incipient stage (probably with the exception of autonomous driving) of developing transferable models, and evaluating this feature, and some machine learning paradigms can help improve this characteristic.

4. Emerging AI Areas towards Actionable ITS

We have hitherto elaborated on the requisites that a model should meet towards leading to actionable data-based insights in ITS applications and processes. Some of these requirements can be fulfilled by properly designing the data-based workflow (e.g., interpretability can be straightforward for certain prediction models, whereas adaptability can be enforced by periodically scheduling the learning algorithm under use and feeding it with new data). However, several research areas have stemmed in the last years from the wide fields of Data Science and Artificial Intelligence that may serve to catalyze the compliance of data-based ITS workflows with the prescribed requisites, and thereby attain the sought actionability of their produced insights.

The main AI areas that have been identified as potentially appropriate for addressing the requirements can be summarized briefly as follows:

- Real-time data processing and online learning, which are not brand new research avenues in ITS, as we can find advanced developments in the literature. However, as we will later show, emerging fields with great potential such as dynamic data fusion and dynamic optimization can expedite and proliferate the adoption of incremental data-based models in more ITS-related applications.

- Transfer learning and domain adaptation, that could allow to develop models for certain contexts and export them to others, linking directly to the transferability requirement, but also to the integration of transportation theories and physical models to data-based models.
- Gray-box modeling, a paradigm halfway between white-box (physical) and black-box (data-based) models. Gray-box modeling represents a promising area to bring awareness to traffic theory and other physical modeling when developing data-based models, with the potential to increase the performance, usability and comprehensibility of the latter.
- Green AI, a trend in Artificial Intelligence research that connects directly with energy and cost efficiency. Developing efficient models has a relevant impact in their sustainability and context awareness.
- Fairness, Accountability, Transparency and Ethics: Data-based models—specially those learning from large amounts of diverse data from many sources—are fragile to biases, and can compromise aspects such as the fairness of decisions or the differential privacy of data. In this context of growing sources of data, including those gathered from people, and increasingly opaque data-based models, it has become essential to understand what models have learned from data, and to analyze them beyond their predictive performance to consider ethical, societal and legal aspects. These aspects have been scarcely considered in ITS research.
- Other Artificial Intelligence areas such as imbalanced learning, reinforcement learning, adversarial machine learning are later highlighted for their noted relevance in ITS.

We next discuss on the research opportunities spurred by the above research lines, their connections with the requirements presented in Section 3 (shown in Figure 3), as well as the challenges that stem from the consideration of these AI areas in the context of ITS.

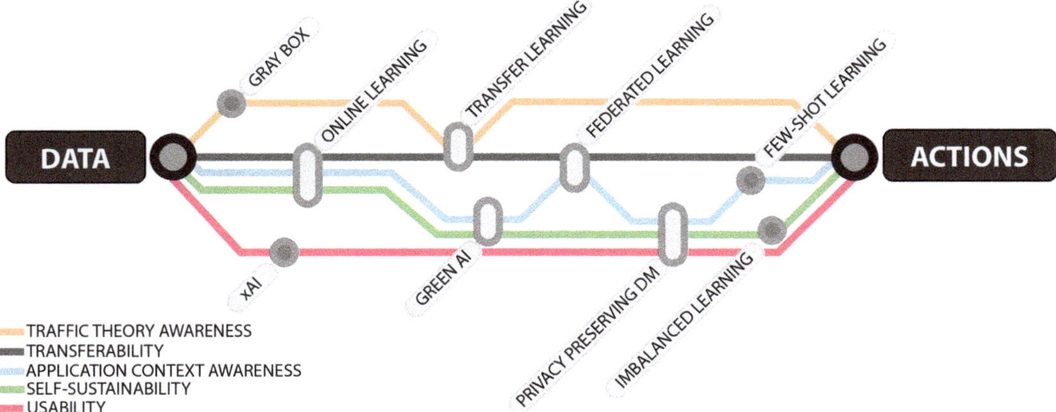

Figure 3. Schematic diagram showing how avant-garde AI subareas can promote actionability in ITS data-based modeling workflows. Subareas contributing with particular emphasis to different functional requirements are connected together along the way from data to actions.

4.1. Online Learning and Dynamic Data Fusion/Optimization

Previously sketched in Section 3.2, by online learning we refer to the capability of the learning model and in general, of the entire workflow, to learn from fastly arriving data possibly produced by non-stationary phenomena that enforces a need for adapting the knowledge captured by the model along time. Changes over data streams can make the data pipeline obsolete, thus demanding active or passive techniques to update it with the characteristics of the stream [7,97].

Although activity around online learning has mostly revolved on certain clustering and classification paradigms (the latter giving rise to the so-called concept drift term to refer to pattern changes), it is important to note that adaptation can be also needed in other stages of the actionable data-based workflow, from data fusion to the prescription of actions. This being said, research areas such as dynamic optimization and dynamic multi-sensor data fusion should be also investigated deeply in future studies related to actionable data-based models, specially when the scenario under analysis can produce information with non-stationary statistical characteristics. When merging different data sources, fusion strategies at different levels can be designed and implemented, from traditional means (data-level fusion, knowledge-level fusion) to modern methods (corr. model-based fusion, federated learning or multiview learning) [164,165]. Fusion of correlated data sources can compensate for missing entries or noisy instances in static environments. However, when data evolve over time as a result of their non-stationarity, new challenges may arise in regards to the inconsistency among multiple information sources, including measurement discrepancy, inconsistent spatial and temporal resolutions, or the timeliness/obsolescence of the data flows to be merged, among other issues. For this reason, close attention should be paid to advances reported around adaptive fusion methods capable of detecting, counteracting and correcting misalignments between data flows that occur and evolve over time. This branch of dynamic data fusion schemes aims at combining together information flows produced by non-stationary sources, synthesizing a representation of the recent history of each of the flows to be merged into a set of more coherent, useful data inputs to the rest of the data-based pipeline [166,167]. On the other hand, dynamic optimization techniques can efficiently deliver optimized actionable policies when the objectives and/or constraints of the underlying optimization problem varies [168,169]. We energetically advocate for a widespread embrace of advances in these fields by the ITS community, emphasizing on those scenarios whose dynamic nature can make the obtained actionable insights eventually obsolete. This is the case, for instance, of traffic related modeling problems (e.g., traffic flow forecasting and optimal routing) or driver characterization for consumption minimization, among many others.

Other requirements for actionability can also benefit from the adoption of the above models in dynamic ITS contexts. For instance, cost efficiency in terms of energy consumption can largely harness the incrementality that often features an online learning model. The use of dynamic data fusion can also yield a drastically less usage of communication resources in wireless V2V links, such as those established in cooperative driving scenarios. All in all, the recent literature poses no question around the relevance of adaptation in data-based modeling exercises noted in this work, with an increasing volume of contributions dealing with the extrapolation of adaptation mechanisms to ITS problems [170–172].

4.2. Transfer Learning and Domain Adaptation

In close semantics to its related actionability requirement (*transferability*), transfer learning aims at deriving novel means to export the knowledge captured by a data-based model for a given task to another task with different inputs and/or outputs [156]. Depending on the amount of alikeness between the origin task and the destination task, we may be also referring to *domain adaptation*, by which we adapt the model built to perform a certain task to make it generalize better when processing new unseen inputs that do not follow the same distribution as their original counterparts (only the distribution changes [173]). Techniques such as subspace mapping, representation learning, of feature weighting arise as those methods most used to allow knowledge to be transferred between data-based models used for prediction.

In essence, transfer learning can provide higher prediction accuracy for models whose number of parameters to be learned (e.g., weights in a Neural Network) demands higher amounts of labeled data than those available in practice. However, data augmentation is not the only goal targeted by transfer learning. Domain adaptation may yield a better performance when used between ITS models that can become severely affected by a lack of calibration, different configurations or diverging specifications. An immediate example illustrating this hypothesis is the use of camera sensors for vehicular perception. Models trained to detect and identify objects in the surroundings of the vehicle can fail if the images provided as their inputs are produced by image sensors with new specs. The same holds for car engine prognosis: replaced components can make a data-based characterization of the normal operation of the engine be of no practical use unless a domain adaptation mechanism is applied. Personalization of ITS services can be another problem where domain adaptation can help refine a model trained with data from many sources: a clear example springs from naturalistic driving, where a behavioral characterization model built at first instance from driving data produced by many individuals (source domain) can be progressively specialized to the particular driver of the car where it is deployed [174–176].

In regards to actionability, several functional requisites can be approached by using elements from Transfer Learning over the data-based pipeline. To begin with, it should be clear that the transferability of learned models for their deployment in different locations and contexts could be vastly improved by Transfer Learning, as the purpose of this AI branch is indeed to meet this requirement in data-based learning models. In fact, this approach is currently under study and wide adoption within the ITS community working on vehicular perception: when the capability of the vehicle to sense and identify its surrounding hinges on learning models (e.g., Deep Learning for image segmentation with cameras), a plethora of contributions depart from pretrained models, which are later particularized for the problem/scenario at hand [177]. This exemplified use case supports our advocacy for further efforts to incorporate transfer learning methods in other ITS applications, specially those where data collection and supervision are not straightforward to achieve in practice. Another functional requirement where Transfer Learning can pose a difference in ITS developments to come is cost efficiency. The knowledge transferred between models learned from different contexts can improve their performance, thereby reducing the need for supervising data instances and ultimately, the time, costs and resources required to perform the data annotation.

Finally, the more recent paradigm coined as Federated Learning refers to the privacy-preserving exchange of captured knowledge among models deployed in different contexts [178,179]. Although the main motivation for the initial inception of Federated Learning targeted the mobile sector, techniques supporting the federation of distributed data-based models can be of utmost importance in the future of ITS, specially for V2V communications among autonomous vehicles and in-vehicle ATIS systems. Definitely the enrichment of models with global knowledge about the data-based task(s) at hand will pose a differential breakthrough in vehicular safety and driving experience. For instance, federated models can collectively identify, assess and countermeasure the risk of more complex vehicular scenarios than each of them in isolation [180]. Likewise, ATIS systems can learn from the preferences and habits of other users to better anticipate the preferences of the driver and act accordingly [181]. In a few words: an enhanced and more effective actionability of the data-based workflows built to undertake such tasks.

4.3. Gray-Box Modeling

Gray-box modelling refers to the design of models that combine theoretical developments and structures related to the problem, with data that serve as a complement for such theories to make the overall model match better the scenario under analysis [182,183]. Gray-box models lie in between white-box models, for which the learned structure is deterministic and grounded in theoretical concepts; and black-box models, whose internal structure lacks physical significance and is learned from data. An example of white-box model in ITS systems is the use of computational fluid dynamics for macroscopic traffic flow modeling, whereas Deep Learning models for traffic forecasting can exemplify black-box modeling in this domain. Gray-box models have been lately embraced by the

ITS community in a number of modeling scenarios, such as those combining biological concepts and data-based models for driver characterization [184,185].

Gray-box modeling can contribute to the actionability of data-based workflows for ITS applications in two different albeit interconnected directions. To begin with, the incorporation of theoretical models to data-based pipelines can narrow the gap between engineers and practitioners more acquainted with traditional tools to analyze ITS systems and processes. Indeed, hybrid modeling can tie both worlds together not only without questioning the validity of prevalent theoretical developments, but also evincing the complementarity and synergy of both approaches. On the other hand, using validated theoretical models can help data-based modeling overcome difficult learning contexts such as class imbalance, outlier characterization or the partial interpretability of data clusters, among others.

4.4. Green Artificial Intelligence

A profitable strand of literature has recently stressed on the energy efficiency of data-based models, highlighting the need for redesigning their learning algorithms to minimize their energy consumption and thereby, make them implementable and usable in practice [186–188]. While this issue is particularly relevant for resource-constrained devices (e.g., mobile hand-helds), the concern with energy efficiency goes beyond usability towards environmental friendliness. For this reason many recent contributions are striving for computationally lightweight variants of machine learning models that sacrifice performance for a notable reduction of their energy demand. This is not only the case of predictive models capable of incrementally learning from data, but also of specific Deep Learning architectures tailored for their deployment on embedded devices [189].

Based on the above rationale, cost efficiency is arguably the most evident functional requirement around which energy-aware model designs can pose a breakthrough towards improving the actionability of the overall data-based workflow. In addition, other aspects can be made more actionable by using energy-aware model designs, such as usability [190]. Despite achieving unprecedented levels of predictive accuracy, a data-based workflow may become useless should it deplete the battery of the system on which it is deployed for operation. Therefore, energy efficiency should be under the target of future research efforts, specially when dealing with ITS applications running on battery-powered devices, inspecting interesting paths rooted thereon such as the trade-off between performance and energy consumption, or the adaptation of the model's operation regime depending on the remaining battery life, among others [191].

4.5. Fairness, Accountability, Transparency and Ethics

To end with, the prescription of actions based on the insights provided by a data-based pipeline must be buttressed by a thorough understanding of the mechanisms behind its provided decisions [192]. Extended information about the model must be presented to the end user for several reasons:

- To gauge as many consequences of the actions as possible, identifying situations where decision making based on the outputs of the data-based workflow gives rise to socially unfair scenarios due to the propagation of inadvertently encoding bias to the automated decisions of the model.
- To ensure him/her that the output of the model is reliable and invariant under the same data stimuli, maintaining a record of the intermediate decisions made along the pipeline, allowing for the post-mortem, potentially correcting analysis of bad decision paths, and thereby maximizing the trust and certainty of the user when embracing its output.
- To make the user understand why the developed model produces its prescriptive output when fed with a set of data inputs, shedding light on which inputs correlate more significantly with the prescribed actions, tracing back causal relations between intermediate data inputs, and discriminating extreme cases where decisions can change radically under slight modifications of the model inputs.

- To supervise the ethics of data-based workflows, identifying potentially illegal uses of unlawful data given the prevailing legislation, guaranteeing the privacy and governance of personal data by third-party data-based ITS applications and processes, and certifying that the output of the model's output does not favor inequalities in terms of gender, religion, race or any other aspect alike.

The above requirements have been lately collectively compiled under the FATE (Fairness, Accountability, Transparency and Ethics) concept, which refers to the design of actionable data-based pipelines whose internal operations can be explained, accounted and critically examined in regards to the consequences of their eventual bias in privacy, fairness and ethical issues [193–195]. This recent concern with the operation of machine learning models spawns from the proliferation of real cases where practical model installments have unveiled deficiencies of different kind, from differential privacy breaks (data revealing the identity of the persons to whom they belong) to unnoticed output bias that caused racist discriminatory issues [196]. For instance, data-based models for vehicular perception, obstacle detection and avoidance must be also endowed with ethics and legal design factors to make the overall decision not just drifted by the data themselves. Another clear domain where FATE can be crucial is modeling with crowd-sourced Big Data, where aspects like privacy preservation [197] and bias avoidance [198] are arguably more critical [199,200]. The construction of the data-based modeling workflow must (i) ensure that protected features remain as such once the workflow has been built, without any chance for reverse engineering (via e.g., XAI techniques [90]) that could compromise the differential privacy of data; and (ii) that learning algorithms along the workflow counteract hidden bias in data that could eventually lead to discriminatory decisions (due to skewed samples, tainted annotation, limited data sizes or imbalanced data). From our perspective, these are among the most concerning challenges in the exploitation of Big Data in ITS, and the main source of motivation for a number of recent studies in areas related to data-driven transportation systems such as pedestrian detection [201], autonomous vehicles [202,203] or urban computing [204]. Bias-related issues can be identified by a proper analysis of the decisions made by the workflow, which in turn requires models to be accountable and transparent enough to thoroughly characterize their sensitivity to bias, and how inputs and outputs (decisions) correlate in regards to protected features. It is also remarkable to note that several proposals have been made to quantify fairness in machine learning pipelines, yielding useful metrics that account for the parity of models when processing groups of inputs [205,206]. Without these aspects being considered jointly with performance measures, data-based ITS developments in years to come are at the risk of being restricted to the academia playground [207].

4.6. Other AI Research Areas Connected to Actionability

The above areas have been highlighted as the main propellers for model actionability in ITS systems. However, it is worthwhile to mention other research areas from the AI realm that can also help completing the chain from data to actions:

- Few-shot learning [208], which aims at overcoming the lack of reliably annotated data and the practical difficulty of performing annotation in certain application scenarios. For instance, accident prevention models cannot be enriched with positive samples unless a fatality occurs and the data captured in place is fed into the model. Few shots learning and related subareas (zero-shot, one-shot) deriving solutions that can automatically learn from very small amounts of training data, incorporating mechanisms (e.g., generative models, regularization techniques, guided simulation) to prevent the overall model from overfitting [209]. In regards to actionability, this family of learning techniques can be helpful to make data-based ITS models deployable in situations lacking data supervision, specially when such a data annotation cannot be guaranteed to be achievable over time.
- Imbalanced and cost-sensitive learning [210,211], which link to the need for avoiding model bias, not only to ensure the generalization of its output, but also to reduce the

likeliness of the workflow to cause discriminatory issues as the ones exemplified above. The history of these AI areas in the ITS community has been going for years now [3]. However, we here emphasize the crucial role of these techniques beyond performance boosting: the techniques originally aimed to counteract the effects of class imbalance in the output of data-based models could be also leveraged to reflect legal impositions that not necessarily relate to the model's performance nor can they be inferred easily for the attributes within the data themselves. The lack of compliance of the model with fairness and ethics standards does not necessarily render a performance degradation observed at its output, nor can it be inferred easily from the available data.

- Hybrid models encompassing linguistic rules and data-based learning techniques, capable of supporting the transition from the traditional way of doing to the new data-based modeling era in the management of ITS systems. We foresee that the community will witness a renaissance of data mining methods incorporating methods such as fuzzy logic not only to implement human knowledge to decision workflows, but also to explain and describe the internal structure of learned models, as it is currently under investigation in many contributions under the XAI umbrella [212,213].

- New prescriptive data-based techniques such as Deep Reinforcement Learning [214] and Algorithmic Game Theory [215] will also grasp interest in the near future for their close connection to actionable data science. The interaction of data-based workflows with humans will require techniques capable of learning actions from experience, and eventually orchestrating the interaction and negotiation among users when their actions are governed by interrelated yet conflicting objectives. In fact such new prescriptive elements are progressively entering the literature in certain ITS applications that target machine autonomy (e.g., autonomous vehicle [216,217] or automated signaling [51]), but it is our vision that they will gain momentum in many other ITS setups.

- Privacy-preserving Data Mining [218,219], which has garnered a great interest in the last year with major breakthroughs reported in the intersection between machine learning, cryptography, homomorphic encryption, secure enclaves and blockchains [220]. The use of personal data and the stringent pressure placed by governments and agencies on differential privacy preservation has spurred a flurry of research to prevent models from revealing sensitive data from their training instances [197,221]. Within the ITS domain, it is possible to find many areas in which privacy preservation has recently been a subject of intense research: from origin-destination flow estimation [222] to route planners [223,224], or pattern mining [225], a glance at recent literature reveals the momentum this topic has acquired lately. In any of these examples data are available as a result of the sensing pervasiveness (specially in the case of VANETs) and the capture of user data. While previous works explored how to used these data in a proper way with respect to privacy matters, it is straightforward to think that the natural evolution of this research line arrives at how protected data is preserved through the modeling workflow.

- Furthermore, the proven vulnerability of data-based models against adversarial attacks has also motivated the community to lay the foundations of an entirely new research area—Adversarial Machine Learning—, committed to the design of robust models against attacks crafted to confuse their outputs [226,227]. Interestingly, one of the most widely exemplified scenarios in this research area relates to ITS: automated traffic signal classification models were proven to be vulnerable to adversarial attacks by placing a simple, intelligently designed sticker on the traffic sign itself [228]. Likewise, the rationale behind Federated Learning (discussed in Section 4.2) also spans beyond the efficient distribution of locally captured knowledge among models: since no raw data instances are involved in the information transfer, privacy of local data is consequently preserved. In short: security also matters in actionable data-based pipelines.

- Finally, the ever-growing scales of ITS scenarios demand more research invested in scaling up learning algorithms in a computationally efficient manner [229]. Automated traffic, smart cities, mobility as a service constitute ITS scenarios where a plethora of information sources interact with each other. Definitely more efforts must be invested in aggregation strategies for data-based models learned from different interrelated data ecosystems, either in a distributed fashion (e.g., federated learning) or in a centralized system (correspondingly, Map-Reduce implementations of data-based models, cloud-based architectures, etc). Computational aspects of large-scale implementations should be also under study due to their implications in terms of actionability, such as the latency of the system when prescribing decisions from data. This latter aspect can be a key for real-time ITS applications for which the gap from data to actions must be shortened to its minimum.

5. Concluding Notes and Outlook

This work has built upon the overabundance of contributions within the ITS community dealing with performance-based comparisons among data-based models. Our claim is that, as in any other domain of application, data-based modeling should bridge the gap between data and actions, providing further value to the ITS application at hand than superior model performance statistics. It is our firm belief that the research community should embrace actionability as the primary design motto, with negligible performance improvements being left behind in favor of relevant aspects such as adaptability, usability, resiliency, scalability or efficiency.

To provide a solid rationale for our postulations, we have first presented a reference model for actionable data-based workflows, placing emphasis on the different phases that should be undertaken to translate data into actions of added value for the decision maker. Adaptation has been highlighted as a necessary albeit often neglected processing step in data-based modeling, which allows models to be effective when deployed on dynamic ITS environments with time-varying data sources. Next, our study has listed the main functional requirements that models along the reference model should meet to guarantee their actionability, followed by an overview of incipient research areas in Data Science and Artificial Intelligence that should progressively enter the ITS arena. Indeed, advances in XAI, Online Learning, Gray-box Modeling and Transfer Learning are currently investigated mostly from an application-agnostic perspective. Their undoubted connection to actionability makes them the core of a promising future for data-based modeling in ITS systems, processes and applications.

Other research areas related to Artificial Intelligence beyond those covered in our reflections will surely spawn further opportunities for actionability in ITS, provided that they fully embrace their ultimate goal: to effectively support decision making. Among them, the use of Automated Machine Learning (AutoML [230]) for tuning data-based models should not only optimize performance-based metrics (e.g., finding a model that attains maximum accuracy for image segmentation in vehicular perception cameras), but also comply with other objectives and constraints that closely link to actionability (e.g., robustness against adversarial attacks, or a lower epistemic uncertainty of the model induced in its output). Unless all such actionability constraints are regarded as design objectives and accounted for as such in the automated discovery of new data-based pipelines, any incursion of AutoML in ITS will be of no practical value. For this to occur, it is our belief that the confluence of multiple functional and non-functional requirements in this automated design process will pave the way towards the massive adoption of multi-objective optimization algorithms as a massive framework to infer and analyze all trade-offs existing among the design objectives.

Data-based modeling has brought a deep transformation to ITS. A vast amount of research works in the field are produced by data-based modeling specialists attracted by the profusion of available data, and with limited knowledge of transportation. Data-based models are getting progressively more complex, increasing the gap between research and practice. This situation calls for a change of paradigm, to a one in which actionability requirements of models is desired by researchers, and practitioners are aware of the technologies available to provide it. Model actionability is a great whole that can act as an incentive to perform smaller steps towards its realization. It is probably unthinkable to develop, in a research environment, a data-based model that meets all proposed requirements. However, addressing some of the postulated requirements while developing a competing data-based ITS model will make it closer to actionability. There is, therefore, a long road to be travelled in ITS model actionability, with interesting avenues around the thorough understanding of models, and the adoption of emerging AI technologies to endow data-based workflows with the requirements needed to make them actionable in practice. As exposed in our study, there is a germinal interest in these research topics. Nevertheless, we foresee vast opportunities for future work when model actionability is set as a design priority.

On a closing note, we advocate for a new dawn of Data Science in the ITS domain, where advances in modeling performance concurrently emerge along with histories and reports about how such models have helped decision making in practical scenarios. Data mining has limited merit without actions prescribed from its outputs, always in compliance and close match with the specificities of its context.

Author Contributions: I.L.: Conceptualization, Methodology, Investigation, Writing—Original Draft, Writing—Review & Editing. J.J.S.-M.: Conceptualization, Methodology, Investigation, Validation, Writing—Original Draft, Funding acquisition. E.I.V.: Conceptualization, Methodology, Writing—Original Draft, Funding acquisition. J.D.S.: Conceptualization, Methodology, Investigation, Supervision, Writing—Review & Editing, Funding acquisition. All authors have read and agreed to the published version of the manuscript.

Funding: This work was supported in part by the Basque Government for its funding support through the EMAITEK program (3KIA, ref. KK-2020/00049). It has also received funding support from the Consolidated Research Group MATHMODE (IT1294-19) granted by the Department of Education of the Basque Government.

Institutional Review Board Statement: Not applicable.

Informed Consent Statement: Not applicable.

Data Availability Statement: Data sharing not applicable.

Conflicts of Interest: The authors declare no conflict of interest.

References

1. Zhu, L.; Yu, F.R.; Wang, Y.; Ning, B.; Tang, T. Big data analytics in intelligent transportation systems: A survey. *IEEE Trans. Intell. Transp. Syst.* **2018**, *20*, 383–398. [CrossRef]
2. Albino, V.; Berardi, U.; Dangelico, R.M. Smart cities: Definitions, dimensions, performance, and initiatives. *J. Urban Technol.* **2015**, *22*, 3–21. [CrossRef]
3. Zhang, J.; Wang, F.Y.; Wang, K.; Lin, W.H.; Xu, X.; Chen, C. Data-driven intelligent transportation systems: A survey. *IEEE Trans. Intell. Transp. Syst.* **2011**, *12*, 1624–1639. [CrossRef]
4. Karlaftis, M.G.; Vlahogianni, E.I. Statistical methods versus neural networks in transportation research: Differences, similarities and some insights. *Transp. Res. Part C Emerg. Technol.* **2011**, *19*, 387–399. [CrossRef]
5. Del Ser, J.; Osaba, E.; Sanchez-Medina, J.J.; Fister, I. Bioinspired Computational Intelligence and Transportation Systems: A Long Road Ahead. *IEEE Trans. Intell. Transp. Syst.* **2019**, *21*, 466–495. [CrossRef]
6. Eiben, A.E.; Smith, J.E. *Introduction to Evolutionary Computing*; Springer: Berlin/Heidelberg, Germany, 2003; Volume 53.
7. Ditzler, G.; Roveri, M.; Alippi, C.; Polikar, R. Learning in nonstationary environments: A survey. *IEEE Comput. Intell. Mag.* **2015**, *10*, 12–25. [CrossRef]
8. Said, H.; Nicoletti, T.; Perez-Hernandez, P. Utilizing telematics data to support effective equipment fleet-management decisions: Utilization rate and hazard functions. *J. Comput. Civ. Eng.* **2016**, *30*, 04014122. [CrossRef]
9. Urbahs, A.; Žavtkēvičs, V. Remotely Piloted Aircraft route optimization when performing oil pollution monitoring of the sea aquatorium. *Aviation* **2017**, *21*, 70–74. [CrossRef]
10. Khaksar, H.; Sheikholeslami, A. Airline delay prediction by machine learning algorithms. *Sci. Iran.* **2019**, *26*, 2689–2702. [CrossRef]

11. Mott, J.H.; Bullock, D.M.; McNamara, M.L. Estimating Aircraft Operations at Airports Using Transponder Data. US Patent Application No. 15/248,581, 18 May 2017.

12. Herring, R.; Hofleitner, A.; Abbeel, P.; Bayen, A. Estimating arterial traffic conditions using sparse probe data. In Proceedings of the 13th International IEEE Conference on Intelligent Transportation Systems, Madeira, Portugal, 19–22 September 2010; pp. 929–936.

13. Kujala, R.; Aledavood, T.; Saramäki, J. Estimation and monitoring of city-to-city travel times using call detail records. *EPJ Data Sci.* **2016**, *5*, 1–16. [CrossRef]

14. Sun, Z.; Hao, P.; Ban, X.J.; Yang, D. Trajectory-based vehicle energy/emissions estimation for signalized arterials using mobile sensing data. *Transp. Res. Part D Transp. Environ.* **2015**, *34*, 27–40. [CrossRef]

15. Rodrigues, J.G.; Aguiar, A.; Vieira, F.; Barros, J.; Cunha, J.P.S. A mobile sensing architecture for massive urban scanning. In Proceedings of the 2011 14th International IEEE Conference on Intelligent Transportation Systems (ITSC), Washington, DC, USA, 5–7 October 2011; pp. 1132–1137.

16. Laña, I.; Del Ser, J.; Velez, M.; Vlahogianni, E.I. Road Traffic Forecasting: Recent Advances and New Challenges. *IEEE Intell. Transp. Syst. Mag.* **2018**, *10*, 93–109. [CrossRef]

17. García, S.; Luengo, J.; Herrera, F. *Data Preprocessing in Data Mining*; Springer: Berlin/Heidelberg, Germany, 2015.

18. Lopes, J.; Bento, J.; Huang, E.; Antoniou, C.; Ben-Akiva, M. Traffic and mobility data collection for real-time applications. In Proceedings of the 13th International IEEE Conference on Intelligent Transportation Systems, Madeira, Portugal, 19–22 September 2010; pp. 216–223.

19. Vlahogianni, E.I.; Golias, J.C.; Karlaftis, M.G. Short-term traffic forecasting: Overview of objectives and methods. *Transp. Rev.* **2004**, *24*, 533–557. [CrossRef]

20. Chen, H.; Grant-Muller, S.; Mussone, L.; Montgomery, F. A study of hybrid neural network approaches and the effects of missing data on traffic forecasting. *Neural Comput. Appl.* **2001**, *10*, 277–286. [CrossRef]

21. Qu, L.; Li, L.; Zhang, Y.; Hu, J. PPCA-based missing data imputation for traffic flow volume: A systematical approach. *IEEE Trans. Intell. Transp. Syst.* **2009**, *10*, 512–522.

22. Tan, H.; Feng, G.; Feng, J.; Wang, W.; Zhang, Y.J.; Li, F. A tensor-based method for missing traffic data completion. *Transp. Res. Part C Emerg. Technol.* **2013**, *28*, 15–27. [CrossRef]

23. Li, Y.; Li, Z.; Li, L. Missing traffic data: Comparison of imputation methods. *IET Intell. Transp. Syst.* **2014**, *8*, 51–57. [CrossRef]

24. Ran, B.; Tan, H.; Wu, Y.; Jin, P.J. Tensor based missing traffic data completion with spatial–temporal correlation. *Phys. A Stat. Mech. Appl.* **2016**, *446*, 54–63. [CrossRef]

25. Laña, I.; Olabarrieta, I.I.; Vélez, M.; Del Ser, J. On the imputation of missing data for road traffic forecasting: New insights and novel techniques. *Transp. Res. Part C Emerg. Technol.* **2018**, *90*, 18–33. [CrossRef]

26. Krempl, G.; Żliobaite, I.; Brzeziński, D.; Hüllermeier, E.; Last, M.; Lemaire, V.; Noack, T.; Shaker, A.; Sievi, S.; Spiliopoulou, M.; et al. Open challenges for data stream mining research. *ACM SIGKDD Explor. Newsl.* **2014**, *16*, 1–10. [CrossRef]

27. Etemad, M.; Soares Júnior, A.; Matwin, S. Predicting transportation modes of GPS trajectories using feature engineering and noise removal. In Proceedings of the Advances in Artificial Intelligence: 31st Canadian Conference on Artificial Intelligence, Canadian AI 2018, Toronto, ON, Canada, 8–11 May 2018; pp. 259–264.

28. Zheng, C.; Chen, S.; Wang, W.; Lu, J. Using principal component analysis to solve a class imbalance problem in traffic incident detection. *Math. Probl. Eng.* **2013**, *2013*, 524861. [CrossRef]

29. Smith, B.L.; Babiceanu, S. Investigation of extraction, transformation, and loading techniques for traffic data warehouses. *Transp. Res. Rec.* **2004**, *1879*, 9–16. [CrossRef]

30. El Faouzi, N.E.; Leung, H.; Kurian, A. Data fusion in intelligent transportation systems: Progress and challenges—A survey. *Inf. Fusion* **2011**, *12*, 4–10. [CrossRef]

31. Choi, K.; Chung, Y. A data fusion algorithm for estimating link travel time. *ITS J.* **2002**, *7*, 235–260. [CrossRef]

32. Chang, B.R.; Tsai, H.F.; Young, C.P. Intelligent data fusion system for predicting vehicle collision warning using vision/GPS sensing. *Expert Syst. Appl.* **2010**, *37*, 2439–2450. [CrossRef]

33. Han, L.; Wu, K. Radar and radio data fusion platform for future intelligent transportation system. In Proceedings of the 7th European Radar Conference, Paris, France, 30 September–1 October 2010; pp. 65–68.

34. Treiber, M.; Kesting, A.; Wilson, R.E. Reconstructing the traffic state by fusion of heterogeneous data. *Comput. Aided Civ. Infrastruct. Eng.* **2011**, *26*, 408–419. [CrossRef]

35. Vlahogianni, E.I. Enhancing predictions in signalized arterials with information on short-term traffic flow dynamics. *J. Intell. Transp. Syst.* **2009**, *13*, 73–84. [CrossRef]

36. Laña, I.; Lobo, J.L.; Capecci, E.; Del Ser, J.; Kasabov, N. Adaptive long-term traffic state estimation with evolving spiking neural networks. *Transp. Res. Part C Emerg. Technol.* **2019**, *101*, 126–144. [CrossRef]

37. Liu, T.; Hu, J.; Pei, X. Mining the Temporal-Spatial Patterns of Urban Traffic Demands Based on Taxi Mobility Data. In Proceedings of the 19th COTA International Conference of Transportation Professionals, Nanjing, China, 6–8 July 2019; pp. 2716–2728.

38. Moretti, F.; Pizzuti, S.; Panzieri, S.; Annunziato, M. Urban traffic flow forecasting through statistical and neural network bagging ensemble hybrid modeling. *Neurocomputing* **2015**, *167*, 3–7. [CrossRef]

39. Cong, Y.; Wang, J.; Li, X. Traffic flow forecasting by a least squares support vector machine with a fruit fly optimization algorithm. *Procedia Eng.* **2016**, *137*, 59–68. [CrossRef]

40. Kim, Y.J.; Hong, J.S. Urban traffic flow prediction system using a multifactor pattern recognition model. *IEEE Trans. Intell. Transp. Syst.* **2015**, *16*, 2744–2755.
41. Fusco, G.; Colombaroni, C.; Comelli, L.; Isaenko, N. Short-term traffic predictions on large urban traffic networks: Applications of network-based machine learning models and dynamic traffic assignment models. In Proceedings of the 2015 International Conference on Models and Technologies for Intelligent Transportation Systems (MT-ITS), Budapest, Hungary, 3–5 June 2015; pp. 93–101.
42. Montanino, M.; Punzo, V. Trajectory data reconstruction and simulation-based validation against macroscopic traffic patterns. *Transp. Res. Part B Methodol.* **2015**, *80*, 82–106. [CrossRef]
43. Chaulwar, A.; Botsch, M.; Utschick, W. A hybrid machine learning approach for planning safe trajectories in complex traffic-scenarios. In Proceedings of the 2016 15th IEEE International Conference on Machine Learning and Applications (ICMLA), Anaheim, CA, USA, 18–20 December 2016; pp. 540–546.
44. Vlahogianni, E.I. Optimization of traffic forecasting: Intelligent surrogate modeling. *Transp. Res. Part C Emerg. Technol.* **2015**, *55*, 14–23. [CrossRef]
45. Teodorović, D. Swarm intelligence systems for transportation engineering: Principles and applications. *Transp. Res. Part C Emerg. Technol.* **2008**, *16*, 651–667. [CrossRef]
46. Kumar, S.N.; Panneerselvam, R. A survey on the vehicle routing problem and its variants. *Intell. Inf. Manag.* **2012**, *4*, 66. [CrossRef]
47. Osaba, E.; Yang, X.S.; Diaz, F.; Lopez-Garcia, P.; Carballedo, R. An improved discrete bat algorithm for symmetric and asymmetric traveling salesman problems. *Eng. Appl. Artif. Intell.* **2016**, *48*, 59–71. [CrossRef]
48. Mendes-Moreira, J.; Moreira-Matias, L.; Gama, J.; de Sousa, J.F. Validating the coverage of bus schedules: A machine learning approach. *Inf. Sci.* **2015**, *293*, 299–313. [CrossRef]
49. Szeto, W.; Jiang, Y.; Wang, D.; Sumalee, A. A sustainable road network design problem with land use transportation interaction over time. *Netw. Spat. Econ.* **2015**, *15*, 791–822. [CrossRef]
50. Van Winden, K.; Biljecki, F.; Van der Spek, S. Automatic update of road attributes by mining GPS tracks. *Trans. GIS* **2016**, *20*, 664–683. [CrossRef]
51. Mannion, P.; Duggan, J.; Howley, E. An experimental review of reinforcement learning algorithms for adaptive traffic signal control. In *Autonomic Road Transport Support Systems*; Springer: Berlin/Heidelberg, Germany, 2016; pp. 47–66.
52. Osaba, E.; Yang, X.S.; Diaz, F.; Onieva, E.; Masegosa, A.D.; Perallos, A. A discrete firefly algorithm to solve a rich vehicle routing problem modelling a newspaper distribution system with recycling policy. *Soft Comput.* **2017**, *21*, 5295–5308. [CrossRef]
53. Imprialou, M.I.M.; Orfanou, F.P.; Vlahogianni, E.I.; Karlaftis, M.G. Methods for defining spatiotemporal influence areas and secondary incident detection in freeways. *J. Transp. Eng.* **2013**, *140*, 70–80. [CrossRef]
54. Yu, C.; Wang, X.; Xu, X.; Zhang, M.; Ge, H.; Ren, J.; Sun, L.; Chen, B.; Tan, G. Distributed Multiagent Coordinated Learning for Autonomous Driving in Highways Based on Dynamic Coordination Graphs. *IEEE Trans. Intell. Transp. Syst.* **2019**, *21*, 735–748. [CrossRef]
55. Lécué, F.; Tallevi-Diotallevi, S.; Hayes, J.; Tucker, R.; Bicer, V.; Sbodio, M.L.; Tommasi, P. Star-city: Semantic traffic analytics and reasoning for city. In Proceedings of the 19th International Conference on Intelligent User Interfaces, Haifa, Israel, 24–27 February 2014; pp. 179–188.
56. Gindele, T.; Brechtel, S.; Dillmann, R. Learning driver behavior models from traffic observations for decision making and planning. *IEEE Intell. Transp. Syst. Mag.* **2015**, *7*, 69–79. [CrossRef]
57. Kammoun, H.M.; Kallel, I.; Casillas, J.; Abraham, A.; Alimi, A.M. Adapt-Traf: An adaptive multiagent road traffic management system based on hybrid ant-hierarchical fuzzy model. *Transp. Res. Part C Emerg. Technol.* **2014**, *42*, 147–167. [CrossRef]
58. Mesbah, A. Stochastic model predictive control: An overview and perspectives for future research. *IEEE Control. Syst. Mag.* **2016**, *36*, 30–44.
59. Hrovat, D.; Di Cairano, S.; Tseng, H.E.; Kolmanovsky, I.V. The development of model predictive control in automotive industry: A survey. In Proceedings of the 2012 IEEE International Conference on Control Applications, Dubrovnik, Croatia, 3–5 October 2012; pp. 295–302.
60. Buchanan, C. *Traffic in Towns: A Study of the Long Term Problems of Traffic in Urban Areas*; Routledge: Abingdon, UK, 2015.
61. Pan, B.; Zheng, Y.; Wilkie, D.; Shahabi, C. Crowd sensing of traffic anomalies based on human mobility and social media. In Proceedings of the 21st ACM SIGSPATIAL International Conference on Advances in Geographic Information Systems, Orlando, FL, USA, 5–8 November 2013; pp. 344–353.
62. Davison, L.J.; Knowles, R.D. Bus quality partnerships, modal shift and traffic decongestion. *J. Transp. Geogr.* **2006**, *14*, 177–194. [CrossRef]
63. Nielsen, J. *Usability Engineering*; Elsevier: Amsterdam, The Netherlands, 1994.
64. Nielsen, J. Usability inspection methods. In Proceedings of the Conference Companion on Human Factors in Computing Systems, Boston, MA, USA, 24–28 April 1994; pp. 413–414.
65. Brooke, J. SUS-A quick and dirty usability scale. *Usability Eval. Ind.* **1996**, *189*, 4–7.
66. Nielsen, J. 10 Usability Heuristics for User Interface Design. 1995. Available online: https://www.nngroup.com/articles/ten-usability-heuristics/ (accessed on 1 December 2018).

67. Nielsen, J. Usability 101: Introduction to Usability. 2003. Available online: http://www.nngroup.com/articles/usability-101-introduction-to-usability/ (accessed on 1 December 2018).

68. Noy, Y.I. Human factors in modern traffic systems. *Ergonomics* **1997**, *40*, 1016–1024. [CrossRef]

69. Green, P. Estimating compliance with the 15-second rule for driver-interface usability and safety. In Proceedings of the Human Factors and Ergonomics Society Annual Meeting, Houston, TX, USA, 27 September–1 October 1999; Volume 43, pp. 987–991.

70. Green, P. Navigation System Data Entry: Estimation of Task Times. 1999. Available online: https://deepblue.lib.umich.edu/bitstream/handle/2027.42/1288/94020.0001.001.pdf?sequence=2 (accessed on 2 September 2018).

71. Burns, P.; Harbluk, J.; Foley, J.P.; Angell, L. The importance of task duration and related measures in assessing the distraction potential of in-vehicle tasks. In Proceedings of the 2nd International Conference on Automotive User Interfaces and Interactive Vehicular Applications, Pittsburgh, PA, USA, 11–12 November 2010; pp. 12–19.

72. Burnett, G. 'Turn right at the traffic lights': The requirement for landmarks in vehicle navigation systems. *J. Navig.* **2000**, *53*, 499–510. [CrossRef]

73. Dos Santos, C.; Botura, G. Proposal of Ergonomic Intervention in Horizontal Traffic Signaling. In Proceedings of the International Conference on Applied Human Factors and Ergonomics, Los Angeles, CA, USA, 17–21 July 2017; pp. 1121–1130.

74. Avelar, S.; Hurni, L. On the design of schematic transport maps. *Cartogr. Int. J. Geogr. Inf. Geovis.* **2006**, *41*, 217–228. [CrossRef]

75. Roberts, M.J.; Newton, E.J.; Canals, M. Radi (c) al departures: Comparing conventional octolinear versus concentric circles schematic maps for the Berlin U-Bahn/S-Bahn networks using objective and subjective measures of effectiveness. *Inf. Des. J.* **2016**, *22*, 92–115.

76. Lyons, G. From Advanced Towards Effective Traveller Information Systems. In *Travel Behaviour Research The Leading Edge*; Hensher, D., Ed.; International Association for Travel Behaviour Research: Pergamon, Turkey, 2001; Volume 47, pp. 813–826.

77. Barfield, W.; Dingus, T.A. *Human Factors in Intelligent Transportation Systems*; Psychology Press: Hove, UK, 2014.

78. Vlahogianni, E.I.; Karlaftis, M.G.; Golias, J.C. Short-term traffic forecasting: Where we are and where we're going. *Transp. Res. Part C Emerg. Technol.* **2014**, *43*, 3–19. [CrossRef]

79. Horan, T.A.; Abhichandani, T.; Rayalu, R. Assessing user satisfaction of e-government services: Development and testing of quality-in-use satisfaction with advanced traveler information systems (ATIS). In Proceedings of the 39th Annual Hawaii International Conference on System Sciences (HICSS'06), Kauai, Hawaii, 4–7 January 2006; Volume 4, p. 83b.

80. Ross, T.; Burnett, G. Evaluating the human–machine interface to vehicle navigation systems as an example of ubiquitous computing. *Int. J. Hum. Comput. Stud.* **2001**, *55*, 661–674. [CrossRef]

81. Fischer, G.; Sullivan, J., Jr. Human-centered public transportation systems for persons with cognitive disabilities. In Proceedings of the Participatory Design Conference, Malmo, Sweden, 23–25 June 2002; pp. 194–198.

82. Dingus, T.A.; Hulse, M.C.; Jahns, S.K.; Alves-Foss, J.; Confer, S.; Rice, A.; Roberts, I.; Hanowski, R.J.; Sorenson, D. Development of Human Factors Guidelines for Advanced Traveler Information Systems and Commercial Vehicle Operations: Literature Review. 1996. Available online: https://www.fhwa.dot.gov/publications/research/safety/95153/index.cfm (accessed on 3 August 2018).

83. Beul-Leusmann, S.; Samsel, C.; Wiederhold, M.; Krempels, K.H.; Jakobs, E.M.; Ziefle, M. Usability evaluation of mobile passenger information systems. In Proceedings of the International Conference of Design, User Experience, and Usability, Heraklion, Greece, 22–27 June 2014; pp. 217–228.

84. Mazloumi, E.; Rose, G.; Currie, G.; Moridpour, S. Prediction intervals to account for uncertainties in neural network predictions: Methodology and application in bus travel time prediction. *Eng. Appl. Artif. Intell.* **2011**, *24*, 534–542. [CrossRef]

85. Van Hinsbergen, C.I.; Van Lint, J.; Van Zuylen, H. Bayesian committee of neural networks to predict travel times with confidence intervals. *Transp. Res. Part C Emerg. Technol.* **2009**, *17*, 498–509. [CrossRef]

86. Liu, Y.; Liu, Z.; Li, X.; Huang, W.; Wei, Y.; Cao, J.; Guo, J. Dynamic traffic demand uncertainty prediction using radio-frequency identification data and link volume data. *IET Intell. Transp. Syst.* **2019**, *13*, 1309–1317. [CrossRef]

87. Tsekeris, T.; Stathopoulos, A. Short-term prediction of urban traffic variability: Stochastic volatility modeling approach. *J. Transp. Eng.* **2009**, *136*, 606–613. [CrossRef]

88. Khosravi, A.; Mazloumi, E.; Nahavandi, S.; Creighton, D.; Van Lint, J. Prediction intervals to account for uncertainties in travel time prediction. *IEEE Trans. Intell. Transp. Syst.* **2011**, *12*, 537–547. [CrossRef]

89. Gunning, D. Explainable artificial intelligence (xai). In *Defense Advanced Research Projects Agency (DARPA), nd Web*; 2017; Available online: https://www.cc.gatech.edu/~alanwags/DLAI2016/(Gunning)%20IJCAI-16%20DLAI%20WS.pdf (accessed on 4 April 2018).

90. Barredo Arrieta, A.; Díaz-Rodríguez, N.; Del Ser, J.; Bennetot, A.; Tabik, S.; Barbado, A.; García, S.; Gil-López, S.; Molina, D.; Benjamins, R.; et al. Explainable Artificial Intelligence (XAI): Concepts, taxonomies, opportunities and challenges toward responsible AI. *Inf. Fusion* **2020**, *58*, 82–115. [CrossRef]

91. Samek, W.; Wiegand, T.; Müller, K.R. Explainable artificial intelligence: Understanding, visualizing and interpreting deep learning models. *arXiv* **2017**, arXiv:1708.08296.

92. Ras, G.; van Gerven, M.; Haselager, P. Explanation methods in deep learning: Users, values, concerns and challenges. In *Explainable and Interpretable Models in Computer Vision and Machine Learning*; Springer: Berlin/Heidelberg, Germany, 2018; pp. 19–36.

93. Vlahogianni, E.I.; Karlaftis, M.G.; Orfanou, F.P. Modeling the effects of weather and traffic on the risk of secondary incidents. *J. Intell. Transp. Syst.* **2012**, *16*, 109–117. [CrossRef]

94. Thiagarajan, A.; Ravindranath, L.; LaCurts, K.; Madden, S.; Balakrishnan, H.; Toledo, S.; Eriksson, J. VTrack: Accurate, energy-aware road traffic delay estimation using mobile phones. In Proceedings of the 7th ACM Conference on Embedded Networked Sensor Systems, Berkeley, CA, USA, 4–6 November 2009; pp. 85–98.

95. Thiagarajan, A. Probabilistic Models for Mobile Phone Trajectory Estimation. Ph.D. Thesis, Massachusetts Institute of Technology, Cambridge, MA, USA, 2011.

96. Geisler, S.; Quix, C.; Schiffer, S.; Jarke, M. An evaluation framework for traffic information systems based on data streams. *Transp. Res. Part C Emerg. Technol.* **2012**, *23*, 29–55. [CrossRef]

97. Gama, J.; Žliobaitė, I.; Bifet, A.; Pechenizkiy, M.; Bouchachia, A. A survey on concept drift adaptation. *ACM Comput. Surv. (CSUR)* **2014**, *46*, 44. [CrossRef]

98. Žliobaitė, I.; Pechenizkiy, M.; Gama, J. An overview of concept drift applications. In *Big Data Analysis: New Algorithms for a New Society*; Springer: Berlin/Heidelberg, Germany, 2016; pp. 91–114.

99. Delany, S.J.; Cunningham, P.; Tsymbal, A.; Coyle, L. A case-based technique for tracking concept drift in spam filtering. In *Applications and Innovations in Intelligent Systems XII*; Springer: Berlin/Heidelberg, Germany, 2005; pp. 3–16.

100. Méndez, J.R.; Fdez-Riverola, F.; Iglesias, E.L.; Díaz, F.; Corchado, J.M. Tracking concept drift at feature selection stage in spamhunting: An anti-spam instance-based reasoning system. In Proceedings of the European Conference on Case-Based Reasoning, Fethiye, Turkey, 4–7 September 2006; pp. 504–518.

101. Stiglic, G.; Kokol, P. Interpretability of sudden concept drift in medical informatics domain. In Proceedings of the 2011 IEEE 11th International Conference on Data Mining Workshops (ICDMW), Vancouver, BC, Canada, 11–14 December 2011; pp. 609–613.

102. Moreira-Matias, L.; Mendes-Moreira, J.; Gama, J.; Ferreira, M. On Improving Operational Planning and Control in Public Transportation Networks Using Streaming Data: A Machine Learning Approach. Ph.D. Thesis, Porto University, Porto, Porgutal, 2014.

103. Laña, I.; Del Ser, J.; Padró, A.; Vélez, M.; Casanova-Mateo, C. The role of local urban traffic and meteorological conditions in air pollution: A data-based case study in Madrid, Spain. *Atmos. Environ.* **2016**, *145*, 424–438. [CrossRef]

104. Moreira-Matias, L.; Alesiani, F. Drift3flow: Freeway-incident prediction using real-time learning. In Proceedings of the 2015 IEEE 18th International Conference on Intelligent Transportation Systems (ITSC), Gran Canaria, Spain, 15–18 September 2015; pp. 566–571.

105. Osekowska, E.; Johnson, H.; Carlsson, B. Maritime vessel traffic modeling in the context of concept drift. *Transp. Res. Procedia* **2017**, *25*, 1457–1476. [CrossRef]

106. Wibisono, A.; Jatmiko, W.; Wisesa, H.A.; Hardjono, B.; Mursanto, P. Traffic big data prediction and visualization using fast incremental model trees-drift detection (FIMT-DD). *Knowl. Based Syst.* **2016**, *93*, 33–46. [CrossRef]

107. Wu, T.; Xie, K.; Xinpin, D.; Song, G. A online boosting approach for traffic flow forecasting under abnormal conditions. In Proceedings of the 2012 9th International Conference on Fuzzy Systems and Knowledge Discovery (FSKD), Chongqing, China, 29–31 May 2012; pp. 2555–2559.

108. Procopio, M.J.; Mulligan, J.; Grudic, G. Learning terrain segmentation with classifier ensembles for autonomous robot navigation in unstructured environments. *J. Field Robot.* **2009**, *26*, 145–175. [CrossRef]

109. Zhang, Y.; Ye, Z. Short-term traffic flow forecasting using fuzzy logic system methods. *J. Intell. Transp. Syst.* **2008**, *12*, 102–112. [CrossRef]

110. Isaacson, D.; Robinson, J.; Swenson, H.; Denery, D. A concept for robust, high density terminal air traffic operations. In Proceedings of the 10th AIAA Aviation Technology, Integration, and Operations (ATIO) Conference, Fort Worth, TX, USA, 13–15 September 2010; p. 9292.

111. Chen, J.; Chen, L.; Sun, D. Air traffic flow management under uncertainty using chance-constrained optimization. *Transp. Res. Part B Methodol.* **2017**, *102*, 124–141. [CrossRef]

112. Wechsler, S.P.; Ban, H.; Li, L. The Pervasive Challenge of Error and Uncertainty in Geospatial Data. In *Geospatial Challenges in the 21st Century*; Springer: Berlin/Heidelberg, Germany, 2019; pp. 315–332.

113. Adar, E.; Re, C. Managing uncertainty in social networks. *IEEE Data Eng. Bull.* **2007**, *30*, 15–22.

114. De Lara, M. A mathematical framework for resilience: Dynamics, strategies, shocks and acceptable paths. *arXiv* **2017**, arXiv:1709.01389.

115. Colpaert, P.; Ballieu, S.; Verborgh, R.; Mannens, E. The Impact of an Extra Feature on the Scalability of Linked Connections. In Proceedings of the 7th International Workshop on Consuming Linked Data co-located with 15th International Semantic Web Conference, COLD@ISWC 2016, Kobe, Japan, 18 October 2016; p. 10.

116. Basu, A.; Vanajakshi, L. Genetic Algorithm Based Dynamic Route Planner for Public Transport 2. *Transport* **2016**, *2*, 3.

117. Schmitt, F.; Schulte, A. Experimental evaluation of a scalable mixed-initiative planning associate for future military helicopter missions. In Proceedings of the International Conference on Engineering Psychology and Cognitive Ergonomics, Las Vegas, NV, USA, 15–20 July 2018; pp. 649–663.

118. Zhang, W.; Zhu, B.; Zhang, L.; Yuan, J.; You, I. Exploring urban dynamics based on pervasive sensing: Correlation analysis of traffic density and air quality. In Proceedings of the 2012 Sixth International Conference on Innovative Mobile and Internet Services in Ubiquitous Computing, Palermo, Italy, 4–6 July 2012; pp. 9–16.

119. Zhang, Z.; Ni, M.; He, Q.; Gao, J.; Gou, J.; Li, X. Exploratory study on correlation between Twitter concentration and traffic surges. *Transp. Res. Rec.* **2016**, *2553*, 90–98. [CrossRef]

120. Zhang, S.; Wu, G.; Costeira, J.P.; Moura, J.M. Understanding traffic density from large-scale web camera data. In Proceedings of the IEEE Conference on Computer Vision and Pattern Recognition, Honolulu, HI, USA, 21–26 July 2017; pp. 5898–5907.
121. Kerner, B.S. Three-phase traffic theory and highway capacity. *Phys. A Stat. Mech. Its Appl.* **2004**, *333*, 379–440. [CrossRef]
122. Kerner, B.S. The physics of traffic. *Phys. World* **1999**, *12*, 25. [CrossRef]
123. Karpatne, A.; Atluri, G.; Faghmous, J.H.; Steinbach, M.; Banerjee, A.; Ganguly, A.; Shekhar, S.; Samatova, N.; Kumar, V. Theory-guided data science: A new paradigm for scientific discovery from data. *IEEE Trans. Knowl. Data Eng.* **2017**, *29*, 2318–2331. [CrossRef]
124. Vlahogianni, E.I.; Karlaftis, M.G.; Golias, J.C. Spatio-temporal short-term urban traffic volume forecasting using genetically optimized modular networks. *Comput. Aided Civ. Infrastruct. Eng.* **2007**, *22*, 317–325. [CrossRef]
125. Ramezani, M.; Geroliminis, N. On the estimation of arterial route travel time distribution with Markov chains. *Transp. Res. Part B Methodol.* **2012**, *46*, 1576–1590. [CrossRef]
126. Chen, B.; Cheng, H.H. A review of the applications of agent technology in traffic and transportation systems. *IEEE Trans. Intell. Transp. Syst.* **2010**, *11*, 485–497. [CrossRef]
127. Shahrbabaki, M.R.; Safavi, A.A.; Papageorgiou, M.; Papamichail, I. A data fusion approach for real-time traffic state estimation in urban signalized links. *Transp. Res. Part C Emerg. Technol.* **2018**, *92*, 525–548. [CrossRef]
128. Mintsis, E.; Vlahogianni, E.I.; Mitsakis, E.; Ozkul, S. Evaluation of a cooperative speed advice service implemented along an urban arterial corridor. In Proceedings of the 2017 5th IEEE International Conference on Models and Technologies for Intelligent Transportation Systems (MT-ITS), Naples, Italy, 26–28 June 2017; pp. 232–237.
129. Gupta, A.; Ong, Y.S. *Memetic Computation: The Mainspring of Knowledge Transfer in a Data-Driven Optimization Era*; Springer: Berlin/Heidelberg, Germany, 2018; Volume 21.
130. Neri, F.; Cotta, C. Memetic algorithms and memetic computing optimization: A literature review. *Swarm Evol. Comput.* **2012**, *2*, 1–14. [CrossRef]
131. Gonzalez, M.C.; Hidalgo, C.A.; Barabasi, A.L. Understanding individual human mobility patterns. *Nature* **2008**, *453*, 779. [CrossRef]
132. Chatterjee, S.; Mitra, B.; Chakraborty, S. Type2Motion: Detecting Mobility Context from Smartphone Typing. In Proceedings of the 24th Annual International Conference on Mobile Computing and Networking, New Delhi, India, 29 October–2 November 2018; pp. 753–755.
133. Tselentis, D.I.; Yannis, G.; Vlahogianni, E.I. Innovative motor insurance schemes: A review of current practices and emerging challenges. *Accid. Anal. Prev.* **2017**, *98*, 139–148. [CrossRef]
134. Andreasson, H.; Bouguerra, A.; Cirillo, M.; Dimitrov, D.N.; Driankov, D.; Karlsson, L.; Lilienthal, A.J.; Pecora, F.; Saarinen, J.P.; Sherikov, A.; et al. Autonomous transport vehicles: Where we are and what is missing. *IEEE Robot. Autom. Mag.* **2015**, *22*, 64–75. [CrossRef]
135. Kulik, A.; Dergachev, K. Intelligent transport systems in aerospace engineering. In *Intelligent Transportation Systems–Problems and Perspectives*; Springer: Berlin/Heidelberg, Germany, 2016; pp. 243–303.
136. Barmpounakis, E.N.; Vlahogianni, E.I.; Golias, J.C. Unmanned Aerial Aircraft Systems for transportation engineering: Current practice and future challenges. *Int. J. Transp. Sci. Technol.* **2016**, *5*, 111–122. [CrossRef]
137. Lin, M.; Hsu, W.J. Mining GPS data for mobility patterns: A survey. *Pervasive Mob. Comput.* **2014**, *12*, 1–16. [CrossRef]
138. Khodaei, M.; Papadimitratos, P. The key to intelligent transportation: Identity and credential management in vehicular communication systems. *IEEE Veh. Technol. Mag.* **2015**, *10*, 63–69. [CrossRef]
139. Menouar, H.; Guvenc, I.; Akkaya, K.; Uluagac, A.S.; Kadri, A.; Tuncer, A. UAV-enabled intelligent transportation systems for the smart city: Applications and challenges. *IEEE Commun. Mag.* **2017**, *55*, 22–28. [CrossRef]
140. Sucasas, V.; Mantas, G.; Saghezchi, F.B.; Radwan, A.; Rodriguez, J. An autonomous privacy-preserving authentication scheme for intelligent transportation systems. *Comput. Secur.* **2016**, *60*, 193–205. [CrossRef]
141. Yuan, Y.; Wang, F.Y. Towards blockchain-based intelligent transportation systems. In Proceedings of the 2016 IEEE 19th International Conference on Intelligent Transportation Systems (ITSC), Rio de Janeiro, Brazil, 1–4 November 2016; pp. 2663–2668.
142. Lei, A.; Cruickshank, H.; Cao, Y.; Asuquo, P.; Ogah, C.P.A.; Sun, Z. Blockchain-based dynamic key management for heterogeneous intelligent transportation systems. *IEEE Internet Things J.* **2017**, *4*, 1832–1843. [CrossRef]
143. Zheng, X.; Chen, W.; Wang, P.; Shen, D.; Chen, S.; Wang, X.; Zhang, Q.; Yang, L. Big data for social transportation. *IEEE Trans. Intell. Transp. Syst.* **2015**, *17*, 620–630. [CrossRef]
144. He, J.; Shen, W.; Divakaruni, P.; Wynter, L.; Lawrence, R. Improving traffic prediction with tweet semantics. In Proceedings of the Twenty-Third International Joint Conference on Artificial Intelligence, Beijing, China, 3–9 August 2013.
145. Ni, M.; He, Q.; Gao, J. Forecasting the subway passenger flow under event occurrences with social media. *IEEE Trans. Intell. Transp. Syst.* **2016**, *18*, 1623–1632. [CrossRef]
146. Evans-Cowley, J.S.; Griffin, G. Microparticipation with social media for community engagement in transportation planning. *Transp. Res. Rec.* **2012**, *2307*, 90–98. [CrossRef]
147. Kuflik, T.; Minkov, E.; Nocera, S.; Grant-Muller, S.; Gal-Tzur, A.; Shoor, I. Automating a framework to extract and analyse transport related social media content: The potential and the challenges. *Transp. Res. Part C Emerg. Technol.* **2017**, *77*, 275–291. [CrossRef]

148. Woodcock, J.; Edwards, P.; Tonne, C.; Armstrong, B.G.; Ashiru, O.; Banister, D.; Beevers, S.; Chalabi, Z.; Chowdhury, Z.; Cohen, A.; et al. Public health benefits of strategies to reduce greenhouse-gas emissions: Urban land transport. *Lancet* **2009**, *374*, 1930–1943. [CrossRef]

149. Chen, Z.; Xia, J.C.; Irawan, B. Development of fuzzy logic forecast models for location-based parking finding services. *Math. Probl. Eng.* **2013**, *2013*, 473471. [CrossRef]

150. Kramers, A. Designing next generation multimodal traveler information systems to support sustainability-oriented decisions. *Environ. Model. Softw.* **2014**, *56*, 83–93. [CrossRef]

151. Zhang, S.; Lee, C.K.; Chan, H.K.; Choy, K.L.; Wu, Z. Swarm intelligence applied in green logistics: A literature review. *Eng. Appl. Artif. Intell.* **2015**, *37*, 154–169. [CrossRef]

152. Feigon, S.; Murphy, C. *Shared Mobility and the Transformation of Public Transit*; TRID: Washington, DC, USA, 2016.

153. Vlahogianni, E.I.; Barmpounakis, E.N. Driving analytics using smartphones: Algorithms, comparisons and challenges. *Transp. Res. Part C Emerg. Technol.* **2017**, *79*, 196–206. [CrossRef]

154. Huang, Y.; Ng, E.C.; Zhou, J.L.; Surawski, N.C.; Chan, E.F.; Hong, G. Eco-driving technology for sustainable road transport: A review. *Renew. Sustain. Energy Rev.* **2018**, *93*, 596–609. [CrossRef]

155. Adamidis, F.K.; Mantouka, E.G.; Barmpounakis, E.N.; Vlahogianni, E.I. *Impacts of Eco Driving on Traffic Flow and Emissions in Large Scale Urban Networks*; Technical Report; TRID: Washington, DC, USA, 2019.

156. Pan, S.J.; Yang, Q. A survey on transfer learning. *IEEE Trans. Knowl. Data Eng.* **2009**, *22*, 1345–1359. [CrossRef]

157. Bajwa, S.I.; Chung, E.; Kuwahara, M. Performance evaluation of an adaptive travel time prediction model. In Proceedings of the Intelligent Transportation Systems, Vienna, Austria, 13–16 September 2005; pp. 1000–1005.

158. Getachew, A.; Obrien, E.J. Simplified site-specific traffic load models for bridge assessment. *Struct. Infrastruct. Eng.* **2007**, *3*, 303–311. [CrossRef]

159. Habibzadeh, H.; Boggio-Dandry, A.; Qin, Z.; Soyata, T.; Kantarci, B.; Mouftah, H.T. Soft sensing in smart cities: Handling 3vs using recommender systems, machine intelligence, and data analytics. *IEEE Commun. Mag.* **2018**, *56*, 78–86. [CrossRef]

160. Shew, C.; Pande, A.; Nuworsoo, C. Transferability and robustness of real-time freeway crash risk assessment. *J. Saf. Res.* **2013**, *46*, 83–90. [CrossRef] [PubMed]

161. Xu, C.; Wang, W.; Liu, P.; Guo, R.; Li, Z. Using the Bayesian updating approach to improve the spatial and temporal transferability of real-time crash risk prediction models. *Transp. Res. Part C Emerg. Technol.* **2014**, *38*, 167–176. [CrossRef]

162. Ibisch, A.; Stümper, S.; Altinger, H.; Neuhausen, M.; Tschentscher, M.; Schlipsing, M.; Salinen, J.; Knoll, A. Towards autonomous driving in a parking garage: Vehicle localization and tracking using environment-embedded lidar sensors. In Proceedings of the 2013 IEEE Intelligent Vehicles Symposium (IV), Gold Coast City, Australia, 23–26 June 2013; pp. 829–834.

163. Shen, M.; Habibi, G.; How, J.P. Transferable Pedestrian Motion Prediction Models at Intersections. *arXiv* **2018**, arXiv:1804.00495.

164. Smirnov, A.; Levashova, T. Knowledge fusion patterns: A survey. *Inf. Fusion* **2019**, *52*, 31–40. [CrossRef]

165. Wang, P.; Yang, L.T.; Li, J.; Chen, J.; Hu, S. Data fusion in cyber-physical-social systems: State-of-the-art and perspectives. *Inf. Fusion* **2019**, *51*, 42–57. [CrossRef]

166. Khaleghi, B.; Khamis, A.; Karray, F.O.; Razavi, S.N. Multisensor data fusion: A review of the state-of-the-art. *Inf. Fusion* **2013**, *14*, 28–44. [CrossRef]

167. Ramachandran, U.; Kumar, R.; Wolenetz, M.; Cooper, B.; Agarwalla, B.; Shin, J.; Hutto, P.; Paul, A. Dynamic data fusion for future sensor networks. *ACM Trans. Sens. Networks (TOSN)* **2006**, *2*, 404–443. [CrossRef]

168. Nguyen, T.T.; Yang, S.; Branke, J. Evolutionary dynamic optimization: A survey of the state of the art. *Swarm Evol. Comput.* **2012**, *6*, 1–24. [CrossRef]

169. Mavrovouniotis, M.; Li, C.; Yang, S. A survey of swarm intelligence for dynamic optimization: Algorithms and applications. *Swarm Evol. Comput.* **2017**, *33*, 1–17. [CrossRef]

170. Chang, W.C.; Cho, C.W. Online boosting for vehicle detection. *IEEE Trans. Syst. Man Cybern. Part B (Cybern.)* **2009**, *40*, 892–902. [CrossRef] [PubMed]

171. Saadallah, A.; Moreira-Matias, L.; Sousa, R.; Khiari, J.; Jenelius, E.; Gama, J. BRIGHT-Drift-Aware Demand Predictions for Taxi Networks. *IEEE Trans. Knowl. Data Eng.* **2019**, in press. [CrossRef]

172. Moreira-Matias, L.; Gama, J.; Ferreira, M.; Mendes-Moreira, J.; Damas, L. Predicting taxi–passenger demand using streaming data. *IEEE Trans. Intell. Transp. Syst.* **2013**, *14*, 1393–1402. [CrossRef]

173. Sun, S.; Shi, H.; Wu, Y. A survey of multi-source domain adaptation. *Inf. Fusion* **2015**, *24*, 84–92. [CrossRef]

174. Ou, C.; Ouali, C.; Karray, F. Transfer Learning Based Strategy for Improving Driver Distraction Recognition. In Proceedings of the International Conference Image Analysis and Recognition, Póvoa de Varzim, Portugal, 27–29 June 2018; pp. 443–452.

175. Xing, Y.; Lv, C.; Wang, H.; Cao, D.; Velenis, E.; Wang, F.Y. Driver activity recognition for intelligent vehicles: A deep learning approach. *IEEE Trans. Veh. Technol.* **2019**, *68*, 5379–5390. [CrossRef]

176. Xing, Y.; Tang, J.; Liu, H.; Lv, C.; Cao, D.; Velenis, E.; Wang, F.Y. End-to-End Driving Activities and Secondary Tasks Recognition Using Deep Convolutional Neural Network and Transfer Learning. In Proceedings of the 2018 IEEE Intelligent Vehicles Symposium (IV), Changshu, China, 26 June–1 July 2018; pp. 1626–1631.

177. Ye, H.; Liang, L.; Li, G.Y.; Kim, J.; Lu, L.; Wu, M. Machine learning for vehicular networks: Recent advances and application examples. *IEEE Veh. Technol. Mag.* **2018**, *13*, 94–101. [CrossRef]

178. Konečnỳ, J.; McMahan, H.B.; Ramage, D.; Richtárik, P. Federated optimization: Distributed machine learning for on-device intelligence. *arXiv* **2016**, arXiv:1610.02527.

179. McMahan, B.; Moore, E.; Ramage, D.; Hampson, S.; y Arcas, B.A. Communication-Efficient Learning of Deep Networks from Decentralized Data. In Proceedings of the Artificial Intelligence and Statistics, Fort Lauderdale, FL, USA, 20–22 April 2017; pp. 1273–1282.

180. Ferdowsi, A.; Challita, U.; Saad, W. Deep Learning for Reliable Mobile Edge Analytics in Intelligent Transportation Systems: An Overview. *IEEE Veh. Technol. Mag.* **2019**, *14*, 62–70. [CrossRef]

181. Vögel, H.J.; Süß, C.; Hubregtsen, T.; André, E.; Schuller, B.; Härri, J.; Conradt, J.; Adi, A.; Zadorojniy, A.; Terken, J.; et al. Emotion-awareness for intelligent vehicle assistants: A research agenda. In Proceedings of the IEEE/ACM 1st International Workshop on Software Engineering for AI in Autonomous Systems (SEFAIAS), Gothenburg, Sweden, 28 May 2018; pp. 11–15.

182. Kroll, A. Grey-box models: Concepts and application. *New Front. Comput. Intell. Its Appl.* **2000**, *57*, 42–51.

183. Oussar, Y.; Dreyfus, G. How to be a gray box: Dynamic semi-physical modeling. *Neural Netw.* **2001**, *14*, 1161–1172. [CrossRef]

184. Inga, J.; Flad, M.; Diehm, G.; Hohmann, S. Gray-Box Driver Modeling and Prediction: Benefits of Steering Primitives. In Proceedings of the 2015 IEEE International Conference on Systems, Man, and Cybernetics, Kowloon Tong, Hong Kong, 9–12 October 2015; pp. 3054–3059.

185. Flad, M.; Fröhlich, L.; Hohmann, S. Cooperative shared control driver assistance systems based on motion primitives and differential games. *IEEE Trans. Hum. Mach. Syst.* **2017**, *47*, 711–722. [CrossRef]

186. Mittal, S. A survey of techniques for approximate computing. *ACM Comput. Surv. (CSUR)* **2016**, *48*, 62. [CrossRef]

187. Alwadi, M.; Chetty, G. Energy Efficient Data Mining Scheme for High Dimensional Data. *Procedia Comput. Sci.* **2015**, *46*, 483–490. [CrossRef]

188. Han, J.; Orshansky, M. Approximate computing: An emerging paradigm for energy-efficient design. In Proceedings of the 2013 18th IEEE European Test Symposium (ETS), Avignon, France, 27–30 May 2013; pp. 1–6.

189. Lane, N.D.; Georgiev, P. Can deep learning revolutionize mobile sensing? In Proceedings of the 16th International Workshop on Mobile Computing Systems and Applications, Santa Fe, NM, USA, 12–13 February 2015; pp. 117–122.

190. Faisal, S. Towards Energy Efficient Data Mining & Graph Processing. Ph.D. Thesis, The Ohio State University, Columbus, OH, USA, 2015.

191. Zliobaite, I.; Hollmen, J.; Koskinen, L.; Teittinen, J. Towards hardware-driven design of low-energy algorithms for data analysis. *ACM SIGMOD Rec.* **2015**, *43*, 15–20. [CrossRef]

192. Arrieta, A.B.; Díaz-Rodríguez, N.; Ser, J.D.; Bennetot, A.; Tabik, S.; Barbado, A.; García, S.; Gil-López, S.; Molina, D.; Benjamins, R.; et al. Explainable Artificial Intelligence (XAI): Concepts, Taxonomies, Opportunities and Challenges toward Responsible AI. *arXiv* **2019**, arXiv:1910.10045.

193. Martin, K. Ethical implications and accountability of algorithms. *J. Bus. Ethics* **2018**, *160*, 835–850. [CrossRef]

194. Veale, M.; Binns, R. Fairer machine learning in the real world: Mitigating discrimination without collecting sensitive data. *Big Data Soc.* **2017**, *4*, 2053951717743530. [CrossRef]

195. Stoyanovich, J.; Howe, B.; Abiteboul, S.; Miklau, G.; Sahuguet, A.; Weikum, G. Fides: Towards a platform for responsible data science. In Proceedings of the 29th International Conference on Scientific and Statistical Database Management, Chicago, IL, USA, 27–29 June 2017; p. 26.

196. Whittaker, M.; Crawford, K.; Dobbe, R.; Fried, G.; Kaziunas, E.; Mathur, V.; West, S.M.; Richardson, R.; Schultz, J.; Schwartz, O. *AI Now Report 2018*; AI Now Institute at New York University: New York, NY, USA, 2018.

197. Victor, N.; Lopez, D.; Abawajy, J.H. Privacy models for big data: A survey. *Int. J. Big Data Intell.* **2016**, *3*, 61–75. [CrossRef]

198. Rashidi, T.H.; Abbasi, A.; Maghrebi, M.; Hasan, S.; Waller, T.S. Exploring the capacity of social media data for modelling travel behaviour: Opportunities and challenges. *Transp. Res. Part C Emerg. Technol.* **2017**, *75*, 197–211. [CrossRef]

199. Boyd, D.; Crawford, K. Critical questions for big data: Provocations for a cultural, technological, and scholarly phenomenon. *Information, Commun. Soc.* **2012**, *15*, 662–679. [CrossRef]

200. Chen, Y.; Guizani, M.; Zhang, Y.; Wang, L.; Crespi, N.; Lee, G.M. When traffic flow prediction meets wireless big data analytics. *arXiv* **2017**, arXiv:1709.08024.

201. Wilson, B.; Hoffman, J.; Morgenstern, J. Predictive inequity in object detection. *arXiv* **2019**, arXiv:1902.11097.

202. Lim, H.S.M.; Taeihagh, A. Algorithmic decision-making in AVs: Understanding ethical and technical concerns for smart cities. *Sustainability* **2019**, *11*, 5791. [CrossRef]

203. Bigman, Y.E.; Gray, K. Life and death decisions of autonomous vehicles. *Nature* **2020**, *579*, E1–E2. [CrossRef]

204. Fu, K.; Alhamadani, A.; Ji, T.; Lu, C.T. Batman or the joker? the powerful urban computing and its ethics issues. *SIGSPATIAL Spec.* **2019**, *11*, 16–25. [CrossRef]

205. Leben, D. Normative Principles for Evaluating Fairness in Machine Learning. In Proceedings of the AAAI/ACM Conference on AI, Ethics, and Society, New York, NY, USA, 7–8 February 2020; pp. 86–92.

206. Verma, S.; Rubin, J. Fairness definitions explained. In Proceedings of the IEEE/ACM International Workshop on Software Fairness (FairWare), Gothenburg, Sweden, 29 May 2018; pp. 1–7.

207. Zook, M.; Barocas, S.; Crawford, K.; Keller, E.; Gangadharan, S.P.; Goodman, A.; Hollander, R.; Koenig, B.A.; Metcalf, J.; Narayanan, A.; et al. Ten Simple Rules for Responsible Big Data Research. 2017. Available online: https://journals.plos.org/ploscompbiol/article?id=10.1371/journal.pcbi.1005399 (accessed on 5 January 2019).

208. Fei-Fei, L.; Fergus, R.; Perona, P. One-shot learning of object categories. *IEEE Trans. Pattern Anal. Mach. Intell.* **2006**, *28*, 594–611. [CrossRef]
209. Wang, Y.; Yao, Q.; Kwok, J.; Ni, L.M. Generalizing from a Few Examples: A Survey on Few-Shot Learning. *arXiv* **2019**, arXiv:1904.05046.
210. Krawczyk, B. Learning from imbalanced data: open challenges and future directions. *Prog. Artif. Intell.* **2016**, *5*, 221–232. [CrossRef]
211. Branco, P.; Torgo, L.; Ribeiro, R.P. A survey of predictive modeling on imbalanced domains. *ACM Comput. Surv. (CSUR)* **2016**, *49*, 31. [CrossRef]
212. Fernandez, A.; Herrera, F.; Cordon, O.; del Jesus, M.J.; Marcelloni, F. Evolutionary Fuzzy Systems for Explainable Artificial Intelligence: Why, When, What for, and Where to? *IEEE Comput. Intell. Mag.* **2019**, *14*, 69–81. [CrossRef]
213. Mencar, C.; Alonso, J.M. Paving the Way to Explainable Artificial Intelligence with Fuzzy Modeling. In Proceedings of the International Workshop on Fuzzy Logic and Applications, Genoa, Italy, 6–7 September 2018; pp. 215–227.
214. Li, Y. Deep Reinforcement Learning: An Overview. *arXiv* **2017**, arXiv:1701.07274.
215. Nisan, N.; Roughgarden, T.; Tardos, E.; Vazirani, V.V. *Algorithmic Game Theory*; Cambridge University Press: Cambridge, UK, 2007.
216. Sallab, A.E.; Abdou, M.; Perot, E.; Yogamani, S. Deep reinforcement learning framework for autonomous driving. *Electron. Imaging* **2017**, *2017*, 70–76. [CrossRef]
217. Ruch, C.; Richards, S.; Frazzoli, E. The Value of Coordination in One-Way Mobility-on-Demand Systems. *IEEE Trans. Netw. Sci. Eng.* **2019**, in press. [CrossRef]
218. Aldeen, Y.A.A.S.; Salleh, M.; Razzaque, M.A. A comprehensive review on privacy preserving data mining. *SpringerPlus* **2015**, *4*, 694. [CrossRef]
219. Agrawal, R.; Srikant, R. Privacy-preserving data mining. In Proceedings of the ACM SIGMOD International Conference on Management of Data, Dallas, TX, USA, 16–18 May 2000; Volume 29, pp. 439–450.
220. Mendes, R.; Vilela, J.P. Privacy-preserving data mining: methods, metrics, and applications. *IEEE Access* **2017**, *5*, 10562–10582. [CrossRef]
221. Ding, W.; Jing, X.; Yan, Z.; Yang, L.T. A survey on data fusion in internet of things: Towards secure and privacy-preserving fusion. *Inf. Fusion* **2019**, *51*, 129–144. [CrossRef]
222. Zhou, Y.; Chen, S.; Mo, Z.; Yin, Y. Privacy preserving origin-destination flow measurement in vehicular cyber-physical systems. In Proceedings of the 2013 IEEE 1st International Conference on Cyber-Physical Systems, Networks, and Applications (CPSNA), Taipei, Taiwan, 19–20 August 2013; pp. 32–37.
223. Florian, M.; Finster, S.; Baumgart, I. Privacy-preserving cooperative route planning. *IEEE Internet Things J.* **2014**, *1*, 590–599. [CrossRef]
224. Rabieh, K.; Mahmoud, M.M.; Younis, M. Privacy-preserving route reporting scheme for traffic management in VANETs. In Proceedings of the 2015 IEEE International Conference on Communications (ICC), London, UK, 8–12 June 2015; pp. 7286–7291.
225. Kim, S.W.; Park, S.; Won, J.I.; Kim, S.W. Privacy preserving data mining of sequential patterns for network traffic data. *Inf. Sci.* **2008**, *178*, 694–713. [CrossRef]
226. Huang, L.; Joseph, A.D.; Nelson, B.; Rubinstein, B.I.; Tygar, J. Adversarial machine learning. In Proceedings of the 4th ACM Workshop on Security and Artificial Intelligence, Chicago, IL, USA, 21 October 2011; pp. 43–58.
227. Szegedy, C.; Zaremba, W.; Sutskever, I.; Bruna, J.; Erhan, D.; Goodfellow, I.; Fergus, R. Intriguing properties of neural networks. *arXiv* **2013**, arXiv:1312.6199.
228. Akhtar, N.; Mian, A. Threat of adversarial attacks on deep learning in computer vision: A survey. *IEEE Access* **2018**, *6*, 14410–14430. [CrossRef]
229. Nguyen, G.; Dlugolinsky, S.; Bobák, M.; Tran, V.; García, Á.L.; Heredia, I.; Malík, P.; Hluchý, L. Machine Learning and Deep Learning frameworks and libraries for large-scale data mining: A survey. *Artif. Intell. Rev.* **2019**, *52*, 77–124 [CrossRef]
230. Hutter, F.; Kotthoff, L.; Vanschoren, J. *Automated Machine Learning: Methods, Systems, Challenges*; Springer Nature: Berlin/Heidelberg, Germany, 2019.

Article

How Does the Location of Transfer Affect Travellers and Their Choice of Travel Mode?—A Smart Spatial Analysis Approach

Jason Chia [1], Jinwoo (Brian) Lee [2,*] and Hoon Han [2]

[1]   School of Civil Engineering and Built Environment, Science and Engineering Faculty,
     Queensland University of Technology, Brisbane, QLD 4000, Australia; jasonchiaqut@gmail.com
[2]   City Planning, City Planning Discipline, School of Architecture and Built Environment,
     Faculty of Built Environment, The University of New South Wales, Sydney, NSW 2052, Australia;
     h.han@unsw.edu.au
*   Correspondence: brian.j.lee@unsw.edu.au; Tel.: +61-2-9385-4799

Received: 15 June 2020; Accepted: 4 August 2020; Published: 7 August 2020

Abstract: This study explores the relationship between the spatial distribution of relative transfer location (i.e., the location of the transfer point in relation to the trip origin and destination points) and the attractiveness of the transit service using smart card data. Transfer is an essential component of the transit trip that allows people to reach more destinations, but it is also the main factor that deters the smartness of the public transit. The literature quantifies the inconvenience of transfer in terms of extra travel time or cost incurred during transfer. Unlike this conventional approach, the new "transfer location" variable is formulated by mapping the spatial distribution of relative transfer locations on a homogeneous geocoordinate system. The clustering of transfer points is then quantified using grid-based hierarchical clustering. The transfer location factor is formulated as a new explanatory variable for mode choice modelling. This new variable is found to be statistically significant, and no correlation is observed with other explanatory variables, including transit travel time. These results imply that smart transit users may perceive the travel direction (to transfer) as important, in addition to the travel time factor, which would influence their mode choice. Travellers may disfavour even adjacent transfer locations depending on their relative location. The findings of this study will contribute to improving the understanding of transit user behaviour and impact of the smartness of transfer, assist smart transport planning and designing of new transit routes and services to enhance the transfer performance.

Keywords: transit; bus; transfer; smart card; spatial analysis; mode choice

---

1. Introduction

Smart, reliable and connected service has been a long-standing goal of transit agencies. Smart transit network and service design must consider the service connectivity to allow users to travel to spatially diverse destinations [1,2]. However, providing direct connectivity for all origin—destination pairs is simply infeasible and impractical. Smart transfer is becoming an essential component of the transit trip. The extra effort in making transfers efficient and convenient has deemed to be necessary to expand service coverage and to provide competitive area-wide connectivity [1,2]. Ironically, transfer is recognised by travellers as a significant impeding factor that disrupts the transit travel experience and deters the use of transit [3–5]. The literature formulates the effect of transfer in terms of the extra travel time such as the additional walking time, waiting time and in-vehicle travel time, incurred during transfer [5–7]. Another type of transfer penalty encapsulates subjective and psychological factors based on preferences, attitudes and perceptions of transit users [1,3,8].

The literature formulates the effect of transfer in the scalar form such as the extra walking time, waiting time, in-vehicle travel time and monetary transfer cost incurred during transfer [5–7]. Another type of transfer penalty encapsulates subjective and psychological factors based on preferences, attitudes and perceptions of transit users [1,3,8].

This study builds on the hypothesis that transfer location (i.e., the relative location of the transfer point to the trip origin and destination) has an impact on the attractiveness of the transit trip involving the transfer for travellers. This variable is quantified by the level of deviation of the transfer point from the straight path from the trip origin to destination. The deviation could be interpreted as intrinsic factors that reflect subjective or psychological impedance imposed by the transfer location. Whereas the burden of transfer has been quantified in terms of travel time and/or cost in the literature, we propose a new "transfer location" variable to represent the deviation in the travel direction. We examine if transit users tend to disfavour the transfer locations that deviate from a direct path to the trip destination and test this hypothesis by incorporating this novel explanatory variable to represent its underlying effect on the travel mode choice. To control the distance effect, this study proposes a combination of transformation techniques to keep only the deviation of transfer points. Despite an extensive range of research on transfer, the current literature has neglected the potential implication of transfer location in the decision-making of mode choice.

In the present study, we aim to improve the smartness of transfer and the explanatory ability of mode choice models by incorporating the transfer location variable. The findings of this study could contribute to smart transport planning and designing of new transit routes and smart services to enhance the transfer quality and to a more realistic assessment of transit service accessibility and connectivity. The spatial distribution of transfer points is analysed by their relative location with respect to the destination point. This study presents a transformation approach to convert the actual coordinates of the transit journey itineraries (i.e., origin, destination and transfer points) on a two-dimensional homogeneous geocoordinate. This approach may be useful for transit route choice and further transfer location analysis.

2. Literature Review

Transfer is an essential and inevitable component of the transit journey (a chain of trips). It allows passengers to reach more destinations by switching to different routes and modes, hence enhances the smartness of the transfer. In major cities with a multimodal transit system, the role of the transfer is more prevalent. In an integrated transit system, the focus is to provide seamless transfers between different trips in a journey [2,9]. Smart transfers at strategic locations improve transit connectivity and expand spatial coverage of transit systems [10]. Despite its essential role, transfers are often seen as a burden in using transit [11]. Inconvenient transfers deter the use of transit for potential transit users and reduce the satisfaction level of existing transit users, which ultimately leads to reduction in the ridership.

The conventional way of quantifying the inconvenience of transfer has been by incorporating it into a generalised cost term to account for the extra monetary costs, travel time and discomfort incurred during the service transferring [5,12,13]. Transfer penalty can be measured as an equivalence of the travel time or money saving by taking the ratio between the coefficients of transfer variables and time or cost variables. This ratio shows how much further people are willing to travel (time without transfer) or how much they are willing to pay (cost), to save one transfer, demonstrating the time and money that must be saved in order to justify one transfer [3,5]. The literature suggests that out-of-vehicle travel time is perceived as more onerous than in-vehicle travel time by transit users when making transfers [14,15]. In practice, the general rule of thumb is that walking and wait times are valued twice as much as the in-vehicle travel time [7,12]. Wardman, et al. (2001) suggest that bus users value the wait time about 1.2 times higher than the in-vehicle travel time and the walk time 1.6 times higher than the in-vehicle travel time [7]. Generally, the wait time during transfer is also valued higher than the walking time during transfer [12,16].

Operational factors such as service reliability, headways regularity, on-time performance and the availability of adequate information affect the quality of transfers [5,12,17]. Providing a guaranteed connection and a through ticket for transfer could significantly reduce the perceived penalty of transfers [7]. An empirical study conducted in Haifa, Israel demonstrated that waiving a transfer fee resulted in a significant increase in the transit ridership [6]. Another study conducted in metropolitan Los Angeles showed that the users' satisfaction with the transit service transfer has little to do with the physical characteristics of the facility, but service frequency and reliability have more impact [18]. A study by Currie and Loader found that the volume of transfers could significantly increase along a major transit route when the service headway is 10 min or shorter [2].

Physical environmental factors such as stop and station amenities may affect the smartness of transfer services. Guo and Wilson reported that transit users are more likely to use the transfer service if escalators are available at the transfer station to assist with changing of levels [11]. Providing amenities such as benches, shades, water fountains and rest rooms would increase the comfort and convenience of transit users while waiting and transferring [5,12]. Security and safety, such as the presence of security staffs and the actual crime rates within the transit facilities would affect the perception of transfer quality [19]. A case study of the London underground train found that the worst transfer locations were the stations with the largest and most complex transfer environments, and the best transfer locations perceived were those stations with simple transfer environments [20]. In the case of whether to take a transfer or walk a longer distance to a destination, Guo and Wilson found that the demand of transfer decreases if walking environments are improved [11]. For example, if wider sidewalks exist along the non-transfer path, transit users are less likely to use a transfer service.

In an integrated transit system, more research seeks to understand and minimise the real cost of transfer inconvenience [15]. Much effort has been devoted to understand and minimise the cost of different time components (e.g., walk and wait time) during transfer, such as the timed transfer concept. This concept optimises the slack time between the arrival of incoming vehicles with the departure of outgoing vehicles [21–23]. Ceder et al. developed a synchronised timetable by maximising the number of simultaneous bus arrivals at transfer nodes [14]. Shih et al. employed the heuristic model for the design of a coordinated network with transfer centres [24]. Similarly, Ting and Schonfeld used a heuristic algorithm to optimise the headways and slack times jointly for all coordinated routes, as the optimised slack times vary with different variables such as headways, vehicle arrival time variance and transfer volumes [25].

As much as minimising the transfer time is important, transit users could also consider the travel direction towards the transfer point. Conventionally, the inconvenience to transfer caused by transfer location is considered as an increase in transit travel time, in the scalar form. This concept is similar to the "angular cost" concept presented by Raveau et al. to measure the directness of a chosen transit route [26]. The conventional route choice models account for the service level of the route alternatives and the socioeconomic and demographic characteristics of users [27]. Raveau et al. found that transit users tend to penalise routes that deviate from a direct path to their destination [26]. The "angular cost" is measured as a function of $\sin\left(\frac{\theta}{2}\right)$, where θ is the angle formed between the direct path to the destination (OD) with the origin-transfer (OT) straight route, weighted by the Euclidean distance to transfer point (d).

3. Study Area and Data

The city of Brisbane accounts for approximately 70% of the total daily weekday trips in South East Queensland [28]. Brisbane has an extensive transit network of bus, rail and ferry systems, covering more than 10,000 km^2. The recent report by the Queensland Government revealed that from January to March 2016, 27.38 million trips were conducted by bus, followed by 12.21 million trips by train, 1.71 million trips by ferry and 1.93 million trips by tram [29]. Bus ridership consisted of more than 63% of total transit ridership. This shows that the bus is the dominant transit mode in Brisbane. The benefit of the bus, in comparison to the train, tram and ferry, is that it has the flexibility to access almost all locations where a road network is present. The nature of buses travelling on existing road networks gives more feasibility of adapting to change, such as the addition of new bus routes to serve more destinations. These considerations have steered the scope of this research towards bus ridership in Brisbane.

Brisbane's bus network may be characterised as a typical radial structure where more than 66% of the bus services operating to the city centre [30]. There are many routes heading in the same direction with very minor variations and no feeder or trunk services are currently provided. The CBD is the central hub for the bus system, where three grade-separated bus only corridors (busways) provide high-speed, high-capacity services to regional centres.

This study relies on two main data sources. First is the smart card data (big data), which is used to develop the transfer map of bus users in the study area. The one-day "go-card" data of Brisbane (24 November 2014, Monday) was used for the mapping. The data encapsulates the entire Brisbane area. The go-card is an electronic ticket for use on transit services throughout the network and records travel data when a traveller touches on at the start of any trip stage, and touches off at the end of the trip stage. This dataset contains information such as go-card ID, date of service, route ID, service ID, direction (inbound or outbound), boarding time and alighting time, boarding stop ID and alighting stop ID, ticket type, journey ID and trip ID. If it is a transfer journey, it would have a consecutive trip ID for each trip stage with the identical journey ID. According to TransLink, a journey is defined as the set of trip stages taken under one fare basis, while a trip is a ride on a single transit vehicle. This study adopts the same convention for the terms "journey" and "trip".

The second dataset used in this research is the 2009 Southeast Queensland Household Travel Survey (SQHTS). This single cross-sectional survey provides information on daily travel behaviour of all members of participating households, from 20 April through 28 June 2009. This includes how and why they travel, at what time of day journeys are made and the average journey distance and duration [28]. Respondents were also asked to report a range of personal information (e.g., age, gender, individual income, driver's license, etc.), and household related information (e.g., household size, number of vehicles, etc.).

4. Transformation Mapping of Transfer Coordinate

This section presents a transformation approach to project the transfer locations on a homogeneous coordinate to examine the spatial distribution pattern of transfers. Travellers could be guided by the geographical images formed in minds, rather than the external maps, especially when individuals are familiar with the settings. These mental constructions suffice as the sole source of spatial information. At instances when individuals are unfamiliar with the surroundings and need to rely on an external map, they still have to transcribe the cartographic information into their minds before they can act on the information [31,32]. In both instances, it is the spatial images in minds that best explain travellers' spatial behaviours.

4.1. Processing for Single-Transfer Journey Itineraries

The smart card data was processed to filter out direct bus journeys and the journeys with two or more service transfers. The single transfer bus journeys account for about 20% of the total bus journeys. The journeys with two or more transfers are negligible less than 1% of the total bus journeys. The analytical framework of the transfer impact in this paper was developed applicable to only single-transfer journeys. The first step of the transformation mapping is to reconstruct travel itineraries by combining related trips from each smart card holder to form complete journeys from origins to destinations, including transfers. The data processing to construct single-transfer journeys is shown in Figure 1.

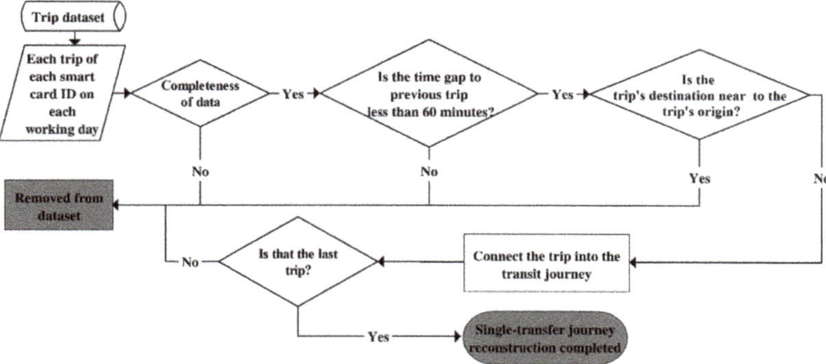

Figure 1. Process to construct single-transfer travel journeys.

The process starts by filtering out noise data such as the incomplete data of origin or destination information. A threshold of 60-min time gap (from the time when travellers alight a stop, to their next boarding time) is applied to identify whether two transactions are connected as a transfer journey. A different threshold has been chosen differently in the literature, ranging from 30 to 90 min [33,34], or a set of thresholds for different transit modes [35]. The threshold of the 60-min time gap is recommended in accordance with Brisbane's transit authority, based on the observed transfer behaviour and transit service characteristics [29]. If the transit user stays at a place for more than 60 min before making the next trip, those two trips are counted as separate trips, rather than a continuous journey through a transfer.

The next process is to distinguish return trips from single-transfer journeys. Studies have shown that transit users are willing to walk on average 400 or 500 m to bus stops [36–39]. A maximum distance threshold of 1 km from origin and destination was used to distinguish single-transfer journeys from return trips. To illustrate, the first bus stop could be located 500 m to the left of the journey's origin (e.g., the residence) and the last bus stop could be located 500 m to the right of the journey's destination (e.g., the residence). If the first and last bus stops are located less than 1 km apart, for the purpose of this study, it was assumed to be a return trip. This study was only interested in single-transfer journeys, so if there was any journey that had more than one transfer, the whole journey was removed from the dataset. After the reconstruction process, a total of 10,083 journeys were identified.

4.2. Transformation

The transit journey data may be illustrated as a triangle where each point of triangle represents the coordinate of the trip origin, destination and transfer. The size of the journey triangles varies by the actual trip distance and therefore the journey data needs to be converted into a homogeneous coordinate system to analyse the spatial distribution pattern. The conversion is done by applying a series of transformation techniques in this study. The first step of the transformation is to transform the journey triangle OTD (origin–transfer–destination) on a spherical Earth's surface to a 2D plan, given the latitudes and longitudes of each point of interest, as shown in Figure 2.

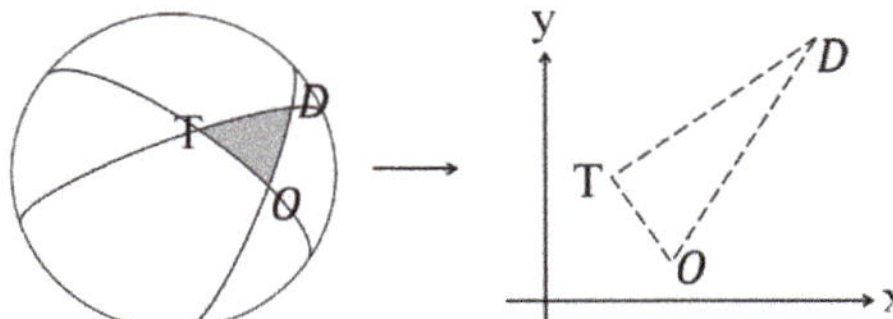

Figure 2. Transformation from a spherical Earth's surface to a 2D plan.

The great-circle distance between two points, which is the shortest distance over the Earth's surface, is calculated based on the spherical law of cosines. The spherical law of cosines states that, for a spherical triangle,

$$\cos OD = \cos OT \cos TD + \sin OT \sin TD \cos T \tag{1}$$

where,

O, T, D = Interest points of the journey triangle, OTD
OD = Distance between origin point and destination point
OT = Distance between origin point and transfer point
TD = Distance between transfer point and destination point

The location of any point on the earth can be defined by its latitude and longitude. In reference to Equation (1), the OD distance can be calculated as the arccosine of cos OD, as shown in Equation (2).

$$\cos OD = \cos OT \cos TD + \sin OT \sin TD \cos T$$
$$OD \text{ (in rad.)} = \cos^{-1}\left[\cos OT \cos TD + \sin OT \sin TD \cos T\right] \tag{2}$$

The unit used for angles is in radians, which gives the distance between origin and destination in radians. Given the convenient mean radius of the earth to be equivalent to 6371 km, the distance between origin and destination, in km, can be calculated by multiplying the OD distance (in radians) with 6371 km, as shown in Equation (3).

$$OD \text{ (in km)} = \cos^{-1}\left[\cos OT \cos TD + \sin OT \sin TD \cos T\right] * 6371 \text{ km} \tag{3}$$

The same technique was applied to calculate the great-circle distance of OT and TD. With the great-circle distance of OT, TD and OD, the respective angles of any triangle on a 2D plane could be calculated using the law of cosines, as shown in Equation (4).

$$\cos O = (OT^2 + OD^2 - TD^2)/2(OT * OD)$$
$$O = \cos^{-1}\left[(OT^2 + OD^2 - TD^2)/2(OT * OD)\right] \tag{4}$$

After the journey triangle OTD was obtained, it needs to undergo a series of Euclidean transformations to display all the origin, destination and transfer points in a standardised Euclidean space, as illustrated in Figure 3.

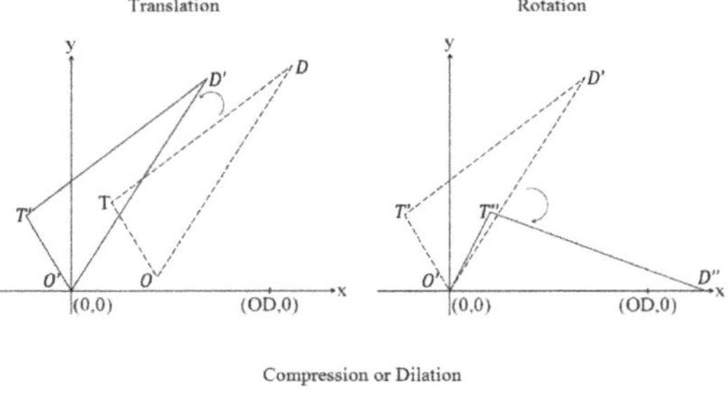
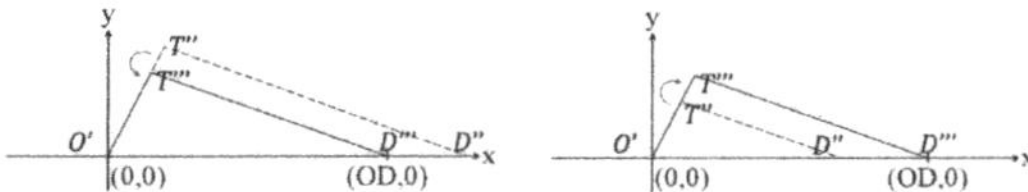

Figure 3. Euclidean transformations.

The first step of Euclidean transformation is translation. Translation relocates the journey triangle *OTD* to set the triangle's origin point, *O*, at (0, 0). This transformation preserves the congruence and distance of the journey triangle *OTD*. Applying the translation process to the single-transfer journeys results in all the journey triangles originating from the same point at (0, 0). The notation for translation ($T_{h,k}$) is shown in Equation (5). The origin and destination points will undergo the same transformation.

$$T_{h,k}\left(T'_x, T'_y\right) = \left(T_x + h,\ T_y + k\right) \tag{5}$$

Preserving the congruence and distance, the journey triangle *OTD* is rotated at *O* (0, 0) until the triangle plane, *OD*, rests on the *x*-axis. This transformation rotates all the journey triangles to lie along the *x*-axis for the destination point, *D*, to have the coordinate of (*x*, 0). The notation for rotation is shown in Equation (6).

$$\begin{bmatrix} T''_x \\ T''_y \end{bmatrix} = \begin{bmatrix} \cos\theta & -\sin\theta \\ \sin\theta & \cos\theta \end{bmatrix} \begin{bmatrix} T'_x \\ T'_y \end{bmatrix} \tag{6}$$

At this stage, all journey triangles *OTD* lie on the same plane (*x*-axis). The next step of the transformation is to loosen up the restriction to consider bijection, which preserves the shape and angles of the triangle, but not distance. The aim of this step is to transform all journey triangles *OTD* to have the same *OD* unit distance, as shown in Figure 4. The notation for compression and dilation (CD_k) is shown in Equation (7).

$$CD_k\left(T'''_x, T'''_y\right) = \left(kT''_x, kT''_y\right) \tag{7}$$

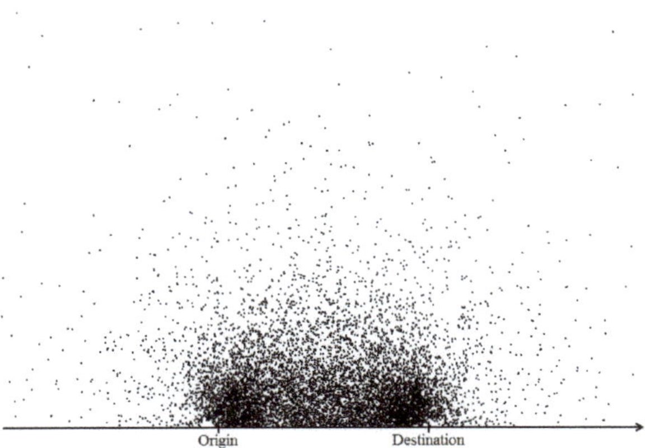

Figure 4. Transfer locations on a homogeneous coordinate.

4.3. Transfer Location Map

Figure 4 illustrates the transfer points of the single-transfer bus journeys, transformed to the scale of *OD* unit distance for both the *x* and *y* axis. In the figure, the scale is not the actual distance, but is adjusted through either compression or dilation. This study assumes that the plot represents the "acceptable" or "viable" transfer locations in relation to the straight path to destination. Consequent analysis quantifies and ranks the viability of transfer points and validates its impact on the travel mode choice.

The distribution of transfer points in Figure 4 may have been influenced by the availability and quality of transit services in the area. They are chosen transfer locations possibly among multiple alternatives, available to the travellers. The transit network structure must be an important determinant of the distribution pattern. The transit network of Brisbane takes the typical radial structure, with no trunk or feeder service. It is common that a transfer requires a significant deviation from the direct path to the destination or even in the opposite direction from the destination. The distribution pattern in the figure may reflect the inconvenience factor of the service transfer under the existing network structure.

4.4. Grid-Based Hierarchical Clustering

To analyse the spatial distribution pattern of the transfer points, this study used the grid-based hierarchical clustering method, which combines the grid-based clustering and hierarchical clustering methods. Cluster analysis is a data reduction tool that partitions a sample dataset into clusters, where objects within a specific cluster share many characteristics, but are very dissimilar to objects not belonging to that cluster [40]. The grid-based clustering (also known as density-based clustering) is one of the most efficient approaches for mining large data sets. Unsupervised clustering such as K-means was inappropriate for this study because it clusters the data in similar sizes (i.e., point densities). The underlying assumption of the choice modelling in the next step is that the cells with higher point densities are considered as more viable transfer locations by travellers. This method adopts algorithms that partition the data space into a finite number of cells to form a grid structure [41] as shown in Figure 5.

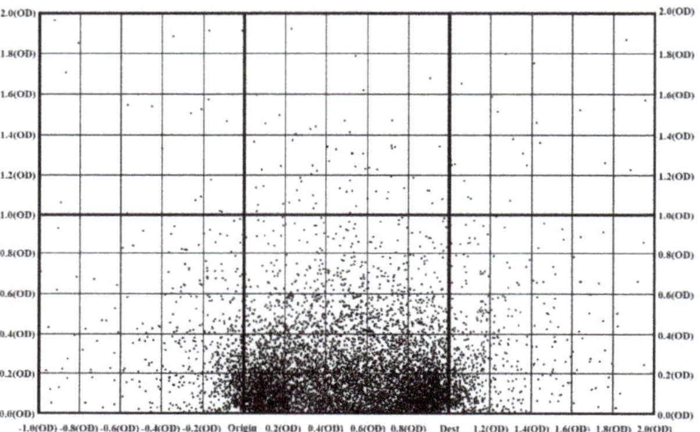

Figure 5. Grid structure on the transformed transfer location map.

For grid-clustering, each grid is defined as $0.2 \times OD$ (origin to destination) unit distance increment. These transfer points are plotted in reference to $1.0 \times OD$ unit distance. Figure 5 shows the clear concentration of transfer points in the cells, along with the direct path between the origin and destination. For cell clustering, the cell density is calculated for each cell as follows:

$$\text{Cell density} = \frac{\text{Total number of transfer points in grid } x}{\text{Total number of transfer points}} \tag{8}$$

The hierarchical clustering method was applied to sort the cells into clusters. Hierarchical clustering is useful for finding relatively homogenous clusters of cases based on measured characteristics. It starts with each case as a separate cluster. Next, these clusters are combined sequentially until only one cluster is left. The algorithm for this clustering method uses the dissimilarities or distances between objects when forming the clusters [40]. Figure 6 shows the cell-density for each cell, and to which cluster each cell is assigned by the hierarchical clustering method.

Figure 6 presents the preferred transfer locations with relatively high cell densities. Figure 7 shows the result of hierarchical clustering. Some interesting results were observed in the travellers' transfer selection. The majority of bus journeys conducted had made a transfer located in the cells F1 and J1. These two cells were identified to have the highest transfer point density at 13.78% and 15.47%, respectively (Cluster A). These two cells may be regarded as the most preferred transfer location and by having a transfer service in those cell locations will increase the likelihood of making a transfer and eventually taking a transit, compared to other cells.

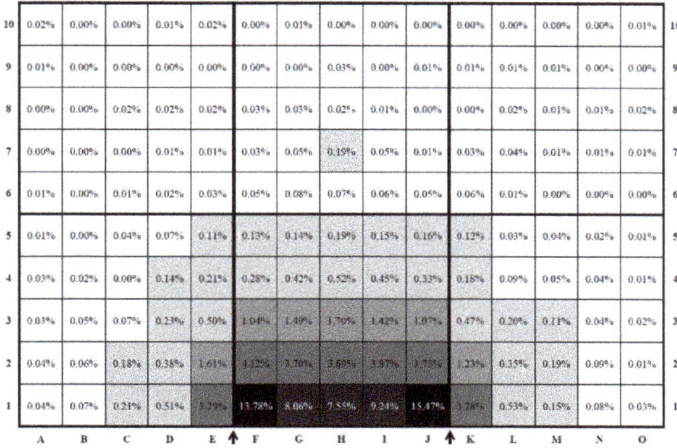

Figure 6. Cell-density in respective clusters.

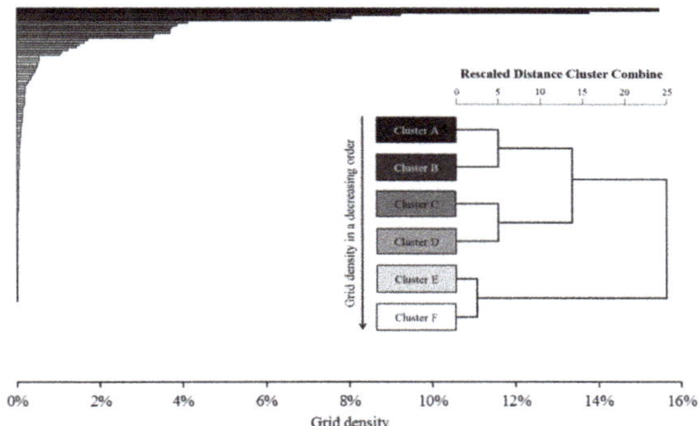

Figure 7. Cell density dendrogram.

All the other cells are categorised into five different clusters by the cell's grid density. The hierarchical clustering uses the Ward's method to measure the dissimilarity among clusters. Ward's method uses an analysis of variance approach, instead of distance metrics to evaluate the distances between clusters, where cluster membership is assessed by calculating the total sum of squared deviations from the mean of a cluster [42]. The dendrogram allows the tracing backward and forward to any cluster at any level. It gives an idea of how great the distance is between clusters in a particular step, using the 0–25 scale along the top of the chart.

Cluster B includes G1, H1 and I1. Transfer points in those five cells of Cluster A and Cluster B account for 54.1% out of the total 150 cells in the map. This implies that most travellers prefer the transfer point to be closely located along the direction to their trip destination. Cluster C consists of seven cells, E1, F2 to J2 and K1. The average cell density significantly declined to 3.67%. Some bus users travelled to transfer points in the opposite direction from their destination, but not too far from their origination location. Similarly, some travellers made a transfer farther from their destination location. The cell density further decreased for the cells in Cluster D with the average density value at 1.37%. The transfer points located in the Cluster A to D groups accounted for 89.35% of the total transfers. The average density value of the cells in Cluster E and Cluster F was negligible at 0.27% and

0.02%, respectively although they accounted for more than 87.33% of the total map area (131 out of 150 cells).

In general, based on the transfer point density in each cell, it was observed that the majority of transfers are conducted at the locations along the direct path to the destination. Transit users occasionally travel to the opposite direction from the destination, or slightly farther away from the destination to make a transfer. When a transit journey is required to make a transfer that is deviated from the direct path between origin and destination, the realisation of such trips is unlikely. This demonstrates the impedance of transfer location, and the interpretation must take into account the transit network structure.

5. Mode Choice Analysis

Two binomial logistic regression models (base and expanded model) are drawn on two travel modes: private vehicles and the bus. For a mode choice analysis, the information of mode specific variables of the alternative (unchosen) mode is necessary. Information of the alternative mode only can be inferred. This study used the GTFS (General Transit Feed Specification) data to infer the bus journey information for those who have chosen a private vehicle as their travel mode; and Google Maps to infer the private vehicle travel time for those who have chosen the bus as their travel mode. This analysis considers only the home-based work journeys. If a traveller has used the bus as the mode of transport, the journey must include one service transfer using the same travel mode (bus). Due to the strict criteria, only 330 private vehicle journeys and 63 bus journeys were used for the analysis. The 2009-10 SEQHTS is the most recent and detailed dataset available to demonstrate travellers' travel patterns and mode choice.

The dependent variables of the model are dichotomous, representing travel mode choice (transit or private vehicle). The independent variables tested in this analysis include individual characteristics (gender, age, individual weekly income, household size and number of cars in the household), journey attributes (travel time, initial wait time, first mile walk time and last mile walk time) and transfer attributes (proportion of in-vehicle bus travel time, proportion of transfer walk time, proportion of transfer wait time, type of transfer and transfer location). Table 1 presents the list of independent variables with brief descriptions.

Table 1. List of independent variables.

Variable	Description
Socioeconomic Attributes	
Gender	Nominal variables: 0—male; 1—female
Age	Age of individuals
Individual weekly income	Individuals' weekly income, given in different income bracket
Number of cars	Total number of cars per household
Household size	Number of persons in the household
Journey Attributes	
Car travel time (minutes)	Total time taken to travel from origins to destinations using private vehicle
Bus travel time (minutes)	Total time taken to travel from origins to destinations using bus
Initial wait time (minutes)	Total wait time for the next available bus service
First mile walk time (minutes)	Walk time taken to access bus station from origination
Last mile walk time (minutes)	Walk time taken from bus station to destination
Transfer Attributes	
Proportion of in-vehicle bus travel time	Proportion of a journey spent on two buses
Proportion of transfer walk time	Proportion of a journey spent on walking for a transfer
Proportion of transfer wait time	Proportion of a journey spent on waiting for a transfer
Type of transfer	Nominal variables: 0—non-walking transfer; 1—otherwise
Transfer location	Ordinal and nominal variables: The cluster developed using smart card data (i.e., Cluster A–F encoded to 0–5), of which individual transfer location falls into

Two different models were developed to test the effectiveness of the transfer location variable. The base model (Model I) takes the conventional approach to account for the effect of transfer by incorporating the proportion of the in-vehicle bus travel time, proportion of transfer walk time and proportion of transfer wait time variables. The expanded model (Model II) used the same set of independent variables and an additional the "transfer location" variable. The test results of those two models are presented in Table 2.

Table 2. Binomial logit model results: transfer location as an ordinal variable.

Variables	Model I			Model II		
	Base Model			Expanded Model		
	Coefficient	Std. Err.	Exp. β	Coefficient	Std. Err.	Exp. β
Constant	−0.85	1.27	0.43	0.35	1.39	1.42
Socioeconomic Attributes						
Individual weekly income	−0.00 ***	0.00	1.00	−0.00 ***	0.00	1.00
Household size	0.43 ***	0.14	1.54	0.40 ***	0.14	1.49
Number of cars	−1.22 ***	0.24	0.29	−1.31 ***	0.25	0.27
Journey Attributes						
Car travel time (minutes)	0.04 ***	0.02	1.04	0.06 ***	0.02	1.07
Initial wait time (minutes)	−0.03	0.02	0.97	−0.03 *	0.02	0.97
First mile walk time (minutes)	−0.08 *	0.04	0.93	−0.09 **	0.04	0.92
Last mile walk time (minutes)	−0.07 *	0.04	0.93	−0.08 **	0.04	0.92
Transfer Attributes						
Proportion of in-vehicle bus travel time	3.90 **	1.54	49.45	4.24 **	1.59	69.27
Transfer location	Not included			−0.28 **	0.13	0.75
Number of observation			393			393
Log-likelihood function value: Constant only model			−172.99			−172.99
Log-likelihood function value: Parameterised model			−122.40			−119.88
Goodness of fit (McFadden rho squared)			0.29			0.31
Model Improvement Test: −2 *(log-likelihood of basic model—log-likelihood of expanded model)						5.04
Chi-critical based on 1 degree of freedom						3.84

Notes: ***: $p < 0.01$; **: $p < 0.05$; *: $p < 0.1$. Coefficients that are statistically insignificant ($p \geq 0.1$) are not shown in this table.

Table 2 shows only the variables that provided the best fitting model fit. For instance, gender, age, network distance, transfer walking time, transfer wait time and transfer type were found not to be significant. The best-fitting basic model (Model I) incorporated six independent variables including: individual weekly income, household size, number of cars, car travel time, bus travel time, initial wait time for the first bus service, first mile walk time, last mile walk time and proportion of in-vehicle bus travel time. In Model II, the transfer location variable was found significant at the 0.05 level. This is notable as the new variable was found to make substantial influence on the mode choice, significant at the 0.05 level. As for socioeconomic variables, the household size had a positive effect on the utility of transit, whereas the individual weekly income and number of cars in the household had a negative effect on the transit utility. The car travel time factor was found significant (at the 0.01 level) among other journey attributes. Other journey attributes such as the first mile and last mile walk time were found significant at the 0.1 level for both the base and expanded models, which are consistent with the literature. As the access and egress increases, the use of transit decreases [38,43]. The initial wait time for the first bus service was found significant at the 0.1 level, only for the expanded models. If transit is not available at the time when individuals needed to travel, it decreases the attractiveness of transit.

As for the transfer-related variables, only the proportion of the in-vehicle bus travel time factor was found to be significant in both the base and expanded models (significant at the 0.05 level). Exp. β shows the effect of the independent variable on the odds ratio. The Exp. β coefficient relating the proportion of in-vehicle transit travel to the likelihood of using transit was 49.45 and 69.27 in the base and expanded model, respectively. These results implied that travellers are more likely to use transit as the proportion of in-vehicle bus travel time increases. This finding is consistent with the literature that shorter in-vehicle transit travel times could lead travellers to perceive the walking and wait times during transfer more onerous and eventually increases the relative attractiveness of a private vehicle [1,20,44,45].

The transfer location variable in the expanded model was found to be significant at the 95% confidence level. The negative coefficient suggests that a transfer location farther from the OD path will decrease the utility of bus and the probability to choose the bus mode. In fact, it turns out that the transfer location factor is one of the most important determinants of travel mode choice. This variable has the Exp. β (the odds ratio) value of 0.75, which shows that a change in the transfer location from a more preferred cluster to a less preferred cluster (e.g., from Cluster A to Cluster B) would decrease the probability of choosing the bus to 0.43, and increase the probability of choosing a private vehicle to 0.57.

The prediction capability of Model I and Model II was compared using McFadden rho squared to demonstrate the effectiveness of the new transfer location variable and its impact on the travel mode choice. Model I resulted in the pseudo R-squared, ρ^2 at 0.29, whereas Model II increased it to 0.31. McFadden suggested ρ^2 values of between 0.2 and 0.4 should represent a very good fit of the model [46]. The increase in ρ^2 by Model II demonstrates that with the inclusion of the new variable, Model II has a better explanatory power on mode choice as compared to Model I.

The chi-squared (χ^2) test was conducted to investigate the statistical improvement between Model I and Model II, by gauging the change in the log-likelihood function relative to the change in degrees of freedom. The chi-squared, χ^2 value of 5.04 exceeds the critical chi-squared of 1 degree of freedom of 3.84, at the 0.05 significant level. This gives a sufficient evidence to reject the null hypothesis that Model II is no better than Model I. With the inclusion of the transfer location variable into Model II, it outperforms Model I (base model).

The transfer location variable in Table 2 is ordered from the most preferred cluster to the least preferred cluster, in an ordinal-scale. This approach is effective to study the impact of transfer location as a variable, based on the assumption that the distance between clusters is equal. To study the relationship between the clusters, an additional binomial logistic regression model is conducted to include the transfer location variable as nominal variables. The result is shown in Table 3.

Table 3. Binomial logit model results: transfer location as nominal variables.

Variables	Model I			Model II		
	Base Model			**Expanded Model**		
	Coefficient	Std. Err.	Exp. β	Coefficient	Std. Err.	Exp. β
Constant	0.16	1.37	1.18	−0.98	1.53	0.38
Socioeconomic Attributes						
Age	−0.03 **	0.01	0.97	−0.03 ***	0.01	0.97
Individual weekly income	−0.00 ***	0.00	1.00	−0.00 ***	0.00	1.00
Household size	0.34 **	0.14	1.41	0.32 **	0.15	1.37
Number of cars	−1.32 ***	0.24	0.27	−1.48 ***	0.26	0.23
Journey Attributes						
Car travel time (minutes)	−0.05 ***	0.02	0.95	−0.07 ***	0.02	0.94
Initial wait time (minutes)	−0.02 *	0.01	0.98	−0.02 **	0.01	0.98
Last mile walk time (minutes)	−0.07 *	0.04	0.93	−0.08 **	0.04	0.92
Transfer Attributes						
Proportion of in-vehicle bus travel time	4.00 ***	1.51	54.52	4.52 ***	1.59	91.84
Transfer Location **The reference category: Cluster F**						
Cluster A		Not included		1.79 *	1.05	5.99
Cluster B				2.34 **	0.98	10.40
Cluster C				2.31 **	0.96	10.04
Cluster D				1.99 **	1.01	7.30
Cluster E				1.21	0.95	3.36

Table 3. *Cont.*

| Variables | Model I | | | Model II | | |
| | Base Model | | | Expanded Model | | |
	Coefficient	Std. Err.	Exp. β	Coefficient	Std. Err.	Exp. β
Number of observation			393			393
Log-likelihood function value: Constant only model			−172.99			−172.99
Log-likelihood function value: Parameterised model			−121.58			−116.20
Goodness of fit (Nagelkerke R Square)			0.39			0.43
Goodness of fit (McFadden R Square)			0.30			0.33
Model improvement test (Chi-squared test, χ^2): −2 *(log-likelihood of basic model—log-likelihood of expanded model)						10.76
The critical chi-squared value with 5 degrees of freedom at the 0.10 α-level						9.24
The critical chi-squared value with 5 degrees of freedom at the 0.05 α-level						11.07

Notes: ***: $p < 0.01$; **: $p < 0.05$; *: $p < 0.10$. Coefficients that are statistically insignificant ($p \geq 0.10$) are not shown in this table.

The result from Tables 2 and 3 did not differ much. The age of the travellers became significant at the 0.05 confidence level, with a negative effect on transit utility. Having the transfer location as nominal variables, Cluster F was assigned to be the reference category. The exponential β coefficient shows that if a transfer location is in Cluster A, it will have 5.99 times more chance to use the bus over Cluster F. Transfer locations located in Cluster B, C and D were found to be significant at the 0.05 level, but not Cluster E. This implies that as the transfer location changes from a less preferred cluster to a more preferred cluster (e.g., from Cluster F to Cluster A), it will increase the probability of choosing the bus over an automobile.

Transfer Location and Transit Travel Time

The analysis results indicate that the chance to make a transit trip was likely to decrease if the trip involves a transfer at the location that deviates from the direct path to the destination. In the conventional mode choice analysis, the level of deviation is quantified in terms of travel time incurred during the transfer. The new variable was created to capture the impact of deviation in the travel direction. We take two approaches to examine the potential collinearity between transfer location and transit travel time. Firstly, the Spearman's correlation coefficient (ρ) was calculated between the bus travel time (continuous variable) and the transfer location (categorical variable) of the Household Travel Survey bus trip data. A weak correlation ($\rho = -0.211$) was found between two variables, which implies that the location of transfer may play as an independent factor for the travel mode choice.

The second approach presents three plots of the transfer point distribution by the length of the bus journey time (less than 30 min, between 30 and 45 min and more than 45 min), as shown in Figure 8.

The level of deviation of transfer points was derived using Equation (9). An arbitrary cell length of 4 was used—for example, the deviation of a transfer point (x, y) was calculated as the sum of the distance from the origination point $(0, 0)$ and the distance from the destination point $(20, 0)$.

$$\text{Level of deviation} = \sqrt{x^2 + y^2} + \sqrt{(x - 20)^2 + y^2} \tag{9}$$

The average level of deviation of a short (less than 30 min), medium (between 30 and 45 min) and long journey (longer than 45 min) was found at 24.7, 24.3 and 25.7, respectively. Although more deviation was found among the longer bus trips, the distribution pattern is largely unchanged regardless of the length of travel time. This suggests that the preference for transfer location was not affected by the travel time (or distance) and travellers might disfavour adjacent transfer services depending on their relative location with respect to destination. The conventional approach of using door-to-door travel time to capture the transfer cost is not sufficient. Transit travel time may be able to capture the effect of deviation in the travel distance, but it is not capable to capture the effect of deviation in the travel direction towards transfer services.

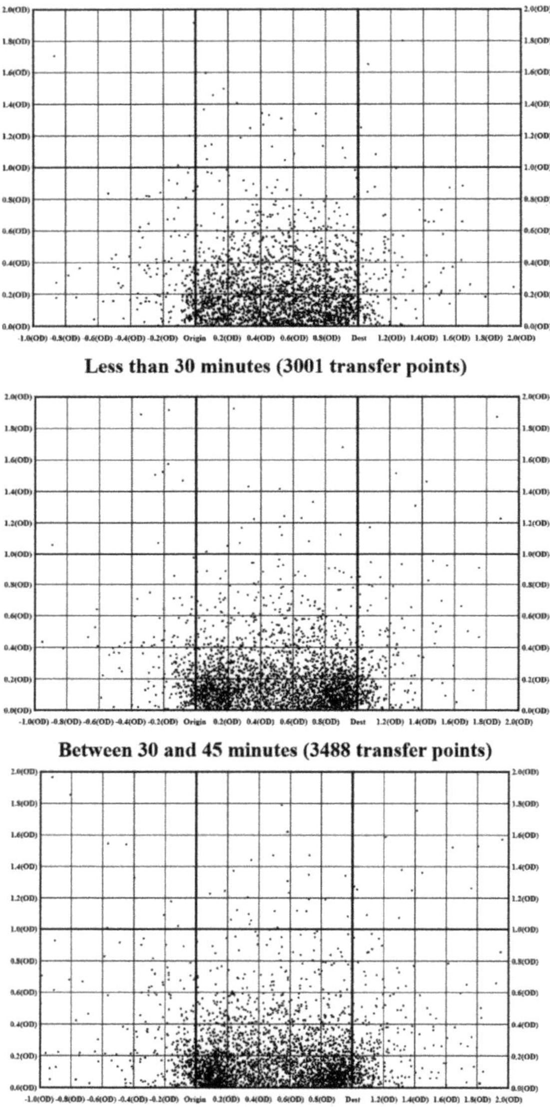

Less than 30 minutes (3001 transfer points)

Between 30 and 45 minutes (3488 transfer points)

More than 45 minutes (3594 transfer points)

Figure 8. Distribution of transfer points.

6. Conclusions

This paper proposed a new approach to take into account the smart distribution of the transfer location impact on travel mode choice. A transformation method was proposed for mapping of the transfer locations on a two-dimensional homogeneous geocoordinate. Transformed transfer locations were grouped into six classes by the level of point density. The novel transfer location variable was included in a mode choice model to demonstrate its underlying effect. The new variable was found as one of the driving determinants of mode choice. The study revealed that the transfer services in the "preferred locations" are likely to increase the smartness of the transit journey including the transfer. The transfer location and the in-vehicle travel time variables were found to be significant and uncorrelated to each other. This implies that travel direction towards transfer points may be an important factor pertaining to mode choice in addition to the travel time factor. The conventional approach of using door-to-door travel time may be not sufficient to capture the real cost of transfer.

This study provides a new approach to analyse the spatial distribution of transfer locations in relation to trip origin and destination points. This new geocoordinate technique may be useful for many applications in smart transport research including transit accessibility and connectivity studies. The conventional methods define the transit service accessibility and connectivity using a travel time constraint, where accessible areas by transit are simply defined as the travel boundary within a specific travel time period (e.g., 45 min). In a radial transit network structure, travelling to neighbouring suburbs often require a transfer at the opposite direction from the destination if there is no direct transit route connecting two suburbs, which is not so smart. For choice users (a private vehicle is available), such locations may be deemed as inaccessible by using transit. Integrating the transfer location variable to the traditional accessibility and connectivity measures may provide a more realistic representation of the service coverage of transit systems.

The findings of this study may contribute to improving the smartness of the public transit and the prediction capability of the mode choice analysis for the future transport demand. The findings presented in this study should be viewed as an exploratory effort to developing a new approach to account for the smartness of transfer and to test is effect on mode choice. The main findings will assist the transit service and performance assessment to identify service gaps and underserved areas. Identifying convenient and strategic transfer locations is essential so that scarce resources can be channelled effectively to improve the quality and smartness of transit service. Minimising the perceived transfer penalty will assist in increasing the competitiveness of public transport, and eventually the transit ridership. In this study, the emphasis is only given to bus journeys with a single transfer. Future research could build upon this concept to consider multimodal transit journeys and those journeys with more than a single transfer.

Author Contributions: Formal analysis, J.C.; Methodology, J.(B.)L. and H.H.; Writing—review and editing, J.(B.)L. and H.H. All authors have read and agreed to the published version of the manuscript.

Funding: This research received no external funding.

Conflicts of Interest: The authors declare no conflict of interest.

References

1. Ceder, A.; Chowdhury, S.; Taghipouran, N.; Olsen, J. Modelling public-transport users' behaviour at connection point. *Transp. Policy* **2013**, *27*, 112–122. [CrossRef]
2. Currie, G.; Loader, C. Bus network planning for transfers and the network effect in Melbourne, Australia. *J. Transp. Res. Board* **2010**, *2145*, 8–17. [CrossRef]
3. Guo, Z.; Ferreira, J. Pedestrian environments, transit path choice, and transfer penalties: Understanding land-use impacts on transit travel. *Environ. Plan. B Plan. Des.* **2008**, *35*, 461–479. [CrossRef]
4. Hadas, Y.; Ceder, A. Public transit network connectivity. *J. Transp. Res. Board* **2012**, *2143*, 1–8. [CrossRef]
5. Schakenbos, R.; Paix, L.L.; Nijenstein, S.; Geurs, K.T. Valuation of a transfer in a multimodal public transport trip. *Transp. Policy* **2016**, *46*, 72–81. [CrossRef]
6. Sharaby, N.; Shiftan, Y. The impact of fare integration on travel behavior and transit ridership. *Transp. Policy* **2012**, *21*, 63–70. [CrossRef]
7. Wardman, M.; Hine, J.; Stradling, S. *Interchange and Travel Choice*; Scottish Executive Central Research Unit: Edinburgh, UK, 2001.
8. Liu, R.; Pendyala, R.; Polzin, S. Assessment of intermodal transfer penalties using stated preference data. *J. Transp. Res. Board* **1997**, *1607*, 74–80. [CrossRef]

9. Givoni, M.; Banister, D. The need for integration in transport policy and practice. In *Integrated Transport: From poLicy to Practice*; Routledge: London, UK, 2010; p. 2.

10. Luk, J.; Olszewski, P. Integrated public transport in Singapore and Hong Kong. *Road Transp. Res.* **2003**, *12*, 41–51.

11. Guo, Z.; Wilson, N. Assessment of the transfer penalty for transit trips geographic information system-based disaggregate modeling approach. *J. Transp. Res. Board* **2004**, *1872*, 10–18. [CrossRef]

12. Iseki, H.; Taylor, B.D. Not all transfers are created equal: Towards a framework relating transfer connectivity to travel behaviour. *Transp. Rev.* **2009**, *29*, 777–800. [CrossRef]

13. Kittelson Associates Inc.; KFH Group Inc.; Parsons Brinckerhoff Quade Douglass, Inc.; Hunter-Zaworski, K. *Transit Capacity and Quality of Service Manual*; Texas Transport Institute and Transport Consulting Limited: Washington, DC, USA, 2013.

14. Ceder, A.; Golany, B.; Tal, O. Creating bus timetables with maximal synchronization. *Transp. Res. Part A Policy Pract.* **2001**, *35*, 913–928. [CrossRef]

15. Garcia-Martinez, A.; Cascajo, R.; Jara-Diaz, S.R.; Chowdhury, S.; Monzon, A. Transfer penalties in multimodal public transport networks. *Transp. Res. Part A Policy Pract.* **2018**, *114*, 52–66. [CrossRef]

16. Vande Walle, S.; Steenberghen, T. Space and time related determinants of public transport use in trip chains. *Transportation Research Part A Policy and Practice* **2006**, *40*, 151–162. [CrossRef]

17. Mishalani, R.G.; McCord, M.M.; Wirtz, J. Passenger wait time perceptions at bus stops: Empirical results and impact on evaluating real-time bus arrival information. *J. Public Transp.* **2006**, *9*, 89–106. [CrossRef]

18. Iseki, H.; Taylor, B.D. Style versus service? An analysis of user perceptions of transit stops and stations. *J. Public Transp.* **2010**, *13*, 2. [CrossRef]

19. Loukaitou-Sideris, A.; Liggett, R.; Iseki, H.; Thurlow, W. Measuring the effects of built environment on bus stop crime. *Environ. Plan. B Plan. Des.* **2001**, *28*, 255–280. [CrossRef]

20. Guo, Z.; Wilson, N. Assessing the cost of transfer inconvenience in public transport systems: A case study of the London Underground. *Transp. Res. Part A Policy Pract.* **2011**, *45*, 91–104. [CrossRef]

21. Chien, S.; Schonfeld, P. Joint optimization of a rail transit line and its feeder bus system. *J. Adv. Transp.* **1998**, *32*, 253–284. [CrossRef]

22. Hall, R.W. Vehicle scheduling at a transportation terminal with random delay en route. *Transp. Sci.* **1985**, *19*, 308–320. [CrossRef]

23. Lee, K.K.T.; Schonfeld, P. Optimal slack time for timed transfers at a transit terminal. *J. Adv. Transp.* **1991**, *25*, 281–308. [CrossRef]

24. Shih, M.C.; Mahmassani, H.; Baaj, M. Planning and design model for transit route networks with coordinated operations. *J. Transp. Res. Board* **1998**, *1623*, 16–23. [CrossRef]

25. Ting, C.J.; Schonfeld, P. Schedule coordination in a multiple hub transit network. *J. Urban Plan. Dev.* **2005**, *131*, 112–124. [CrossRef]

26. Raveau, S.; Muñoz, J.C.; de Grange, L. A topological route choice model for metro. *Transp. Res. Part A Policy Pract.* **2011**, *45*, 138–147. [CrossRef]

27. De Dios Ortúzar, J.; Willumsen, L.G.; Wiley, I. *Modelling Transport*, 4th ed.; Wiley-Blackwell: Oxford, UK, 2011.

28. Queensland Government. *Travel in South-East Queensland*; Queensland Government: Brisbane, Australia, 2012.

29. Queensland Government. *TransLink Tracker January-March 2016 Q3*; Department of Transport and Main Roads: Brisbane, Australia, 2016.

30. Devney, J. Redesigning bus networks to be simpler, faster and more connected. In Proceedings of the Australian Institute of Traffic Planning and Management (AITPM) National Conference, Adelaide, Australia, 13 August 2014.

31. Downs, R.M.; Stea, D. *Maps in Minds: Reflections on Cognitive Mapping*; HarperCollins Publishers: New York, NY, USA, 1977.

32. Schönfelder, S.; Axhausen, K.W. *Urban Rhythms and Travel Behaviour: Spatial and Temporal Phenomena of daily Travel*; Ashgate: Burlington, NJ, USA, 2010.

33. Bagchi, M.; White, P. What role for smart-card data from bus systems? *Munic. Eng.* **2004**, *157*, 39–46. [CrossRef]

34. Hofmann, M.; O'Mahony, M. Transfer journey identification and analyses from electronic fare collection data. In Proceedings of the 2005 IEEE Intelligent Transportation Systems Conference, Vienna, Austria, 13–16 September 2005; pp. 34–39.

35. Seaborn, C.; Attanucci, J.; Wilson, N.H.M. Analyzing multimodal public transport journeys in London with smart card fare payment data. *Transp. Res. Rec.* **2009**, *2121*, 55–62. [CrossRef]

36. Chia, J.; Lee, J.; Kamruzzaman, M. Walking to public transit—Exploring variations by socio-economic status. *Int. J. Sustain. Transp.* **2016**, *10*, 805–814. [CrossRef]

37. Horner, M.W.; Murray, A.T. Spatial representation and scale impacts in transit service assessment. *Environ. Plan. B* **2004**, *31*, 785–798. [CrossRef]

38. O'Sullivan, S.; Morrall, J. Walking distances to and from light-rail transit stations. *J. Transp. Res. Board* **1996**, *1538*, 19–26. [CrossRef]

39. Weinstein Agrawal, A.; Schlossberg, M.; Irvin, K. How far, by which route and why? A spatial analysis of pedestrian preference. *J. Urban Des.* **2008**, *13*, 81–98. [CrossRef]

40. Sarstedt, M.; Mooi, E. Cluster Analysis. In *A Concise Guide to Market Research: The Process, Data, and Methods Using IBM SPSS Statistics*; Sarstedt, M., Mooi, E., Eds.; Springer: Berlin/Heidelberg, Germany, 2011; pp. 273–324.

41. Cheng, W.; Wang, W.; Batista, S. Grid-based clustering. In *Data Clustering: Algorithms and Applications*; Aggarwal, C.C., Reddy, C.K., Eds.; Society for Industrial and Applied Mathematics: Philadelphia, PA, USA, 2013; Volume 31, pp. 127–148.

42. Ward, J.H. Hierarchical grouping to optimize an objective function. *J. Am. Stat. Assoc.* **1963**, *58*, 236–244. [CrossRef]

43. Cervero, R. Walk-and-ride: Factors influencing pedestrian access to transit. *J. Public Transp.* **2001**, *3*, 1. [CrossRef]

44. Frank, L.; Bradley, M.; Kavage, S.; Chapman, J.; Lawton, T.K. Urban form, travel time, and cost relationships with tour complexity and mode choice. *Transportation* **2008**, *35*, 37–54. [CrossRef]

45. Hadas, Y.; Ranjitkar, P. Modeling public-transit connectivity with spatial quality-of-transfer measurements. *J. Transp. Geogr.* **2012**, *22*, 137–147. [CrossRef]

46. Louviere, J.J.; Hensher, D.A.; Swait, J.D. *Stated Choice Methods: Analysis and Applications*; Cambridge University Press: Cambridge, UK, 2000.

Article

Iktishaf+: A Big Data Tool with Automatic Labeling for Road Traffic Social Sensing and Event Detection Using Distributed Machine Learning

Ebtesam Alomari [1], Iyad Katib [1], Aiiad Albeshri [1], Tan Yigitcanlar [2,3] and Rashid Mehmood [4,*

[1] Faculty of Computing and Information Technology, King Abdulaziz University, Jeddah 21589, Saudi Arabia; EAlomari0011@stu.kau.edu.sa (E.A.); IAKatib@kau.edu.sa (I.K.); aaalbeshri@kau.edu.sa (A.A.)

[2] School of Architecture and Built Environment, Queensland University of Technology, 2 George Street, Brisbane 4000, QLD, Australia; tan.yigitcanlar@qut.edu.au

[3] School of Technology, Federal University of Santa Catarina, Campus Universitario, Trindade, Florianópolis 88040-900, SC, Brazil

[4] High Performance Computing Center, King Abdulaziz University, Jeddah 21589, Saudi Arabia

* Correspondence: RMehmood@kau.edu.sa

Abstract: Digital societies could be characterized by their increasing desire to express themselves and interact with others. This is being realized through digital platforms such as social media that have increasingly become convenient and inexpensive sensors compared to physical sensors in many sectors of smart societies. One such major sector is road transportation, which is the backbone of modern economies and costs globally 1.25 million deaths and 50 million human injuries annually. The cutting-edge on big data-enabled social media analytics for transportation-related studies is limited. This paper brings a range of technologies together to detect road traffic-related events using big data and distributed machine learning. The most specific contribution of this research is an automatic labelling method for machine learning-based traffic-related event detection from Twitter data in the Arabic language. The proposed method has been implemented in a software tool called Iktishaf+ (an Arabic word meaning discovery) that is able to detect traffic events automatically from tweets in the Arabic language using distributed machine learning over Apache Spark. The tool is built using nine components and a range of technologies including Apache Spark, Parquet, and MongoDB. Iktishaf+ uses a light stemmer for the Arabic language developed by us. We also use in this work a location extractor developed by us that allows us to extract and visualize spatio-temporal information about the detected events. The specific data used in this work comprises 33.5 million tweets collected from Saudi Arabia using the Twitter API. Using support vector machines, naïve Bayes, and logistic regression-based classifiers, we are able to detect and validate several real events in Saudi Arabia without prior knowledge, including a fire in Jeddah, rains in Makkah, and an accident in Riyadh. The findings show the effectiveness of Twitter media in detecting important events with no prior knowledge about them.

Keywords: smart cities; big data; event detection; road traffic; distributed machine learning; automatic labeling; social media; data analytics; social media analytics; Arabic tweets

Citation: Alomari, E.; Katib, I.; Albeshri, A.; Yigitcanlar, T.; Mehmood, R. Iktishaf+: A Big Data Tool with Automatic Labeling for Road Traffic Social Sensing and Event Detection Using Distributed Machine Learning. *Sensors* **2021**, *21*, 2993. https://doi.org/10.3390/s21092993

Academic Editor: Alberto Gotta

Received: 19 March 2021
Accepted: 21 April 2021
Published: 24 April 2021

Publisher's Note: MDPI stays neutral with regard to jurisdictional claims in published maps and institutional affiliations.

1. Introduction

1.1. Smart Cities, Tranportation, and Social Sensing

Smart cities and societies aim to revolutionize our daily lives and improve social, economic, and environmental sustainability through increased technology penetration, participatory governance, and wise use of natural and other resources [1]. Smart urban and rural developments require timely sensing and analysis of diverse data produced by various edge sensors, smart devices, GPS, cameras, and the Internet of Things (IoT) [2]. Social media such as Twitter have become an important class of sensors for smart urban and rural developments [3], and in many sectors of smart cities and societies, it is increasingly being seen as a conveniently available and relatively inexpensive source of information compared to physical sensors [4]. Road transportation that is considered the backbone of modern economies is one such sector. It costs globally 1.25 million deaths and 50 million human injuries annually and therefore it is a research and development area of high significance.

Increased urbanisation is giving rise to the evolution of cities into megacities where traffic congestion is a leading problem causing devastating economic, social, and ecological losses. The annual cost of congestion in the US is USD305 billion, not to mention the damages to health and the number of deaths. Congestion is caused due to the steadily growing traffic in the cities over the years, road damages, roadworks, traffic accidents, bad weather, and other contingencies. There is a need to detect these causes or events to enable timely planning and operations.

Many times, congestion is caused due to events that are beyond the direct scope of physical road sensors, and therefore physical sensors cannot detect these events until the effects of these events are visible on the roads and can be sensed by the on-road sensors. For example, a football match in a city is likely to disrupt the traffic and increase pressure on the road network in certain segments of the city. Such an event can be detected through social media in advance of the event and timely intervention may reduce the aggravation of congestion in the city. Events such as a major football event may have already been known to the authorities. However, social media can also detect events that are being arranged on ad hoc bases—such as social gatherings, small sports gatherings—and, though these are small, there can be many of these in a city and can create an aggregately large pressure on the city roads. Similarly, unpredictable events such as a fire in a city segment may also disrupt the city traffic and such events can also be detected automatically on social media before their effects on the roads are visible. Moreover, historical analysis of social media data can reveal hidden information related to the traffic that may have not known otherwise and can be used for urban planning.

Twitter is one of the most popular microblogging media used for communication and sharing personal status, events, news, etc. [5]. Twitter allows users to post short text messages called tweets. A massive amount of real-time data is posted by millions of users on various topics including transportation and real-time road traffic [4,6–8]. In the recent decade, the use of Twitter and other social media by researchers and practitioners to study different issues in many application domains and sectors has steadily increased [9–14]. Transportation is no exception where social media has been used to study various aspects such as for analysing travel behaviours [15], recognizing mobility patterns [16], congestion detection [17], and event detection [4,6,9,18,19]. Due to the microblogging and real-time nature of Twitter, people are likely to communicate information about small and large-scale social gatherings, sports events, or events such as a fire, weather, allowing such information to be extracted in real-time [20]. Such information can allow the detection of transportation-related events and their causes for timely planning and operations. However, while manifesting great potential, several major challenges need to be overcome before its wide adoption in transportation and other areas.

1.2. Summary of the Proposed Work

The aim of this work is to develop big data technologies for detecting road traffic-related events (i.e., events that may affect road traffic) from Twitter data in the Arabic language with a focus on Saudi Arabia. Over the past few years, we have continued to build a detailed literature review on social media analytics in transportation. We have learnt from the literature review that the cutting-edge on big data-enabled social media analytics for transportation-related studies is limited. Many more studies are needed to improve the breadth and depth of the research on the subject in several aspects to establish maturity in this area. The research gaps relate to the focus of the studies, the size and diversity of the data, the applicability and performance of the machine learning methods, the diversity in terms of the social media languages, the scalability of the computing platforms, and others [13,21]. The maturity of research in this area will allow the development, commercialization, and wide adoption of the tools for transportation planning and operations (for the literature review and research gap, see Section 2).

This paper brings a range of technologies together to detect road traffic-related events using big data and distributed machine learning. The paper contributes to most of the above-mentioned research gaps. The most specific contribution of this research is an automatic labelling method for machine learning-based traffic-related event detection from Twitter data. In principle, the method itself is generic and can be applied for natural language processing (NLP) in any language. However, in this paper, the method is applied to Twitter data in the Arabic language. One of the approaches to detecting events from social media requires text classification using supervised classification algorithms. Supervised classification requires labeling of data for the training phase. For big data, the manual labeling process is time-consuming and labor-intensive [22]. Using the automatic labelling techniques developed in this paper we are able to deal with over an order of magnitude larger dataset compared to our earlier work in [6]. We are able to detect several real events in Saudi Arabia without any prior knowledge, including a fire in Jeddah, rains in Makkah, and an accident in Riyadh. The proposed automatic labeling method uses predefined dictionaries to reduce the effort, time, and cost of manual labeling of tweets. The dictionaries have been generated automatically for each event type using the top vocabularies extracted from the manually labeled dataset. Then, the dictionaries are adjusted manually to add synonyms and make sure that we do not miss any important vocabulary. After that, we divide them into levels based on the importance and the degree of relevance to the event type. Then, we calculate the weight for each labeled tweet (see Section 3 for details of the tool design including the automatic labelling method).

The proposed method has been implemented in a software tool called Iktishaf+ (an Arabic word meaning discovery) that is able to detect traffic events automatically from tweets in the Arabic language using distributed machine learning over Apache Spark. The tool is built using nine components that are used for nine specific functions namely data collection and storage, data pre-processing, tweets labeling, feature extraction, tweets filtering, event detection, spatio-temporal information extraction, reporting and visualization, and internal and external validation. The architectural blocks of the Iktishaf+ system are depicted in Figure 1 (we will describe in detail the system architecture including its nine components in Section 3). Iktishaf+ is built using a range of technologies including Apache Spark, Spark ML, Spark SQL, NLTK, PowerBI, Parquet, and MongoDB.

The Iktishaf+ tool uses Iktishaf Stemmer that we introduced in [6]. It is a light stemmer for the Arabic language developed by us. It is designed to strip affixes based on the length of the tokens. It allows reducing the feature space and minimizing the number of removed letters from the token to prevent changes in meaning or losing important words. We also use in this work a location extractor developed by us that helps to find the location of the detected events. It uses multiple methods to extract the event location. The location is extracted from the tweet text where the place name is explicitly mentioned in the message or is included as hashtags. Additionally, the event locations are identified from the account names using a predefined list of account names that are specialized in posting about traffic conditions in different cities in Saudi Arabia. If no information is found, the other attributes associated with the tweets JSON object such as coordinates and user profiles are checked for the location information. These methods have allowed us to extract and visualize spatio-temporal information about the detected events.

The specific data used in this work comprises 33.5 million tweets collected from Saudi Arabia using the Twitter API for a period of over a year. We have not used this data in any of the earlier works. The findings show the effectiveness of the Twitter media in detecting important events, and other information in time, space, and information-structure with no earlier knowledge about them. The detected events are validated using internal and external sources.

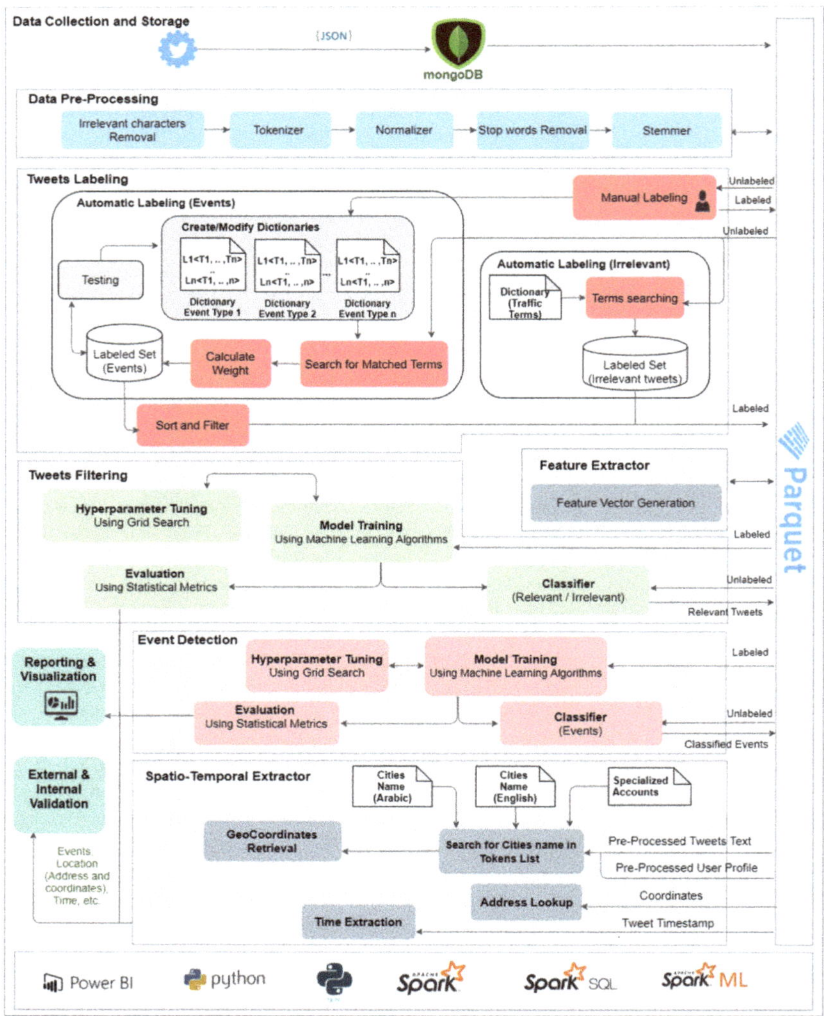

Figure 1. Iktishaf+: The proposed system architecture.

Iktishaf+ is an enhanced version of our tool Iktishaf that was introduced in [6]. The tool Iktishaf+ extends the functionality, capacity, and testing of our earlier work on traffic event detection. The earlier work on event detection has reported analyses of 2.5 million tweets [6]. The number of tweets was limited by our ability to manually label the tweets. We have also applied the Iktishaf tool to detect government measures and public concerns related to CVOID-19 using unsupervised learning [13]. Our earlier work on big data social media analytics has also focused on application areas including public sentiments analysis of government services [23], logistics [24,25], and healthcare [7] in both Arabic and English languages.

The Iktishaf+ tool uses open-source big data distributed computing technologies that enable the scalability and integration of transportation software systems with each other and with other smart city systems such as smart healthcare and urban governance systems. An elaboration of the novelty, contributions, and utilization of this work is given in Section 2.4.

The organization of the paper is as follows: Section 2 highlights the related work, in which we review different techniques for traffic events detection from social media in addition to the existing approaches for labeling large-scale datasets. Section 3 describes the proposed methodology and details the tool design and architecture. Section 4 explains the analysis results, which is followed by conclusions and future work reported in Section 5.

2. Literature Review

Digital societies could perhaps be characterized by their increasing desire to express themselves and interact with others, and this is done through various digital platforms [26]. The core ingredients of these digital societies, or digital platforms that enable these societies include a range of emerging technologies and their convergence. The technologies include big data [27–30], high-performance computing (HPC) [31–34], artificial intelligence [35–38], cloud, fog, and edge computing [39–41], social sensing [42–45], and Internet of Things (IoT) [36,40,46–48]. The applications include transportation [4,49–53], healthcare [7,54–56], and others [57–60]. The pulse for sensing and engaging with the environments are provided by social media and IoT. Sentiment analysis, or opinion mining, is a vital tool in natural language processing (NLP) [61], and many of the notable works on sentiment analysis rely on artificial intelligence, Twitter, and other social media.

We review here the literature related to the topics of this paper, which is detection of events related to road traffic using Twitter in the Arabic language. We begin with the literature about traffic event detection and then we discuss the solutions for automatic labeling. We discuss in Section 2.1 the works that have been developed for event detection in any language whether they use big data or not. Section 2.2 discusses works in the Arabic language and since works for Arabic are very limited, we introduce the studies related to detect any type of events not only traffic events. In Section 2.3, we present the solutions for labeling large datasets. Finally, Section 2.4 reveals the research gap.

2.1. Traffic Events Detection Using Social Data (Any Language)

Sakaki et al. [62] proposed an earthquake reporting system using Japanese tweets. They classified real-time tweets into positive (event-related) and negative (not related to events) classes using an SVM classifier. To prepare the training set, they used three groups of features for each tweet, which are keywords in a tweet, the number of words, and the words before and after the target-event words. Furthermore, they extended their work to extract events from tweets referring to driving information [63]. They collected tweets using a list of keywords about traffic-related events such as heavy traffic, traffic restriction, police checkpoints, parking, and rain mist. As in their previous work, they prepared the features using different methods and then they selected the best features to train a classifier using the SVM algorithm. Moreover, Klaithin and Haruechaiyasak [10] analyzed tweets in the Thai language to extract traffic events. They trained a classifier using Naïve Bayes to classify the tweets into six categories, which are accident, announcement, question, request, sentiment and traffic condition. They applied machine learning classifier based on Naive Bayes Model.

Kumar et al. [64] trained a sentiment classification model to detect negative sentiment about a road hazard from Twitter. The data is collected using search filtering with specific terms that relate to traffic. Then, naïve Bayes, K-nearest-neighbor and the dynamic language model (DLM) are used to build models to classify the tweets into a hazard and not hazard. Semwal et al. [65] applied real-time spatio-temporal analysis on Facebook data to detect traffic insights. They designed a module to detect the occurrence of events based on the spike in the number of posts at a specific time period and location. When the number is more than a threshold, the posts associated with that time and location are analyzed to evaluate their sentiments. Further, a random forest classifier was used to predict the most dominant issue for the next day. To address the problem of having an imbalanced dataset, they used SMOTE. Tejaswin et al. [66] also used random forest classifier to predict traffic incidents. The traffic incidents are clustered and predicted using spatio-temporal data from Twitter. The location information is extracted using NLP and background knowledge by using Freebase API, which is a community-curated structured database containing large number of entities and each one defied by multiple properties and attributes that helps in entity disambiguation.

Moreover, D'Andrea et al. [11] collected real-time Italian tweets and classified them after applying text mining techniques. The tweets are classified into three classes namely, traffic due to an external event, traffic congestion or crash, and non-traffic. They built a set of traffic events detected from official news websites or local newspapers and then they compared the time of detecting an event from these official sites with the time of detection from Twitter's stream fetched by their system.

None of the above-discussed approaches have used big data technologies. Salas et al. [44] used apache spark to process tweets and train model using SVM classification algorithm to classify them into traffic and non-traffic related tweets. To extract location information, they used a combination of name entity recognition (NER) such as Stanford NER and a knowledge base such as Wikipedia.

Suma et al. [67] built a classification model using logistic regression with stochastic gradient descent to detect events related to road traffic from English tweets using Apache Spark. Lau [42] used the latent Dirichlet allocation (LDA) topic modeling module to filter traffic messages. In addition, they used the Spark MLib library and trained classifiers using SVM, KNN and NB to detect traffic events. A detailed survey of event detection techniques using twitter data can be found in [68].

2.2. Traffic Events Detection Using Social Data (Arabic Language)

A very limited number of studies have proposed to analyze Arabic social information for traffic event detection, so first we review the studies about detecting any event not necessarily related to traffic. Then, we review the works that focus on transport and traffic events. Finally, we discuss the works that use big data. Alkouz and Alghbari [69] analyzed English and Arabic data, including standard Arabic and UAE dialectical posts from Twitter and Instagram to detect and predict traffic jams. They filtered the collected data to keep only traffic-related data. They used about 2.4 million tweets and 319,125 traffic-related image captions from Instagram. Further, the text is cleaned, tokenized and then stemmed using the NLTK root stemmer. Then, they used a predefined list of keywords to classify the posts into reporting posts or non-reporting posts where reporting posts contain at least one vocabulary from the list. They developed a tool to identify locations from the text of posts and/or GPS location. Further, they employed a linear regression model to predict future traffic jams. Moreover, Alkhatib et al. [70] analyzed tweets written in Modern Standard Arabic and Dialect Arabic analysis for the purpose of incident and emergency reporting. To detect incidents and disasters occurring in the UAE, they collected tweets in real-time using specific keywords, which are a car accident, earthquake, drought, hailstorm, heatwave, building collapse, riot and civil disorder. They labeled 8000 tweets manually to generate a training set and collected 82,150 tweets as a testing set. They built classification models using five machine learning algorithms, which are Polynomial Networks (PN), NB, KNN, Rachio (RA) and SVM. Moreover, they applied root stemmer and the results showed that it improves the classification accuracy. Further, they built NER corpus using Wikipedia to identify certain types of NEs such as building name, event risk and impact level and number of casualties. To extract location information, they built a dictionary of terms related to location names in the Dubai city.

Other researchers have proposed a solution to detect events but their main focus was not on traffic. AL-Smadi and Qawasmeh [71] extracted events about technology, sports, and politics using unsupervised rule-based technique. Alsaedi and Pete [72] developed a solution using naïve Bayes and online clustering algorithms to detect disruptive events. Alabbas et al. [73] detected a high-risk flood using a SVM classifier.

However, none of the above-discussed approaches for event detection from Arabic have used big data technologies. Alomari and Mehmood [18] developed a dictionary-based approach using SAP HANA, which is an in-memory processing platform to analyze Arabic tweets related to traffic congestion in Jeddah city. Additionally, they extracted traffic congestion causes. Furthermore, they extended their work and applied sentiment analysis on traffic-related tweets [19]. Moreover, they developed a supervised classification models using Apache Spark platform to detect eight types of traffic events, which are accident, roadwork, road closure, road damage, road condition, fire, weather, and social events. The results show that SVM achieves better results compared to logistic regression and Naive Bayes algorithm. Subsequently, they extended their work and validate the ability of the proposed Iktishaf tool [6] in detecting various events, their locations and times, with no earlier knowledge about the events from about 2.5 million tweets. Further, they designed a new light stemmer, Iktishaf Stemmer, for Arabic text and study the effect of using it on the performance. The results show that the performance of the trained model with and without using the proposed stemmer is almost similar. On the other hand, comparing to other light stemmers such as Tashaphyne and ISRI, Iktishaf Stemmer helps to minimize the number of letters removed and eliminate changes on the meaning especially for the words that related to transportation.

2.3. Solution for Labeling Large Scale Dataset (Any Language)

Manual labeling is a very challenging and expensive process especially with having a very large dataset and thus supervised learning is hard to applied on big social data. One of the solutions is crowdsourcing by cooperating with freelancers but one of the issues is the quality of the work [74]. Besides, crowdsourcing is not a fully automatic approach. In this section, we discuss the existing works that are similar to us and enables labeling text automatically to eliminate the need for human experts to label all the training set.

Pandey and Natarajan [75] proposed a system to extract situation awareness (SA) information and location from Twitter during disaster events. They suggested using semi-supervised classification instead of the traditional supervised machine learning approach, which would be tedious and time-consuming in term of labeling. For creating a semi-supervised model, they manually labeled a small set of tweets and then fed them to the SVM to classify them into situation awareness and non-situation awareness. Then, they used the result from this initial classification to self-train the model. However, their model achieved very low precision and recall value for situation awareness class.

Shafiabady et al. [76] suggested using an unsupervised clustering approach such as self-organizing maps (SOM) and correlation coefficient (CorrCoef) to group the unlabelled documents and use them as labelled data to train the SVM for text classification. However, their approach was applied on documents not on short text such as tweets. Ghahreman and Dastjerdi [77] applied semi-automatic labelling by combining co-training algorithms with the similarity evaluation measure. They labelled a small set of data manually, then they used the SVM algorithm to classify the unlabelled document. After that, based on the threshold, part of the output is selected. Then, they calculated the similarity between the selected documents and manually labelled documents

Zewen et al. [78] suggested labeling a few documents automatically using the external semantic resources e.g., HowNet. Then, they combined the labeled data and most of the unlabeled training data to train the classifier by semi-supervised learning. To label the documents automatically, they obtained the knowledge of the category name using lexical databases as external semantic resources and then they generated a set of features for the corresponding category. Further, they extracted features from the documents as a corresponding feature vector. After that, the similarity between each category name and each text document are calculated to rank the documents and classify them into the corresponding category. Triguero et al. [79] provided a taxonomy for the self-labeled techniques. One of the techniques is the addition mechanism. It consists of a variety of schemes, including incremental, batch and amending.

2.4. Research Gap, Novelty, Contributions, and Utilization

It can be seen from the literature review provided in this section that the works that use big data technology for traffic-related event detection are limited. To the best of our knowledge, none of the existing works for Arabic have used big data technology and platforms. Furthermore, none of them have used automatic labeling to address the problem of manual labeling of large datasets. The cutting-edge on big data-enabled social media analytics for transportation-related studies is limited. Many more studies are needed to improve the breadth and depth of the research on the subject in several aspects to establish maturity in this area. The research gaps relate to the focus of the studies, the size and diversity of the data, the applicability and performance of the machine learning methods, the diversity in terms of the social media languages, the scalability of the computing platforms, and others [13,21]. The maturity of research in this area will allow the development, commercialization, and wide adoption of the tools for transportation planning and operations.

The range of technologies that we have incorporated in the Iktishaf+ tool advances the state-of-the-art on big data social media analytics in the Arabic language in a number of ways (some of these contributions related to big data analysis also apply more broadly to English and other languages). Firstly, the extended tool Iktishaf+ has contributed multiple big data pipelines and architectures for event detection (in transportation and other sectors) from social media using cutting-edge technologies including data-driven distributed machine learning and high-performance computing. Secondly, it incorporates a novel pre-processing pipeline for Saudi dialectical Arabic that includes irrelevant characters removal, tokenizer, normalizer, stop words removal, and an Arabic light stemmer to improve event detection and overall performance. This will help many other works in the Arabic language to benefit from our work. Thirdly, the tool incorporates a range of lexicon-based, supervised, and unsupervised machine learning methods for event detection from social media in the Arabic language to enable smarter transportation and smarter societies. Using these methods, we have detected various physical and conceptual events such as congestion, fire, weather, government measures, and public concerns. Fourthly, the extended tool incorporates an automatic labeling method to reduce the effort, time, and cost of manual labeling of large datasets. We are not aware of any automatic labelling work in the Arabic language. Fifthly, we have developed and incorporated methods in the tool for spatial and temporal information from Twitter data to allow spatio-temporal clustering and visualization of detected events. Sixthly, we have developed methods for validating the detected events using internal and external sources. None of the existing works in English and other languages, particularly Arabic, have reported a similar analysis of Twitter data for event detection in terms of the richness of the methods, depth of analysis, and significance of findings. To the best of our knowledge, no work in the Arabic language exists that has used automatic labelling or big data tools or has reported analysis of a large number of tweets such as we have in this paper.

The scalability of the software systems for big data analytics is critical and is being hampered due to the challenges related to the management, integration, and analysis of big data (the 4V challenges). The use of big data distributed computing technologies is important because it will allow the scalability and integration of transportation software systems with each other and with other smart city systems. The ability of the Iktishaf+ tool to execute in parallel could save a month of computing time for the specific dataset size and the problem addressed in our work and speed up the development process [13]. For larger datasets, executing sequential codes may not even be possible, or distributed computing could save years of development time.

The utilization possibilities of our tool are many. For example, governments could learn about the various events, public concerns, and reactions related to certain government policies, measures, and actions (in pandemic and normal times) and develop policies and measures to address these concerns. The public could raise their concerns and give feedback on government policies. The public could learn about various public and industry activities (such as fires, social events, and other events, and economic activities detected by our tool in the earlier work [13]) and get involved in these to address financial, social, and other difficulties. The standardization and adoption of such tools could lead to real-time surveillance and the detection of transportation-related or other events, or disease outbreaks (and other potentially dangerous phenomena) across the globe and allow governments to take timely actions to prevent various risks, the spread of diseases, and other disasters. The international standardization of such tools could allow governments to learn about the impact of policies of various countries and develop best practices for national and international response.

3. Iktishaf+: Methodology and Design

Figure 1 illustrates the Iktishaf+ architecture. It consists of nine components, which are: (1) Data Collection and Storage Component, (2) Data Pre-Processing Component, (3) Tweets Labeling Component, (4) Feature Extractor Component, (5) Tweet Filtering Component, (6) Event Detection Component, (7) Spatio-Temporal Extractor Component, (8) Reporting and Visualization Component and (9) External and Internal Validation Component. The next subsection explains the tools and libraries used to develop Iktishaf+. Sections 3.2–3.10 elaborate each component in detail.

3.1. Tools and Libraries

Iktishaf+ is built over the Apache Spark platform, which enables in-memory processing on distributed data. The main libraries that have been used are Spark ML and Spark SQL. Spark.ML is a new package introduced in Spark 1.2. Unlike the Spark.MLlib package that was built on top of RDDs, Spark.ML contains higher-level API built on top of DataFrames creating and tuning practical machine learning pipelines. Moreover, the script was written using Python and runs on the Aziz supercomputer, a Fujitsu 230 TFLOPS machine comprising around 500 nodes, each with 24 cores. Besides, Fujitsu Exabyte File System (FEFS) has been used to provides high performance storage space as well as Scalable I/O performance. FEFS is a scalable parallel file system based on Lustre. The Aziz supercomputer supports running Spark with YARN, which allocates resources across applications.

Figure 2 shows the architecture of Apache Spark with YARN. Spark applications can run as independent sets of processes on a cluster. It acquires *Executors* on cluster nodes. The SparkContext is responsible for coordinating the application and enable connecting YARN. Then, it sends *tasks* to the executors to run computations for the application.

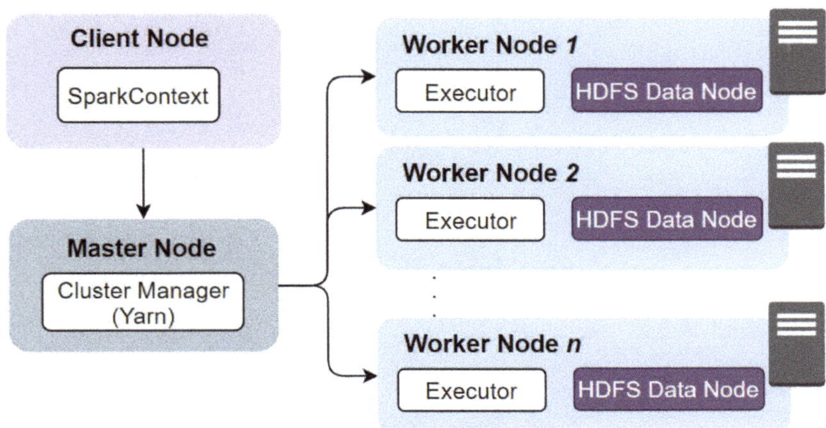

Figure 2. Spark Application Run on Yarn.

3.2. Data Collection and Storage Component (DCSC)

We collected the data using the Twitter REST API, which enables collecting historical data. The returned data by the Twitter API are encoded using JavaScript Object Notation (JSON). Each tweet object includes a unique ID, the text content itself, a timestamp that represents when it was posted, and many child objects such as 'user' and 'place'. Based on their documentation [80], a tweet can have over 150 attributes associated with it. Each child object encapsulates attributes to describe it. For instance, the 'user' object contains 'name', 'screen_ name', 'id', 'followers_count', and others. Some of the attributes belong to 'user' object can be filled in manually by the user such as 'description' and 'location'. Additionally, each tweet includes the 'entities' object, which encapsulates several attributes such as 'hashtags', 'user_mentions', 'media', and 'links'. The example in Figure 3 shows the core attributes of a tweet object.

```
"tweet" : {
        "created_at" : "Fri Oct 05 20:36:52 +0000 2018",
        "id" : NumberLong(1048311077195468801 0),
        "id_str" : "1048311077195468801 0",
        "full_text" : "@JeddahAmanah  ... ‎الأهم من المناظر الجمالية‎ ‎مافي خطوط سريعة كفاية من
‎الشرق للغرب .. طفشنا من رحمة النحلية و فلسطين‎",
        "entities" : {
            "hashtags" : [],
            "symbols" : [],
            "user_mentions" : [
                    "screen_name" : "JeddahAmanah",
                    "name" : "‎أمانة محافظة جدة‎",
                    "id" : 58427580,
                    "id_str" : "58427580",              },
            "urls" : []
        },
        "user" : {
            "id" : 720346162200,
            "id_str" : "720346162200",
            "name" : "ahmadd",
            "screen_name" : "ahd1109",
            "location" : "‎جدة‎",
            "description" : "",
            "url" : null,
            "entities" : {
                "description" : {
                    "urls" : []
                }
            },
        "coordinates" : null,
        "place" : null,

    }
}
```

Figure 3. A Tweet Object.

We fetch Arabic tweets posted by the users in Saudi Arabia by using geo-filtering. Besides, we searched for tweets using hashtags that included cities name. Both methods ensure collecting tweets about Saudi Arabia or posted from any place inside it, but not necessarily related to road traffic. Moreover, we created a list of specialized accounts that post about transportation and traffic condition in Saudi Arabia and then use them to obtain traffic-related tweets. We collected data in the period between September 2018 and October 2019. The total number of the collected tweets is 33.5 million. After that, we clean the data by removing duplicates and the retweets.

For storing the collected tweets, we need a storage method that provides flexible schemas to store/retrieve the data. So, we found that the NoSQL databases are more appropriate comparing to the traditional table structures in relational databases. One of the common types of NoSQL databases is a document-oriented database. In this type, each key is paired with the document of various document data types, such as XML and JSON. One of the most widely used document-oriented databases is MongoDB. Thus, we used MongoDB to store the fetched tweets using Twitter API.

Moreover, we use parquet file storage. The Parquet is a column-oriented format. It is supported by many data processing systems including Apache Spark. Furthermore, it enables very efficient compression. We select it to store the output after each stage because it is efficient and provides good performance for both storage and processing. Besides, Spark SQL supports both reading and writing Parquet files that automatically preserves the schema of the original data. After reading data from the Parquet file, it is stored in Spark DataFrames, which is equivalent to a table in a relational database or a data frame in R/Python. However, it provides richer optimizations.

3.3. Data Pre-Processing Component (DPC)

We use pre-processing component proposed in our earlier work, Iktishaf [6]. Algorithm 1 shows the Iktishaf+ pre-processing algorithm. It received the collected tweets, the Arabic diacritics [D], punctuations [P], and Arabic stop words [SW] as an input while the output is clean, normalized, and stemmed tokens. The collected tweets are exported from MongoDB and stored in Apache Spark DataFrame. The next subsections explain the main pre-processing steps.

3.3.1. Irrelevant Characters Removal

We removed Arabic diacritics, punctuation marks. English letters and numbers. For Arabic diacritics, we created a list of all the three forms of diacritics suggested by Diab et al. [81]. The first form is vowel diacritics. It refers to the three main short vowels, named in Arabic as Fatha (ـَ), Damma (ـُ) and Kasra (ـِ) as well as the Sukun diacritic (ـْ), which indicates the absence of any vowel. The second form is nunation diacritics, which are named in Arabic as Fathatan (ـً), Dammatan (ـٌ) and Kasratan (ـٍ). They represent the doubled version of the short vowels. The third form is called Shadda (germination) and refers to the consonant-doubling diacritical (ـّ). It also can be merged with diacritics from the two previous types and result in a new diacritic such as (ـ), (ـ). Therefore, the total number of Arabic diacritics is thirteen diacritical marks. All of them will be removed from the text.

Furthermore, we created a list of all the punctuation marks such as commas, period, colons, both Arabic and English semi-colons, and question marks, in addition to the different types of brackets, slashes and mathematical symbols as well as the other signs such as $, %, &, and @. For the hashtags, we strip only the hash (#) and underscore (_) symbols and keep the keywords because it may contain useful information such as the place or event name.

Algorithm 1 Pre-Processing.

Input: tweets; [D]; [P]; [SW]
Output: Clean, normalized and stemmed tokens

```
1      $spark ← createSparkSession()
2      tweets_DF ← spark.read(tweets)
3      ForEach tweet in tweets_DF['text'] do
//        Remove Irrelevant Characters
4         clean_tweets ← ""
5      For char in tweet do
6      If char not in [D] AND char not in [P] AND not char.isdigit() AND not char.isEnglishChar()
       then
7      clean_tweets.append(char )
8      end
//     Tokenize
9         tokens ← [ ]
10     tokens ← clean_tweets.split()
//     Normalize
11     normalized_tokens ← [ ]
12     For token in tokens do
13       For alif in ['أ', 'إ', 'آ'] do
14         If alif in token then
15           token ← token.replace(alif, 'ا')
16         end
17       If token.endswith(' ي') or token.endswith('ئ') then
18         token ← token.replaceLastCharWith('ى')
19       If token.endswith(' ة') then
20         token ← token.replaceLastCharWith('ه'}
// Remove Stop Words
21       If token not in [SW] then
22         normalized_tokens.append( token)
23     end
//     Stemmer
24     stemm_tokens ← stemmer(normalized_tokens)
25     End
```

3.3.2. Tokenizer and Normalizer

To divide the text into a list of words (tokens), we used a split() method in python, which returns a list of substrings after breaking the giving string by a specified separator, in our case the separator is any white space in the text. After that, the tokens are passed to the normalizer to replace letters that have different forms into the basic shape. The letter (ا) pronounced Alif had three forms (أ, إ, آ). It will be normalized to bare Alif (ا). Besides, the letter (ي) pronounced Yaa will be normalized to dotless Yaa (ى). In addition, the letter (ة) pronounced Taa marbutah will be normalized to (ه).

3.3.3. Stop-Words Removal

The Natural Language Toolkit (NLTK) [82] provided a stop-words list for Arabic Language. However, it was designed for the formal Modern Standard Arabic. Therefore, we modified the list and added new stop-words that usually used in dialectical Arabic such as "ليش", "اللي" and others. Subsequently, we considered common grammar mistakes. For instance, the preposition "متى" might be written "متا" and "لكن" might be written "لاكن". Besides, we added the common words that are used in Du'aa (prayer) such as "يارب", "اللهم", "الله" because they are frequently used and keeping them is not necessary, in our particular case. Before using the final generated stop-words list, we normalized them because they will be stripped from normalized tweets.

3.3.4. Stemmer

The existing Arabic stemmer tools can be categorized into two types, which are root-based stemmer and light stemmer. The first type extracts the root of the words while the light stemmers strip affixes (prefixes and suffixes). However, root-based stemmers are a heavy stemmer and known to have some weaknesses such as increasing word ambiguity. Thus, in this work, we decided to use a light stemmer. However, most of the existing Arabic light stemmers can lead to removing important parts of the word which results in a few letters with no meaning. Therefore, we used the developed Arabic Light Stemmer, Iktishaf Stemmer [6]. It is designed to strip affix based on the length of the word to reduce the chance of stripping important letters which could lead to change the meaning or losing important words particularly the words that are related to transportation. Algorithm 2 shows the algorithm of the proposed stemmer. It gets the normalized tokens, the lists of prefixes [P], and suffixes [S] as input. For prefix, we divide them into three lists, each list contains specific prefixes that do not usually come together in one word. P1 includes prefixes such as "ك", "و", "ف", "ب" whereas P2 contains "ما", "يا" and P3 includes prefixes such as "هال", "كال", "فال", "وال", "لل", "ال". Both P1 and P2 will be removed only if the token length is greater than 5 to decrease the chance of making mistakes and stripping important letters that are part of the token. For instance, "بريده" city starts with "ب" but the length is greater than 5 so it will not be affected by stemmer. On the other side, any prefixes in P3 will be stripped if the length of the word is greater than 4. Moreover, we take into consideration that the word may contain more than one prefix so the first list of prefixes contains the prefixes that come at the beginning of the word. For instance, the word "بمايناسب" contains two prefixes: 'ب', which is in P1 and 'ما', which is in P2 so 'ما' will be stripped after 'ب'.

Similarly, for suffix, we have three lists. S1 contains suffixes such as, "ها", "كم", "كن", while S2 contains "ين", "ون", "ان", "وا". To clarify, we give an example of the verb "يعرف". It can be ended with any of the following suffixes: كم, ون "ون". It may also contain two suffixes like in "يعرفونهم", which contains "هم, ني, ها, نا, هن" and "هم" So, "هم" will be removed first because it is in S1 and then "ون" will be removed in the next step.

After removing any suffix from the previous lists, we check the last letter in the word to see if it becomes end with 'ت' to replace it with 'ه'. For example, the word "سيارتهم" (their car) will become "سيارت" after removing "هم" but the correct spelling is "سياره" so we need to replace 'ت' with 'ه'.

Subsequently, the stemmer removes suffix in S3, which are 'ه', 'ي' only if the length of the word is greater than five and thus, we reduce the chance of stripping them if they are part of the word like the 'ه' in the previous example "سياره". It will not be removed because the word consists of 5 letters. After that, we check if the new stemmed token ends with 'ت' to replace it with 'ه'. For example, the words "سيارتي" (my car) or "سيارته" (his car) will become after stemming "سيارت", so we need to replace the last letter to make it "سياره". The final suffix is 'ات' and it will be replaced with 'ه'. For instance, "سيارات" (cars) will become "سياره". Finally, we check the length of the word after striping suffixes and prefixes and keep only words that have at least two characters.

Algorithm 2 Stemmer.

Input: normalized_tokens; [P1]; [P2]; [P3]; [S1]; [S2]; [S3]
Output: Stemmed tokens

```
1     stemm_tokens ← [ ]
2     ForEach nToken in normalized_tokens do
// Remove Prefix
3         If nToken.length() > 4 then
4             If nToken.length() > 5 and nToken.startWithany([P1]) then
5                 nToken ← nToken.removeStartWith([P1])
6             If nToken.length() > 5 and nToken.startWithany([P2]) then
7                 nToken ← nToken.removeStartWith([P2])
8             If nToken.startWithany([P3]) then
9                 nToken ← nToken.removeStartWith([P3])
\\ Remove Suffix
10        If nToken.length() > 5 then
11            If nToken.endWithany([S1]) then
12                nToken ← nToken.removeEndsWith([S1])
13                If nToken.endWith('ا') then
14                    nToken ← nToken.removeEndWith(' ا' )
15                If nToken.endWith(' ت') then
16                    nToken← nToken.replaceWith(' ه' )
17            If nToken.endWithany([S2]) then
18                nToken ← nToken.removeEndWith([S2])
19                If nToken.endWith(' ا') then
20                    nToken ← nToken.removeEndWith(' ا' )
21                If nToken.endWith('ت') then
22                    nToken ← nToken.replaceWith( ' ه' )
23            If nToken.endWithany([S3]) then
24                nToken ← nToken.removeEndWith([S3])
25                If nToken.endWithany('ا ت') then
26                    nToken← nToken.replaceWith(' ه' )
27                If nToken.endWith( ' ت') then
28                    nToken ← nToken.replaceWith(' ه' )
29        If nToken.length() >1 then
30            stemm_tokens.append(nToken)
31    End
```

3.4. Tweets Labeling Component (TLC)

To generate a training set for the classifiers, we need labeled tweets. Since we have a very large dataset of around 33.5 million tweets, manual labeling will be very expensive and time-consuming process. We manually labeled approximately twenty thousand tweets of the total 33.5 million tweets and then we combined them with automatically labeled tweets using the automatic labeling approach. The manually labeled tweets also help us to generate dictionaries for automatic labeling since it is a lexical based approach. The following subsections explain the proposed approach for labeling tweets about event classifiers and tweets filtering classifier. Even though we detect events after filtering tweets, we will start with labeling events because the output will be used later on to label the tweets into relevant and irrelevant.

3.4.1. Automatic Labeling for Events Tweets

Creating Dictionaries

For each event type, we automatically generated a dictionary that contains the top frequent terms using the manual labeled tweets. We manually updated each dictionary to include the missing terms related to each event type in addition to add synonyms. Both manually and automatically added terms in the dictionaries are passed to the stemmer since the search for matching terms will be applied to the tweet after pre-processing. We used Iktishaf light stammer (see Section 3.3.4).

The dictionaries contain a group of terms but we cannot use them directly to search for matching tweets because the degree of relevance to the event type is not equal for all the terms. Therefore, for each event type, we created an N number of terms list. Each list is considered as a level. So, we have N number of levels (L_1, L_2, ..., L_n). Each term T in the event dictionary is assigned to a level based on the degree of the importance of this term to the event. Thus, the terms that are highly related to the event and almost exist in each report about this event are assigned to the first level (L_1) while the last level contains terms that are least related to the event.

Furthermore, we gave each list a weight W based on the level it belongs to, which means we have N weights (W_1, W_2, ... , W_n). W_n is the highest weight so it is assigned to L1 which contains the most important terms. In this work, we used 4 levels of terms. To clarify, for Accident event, Level 1 includes terms such as Accident (حادث) and crash (صدم), Level 2 contains terms such as car (سياره), driver (سائق) and, road (طريق), Level 3 include Ambulance (اسعاف) and death(وفاه) while the last level (Level 4) contains the less important/relevant terms such as cause (يسبب).

Algorithm 3 shows the automatic labeling algorithm. It receives the pre-processed tweets (tweets_P), term dictionaries (terms_D), and event types (event_T) as input and provides the labeled tweets as an output. Apache Spark is used and the pre-processed tweets are stored in Spark Dataframe (tweets_DF). For each token in the tweet, it searched for the matched term in the terms dictionary. The "Find Matching Terms" section explains the process of searching for the matching terms. After that, weight is calculated for each labeled tweet. The section "Weight Calculation" clarifies the process of weight calculation. The last step is sorting and filtering the labeled tweets based on the calculated weight, see section "Sort and Filter Automatic Labeled Tweets" for further details.

Find Matching Terms

For each tweet, we applied the pre-processing steps explained in Section 3.3 to remove irrelevant characters, divide the text into tokens, normalize the tokens, remove the stop words, and apply stemmer. The output is N number of clean normalized and stemmed tokens (K1, K2, ... , Kn). Moreover, we iterated over each tweet and for each token K, we searched for the match terms in the term levels (L_1, L_2, ... , L_n). The output is a list of existing terms in each level as shown below where Tx is the matching term:

$$Lx_1 = [Tx_1, Tx_2, \ldots, Tx_n], \ldots, Lx_n = [Tx_1, Tx_2, \ldots, Tx_n] \tag{1}$$

Algorithm 3 Automatic Labeling for Events.

Input: tweets_P; terms_D; event_T
Output: Labeled tweets

1	spark ← createSparkSession()
2	tweets_DF ← spark.read(tweets_P)
3	n ← get_levels_num(terms_D)
4	**ForEach** tweet **in** tweets_DF **do**
5	tokens ← get_tokens(tweet)
6	**forEach** event **in** event_T **do**
7	term_levels ← terms_D[event]
8	**forEach** token **in** tokens **do**
9	l ← 1
10	stopSearch ← False
11	**while** l < n **do**
12	**If** !stopSearch **then**
13	**ForEach** term **in** term_levels[l] **do**
14	**If** term.isEqual(token) **then**
15	matchedTerms ← add(term,l)
16	stopSearch ← True
17	break
18	**end**
19	l ← +1
20	**end**
21	**end**
22	tweet_weight[event] ← calculate_weight(matchedTerms)
23	**end**
24	tweets_DF['weight'] ← tweet_weight
25	**end**
26	tweets_label_DF ← sort_and_filter(tweets_DF['weight'])

Weight Calculation

We filtered the tweets to keep only the tweets contain at least one term from the high-level (L1), except for roadwork/damage event type, we keep tweets that contain at least one term from level (L1) and at least one term from level (L2). For roadwork/damage event, L1 includes terms such as maintenance (صيانه), development (تطوير) while L2 include terms such as road (طريق), street (شارع). So at least on terms from each level should be found to assign a label to a tweet. After that, the weight is calculated using the following equation:

$$W_E = \{(size(Lx_1) \times W_n) + (size(Lx_2) \times W_{n-1}) + \ldots + (size(Lx_n) \times W_1)\} \tag{2}$$

where W_E is the total weight for the event E, Lx is the list of matching terms and W is the weight assigned for this level. Since we have 4 levels, the highest weight is 4, so, each of the term Tx that was found in Level Lx_1 has a weight equal to 4.

Sort and Filter Automatic Labeled Tweets

We used the weight to sort the labeled tweets and then we filtered them. Furthermore, we specified a threshold to discard labeled tweet that has low weight and kept tweets that have high weight because they are most likely related to the event. The same process is repeated for each event type. We take into consideration during this process that the tweet can have multiple labels.

Testing and Evaluation

The testing and evaluation of the proposed automatic labeling tool are performed in two stages. The first stage is in the beginning in order to update and modify the list of terms in each level in the dictionary such as moving terms from level to another level or adding new terms. After each initial labeling iteration, we extracted the top vocabularies to search for the new important terms that are not yet included in the dictionary of that event. For instance, for the first iteration, most of the weather event is about rains so the automatic generated dictionary contains terms related to rains. So, this stage is important to insert missing terms and then we manually update the dictionary to add their synonyms if they exist. The second stage is applied to randomly selected tweets to make sure that the tweets are labeled correctly. The main goal is reducing the number of false positive (labeled as an event but it is not) more than the number of false negatives (event but labeled as not related) to reduce the chance of making mistakes and including none event tweet in the training set of events classifiers. Besides, missing a few events tweets will not have a major effect on the size of the training set.

3.4.2. Automatic Labeling for Irrelevant Tweets

Before detecting the event, we need to train a classifier to classify the tweets into relevant and irrelevant to traffic. The training set contains positive tweets (related to traffic) and negative tweets (not related). Even though we train the filtering classifiers before the events classifiers, we generate the training set for event before filtering classifiers. The output from the automatic event labeling process is used for a positive class for tweet filtering. For negative class, we applied another automatic filtering approach by searching for the tweets that do not contain any terms related to traffic and transportation. We searched in the tweets collected by geo-filtering and we excluded tweets posted by any account related to traffic.

3.5. Feature Extractor Component (FEC)

We used CountVectorizer and IDFModel algorithms provided in the Spark ML package to generate the feature vectors and rescale them. IDFModel applied TF-IDF (Term Frequency-Inverse Document Frequency), which reflects the importance of a token to a document (tweet) in a corpus. The TF-IDF is the product of TF and IDF where TF (t, d) is the frequency of the appearance of token t in document d while the IDF is calculated using Equation (3). A detailed explanation was given in our earlier paper [6]:

$$\text{IDF(t, D)} = log\frac{|D| + 1}{\text{DF}(t, D) + 1} \tag{3}$$

3.6. Tweets Filtering Component (TFC)

3.6.1. Model Training

To filter tweets into related to road traffic and not related, we built a classifier using ML supervised classification algorithms. After labeling the tweets using both automatic and manual labeling approach, we used the Spark ML library to build and train the models. We built three models using SVM, naïve Bayes and logistic regression algorithms. We have an imbalanced dataset because in our work, the number of samples for the negative class (not related to traffic) is much higher than the positive class (traffic-related). This will lead to misleading evaluation results, especially for accuracy. To address this issue, we applied a random under-sampling approach to randomly remove some tweets from the negative class.

3.6.2. Hyperparameter Tuning

After data processing and feature extraction, we need to tune the parameters to obtain the best performance model. Grid search is one of the well-known ways to search for the best tuning parameter values. To do that, we need to specify a set of candidate tuning-parameter values and then evaluate them. Cross-validation can help to generate samples from the training set to evaluate each distinct parameter value combination and see how they perform. After that, we can get the best tuning parameter combination and use them with the entire training set to train the final model.

Spark ML supports model selection using tools such as CrossValidator to select a relatively reasonable parameter setting from a grid of parameters. We used 5-fold cross-validation, so CrossValidator will generate five (training, testing) dataset pairs. Then, the average evaluation metric for the five models will be computed and the best parameter will be founded. In the future, we plan to improve our method and use 10-fold validation.

3.6.3. Classification Model Evaluation

We compare the performance using the common evaluation metrics, which are accuracy, recall, precision and f-score. The model that achieves higher results is selected for the final classification of tweets. Since we are using binary classification to classify into relevant (class 1) and irrelevant (class 0), tuning the prediction threshold is very important. The default threshold is 0.5 and it can be any value in the range [0, 1]. If the estimated probability of class label 1 is greater than the specified threshold, the prediction result will be 1, otherwise, it will be class 0. Thus, specifying a high threshold value will encourage the model to predict 0 more often and vice versa. In our case, we need to minimize the chance of making mistakes and predicting 0 (irrelevant) as 1 (relevant) so we set the threshold to 0.8.

3.7. Events Detection Component (EDC)

We focus in this work on detecting the following event types: Fire, Weather, Social, Traffic Condition, Roadwork/Road Damage, and Accident. The tweets are labeling using the labeling method explained in Section 3.4. The classes are not mutually exclusive where the tweet can be about multi-events at the same time. For instance, the tweet might explain the accident that occurs due to bad weather. Hence, two labels will be assigned to this tweet, which are accident and weather. To address the issue, we used a binary classification. We trained a model for each event type. For each model, we need positive and negative samples. Assume we have event type T, the tweets that are labeled as T considered as positive samples while all the remaining tweets that belong to the other events types are considered negative samples. Moreover, tweet that has more than one label such as accident and weather will be included in the positive class in the training set of accident as well as weather during training both classifiers. However, as the number of tweets on the negative class is very large compared to the positive because it includes all the tweets about the other events types, we have an imbalance dataset problem. To address this problem, we followed the same approach explained in Section 3.6.1 by applying a random under-sampling approach.

3.8. Spatio-Temporal Extractor Component (STEC)

The location is the foremost matter of interest in transportation analysis and event detection domain. Thus, we applied different techniques for location extraction from the Tweet object.

3.8.1. Text, Hashtag and Username

The main approach is extracting location details which are mentioned within the post. It might be explicitly mentioned in the tweet's message or it might exist as part of the hashtags or accounts name especially if the tweets are posted by a specialized account that posts about the events and traffic condition in the cities. We created a list of cities name in Saudi Arabia to search for cities name in the tweet message. We pass the Arabic names list to the stemmer before using them to extract the place name from text because we extract them from pre-processed text. In addition, we searched for the cities name in English to extract them from accounts or hashtags using a predefined list of cities' names in English. We also created a list of specialized accounts that post about traffic in Saudi Arabia cities and does not include a city name. After that, we use this list to find the city name based on the username.

3.8.2. Tweets Geo Attributes

One of the approaches is obtaining coordinates in geotagged tweets by getting latitude and longitude from 'coordinate' or 'place' objects. The 'place' child object consists of several attributes including 'place_type', 'pl ce_name', 'country_code'. The place type is either city or point of interest (poi). However, a small fraction of tweets are geotagged because most of the users used to disable location services in their smartphones for privacy reasons.

3.8.3. User Profile

The location information is also extracted from user profiles where they usually manually write the country and city name. We have to consider that this information might be written in Arabic or English and they use different spelling. For instance, Makkah can be written as Makkah or Mecca. The text is tokenized and then passed to the stemmer before searching for the city name using the created dictionaries.

We cannot rely on the geo attributes alone because geo coordinates information might not have been provided especially for users who disable location services in their smartphones where the value will be 'null' in this case, as shown in the JSON example in Section 3.2. Similarly, we cannot rely on profile information alone because users do not always fill in these fields with accurate information. In addition, they might travel to another city/country so the profile information, does not reflect the current location. Besides, both approaches do not necessarily represent the place of the event because users might post about events that occur in other cities.

In this work, we considered the text as the main source of location information because it is more accurate than the other attributes besides, we need to find the location where the events occur not where they were posted. If the information does not exist, we extract them from coordinates or place attributes. The last option is to find location from the profile because it is less accurate than the other since users specify their information in the profile manually and they do not usually update them whenever they travel to another city. For visualization, the geospatial coordinates of the detected locations are extracted to enable plotting them on the map.

3.9. Reporting and Visualization Component (RVC)

This component supports plotting the output of the spatio-temporal information extraction component and event detection components to show the detected events and their location and time of occurrence. Also, it supports finding peak events based on configurable parameters as well as visualize the results. Algorithm 4 shows the peak events reporting algorithm. It enables searching for hourly, daily and monthly peak events when the tweets intensity exceeds a specific threshold value.

Moreover, this component supports visualizing the results of the model evaluation (see Section 3.6.3) for both tweet filtering and event detection components to illustrate which algorithm achieved better results.

Algorithm 4 Peak Events Reporting.

Input: event_Tweets; event_Types; threshold; duration
Output: Peak events list

```
1      for event in event_Types do
2          for d in duration do
// Duration can be Hours, Days or Months

3              peakEvent← [[[]
4              for tweet in event_Tweets[event] do
5                  if   tweet.intensity() >threshold   then
6                      peakEvent.append( tweet )
7                  end
8          end
9      End
```

3.10. External and Internal Validation Component (EIVC)

To validate Iktishaf+ tool and its ability to detect events and their spatial and temporal nature. We searched against various sources on the web including news media. Then, we compared the information extracted by our tool with the information in the web sources. However, news media do not report all the existing events and even if they report them, they might not mention the time of occurrence. In this case, we searched in the tweets we have, related to the event, to find the validation information we need. We consider this process as an internal validation because it is based on the collected tweets. To find time information, we go back to the earliest tweet we have about the event and if the time is not mentioned explicitly in the tweet text, we refer to the time of posting the tweet as the starting time of the event. The process of searching in the external sources was done manually, but we plan to automate it in the future.

4. Analysis and Results

4.1. Detected Events

4.1.1. Validation of Detected Events

To validate the ability of iktishaf+ and verify if a detected event really happened on the same detected date and location, we searched against external validation sources (news media) or an internal source (Twitter) (see Section 3.10). Table 1 shows a comparison between information extracted by Iktishaf+ and information from external/internal sources. We cannot discuss all the detected events, due to the limited number of pages allowed and the large period we are covering in this work (September 2018–October 2019). So, we selected samples of different event types occurred in different time and location. Column 1 shows the event types. Column 2 lists the location (city name) where the event occurs as we found from searching in various external sources as well as the location extracted by our tool. Column 3 gives the date when the events occur. Column 4 gives the time of occurrence to assess the ability of our tool to detect the time. We compared the starting time mentioned in the web sources with the peak time showing by Iktishaf+. As explained in Section 3.10 the time information may not be mentioned in the news reports. So, in this case, we searched in the collected tweet and get the earliest tweet about this event. Then, we extracted the time from the timestamp attached to the tweet.

Table 1. Example of Validation.

Event Type	Location		Date		Time	
	Validation Sources	Iktishaf+	Validation Sources	Iktishaf+	Validation Sources	Iktishaf+
Fire	Riyadh		1/10/2018		Started at 3 p.m.	Found peak around 4 p.m.
	Jeddah		29/9/2019		Started at 12:35 p.m.	Found peak around 12 p.m.
Weather	Makkah and Jeddah		23/11/2018		Started at dawn (before 5:42 a.m.)	Found peak around 4 a.m.
Accident	Riyadh		8/10/2018		Started around 4:48 a.m.	Found peak around 5 a.m.

Moreover, we drew charts to display the time extracted by Iktishaf+ for each event in the table. Further, the locations of each event are overlayed on top of the Saudi Arabia map.

Row 1 shows the "Fire" event on 1 October 2018. Figure 4 illustrates the locations. Note the largest circle in Riyadh city, this matches the information found in the newspaper, where they reported about a huge fire that broke out at a power plant in Riyadh [83]. They also mentioned that the Saudi Civil Defense received notification about the fire at 3 p.m. Figure 5 shows the time extracted by our tool. It can be seen that the intensity started raising at 3 p.m. and the highest peak was at 4 p.m.

Figure 4. Fire Event on 1 October 2018.

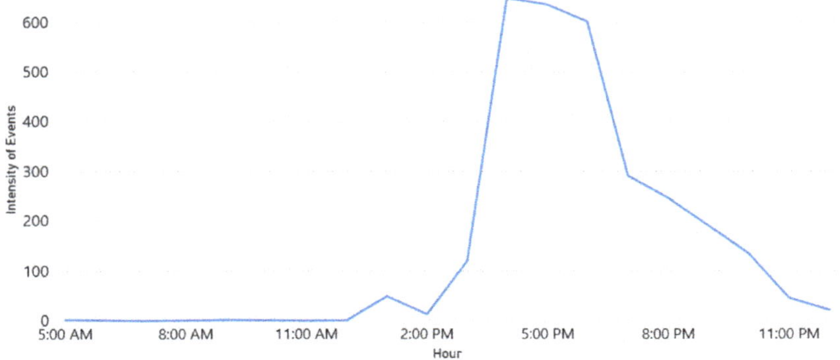

Figure 5. Intensity of Detected Fire Event in Riyadh (1 October 2018).

Row 2 validates another "Fire" event. As we found in the web source [84], it was a massive fire that ripped through the main station of the Haramain high-speed railway in Jeddah city on 29 September 2019. It started at 12:35 p.m. according to the Haramain high-speed railways' Twitter account. Further, it burned for hours before it was brought under control. Figures 6 and 7 show the location and time information extracted by Iktishaf+. Note the largest circle in Jeddah city as well as the peak time around noon as shown in Figure 7.

Figure 6. Fire Event on 29 September 2019.

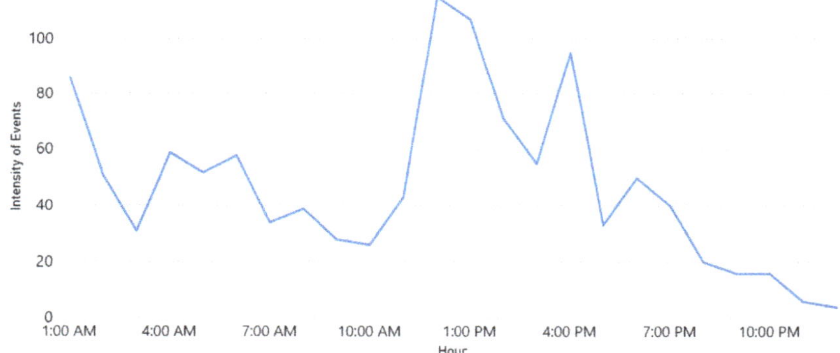

Figure 7. Intensity of Detected Fire Event in Jeddah (29 September 2019).

Furthermore, Row 3 validates the "Weather" event on 23 November 2018. This was due to the rains in Makkah and Jeddah cities as reported in the newspaper [85]. Figure 8 plots the locations of weather events on that date. Note the largest circles are in Jeddah and Makkah cities. The news article we found was posted around 5:42 a.m. and they mentioned that the rains started at dawn. This validates the time extracted by our tool. Figure 9 shows a peak at 4 a.m. As we know, Twitter users almost post about events like rain immediately once they happen and most likely earlier than newspapers.

Finally, Row 4 illustrates the "Accident" event on 8 October 2018. Note the largest circle in Riyadh shown in Figure 10. The time of occurrence is not available on the newspaper website so in this case, we went back to the tweets and searched for the earliest tweet that mentioned information about the same accident we wanted to validate. Then, we extracted the time from the timestamp (created_at attribute) included with the tweet object and assumed that it was the time of occurrence (see Section 3.10). This is the English translation for the earliest tweet we found about this event "Congestion in every street and accident in the Alwashm bridge, stations crowded, crowded everywhere in Riyadh #Riyadh_now". The time attached to this tweet is "Mon Oct. 08 04:48:17 +0000 2018" and the first peak time detected by Iktishaf+ as shown in Figure 11 is at 5 a.m. Therefore, it can be seen from the discussed results above that the information from external or internal validation sources matches the information detected by Iktishaf+, which proves the ability of our tool to automatically detect events and their location and time without prior knowledge.

Figure 8. Weather Event on 23 November 2018.

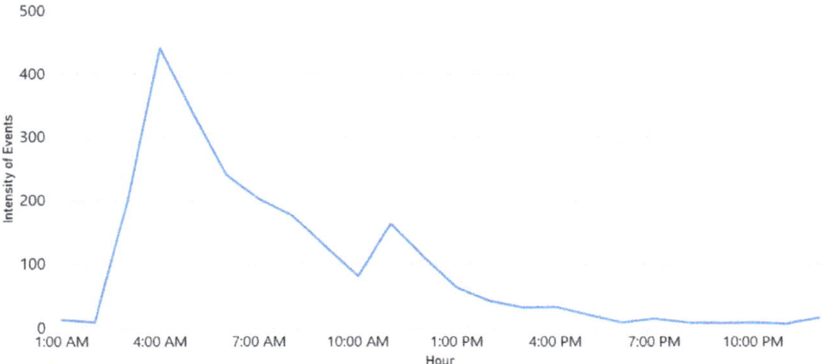

Figure 9. Intensity of Detected Weather Event in Makkah (23 November 2018).

4.1.2. Spatial Analysis

Figure 12 depicts the percentage of the extracted location information using different approaches explained in Section 3.8. As shown in the pie chart, 44% of the information is extracted from tweet text while 16% are extracted using the information in the user's profile. However, 27% of tweets about events did not include any information about the location where it occurs. Besides, only 5% are extracted from geo attributes. This could be because few tweets are geo-tagged because users usually turn off the location service in their smartphones. Also, we only look into the geo attributes if the location does not mention in the text because we mainly focus on where the event occurs not where it has been posted.

Figure 10. Accident Event on 8 October 2018.

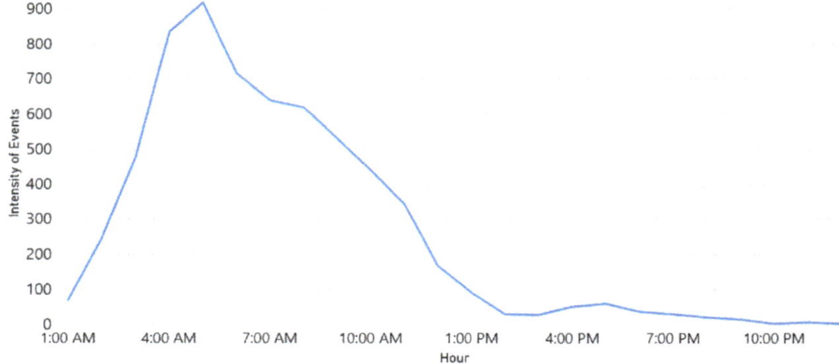

Figure 11. Intensity of Detected Accident Event in Riyadh (8 October 2018).

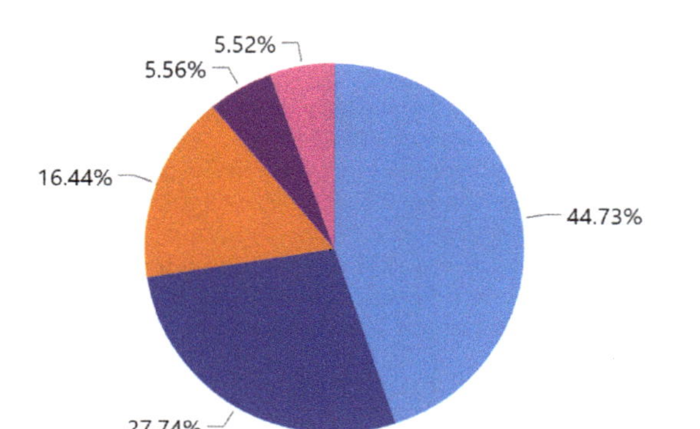

Figure 12. Number of Tweets Using Different Location Extraction Approaches.

After inferring cities' name from the tweets, we group them by province. Figure 13 gives the number of tweets for each event type in the large provinces in Saudi Arabia. It shows the aggregated number of tweets for the whole period (from September 2018 to October 2019). It can be seen that the number of events detected in Riyadh is higher than the events in other provinces. This could be because Riyadh is the capital and one of the largest cities. Besides, based on the latest report published by INREX [86], Riyadh is the most congested city in Saudi Arabia, which may explain the results we got.

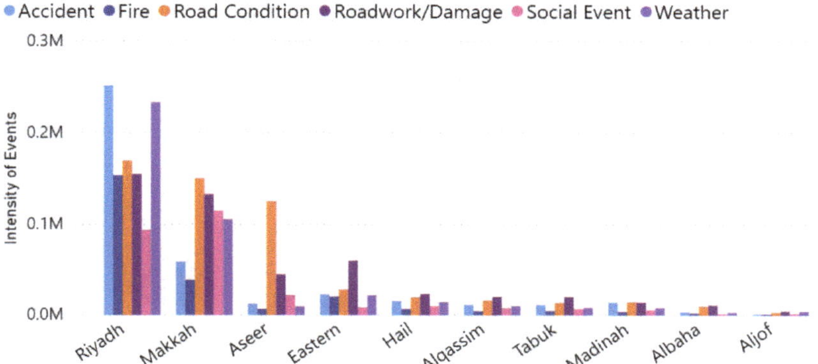

Figure 13. The Number of Detected Events in Different Provinces.

4.1.3. Spatio-Temporal Analysis

Figure 14 shows the hourly distribution for the aggregated number of tweets for the whole period. We plot only the provinces that show the high number of events to eliminates having too much data in the chart since we have 13 provinces in Saudi Arabia. As shown in this figure, the number of tweets starts raising in the morning and becomes very high by the time of coming back from school and work, which usually between 12 p.m. and 5 p.m. Further, the number goes down after 8 pm, which is expected because usually the traffic flow and activities during day-time are higher than the night-time because of work and schools.

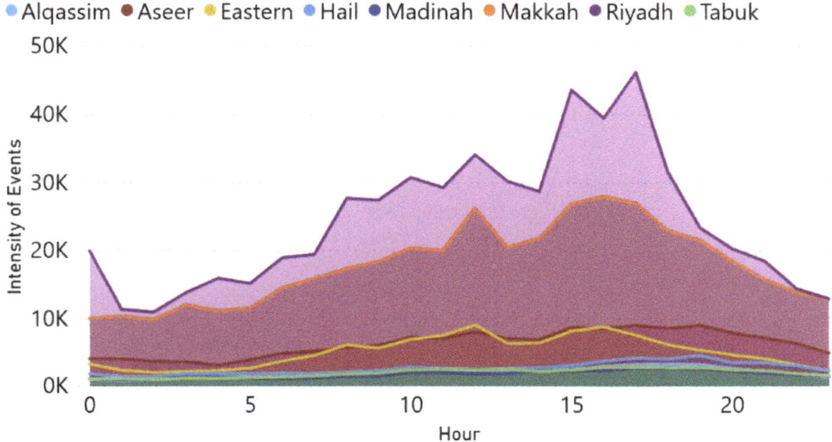

Figure 14. Hourly Distribution of Tweets Divided by Provinces (Aggregated).

4.2. Evaluation: Tweet Filtering Classifiers

To evaluate the trained model for the Tweets Filtering Component (TFC), we used the common statistical metrics, accuracy, precision, recall, f-score (see Section 3.6.3). Most of the tweets we have are irrelevant to traffic, so we have an imbalanced dataset. So, to eliminate the effect on the evaluation results, we have two options either oversampling the majority class that represents the irrelevant tweets or under-sampling the minority class that represents the traffic tweets. In our particular case, it is better to have a large number of samples for both classes. Therefore, we decided to apply oversampling on positive (traffic-related) class. We simply duplicate the number of the tweet in the positive class (see Section 3.6.3). Figure 15 shows that SVM achieved higher results compared to the other algorithms. It achieved 91% for both accuracy and f1-score, 90% for precision and 89% for recall. The difference between the results achieved by SVM and LR is approximately 1%. However, we selected SVM where it performed better.

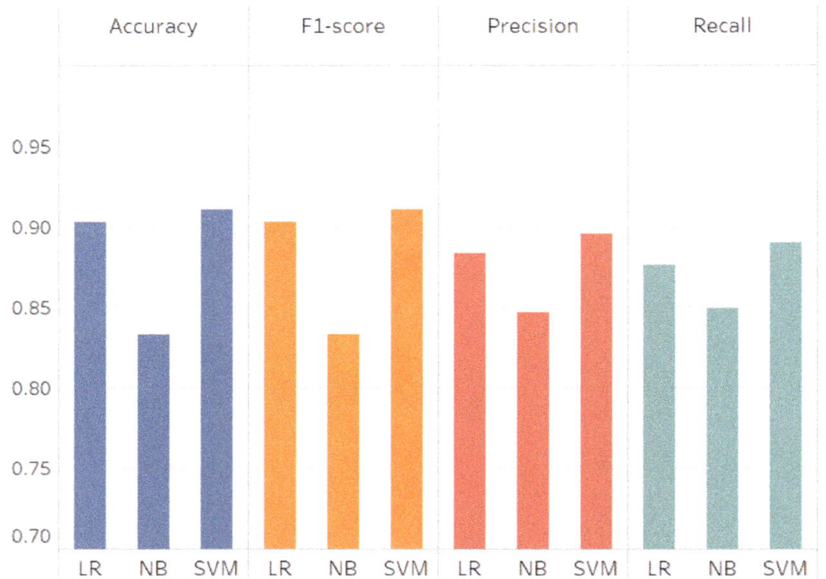

Figure 15. Numerical Evaluation (Tweets Filtering).

4.3. Evaluation: Event Classifiers

We numerically evaluated the built binary classifiers in the Event Detection Component (EDC). For each event type, we trained three models using NB, SVM, and LR algorithms and then we selected the algorithm that achieved higher results in most evaluation metrics (accuracy, precision, recall and f1-score). We have used the four performance metrics as discussed before in the previous section (also see Section 3.6.3).

Figure 16 illustrates the evaluation results for the four performance metrics (left to right: accuracy, precision, recall, and F-score) in four separate figures. The results show that SVM performs better than other algorithms for Weather, Roadwork and Traffic condition events while NB performed the best for Fire event. Besides, for accident events, SVM achieved higher results for accuracy, precision and f1-score while NB performed slightly better for the recall where SVM achieves 86% whereas NB achieves 88%. However, since the SVM performed better for most metrics, we selected SVM for accident event. For Social event, NB achieved higher recall and precision while SVM performed better for accuracy and f1-score. However, we selected NB for Social events since the accuracy and f1-score achieved by SVM are approximately 1% higher than NB. To summarize, SVM has been used for all the event types except Fire and Social where we used NB. Moreover, the highest results we got in all metrics were achieved by SVM for Weather event where it achieved 98% for both accuracy and f1-score and 97% for recall and precision. We assume that the reason is we have a larger number of tweets for the training set of weather event since it occurs more often and a lot of users post about it compared with other event types such as accident and roadwork/damage.

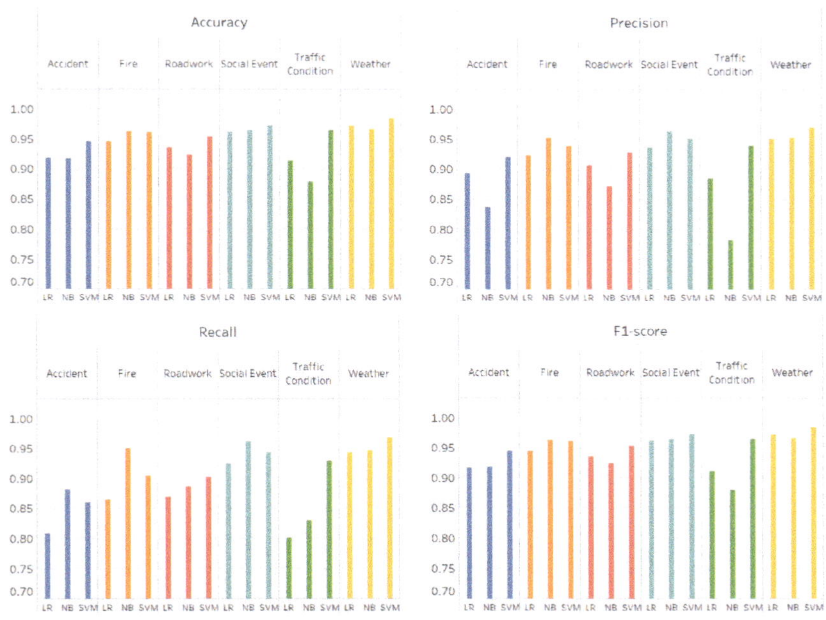

Figure 16. Numerical Evaluation (Events Classification).

5. Conclusions and Outlook

Digital societies could be characterized by their increasing desire to express themselves and interact with others. This is being realized through digital platforms such as social media that have increasingly become convenient and inexpensive sensors compared to physical sensors in many sectors of smart societies. One such major sector is road transportation, which is the backbone of modern economies and costs globally 1.25 million deaths and 50 million human injuries annually. The cutting-edge on big data-enabled social media analytics for transportation-related studies is limited.

In this paper, we introduced the Iktishaf+ tool that uses big data and distributed machine learning to automatically detect road traffic events from Arabic tweets. Manual labeling is a time-consuming process that makes supervised classification hard to apply to big data. In order to address this problem; we proposed an automatic labeling approach to reduce the effort and time of generating a training set for training supervised classification models. The traditional manual labeling for text is usually achieved by looking for specific terms to decide whether the text is relevant to the topic or not. Hence, our tool was designed to follow the same procedure. We built a dictionary for each event type that contains lists of terms that usually used when posting about the events. The dictionaries were generated automatically using the top vocabularies extracted from the manual labeled tweets. Then, we updated them manually to add synonyms and missing vocabulary. After that, we divided them into levels based on the degree of importance and relevance to the event type. Subsequently, the tool looked up the matched terms and labels the tweets based on that. Finally, the tool calculated weight for each labeled tweet, and only tweets that are highly related to the event are included in the training set.

Furthermore, we developed a location extractor to find the location of the events allowing spatio-temporal information extraction and visualization of the events. Moreover, using a stemmer is necessary for our work not only to minimize feature space for model training but it is also very helpful especially for terms searching during automatic labeling and location extraction. The existing Arabic stemmers are not efficient in our case where they might lead to removing an important letter from the word and then cause losing or changing the meaning of important words. Therefore, we designed a light stemmer that enables affix stripping with fewer changes in the word meaning.

We built and trained models to filter out irrelevant tweets to traffic events. We focused on six events that might affect road traffic which are accident, fire, weather, roadwork/damage, road condition, and social events. Furthermore, we built classifiers to automatically classify tweets into different events. We used three machine learning algorithms, which are SVM, NB, and logistic regression. Then, we selected the algorithm that achieves better results in terms of accuracy, recall, precision, and f-score. Moreover, we applied external validation using online sources such as newspapers. We selected the highest peaks from the detected events and find whether they occurred or not. The results show that our tool is able to automatically detect events and their spatial and temporal nature without prior knowledge.

The ability of the Iktishaf+ tool to use big data distributed computing technologies could save days, months, or years of computing time proportional to the size of the data. Moreover, it enables the scalability and interworking of big data analytics software systems. The utilization possibilities of our tool are many such as detection of transportation-related events for planning and operations, detection of causes of road congestion, understanding public concerns and their reactions to government policies and actions, and many more. An elaboration of these aspects of our work (the novelty, contributions, and utilization) was given in Section 2.4.

We have shown good evidence of the use of automatic labelling, machine learning, and other methods. However, more work is needed to improve the breadth and depth of the work with regard to what can be detected, the diversity of data and machine and deep learning methods, the accuracy of detection in space and time, and the real-time analysis of the tweets.

The real-time operation of the proposed system could depend on a number of factors. Firstly, the definition of the term "real-time" per se depends on the application and the requirements at hand. Some applications may require reactions within sub-second periods while others may tolerate a few minutes or more. Moreover, taking preventive actions also depends on the event and the action being taken. In this particular context, and considering the example of a car accident, the Iktishaf system once trained can detect an accident from tweets instantaneously provided the tweets are available in real-time for the software to process. This can be achieved, for instance, by running the software at the edge or fog layers. The reactive actions, in this case, can mean to inform the police and ambulance services, which can be done in real-time by the software automatically through an automatic emergency call to 911, by sending tweets or other messages to the concerned bodies, or by other emergency strategies available in the area. The messages related to the particular actions in this context can also be propagated using vehicular ad hoc networks (VANETs), dedicated short-range communications (DSRC), etc. A more interesting and lucrative work would be to detect certain events (such as car chases or certain patterns in the traffic that may lead to accidents or certain social events that may cause congestion) before these happen and take actions to prevent the events before they happen. These will require further research and adding additional functionalities to the Iktishaf tool. Our future work will look into these areas.

Digitally and data-driven methods while bringing many benefits to the research and practice have their risks and disadvantages as is the case for anything else. These include issues related to security, privacy, data ownership, lack of standards describing ethical requirements from digital methods and compliance to these standards, the safety of the stakeholders involved in data-driven and digital methods, vulnerabilities of digital platforms, and the digital divide. For a detailed discussion of these issues, see Section 11 in [87], and the references therein. As regards the specific privacy issues of Twitter data, the data we use is openly available. The information about the location of these tweets is also public. However, we have not disclosed any personal information through our analysis. The information we detected and published is of general nature and therefore does not infringe on individual privacy. However, generally speaking, it is possible to detect information from Twitter data that affects individuals' privacy. Our earlier works [57,88,89] have looked into privacy and we plan to extend this to investigate Twitter data privacy in the future.

Our focus in this work is on Saudi Arabia. The tool hence currently works with tweets only in the Arabic language. The tool can be used in other Arabic language-speaking countries, such as Egypt, Kuwait, Bahrain, and UAE. The system methodology and design of the tool developed in this paper are generic, and therefore the tool can be extended to other countries globally. This will require the adaptation of the tool with additional languages, such as English, Spanish, or Chinese, by additional modules in the pre-processing and clustering modules.

This line of our work deals with the use of Twitter data as a virtual sensor to detect transportation-related events. It is necessary to look also into other sources and methods of sensing in transportation systems, such as inductive loops, floating car data, automatic vehicle locators, virtual loop detectors, cooperative driving, etc. The real vision and potential of smart transportation systems will be realized when different sensing systems will be integrated within the transportation systems as well as with other urban systems. Our other strands of research have looked into different traffic sensing methods such as GPS [51], inductive loops [49], cooperative decision-making for autonomous vehicles [35], and urban travel data from travel cards and other sources [50]. Our future work will look into integrating these sensing methods along with other urban sensing systems, such as healthcare [7].

Author Contributions: Conceptualization, E.A. and R.M.; methodology, E.A. and R.M.; software, E.A.; validation, E.A. and R.M.; formal analysis, E.A. and R.M.; investigation, E.A., R.M., A.A., I.K. and T.Y.; resources, R.M., I.K., and A.A.; data curation, E.A.; writing—original draft preparation, E.A. and R.M.; writing—review and editing, R.M., A.A., I.K. and T.Y.; visualization, E.A.; supervision, R.M.; project administration, R.M., I.K. and A.A.; funding acquisition, R.M., A.A. and I.K. All authors have read and agreed to the published version of the manuscript.

Funding: This project was funded by the Deanship of Scientific Research (DSR) at King Abdulaziz University, Jeddah, under grant number RG-11-611-40. The authors, therefore, acknowledge with thanks the DSR for their technical and financial support.

Institutional Review Board Statement: Not applicable.

Informed Consent Statement: Not applicable.

Data Availability Statement: Data was obtained from Twitter. Restrictions apply to the availability of these data.

Acknowledgments: The experiments reported in this paper were performed on the Aziz supercomputer at King Abdulaziz University.

Conflicts of Interest: The authors declare no conflict of interest.

References

1. Mehmood, R.; See, S.; Katib, I.; Chlamtac, I. (Eds.) Smart Infrastructure and Applications: Foundations for Smarter Cities and Societies. In *EAI/Springer Innovations in Communication and Computing*; Springer: Cham, Switzerland, 2020; p. 692.
2. Hashem, I.A.T.; Chang, V.; Anuar, N.B.; Adewole, K.; Yaqoob, I.; Gani, A.; Ahmed, E.; Chiroma, H. The role of big data in smart city. *Int. J. Inf. Manag.* **2016**, *36*, 748–758. [CrossRef]
3. Zheng, X.; Chen, W.; Wang, P.; Shen, D.; Chen, S.; Wang, X.; Zhang, Q.; Yang, L. Big Data for Social Transportation. *IEEE Trans. Intell. Transp. Syst.* **2016**, *17*, 620–630. [CrossRef]
4. AlOmari, E.; Mehmood, R.; Katib, I. Road Traffic Event Detection Using Twitter Data, Machine Learning, and Apache Spark. In *2019 IEEE Smart World, Ubiquitous Intelligence & Computing, Advanced & Trusted Computing, Scalable Computing & Communications, Cloud & Big Data Computing, Internet of People and Smart City Innovation (SmartWorld/SCALCOM/UIC/ATC/CBDCom/IOP/SCI), Leicester, UK, 19–23 August 2019*; Institute of Electrical and Electronics Engineers (IEEE): Piscataway, NJ, USA, 2019; pp. 1888–1895.
5. Huang, W.; Xu, S.; Yan, Y.; Zipf, A. An exploration of the interaction between urban human activities and daily traffic conditions: A case study of Toronto, Canada. *Cities* **2019**, *84*, 8–22. [CrossRef]
6. AlOmari, E.; Katib, I.; Mehmood, R. Iktishaf: A Big Data Road-Traffic Event Detection Tool Using Twitter and Spark Machine Learning. *Mob. Netw. Appl.* **2020**, 1–16. [CrossRef]
7. Alotaibi, S.; Mehmood, R.; Katib, I.; Rana, O.; Albeshri, A. Sehaa: A Big Data Analytics Tool for Healthcare Symptoms and Diseases Detection Using Twitter, Apache Spark, and Machine Learning. *Appl. Sci.* **2020**, *10*, 1398. [CrossRef]
8. Yigitcanlar, T.; Kankanamge, N.; Preston, A.; Gill, P.S.; Rezayee, M.; Ostadnia, M.; Xia, B.; Ioppolo, G. How can social media analytics assist authorities in pandemic-related policy decisions? Insights from Australian states and territories. *Health Inf. Sci. Syst.* **2020**, *8*, 37. [CrossRef] [PubMed]
9. Agarwal, S.; Mittal, N.; Sureka, A. Potholes and Bad Road Conditions- Mining Twitter to Extract Information on Killer Roads. In Proceedings of the ACM India Joint International Conference on Data Science and Management of Data, Goa, India, 11–13 January 2018.
10. Klaithin, S.; Haruechaiyasak, C. Traffic information extraction and classification from Thai Twitter. In Proceedings of the 2016 13th International Joint Conference on Computer Science and Software Engineering (JCSSE), Khon Kaen, Thailand, 3–15 July 2016; pp. 1–6. [CrossRef]
11. D'Andrea, E.; Ducange, P.; Lazzerini, B.; Marcelloni, F. Real-Time Detection of Traffic from Twitter Stream Analysis. *IEEE Trans. Intell. Transp. Syst.* **2015**, *16*, 2269–2283. [CrossRef]
12. Kurniawan, D.A.; Wibirama, S.; Setiawan, N.A. Real-time traffic classification with Twitter data mining. In Proceedings of the 2016 8th International Conference on Information Technology and Electrical Engineering (ICITEE), Yogyakarta, Indonesia, 5–6 October 2016; pp. 1–5.
13. AlOmari, E.; Katib, I.; Albeshri, A.; Mehmood, R. COVID-19: Detecting Government Pandemic Measures and Public Concerns from Twitter Arabic Data Using Distributed Machine Learning. *Int. J. Environ. Res. Public Health* **2021**, *18*, 282. [CrossRef] [PubMed]
14. Kankanamge, N.; Yigitcanlar, T.; Goonetilleke, A. How engaging are disaster management related social media channels? The case of Australian state emergency organisations. *Int. J. Disaster Risk Reduct.* **2020**, *48*, 101571. [CrossRef]
15. Agarwal, A.; Toshniwal, D. Face off: Travel Habits, Road Conditions and Traffic City Characteristics Bared Using Twitter. *IEEE Access* **2019**, *7*, 66536–66552. [CrossRef]
16. Assem, H.; Buda, T.S.; O'Sullivan, D. RCMC: Recognizing Crowd-Mobility Patterns in Cities Based on Location Based Social Networks Data. *ACM Trans. Intell. Syst. Technol.* **2017**, *8*, 1–30. [CrossRef]
17. Cárdenas-Benítez, N.; Aquino-Santos, R.; Magaña-Espinoza, P.; Aguilar-Velazco, J.; Edwards-Block, A.; Cass, A.M. Traffic Congestion Detection System through Connected Vehicles and Big Data. *Sensors* **2016**, *16*, 599. [CrossRef] [PubMed]
18. Al'Omari, E.; Mehmood, R. Analysis of Tweets in Arabic Language for Detection of Road Traffic Conditions. In *Lecture Notes of the Institute for Computer Sciences, Social Informatics and Telecommunications Engineering*; Springer Science and Business Media LLC.: Berlin, Germany, 2018; Volume 224, pp. 98–110.
19. Al'Omari, E.; Mehmood, R.; Katib, I. Sentiment Analysis of Arabic Tweets for Road Traffic Congestion and Event Detection. In *Smart Infrastructure and Applications*; Metzler, J.B., Ed.; Springer International Publishing: Cham, Switzerland, 2020; pp. 37–54.
20. Yigitcanlar, T.; Kankanamge, N.; Regona, M.; Maldonado, A.; Rowan, B.; Ryu, A.; DeSouza, K.C.; Corchado, J.M.; Mehmood, R.; Li, R.Y.M. Artificial Intelligence Technologies and Related Urban Planning and Development Concepts: How Are They Perceived and Utilized in Australia? *J. Open Innov. Technol. Mark. Complex.* **2020**, *6*, 187. [CrossRef]
21. Yigitcanlar, T.; Kankanamge, N.; Vella, K. How are Smart City Concepts and Technologies Perceived and Utilized? A Systematic Geo-Twitter Analysis of Smart Cities in Australia. *J. Urban Technol.* **2021**, *28*, 135–154. [CrossRef]
22. Kankanamge, N.; Yigitcanlar, T.; Goonetilleke, A. Kamruzzaman Determining disaster severity through social media analysis: Testing the methodology with South East Queensland Flood tweets. *Int. J. Disaster Risk Reduct.* **2020**, *42*, 101360. [CrossRef]
23. Alsulami, M.; Mehmood, R. Sentiment Analysis Model for Arabic Tweets to Detect Users' Opinions about Government Services in Saudi Arabia: Ministry of Education as a Case Study in Al Yamamah Information and Communication Technology Forum. 2018. Available online: https://www.researchgate.net/publication/324000226_Sentiment_Analysis_Model_for_Arabic_Tweets_to_Detect_Users\T1\textquoteright_Opinions_about_Government_Services_in_Saudi_Arabia_Ministry_of_Education_as_a_case_study (accessed on 1 March 2021).

24. Suma, S.; Mehmood, R.; Albeshri, A. Automatic Detection and Validation of Smart City Events Using HPC and Apache Spark Platforms. In *Smart Infrastructure and Applications*; Metzler, J.B., Ed.; Springer: Cham, Switzerland, 2020; pp. 55–78.

25. Suma, S.; Mehmood, R.; Albugami, N.; Katib, I.; Albeshri, A. Enabling Next Generation Logistics and Planning for Smarter Societies. *Procedia Comput. Sci.* **2017**, *109*, 1122–1127. [CrossRef]

26. Mehmood, R.; Bhaduri, B.; Katib, I.; Chlamtac, I. (Eds.) Smart Societies, Infrastructure, Technologies and Applications. In *Lecture Notes of the Institute for Computer Sciences, Social Informatics and Telecommunications Engineering (LNICST)*; Springer: Cham, Switzerland, 2018; Volume 224, p. 367.

27. Arfat, Y.; Usman, S.; Mehmood, R.; Katib, I. Big Data Tools, Technologies, and Applications: A Survey. In *Smart Infrastructure and Applications*; Metzler, J.B., Ed.; Springer: Cham, Switzerland, 2020; pp. 453–490.

28. Arfat, Y.; Usman, S.; Mehmood, R.; Katib, I. Big Data for Smart Infrastructure Design: Opportunities and Challenges. In *Smart Infrastructure and Applications*; Metzler, J.B., Ed.; Springer: Cham, Switzerland, 2020; pp. 491–518.

29. Arfat, Y.; Suma, S.; Mehmood, R.; Albeshri, A. Parallel Shortest Path Big Data Graph Computations of US Road Network Using Apache Spark: Survey, Architecture, and Evaluation. In *Smart Infrastructure and Applications*; Metzler, J.B., Ed.; Springer: Cham, Switzerland, 2020; pp. 185–214.

30. Usman, S.; Mehmood, R.; Katib, I. Big Data and HPC Convergence for Smart Infrastructures: A Review and Proposed Architecture. In *Smart Infrastructure and Applications*; Springer: Cham, Switzerland, 2020; pp. 561–586.

31. Muhammed, T.; Mehmood, R.; Albeshri, A.; Katib, I. SURAA: A Novel Method and Tool for Loadbalanced and Coalesced SpMV Computations on GPUs. *Appl. Sci.* **2019**, *9*, 947. [CrossRef]

32. Alyahya, H.; Mehmood, R.; Katib, I. Parallel Iterative Solution of Large Sparse Linear Equation Systems on the Intel MIC Architecture. In *Smart Infrastructure and Applications*; Advanced Controllers for Smart Cities; Springer: Cham, Switzerland, 2020; pp. 377–407.

33. Usman, S.; Mehmood, R.; Katib, I.; Albeshri, A.; Altowaijri, S.M. ZAKI: A Smart Method and Tool for Automatic Performance Optimization of Parallel SpMV Computations on Distributed Memory Machines. *Mob. Netw. Appl.* **2019**, 1–20. [CrossRef]

34. Usman, S.; Mehmood, R.; Katib, I.; Albeshri, A. ZAKI+: A Machine Learning Based Process Mapping Tool for SpMV Computations on Distributed Memory Architectures. *IEEE Access* **2019**, *7*, 81279–81296. [CrossRef]

35. Alam, F.; Mehmood, R.; Katib, I.; Altowaijri, S.M.; Albeshri, A. TAAWUN: A Decision Fusion and Feature Specific Road Detection Approach for Connected Autonomous Vehicles. *Mob. Netw. Appl.* **2019**, 1–17. [CrossRef]

36. Mehmood, R.; Alam, F.; Albogami, N.N.; Katib, I.; Albeshri, A.; Altowaijri, S.M. UTiLearn: A Personalised Ubiquitous Teaching and Learning System for Smart Societies. *IEEE Access* **2017**, *5*, 2615–2635. [CrossRef]

37. Yigitcanlar, T.; Corchado, J.; Mehmood, R.; Li, R.; Mossberger, K.; Desouza, K. Responsible Urban Innovation with Local Government Artificial Intelligence (AI): A Conceptual Framework and Research Agenda. *J. Open Innov. Technol. Mark. Complex.* **2021**, *7*, 71. [CrossRef]

38. Alshareef, A.; Albeshri, A.; Katib, I.; Mehmood, R. Road Traffic Vehicle Detection and Tracking using Deep Learning with Custom-Collected and Public Datasets. *IJCSNS Int. J. Comput. Sci. Netw. Secur.* **2020**, *20*. [CrossRef]

39. Schlingensiepen, J.; Schlingensiepen, J.; Mehmood, R.; Mehmood, R.; Nemtanu, F.C.; Nemtanu, F.C.; Niculescu, M.; Niculescu, M. Increasing Sustainability of Road Transport in European Cities and Metropolitan Areas by Facilitating Autonomic Road Transport Systems (ARTS). In *Sustainable Automotive Technologies 2013*; Advanced Microsystems for Automotive Applications 2016; Springer: Cham, Switzerland, 2014; pp. 201–210.

40. Muhammed, T.; Mehmood, R.; Albeshri, A.; Katib, I. UbeHealth: A Personalized Ubiquitous Cloud and Edge-Enabled Networked Healthcare System for Smart Cities. *IEEE Access* **2018**, *6*, 32258–32285. [CrossRef]

41. Janbi, N.; Katib, I.; Albeshri, A.; Mehmood, R. Distributed Artificial Intelligence-as-a-Service (DAIaaS) for Smarter IoE and 6G Environments. *Sensors* **2020**, *20*, 5796. [CrossRef] [PubMed]

42. Lau, R.Y. Toward a social sensor based framework for intelligent transportation. In Proceedings of the 2017 IEEE 18th International Symposium on A World of Wireless, Mobile and Multimedia Networks (WoWMoM), Macau, China, 12–15 June 2017; pp. 1–6. [CrossRef]

43. Pandhare, K.R.; Shah, M.A. Real time road traffic event detection using Twitter and spark. In Proceedings of the 2017 International Conference on Inventive Communication and Computational Technologies (ICICCT), Coimbatore, India, 10–11 March 2017; pp. 445–449. [CrossRef]

44. Salas, A.; Georgakis, P.; Nwagboso, C.; Ammari, A.; Petalas, I. Traffic event detection framework using social media. In *Proceedings of the 2017 IEEE International Conference on Smart Grid and Smart Cities (ICSGSC)*; Institute of Electrical and Electronics Engineers (IEEE): Piscataway, NJ, USA, 2017; pp. 303–307.

45. Garg, S.; Kaur, K.; Kumar, N.; Rodrigues, J.J.P.C. Hybrid Deep-Learning-Based Anomaly Detection Scheme for Suspicious Flow Detection in SDN: A Social Multimedia Perspective. *IEEE Trans. Multimed.* **2019**, *21*, 566–578. [CrossRef]

46. Alam, F.; Mehmood, R.; Katib, I.; Albogami, N.N.; Albeshri, A. Data Fusion and IoT for Smart Ubiquitous Environments: A Survey. *IEEE Access* **2017**, *5*, 9533–9554. [CrossRef]

47. Muhammed, T.; Mehmood, R.; Albeshri, A.; Alzahrani, A. HCDSR: A Hierarchical Clustered Fault Tolerant Routing Technique for IoT-Based Smart Societies. In *Smart Infrastructure and Applications*; Springer: Cham, Switzerland, 2020; pp. 609–628.

48. Lin, C.; He, D.; Kumar, N.; Choo, K.-K.R.; Vinel, A.; Huang, X. Security and Privacy for the Internet of Drones: Challenges and Solutions. *IEEE Commun. Mag.* **2018**, *56*, 64–69. [CrossRef]

49. Aqib, M.; Mehmood, R.; Alzahrani, A.; Katib, I.; Albeshri, A.; Altowaijri, S.M. Smarter Traffic Prediction Using Big Data, In-Memory Computing, Deep Learning and GPUs. *Sensors* **2019**, *19*, 2206. [CrossRef] [PubMed]
50. Aqib, M.; Mehmood, R.; Alzahrani, A.; Katib, I.; Albeshri, A.; Altowaijri, S.M. Rapid Transit Systems: Smarter Urban Planning Using Big Data, In-Memory Computing, Deep Learning, and GPUs. *Sustainability* **2019**, *11*, 2736. [CrossRef]
51. Alsolami, B.; Mehmood, R.; Albeshri, A. Hybrid Statistical and Machine Learning Methods for Road Traffic Prediction: A Review and Tutorial. In *Smart Infrastructure and Applications: Foundations for Smarter Cities and Societies*; Springer: Cham, Switzerland, 2020; pp. 115–133.
52. Kumar, N.; Chilamkurti, N.; Park, J.H. ALCA: Agent learning–based clustering algorithm in vehicular ad hoc networks. *Pers. Ubiquitous Comput.* **2013**, *17*, 1683–1692. [CrossRef]
53. Miglani, A.; Kumar, N. Deep learning models for traffic flow prediction in autonomous vehicles: A review, solutions, and challenges. *Veh. Commun.* **2019**, *20*, 100184. [CrossRef]
54. AlAmoudi, E.; Mehmood, R.; Albeshri, A.; Gojobori, T. A Survey of Methods and Tools for Large-Scale DNA Mixture Profiling. In *Smart Infrastructure and Applications*; Advanced Controllers for Smart Cities; Springer: Cham, Switzerland, 2020; pp. 217–248.
55. Alotaibi, S.; Mehmood, R. Big Data Enabled Healthcare Supply Chain Management: Opportunities and Challenges. In *International Conference on Smart Cities, Infrastructure, Technologies and Applications*; Metzler, J.B., Ed.; Springer: Cham, Switzerland, 2018; Volume 224, pp. 207–215.
56. Mehmood, R.; Graham, G. Big Data Logistics: A health-care Transport Capacity Sharing Model. *Procedia Comput. Sci.* **2015**, *64*, 1107–1114. [CrossRef]
57. Al-Dhubhani, R.; Mehmood, R.; Katib, I.; Algarni, A. Location Privacy in Smart Cities Era. In *Lecture Notes of the Institute for Computer Sciences, Social-Informatics and Telecommunications Engineering, LNICST*; Springer: Cham, Switzerland, 2018; Volume 224, pp. 123–138.
58. Khanum, A.; Alvi, A.; Mehmood, R. Towards a Semantically Enriched Computational Intelligence (SECI) Framework for Smart Farming. In *Lecture Notes of the Institute for Computer Sciences, Social-Informatics and Telecommunications Engineering, LNICST*; Springer: Cham, Switzerland, 2018; Volume 224, pp. 247–257.
59. Alkhamisi, A.O.; Mehmood, R. An Ensemble Machine and Deep Learning Model for Risk Prediction in Aviation Systems. In Proceedings of the 2020 6th Conference on Data Science and Machine Learning Applications (CDMA), Riyadh, Saudi Arabia, 4–5 March 2020; pp. 54–59.
60. Garg, S.; Kaur, K.; Kumar, N.; Kaddoum, G.; Zomaya, A.Y.; Ranjan, R. A Hybrid Deep Learning-Based Model for Anomaly Detection in Cloud Datacenter Networks. *IEEE Trans. Netw. Serv. Manag.* **2019**, *16*, 924–935. [CrossRef]
61. Liu, B. Sentiment Analysis and Opinion Mining. *Synth. Lect. Hum. Lang. Technol.* **2012**, *5*, 1–167. [CrossRef]
62. Sakaki, T.; Okazaki, M.; Matsuo, Y. Earthquake shakes Twitter users. In Proceedings of the 19th International Conference on World Wide Web, Raleigh, NC, USA, 26–30 April 2010; pp. 851–860.
63. Sakaki, T.; Matsuo, Y.; Yanagihara, T.; Chandrasiri, N.P.; Nawa, K. Real-time event extraction for driving information from social sensors. In Proceedings of the 2012 IEEE International Conference on Cyber Technology in Automation, Control, and Intelligent Systems (CYBER), Bangkok, Thailand, 27–31 May 2012; pp. 221–226.
64. Kumar, A.; Jiang, M.; Fang, Y. Where not to go? In *Proceedings of the 37th International ACM SIGIR Conference on Research & Development in Information Retrieval*; ACM: New York, NY, USA, 2014; Volume 2609550, pp. 1223–1226.
65. Semwal, D.; Patil, S.; Galhotra, S.; Arora, A.; Unny, N. STAR. In *Proceedings of the 2nd IKDD Conference on Data Sciences*; ACM: New York, NY, USA, 2015; p. 7.
66. Tejaswin, P.; Kumar, R.; Gupta, S. Tweeting Traffic: Analyzing Twitter for generating real-time city traffic insights and predictions. In *Proceedings of the 2nd IKDD Conference on Data Sciences*; ACM: New York, NY, USA, 2015; pp. 1–4.
67. Suma, S.; Mehmood, R.; Albeshri, A. Automatic Event Detection in Smart Cities Using Big Data Analytics. In *Proceedings of the Communications and Networking*; Metzler, J.B., Ed.; Springer: Cham, Switzerland, 2018; Volume 224, pp. 111–122.
68. Atefeh, F.; Khreich, W. A Survey of Techniques for Event Detection in Twitter. *Comput. Intell.* **2013**, *31*, 132–164. [CrossRef]
69. Alkouz, B.; Al Aghbari, Z. SNSJam: Road traffic analysis and prediction by fusing data from multiple social networks. *Inf. Process. Manag.* **2020**, *57*, 102139. [CrossRef]
70. Alkhatib, M.; El Barachi, M.; Shaalan, K. An Arabic social media based framework for incidents and events monitoring in smart cities. *J. Clean. Prod.* **2019**, *220*, 771–785. [CrossRef]
71. Al-Smadi, M.; Qawasmeh, O. Knowledge-based Approach for Event Extraction from Arabic Tweets. *Int. J. Adv. Comput. Sci. Appl.* **2016**, *1*, 483–490. [CrossRef]
72. Alsaedi, N.; Burnap, P. Arabic Event Detection in Social Media. *LNCS* **2015**, *9041*, 384–401.
73. Alabbas, W.; Al-Khateeb, H.M.; Mansour, A.; Epiphaniou, G.; Frommholz, I. Classification of colloquial Arabic tweets in real-time to detect high-risk floods. In *2017 International Conference on Social Media, Wearable and Web Analytics (Social Media)*; Institute of Electrical and Electronics Engineers (IEEE): Piscataway, NJ, USA, 2017; pp. 1–8.
74. Kankanamge, N.; Yigitcanlar, T.; Goonetilleke, A.; Kamruzzaman, M. Can volunteer crowdsourcing reduce disaster risk? A systematic review of the literature. *Int. J. Disaster Risk Reduct.* **2019**, *35*, 101097. [CrossRef]
75. Pandey, N.; Natarajan, S. How social media can contribute during disaster events? Case study of Chennai floods 2015. In Proceedings of the 2016 International Conference on Advances in Computing, Communications and Informatics (ICACCI), Jaipur, India, 21–24 September 2016; pp. 1352–1356. [CrossRef]

76. Shafiabady, N.; Lee, L.; Rajkumar, R.; Kallimani, V.; Akram, N.A.; Isa, D. Using unsupervised clustering approach to train the Support Vector Machine for text classification. *Neurocomputing* **2016**, *211*, 4–10. [CrossRef]

77. Ghahreman, N.; Dastjerdi, A.B. Semi-Automatic Labeling of Training Data Sets in Text Classification. *Comput. Inf. Sci.* **2011**, *4*, 48. [CrossRef]

78. Xu, Z.; Li, J.; Liu, B.; Bi, J.; Li, R.; Mao, R. Semi-supervised learning in large scale text categorization. *J. Shanghai Jiaotong Univ. (Sci.)* **2017**, *22*, 291–302. [CrossRef]

79. Triguero, I.; García, S.; Herrera, F. Self-labeled techniques for semi-supervised learning: Taxonomy, software and empirical study. *Knowl. Inf. Syst.* **2015**, *42*, 245–284. [CrossRef]

80. Twitter. Tweet Objects. 2019. Available online: https://developer.twitter.com/en/docs/tweets/data-dictionary/overview/intro-to-tweet-json (accessed on 1 March 2021).

81. Messager, S.; Hann, A.C.; Goddard, P.A.; Dettmar, P.W.; Maillard, J.-Y. Use of the 'ex vivo' test to study long-term bacterial survival on human skin and their sensitivity to antisepsis. *J. Appl. Microbiol.* **2004**, *97*, 1149–1160. [CrossRef]

82. Loper, E.; Bird, S. NLTK: The Natural Language Toolkit. *arXiv* **2002**, preprint, arXiv:cs/0205028.

83. WAS, S.P.A. Civil Defense in Riyadh Conducts Cooling Operations for Burnt Transformers in Al-Nafal Neighborhood. Available online: https://www.spa.gov.sa/1821344 (accessed on 1 March 2021).

84. Saudi, A. 5 Injured in Al-Haramain Train Station Fire. 2019. Available online: Alarabiya.net (accessed on 1 March 2021).

85. Salamah, H. Watch rain in Makkah at dawn on Frida and a thunderstorm in Jeddah. Available online: youm7.com/story/2018/11/23/4041930/شاهدسقوطأمطارفالبيتالحرامفجرالجمعةوعاصفةرعدية (accessed on 1 March 2021).

86. INRIX. INRIX Global Traffic Scorecard. 2020. Available online: http://inrix.com/scorecard/ (accessed on 19 March 2021).

87. Alam, F.; Almaghthawi, A.; Katib, I.; Albeshri, A.; Mehmood, R. iResponse: An AI and IoT-Enabled Framework for Autonomous COVID-19 Pandemic Management. *Sustainability* **2021**, *13*, 3797. [CrossRef]

88. Al-Dhubhani, R.S.; Cazalas, J.; Mehmood, R.; Katib, I.; Saeed, F. A Framework for Preserving Location Privacy for Continuous Queries. In *International Conference of Reliable Information and Communication Technology*; Metzler, J.B., Ed.; Springer: Cham, Switzerland, 2019; Volume 1073, pp. 819–832.

89. Ayres, G.; Mehmood, R. LocPriS: A Security and Privacy Preserving Location Based Services Development Framework. *Lect. Notes Comput. Sci.* **2010**, *6279*, 566–575. [CrossRef]

Article

LidSonic V2.0: A LiDAR and Deep-Learning-Based Green Assistive Edge Device to Enhance Mobility for the Visually Impaired

Sahar Busaeed [1], Iyad Katib [2], Aiiad Albeshri [2], Juan M. Corchado [3,4,5], Tan Yigitcanlar [6] and Rashid Mehmood [7,*]

[1] Faculty of Computer and Information Sciences, Imam Mohammad Ibn Saud Islamic University, Riyadh 11564, Saudi Arabia
[2] Department of Computer Science, Faculty of Computing and Information Technology, King Abdulaziz University, Jeddah 21589, Saudi Arabia
[3] Bisite Research Group, University of Salamanca, 37007 Salamanca, Spain
[4] Air Institute, IoT Digital Innovation Hub, 37188 Salamanca, Spain
[5] Department of Electronics, Information and Communication, Faculty of Engineering, Osaka Institute of Technology, Osaka 535-8585, Japan
[6] School of Architecture and Built Environment, Queensland University of Technology, 2 George Street, Brisbane, QLD 4000, Australia
[7] High Performance Computing Center, King Abdulaziz University, Jeddah 21589, Saudi Arabia
* Correspondence: rmehmood@kau.edu.sa

Citation: Busaeed, S.; Katib, I.; Albeshri, A.; Corchado, J.M.; Yigitcanlar, T.; Mehmood, R. LidSonic V2.0: A LiDAR and Deep-Learning-Based Green Assistive Edge Device to Enhance Mobility for the Visually Impaired. *Sensors* **2022**, *22*, 7435. https://doi.org/10.3390/s22197435

Academic Editor: Antonio Guerrieri

Received: 11 August 2022
Accepted: 26 September 2022
Published: 30 September 2022

Publisher's Note: MDPI stays neutral with regard to jurisdictional claims in published maps and institutional affiliations.

Abstract: Over a billion people around the world are disabled, among whom 253 million are visually impaired or blind, and this number is greatly increasing due to ageing, chronic diseases, and poor environments and health. Despite many proposals, the current devices and systems lack maturity and do not completely fulfill user requirements and satisfaction. Increased research activity in this field is required in order to encourage the development, commercialization, and widespread acceptance of low-cost and affordable assistive technologies for visual impairment and other disabilities. This paper proposes a novel approach using a LiDAR with a servo motor and an ultrasonic sensor to collect data and predict objects using deep learning for environment perception and navigation. We adopted this approach using a pair of smart glasses, called LidSonic V2.0, to enable the identification of obstacles for the visually impaired. The LidSonic system consists of an Arduino Uno edge computing device integrated into the smart glasses and a smartphone app that transmits data via Bluetooth. Arduino gathers data, operates the sensors on the smart glasses, detects obstacles using simple data processing, and provides buzzer feedback to visually impaired users. The smartphone application collects data from Arduino, detects and classifies items in the spatial environment, and gives spoken feedback to the user on the detected objects. In comparison to image-processing-based glasses, LidSonic uses far less processing time and energy to classify obstacles using simple LiDAR data, according to several integer measurements. We comprehensively describe the proposed system's hardware and software design, having constructed their prototype implementations and tested them in real-world environments. Using the open platforms, WEKA and TensorFlow, the entire LidSonic system is built with affordable off-the-shelf sensors and a microcontroller board costing less than USD 80. Essentially, we provide designs of an inexpensive, miniature green device that can be built into, or mounted on, any pair of glasses or even a wheelchair to help the visually impaired. Our approach enables faster inference and decision-making using relatively low energy with smaller data sizes, as well as faster communications for edge, fog, and cloud computing.

Keywords: visually impaired; smart mobility; sensors; LiDAR; ultrasonic; deep learning; obstacle detection; obstacle recognition; assistive tools; edge computing; green computing; sustainability; Arduino Uno; smart app

1. Introduction

There are over 1 billion disabled people today around the world, comprising 15% of the world population, and this number is greatly increasing due to ageing, chronic diseases, and poor environments and health according to the World Health Organization (WHO) [1]. WHO defines disability as having three dimensions, "impairment in a person's body structure or function, or mental functioning; activity limitation; and participation restrictions in normal daily activities", and states that disability "results from the interaction between individuals with a health condition with personal and environmental factors" [2]. Cambridge Dictionary defines disability as "not having one or more of the physical or mental abilities that most people have" [3]. Wikipedia defines physical disability as "a limitation on a person's physical functioning, mobility, dexterity, or stamina" [4]. Disabilities can relate to various human functions, including hearing, mobility, communication, intellectual ability, learning, and vision [5]. In the UK, 14.6 million people are disabled [6], forming over 20% of the population. In the US, 13.2% of the population were disabled according to the 2019 statistics, comprising over 43 million people [7]. Similar statistics could be found in various countries around the world, some worse than others, which means that, on average, disabled people make up 15% of the population globally.

With 253 million people affected by visual impairment and blindness around the globe, it is the second most prevalent disability in the world population after hearing loss and deafness [8]. Four terminologies can be used to identify various rates of loss of vision and blindness, namely, partially sighted, low vision, legally blind, and totally blind [9]. People with partial vision in one or both eyes are considered partially sighted. Low vision relates to a serious visual impairment, where visual acuity in the good-seeing eye is 20/70 or lower and cannot be enhanced with glasses or contact lenses. If the best-seeing eye can be corrected to achieve 20/200, then the person is considered legally blind [9]. Finally, people who are totally blind are those with a total loss of vision [10]. Even though vision impairment can happen at any point in life, it is more common among older people. Visual impairment can be hereditary. In these kinds of circumstances, it occurs from birth or in childhood [11].

While visual impairment and blindness are among the most disabling disabilities, we know relatively little about the lives of visually impaired and blind individuals [12]. The WHO predicts that the number of people with visual impairments will increase owing to population growth and aging. Moreover, contemporary lifestyles have spawned a multitude of chronic disorders that degrade vision and other human functions [13]. Diabetes and hyperglycemia, for instance, can cause a range of health issues, including visual impairment. Several tissues of the ocular system can be affected by diabetes, and cataracts are one of the most prevalent causes of vision impairment [14].

Behavioral and neurological investigations relevant to human navigation have demonstrated that the way in which we perceive visual information is a critical part of our spatial representation [15]. It is typically hard for visually impaired individuals to orient themselves and move in an unknown location without help [16]. For example, landplane tracking is a natural mobility task for humans, but it is an issue for individuals with poor or no eyesight. This capacity is necessary for individuals to avoid the risk of falling and to alter their position, posture, and balance [16]. Moving up and down staircases, low and high static movable obstacles, damp flooring, potholes, a lack of information about recognized landmarks, obstacle detection, object identification, and dangers are among the major challenges that visually impaired people confront indoors and outdoors [17,18].

The disability and visual impairment statistics of Saudi Arabia are also alarming. Around 3% of people in Saudi Arabia reported the presence of a disability in 2016 [19]. According to the General Authority for Statistics in Saudi Arabia, 46% of all the disabled in Saudi Arabia who have one disability are visually impaired or blind, and 2.9% of the Saudi population have disabilities amounting to extreme difficulty [20]. The information provided above highlights the urgent need for research in the development of assistive technologies for general disabilities, including visual impairment.

A white cane is the most popular tool used by visually impaired individuals to navigate their environments; nevertheless, it has a number of drawbacks. It requires physical contact with the environment [12], cannot detect barriers above the ground, such as ladders, scaffolding, tree branches, and open windows [21], and generates neuromusculoskeletal overuse injuries and syndromes that may require rehabilitation [22]. Moreover, the user of a white cane is sometimes ostracized for social reasons [12]. In the absence of appropriate assistive technologies, visually impaired individuals must rely on family members or other people [23]. However, human guides can be dissatisfying at times, since they may be unavailable when assistance is required [24]. The use of assistive technologies can help visually impaired and blind people engage with sighted people and enrich their lives [12].

Smart societies and environments are driving extraordinary technical advancements with the promise of a high quality of life [25–28]. Smart wearable technologies are generating numerous new opportunities in order to enhance the quality of life of all people. Fitness trackers, heart rate monitors, smart glasses, smartwatches, and electronic travel aids are a few examples. The same holds true for visually impaired people. Multiple devices have been developed and marketed to aid visually impaired individuals in navigating their environments [29]. An electronic travel aid (ETA) is a regularly used type of mobility-enhancing assistive equipment for the visually impaired and blind. It is anticipated that ETAs will increasingly facilitate "independent, efficient, effective, and safe movement in new environments" [30]. ETAs can provide information about the environment through the integration of multiple electronic sensors and have shown their effectiveness in improving the daily lives of visually impaired people [31]. ETAs are available in a variety of wearable and handheld devices and may be categorized according to their usage of cellphones, sensors, or computer vision [32]. The acceptance rate of ETAs is poor among the visually impaired and blind population [23]. Their use is not common among potential users because they have inadequate user interface designs, are restricted to navigation purposes, are functionally complex, weighty to carry, expensive, and lack functionality for object recognition, even in familiar indoor environments [23]. The low adoption rate does not necessarily indicate that disabled people oppose the use of ETAs; rather, it confirms that additional research is required to investigate the causes of the low adoption rate and to improve the functionality, usability, and adaptability of assistive technologies [33]. In addition, the introduction of unnecessarily complicated ETAs that may necessitate extensive and supplementary training to learn additional and difficult abilities is not a realistic alternative and is not a feasible solution [22]. Robots can assist the visually impaired in navigating from one location to another, but they are costly, along with their other challenges [34]. Augmented reality has been used as a solution for magnifying text and images through a finger wearable applied with a camera to project on a HoloLens [35], but this technology is not suitable for blind people.

We developed a comprehensive understanding of the state-of-the-art requirements of, and solutions for, visually impaired assistive technologies using a detailed literature review (see Section 2) and a survey [36] of this topic. Using this knowledge, we identified the design space for assistive technologies for the visually impaired and the research gaps. We found that the design considerations for assistive technologies for the visually impaired are complex and include reliability, usability, and functionality in indoor, outdoor, and dark environments; transparent object detection; hand-free operations; high-speed, real-time operations; low battery usage and energy consumption; low computation and memory requirements; low device weight; and cost effectiveness. Despite the fact that several devices and systems for the visually impaired have been proposed and developed in academic and commercial settings, the current devices and systems lack maturity and do not completely fulfill user requirements and satisfaction [18,37]. For instance, numerous camera-based and computer-based solutions have been produced. However, the computational cost and energy consumption of image processing algorithms pose a concern for low-power portable or wearable devices [38]. These solutions require large storage and computational resources, including large RAMs to process large volumes of data containing images. This would require substantial processing, communication, and decision-making times, and would also consume energy and battery life. Significantly more research effort is required to bring innovation, intelligence, and user satisfaction to this crucial area.

In this paper, we propose a novel approach that uses a combination of a LiDAR with a servo motor and an ultrasonic sensor to collect data and predict objects using machine and deep learning for environment perception and navigation. We implemented this approach using a pair of smart glasses, called LidSonic V2.0, to identify obstacles for the visually impaired. The LidSonic system consists of an Arduino Uno edge computing device integrated into the smart glasses and a smartphone app that transmits data via Bluetooth. Arduino gathers data, operates the sensors on smart glasses, detects obstacles using simple data processing, and provides buzzer feedback to visually impaired users. The smartphone application collects data from Arduino, detects and classifies items in the spatial environment, and gives spoken feedback to the user on the detected objects. LidSonic uses far less processing time and energy than image-processing-based glasses by classifying obstacles using simple LiDAR data and using several integer measurements.

We comprehensively describe the proposed system's hardware and software design, having constructed their prototype implementations and tested them in real-world environments. Using the open platforms WEKA and TensorFlow, the entire LidSonic system was built with affordable off-the-shelf sensors and a microcontroller board costing less than USD 80. Essentially, we provide the design of inexpensive, miniature green devices that can be built into, or mounted on, any pair of glasses or even a wheelchair so as to help the visually impaired. Our approach affords faster inference and decision-making using relatively low energy with smaller data sizes. Smaller data sizes are also beneficial in communications, such as those between the sensor and processing device, or in the case of fog and cloud computing, because they require less bandwidth and energy and can be transferred in relatively shorter periods of time. Moreover, our approach does not require a white cane (although it can be adapted to be used with a white cane) and, therefore, it allows for handsfree operation.

The work presented in this paper is a substantial extension of our earlier system LidSonic (V1.0) [39]. LidSonic V2.0, the new version of the system, uses both machine learning and deep learning methods for classification, as opposed to V1.0, which uses machine learning alone. LidSonic V2.0 provides a higher accuracy of 96% compared to 92% for LidSonic V1.0, despite the fact that it uses a lower number of data features (14 compared to 45) and a wider vision angle of 60 degrees compared to 45 degrees for LidSonic V1.0. The benefits of a lower number of features are evident in LidSonic V2.0, requiring even lower computing resources and energy than LidSonic V1.0. We have extended the LidSonic system with additional obstacle classes, provided an improved and extended explanation of its various system components, and conducted extensive testing with two new datasets, six machine learning models, and two deep learning models. System V2.0 was implemented using the Weka and TensorFlow platforms (providing dual options for open-source development) compared to the previous system that was implemented using Weka alone. Moreover, this paper provides a much extended, completely new literature review and taxonomy of assistive technologies and solutions for the blind and visually impaired.

Earlier in this section, we noted that, despite the fact that several devices and systems for the visually impaired have been developed in academic and commercial settings, the current devices and systems lack maturity and do not completely fulfill user requirements and satisfaction. Increased research activity in this field will encourage the development, commercialization, and widespread acceptance of devices for the visually impaired. The technologies developed in this paper are of high potential and are expected to open new directions for the design of smart glasses and other solutions for the visually impaired using open software tools and off-the-shelf hardware.

The paper is structured as follows. Section 2 explores relevant works in the field of assistive technologies for the visually impaired and provides a taxonomy. Section 3 gives an overview of the LidSonic V2.0 system, highlighting its user, developer, and system features. Section 4 provides a detailed illustration of the software and hardware design and implementation. The system is evaluated in Section 5 Conclusions and thoughts regarding future work are provided in Section 6.

2. Related Work

This section reviews the literature relating to this paper. Section 2.1 presents the sensor technologies and the types of sensor technologies used in the assistive tools for the visually impaired. Section 2.2 reviews the processing methods. Section 2.3 discusses the feedback techniques. A taxonomy of functions and applications is provided in Section 2.4. Section 2.5 identifies the research gap and justifies the need for this research. A taxonomy of the research on the visually impaired presented in this section is given in Figure 1. An extensive complimentary review of the assistive technologies for the visually impaired and blind can be found in our earlier work [39].

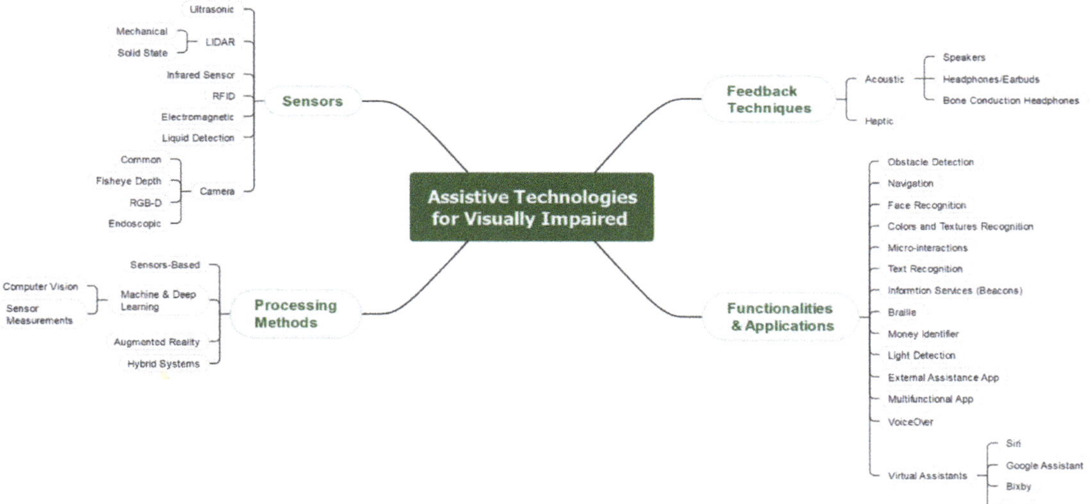

Figure 1. A Taxonomy of Research on Assistive Technologies for the Visually Impaired.

2.1. Sensor Technologies and Types Used in Assistive Tools

Sensors are indistinguishable components of cyberphysical systems. They collect knowledge regarding environmental factors, as well as non-electrical system parameters, and provide the findings as electrical signals. With the development of microelectronics, sensors are available as compact devices at low costs and have a wide range of applications in different fields, especially in control systems [40]. There are two types of sensors, including passive and active sensors. The active sensor needs an incentive to activate. On the contrary, the passive sensor detects inputs and generates output signals directly without an external incentive. The classification may be dependent on the sensor's means of detection, e.g., electric, biological, chemical, radioactive, etc. [41].

A number of sensors have been employed in the field of technologies for visually impaired people. They have been used to solve a wide range of vision issues. The most frequent types of sensors used in assistive devices for the visually impaired are listed in Table 1. It also shows the sensors' functions (functionality), types of wearables, and types of feedback.

2.1.1. Ultrasonic Sensors

An ultrasonic sensor is an electronic device that uses ultrasonic sound waves to detect the distance between the user and a target item and transforms the reflected sound into an electric signal. Fauzul and Salleh [42] developed a visual assistive technology to assist visually impaired individuals in safely and conveniently navigating both indoor and outdoor situations. A smartphone app and an obstacle sensor are the two major components of the system. To deliver auditory cue instructions to the user, the mobile software makes use of the smartphone's microelectromechanical sensors, location services, and Google Maps. Ultrasonic sensors in the obstacle sensor are used to detect objects and offer tactile feedback. The obstacle avoidance gadget attaches to a standard cane and vibrates the handle with varying intensities according to the nature of the obstructions. The spatial distance and direction from the present position to the intended location are used to produce spatial sound cues. Gearhart et al. [43] proposed a technique for identifying the position of the detected object using triangulation by geometric relationships with scalar measurements. The authors placed two ultrasonic sensors one on each shoulder, which were angled towards each other at five degrees from parallel, with a space of 10 inches. However, this technique is too complex to be applied to several objects in front of the sensor. A significant number of research papers in the field of objects detection have depended on ultrasonic sensors. Tudor et al. [44] proposed a wearable belt with two ultrasonic sensors and two vibration motors to direct the visually impaired away from obstacles. The authors used an Arduino Nano Board with an ATmega328P microcontroller to connect and build their system. According to their findings, the authors in [45] found that ultrasonic sensors and vibrator devices are easily operated by Arduino UNO R3Impatto Zero boards. Noman et al. The authors of [46] proposed a robot equipped with several ultrasonic sensors and Arduino Mega (ATMega 2560 processor) to detect obstacles, holes, and stairs. The robots can be utilized in indoor environments; however, their use outdoors is not practical.

2.1.2. LiDAR Sensors

Light detection and ranging, or LiDAR, is a common remote sensing technique used for determining an object's distance. Chitra et al. [47] proposed a handsfree LVU (LiDARs and Vibrotactile Units) discrete wearable gadget that helps blind persons to identify impediments. Proper mobile assistance equipment is required. The proposed gadget consists of a wearable sensor strap. Liu et al. proposed HIDA. This is a lightweight assistance system used for comprehensive indoor detection and avoidance based on 3D point cloud instance segmentation and a solid-state LiDAR sensor. The authors created a point cloud segmentation model with dual lightweight decoders for semantic and offset predictions, ensuring the system's efficiency. The segmented point cloud was post-processed by eliminating outliers and projecting all points onto a top-view 2D map representation after the 3D instance segmentation.

2.1.3. Infrared (IR) Sensors

An infrared (IR) sensor is an electronic device that monitors and senses the infrared radiation in its surroundings [48]. Infrared signals are similar to RFID in that they rely on distant data transfer. As previously stated, the latter uses radio waves, whilst the former uses light signals. Air conditioner remotes and motion detectors, for example, all use infrared technology. Smartphones now come with infrared blasters, allowing users to control any compatible device with an infrared receiver. It is an eye-safe light, which emits pulses and measures the time taken to calculate the distance using the reflected light. Every metric of the IR consists of thousands of separate pulses of light that lead to reliable measurements of rain, snow, fog, or dust and can be obtained by an infrared sensor. These measurements are difficult to capture with cameras [22]. In addition, IR has a long range in both indoor and outdoor environments, high precision, small size, and low latency. An IR sensor can detect obstacles up to 14 m away, with a 0.5 resolution and 4 cm accuracy [12]. IR has medium width among the ultrasonic and laser sensors. The laser has a rather narrow scope, and it gathers a very limited amount of space information, which is not large enough for free paths. On the other hand, ultrasonic sensors have many reflections; thus, they are limited [49].

2.1.4. RFID Sensors

RFID is the abbreviation of radio frequency identification. Data may be transferred and received through radio waves using RFID. In RFID, the sender sends a radio receiver. In this method, the sender is commonly an RFID chip (or an RFID tag) inserted into the object being read or scanned. The receiver, on the other hand, is an electrical device that detects the RFID chip's data. The chip and receiver do not need to be in physical contact, because the data is broadcast and received by radio waves. RFID is appealing because of its remote capabilities, but it is also harmful. Because the chip's RFID signal may be read by anybody with an RFID reader, it may lead to an unethical and harmful situation, namely, data theft. The fact that the person using the scanner does not even have to be near the chip/tag increases the danger. Another disadvantage is that each tag has a certain range, which necessitates extensive individual testing, limiting the scope. In addition, the system may be quickly turned off if the tags are wrapped or covered, preventing them from receiving radio signals [50].

An intelligent walking stick for the blind was proposed by Chaitrali et al. [51]. The proposed navigation system for vision impairment uses infrared sensors, RFID technology, and Android handsets to provide speech output for the purpose of obstacle navigation. The gadget is equipped with proximity infrared sensors, and RFID tags are implanted in public buildings, as well as in the walking sticks of blind people. The gadget is Bluetooth-connected to an Android phone. An Android application that provides voice navigation based on the RFID tag reading and also updates the server with the user's position information is being developed. Another application allows family members to access the location of the blind person via the server at any time. The whereabouts of a blind person may be traced at any time, providing further security. This approach has the disadvantage of not being compact. When the intelligent stick is within the range of the PCB unit, the active RFID tags immediately send location information. It is not necessary for the RFID sensor to read it explicitly.

2.1.5. Electromagnetic Sensors (Microwave Radar)

Adopting a pulsed chirp scheme can reduce the power consumption and preserve a high resolution by managing the spatial resolution in terms of the frequency modulation bandwidth. A pulsed signal enables the transmitter to be turned off in the listening time of the echo, thereby significantly reducing the energy consumption [37]. Using a millimeter-wave radar and a typical white cane, a method of electronic travel assistance for blind and visually impaired persons was proposed [52]. It is a sophisticated system that not only warns the user of possible difficulties but also distinguishes between human and nonhuman targets. Because real-world situations are likely to include moving targets, a novel range alignment approach was developed to detect minute chest movements caused by physiological activity as vital evidence of human presence. The proposed system recognizes humans in complicated situations, with many moving targets, giving the user a comprehensive set of information, including the presence, location, and type of the accessible targets. The authors used a 122 GHz radar board to carry out appropriate measurements, which were used to demonstrate the system's working principle and efficacy.

2.1.6. Liquid Detection Sensors

Research usually involves more than one type of sensor in order to cover most of the prevalent challenges facing the visually impaired. Ikbal et al. [53] proposed a stick that is equipped with various kinds of sensors to assist in detecting obstacles. One of the major obstacles that jeopardize the visually impaired is water on the floor [54]. Therefore, the authors included a water sensor in their solution, in addition to two ultrasonic sensors to detect 180 cm obstacles, including one IR sensor to detect stairway gaps and holes on streets, and a temperature sensor for fire alert. The authors connected all the sensors with an Arduino microcontroller board. This sensor must come into contact with the surface of the water in order to provide the result; thus, this method must consider the appropriate wearable. One of the most important uses of a liquid detector for the blind is the use of a sensor that is placed on the cup to prevent it from spilling. To improve navigation safety, the authors present a polarized RGB-Depth (pRGB-D) framework to detect the traversable area and water hazards while using polarization-color-depth-attitude information [55].

2.1.7. Cameras

The camera is utilized to provide various functions in different technology solutions using machine learning algorithms, such as facial recognition, object recognition, and localization (see Table 1). Research has used various types of cameras. The most frequently used types are the common camera and the RGB-Depth camera. The common camera is used mostly in facial, emotion, and obstacle recognition. On the other hand, the RGB-D camera has been used for detecting and avoiding obstacles and mapping to assist in navigation through indoor environments. A depth image is an image channel in which each pixel is related to a distance between the image plane and the respective point in the RGB picture. Adding depth to standard color camera techniques increases both the precision and density of the map. RGB-D sensors are popular in a variety of visual aid applications due to their low power consumption and inexpensive cost, as well as their resilience and high performance, as they can concurrently sense color and depth information at a smooth video framerate. Because polarization characteristics reflect the physical properties of materials, polarization and associated imaging can be employed for material identification and target detection, in addition to color and depth [30]. Meanwhile, because various polarization states of light act differentially at the interface of an object's surface, polarization has been utilized in a variety of surface measuring techniques. Nevertheless, most industrial RGB-D sensors, such as light-coding sensors and stereo cameras, depend solely on intensity data, with polarization indications either missing or insufficient [55]. The study reported in [56] describes a 3D object identification algorithm and its implementation in a robotic navigation aid (RNA) that allows for the real-time detection of indoor items for blind people utilizing a 3D time-of-flight camera for navigation. Then, using a Gaussian-mixture-model-based plane classifier, each planar patch is classified as belonging to a certain object model. Finally, the categorized planes are clustered into model objects using a recursive plane clustering process. The approach can also identify various non-structural elements in the indoor environment. The authors of the research reported in [57] proposed a new approach to autonomous obstacle identification and classification that combines a new form of sensor, a patterned light field, with a camera. The proposed gadget is compact in terms of size, portable, and inexpensive. As the sensor system is transported in natural interior and outdoor situations over and toward various sorts of barriers, the grid projected by the patterned light source is visible and distinguishable. The proposed solution uses deep learning techniques, including a convolutional neural-network-based categorization of individual frames, to leverage these patterns without calibration. The authors improved their method by smoothing frame-based classifications across many frames using lengthy short-term memory units.

Table 1. Types of Sensors Used in Assistive Devices for VI.

Sensor Name	Works	Purpose of Sensor	No. of Sensors	Weight	Wearable/Assistive	Feedback Method
IR Sensors	[58]	Touch down, touch up sensor	2	Light	Mounted on top of a finger	Acoustic
	[49]	Detect obstacles, stairs	2	Light	Cane	Acoustic
IMU	[58]	Recognize gestures and sense movements	1	Light	Mounted on top of a finger	Acoustic
Ultrasonic Sensors	[23]	Detect obstacles up to the chest level	5	Light	Cane	Acoustic, vibration
	[59]	Detect obstacles	2	High	Guide dog robot and portable robot	Acoustic
	[45]	Detect obstacles	5	Fair	Mounted on the head, legs, and arms	Buzzer, vibration
	[44]	Detect obstacles	2	Fair	Belt	Vibration
ToF Distance Sensors	[12]	Detect obstacles	7	High	Belt	Vibration belt
Microwave Radar	[37]	Detect obstacles	1	Light	Mounted on a cane	Acoustic, vibration
Wet Floor Detection Sensors	[53]	Detect wet floors	1	Light	Cane	Buzzer
Bluetooth	[60]	Informing about indoor environments	3	Light	Beacon transmitter, smartphone	Acoustic
Laser Pointer	[61]	Detect obstacles	1	Light	Belt	Vibration belt
Cameras	[58]	Localize the hand touch	1	Light	Mounted on top of a finger	Acoustic
	[62]	Emotion recognition	1	Light	Clipped on to spectacles	Vibration belt
	[59]	Obstacle recognition (traffic light, cones, bus, etc.)	2	High	Guide dog robot and portable robot	Acoustic
	[63]	Localization system	1	Low	Head level (helmet), chest level (hanged)	Location in a Map
RGB-D Cameras	[61]	Detect obstacles	1	Light	Smartphone simulating a cane	Vibratory belt
	[64]	Avoid obstacles, localization system for indoor navigation	1	Light	Glass, tactile vest, smartphone	Haptic vest (4 vibration motors)
Endoscopic Cameras	[65]	Identify clothing colors, visual texture recognition	1	Light	Mounted on top of a finger	Acoustic
Compass	[66]	Indoor navigation	1	Light	Optical head-mounted (Glass)	Acoustic

2.2. Processing Methods

Researchers have used a range of processing methods for assistive technologies. Recent years have seen an increase in the use of machine learning and deep learning methods in various applications and sectors, including healthcare [67–69], mobility [70–72], disaster management [73,74], education [75,76], governance [77], and many other fields [78]. Assistive technologies are no different and have begun to increasingly rely on machine learning methods. This section reviews some of the works on processing methods for assistive technologies, including both machine learning-based methods and other methods.

There are numerous ideas and methods that have been proposed to solve the problems and challenges facing the blind. Katzschmann et al. [12] incorporated several sensors and feedback motors in a belt to produce an aiding navigation system, called Array of LiDARs and Vibrotactile Units (ALVU), for visually impaired people. The authors developed a secure navigation system, which is effective in providing detailed feedback to a user about the obstacles and free areas around the user. Their technology is made up of two components: a belt with a distance sensor array and a haptic array of feedback modules. The haptic strap that goes around the upper abdomen provides input to the person wearing the ALVU, which allows them to sense the distance between themselves and their surroundings. As the user approaches an impediment, they receive greater pulse rates and a higher vibration force. The vibration and pulses stop once the user has overcome the obstacle. However, this kind of feedback is primitive and cannot define the type of obstacle that the user should avoid. Moreover, it does not determine whether the obstacle should or should not be avoided. In addition, wearing two belts may not be easy and comfortable for the user. Meshram et al. [23] designed a NavCane that detects and avoids obstacles from the floor up to the chest level. It can also identify water on the floor. It has a user button to send auto alerts through SMS and email in emergencies. It provides two kinds of feedback, including tactile feedback using vibration and auditory feedback using the headphones. However, the device cannot identify the nature of the objects and cannot detect obstacles above chest level.

Hong et al. [79] proposed a solution for blind people based on two haptic wristbands used to provide feedback on objects. Using a LiDAR, Chun et al. [80] proposed a detection technique that reads the distances of deferent angles and then measures the predicted obstacles by comparing these reading.

Using the Internet of Things (IoT), machine learning, and embedded technologies, Mallikarjuna et al. [34] developed a low-cost visual aid system for object recognition. The image is acquired by the camera and then forwarded to the Raspberry Pi. To classify the image, the Raspberry Pi is trained using the TensorFlow Machine Learning Framework and Python programming. However, their technique requires a long period of time (5 s to 8 s) to inform the visually impaired individual about the item in front of them.

Gurumoorthy et al. [81] proposed a technique using the rear camera of a mobile phone to capture and analyze the image in front of the visually impaired. To execute tasks related to computer vision, this device uses Microsoft Cognitive Services. Then, image feedback is provided to the user through Google talkback. This technique needs a mobile internet service in order to be performed. Additionally, it is hard for the visually impaired to take a proper picture. A similar solution by means of sending the picture to the cloud to be analyzed was proposed in [33]; however, the authors captured the image through a camera mounted into the white cane. The authors also proposed a solution for improving visually impaired people's mobility, which comprises a smart folding stick that works in tandem with a smartphone app using interconnection mechanisms based on GPS satellites. Navigational feedback is presented to the user as a voice output, as well as to the visually impaired family/guardians via the smartphone application. Rao and Singh [82] developed an obstacle detection method based on computer vision using a fisheye camera that is mounted onto a shoe. The photo is transmitted to a mobile application that uses TensorFlow Lite to classify the picture and alert visually impaired users about potholes, ditches, crowded places, and staircases. The device gives a vibration notification. In addition, an ultrasonic sensor is mounted with a servo on the front of the shoe to detect nearby obstacles. A vibration motor inside the shoe is used for feedback.

2.3. Feedback Techniques

People with standard vision depend on feedback that they gain from vision. They perceive more through vision than through hearing or touch. This is something that the visually impaired lack. Therefore, ETAs must be able to provide sufficient input on the perceived knowledge about the world of the user. Furthermore, feedback should be swift and not conflict with hearing and feeling [61].

2.3.1. Haptics

The haptic methods for the visually impaired person can offer more methods for interaction with the other human senses, such as hearing, and do not interfere with them. It has been noted in studies that the visually impaired have higher memorization abilities and recognition of haptic tasks [83]. The advantages of haptic feedback are high privacy, because only the person can observe the stimuli, as well as usefulness in high-noise environments and the fact that they can expand the person's experience as an additional communicational channel [84]. Buimer et al. [62] presented an experiment of a technique used to recognize facial emotion and send feedback through a vibration belt to the user. The authors conveyed information regarding six emotions by installing six vibration units in a belt. Even though the technique has accuracy problems, the satisfactory results of this method are based on a study of eight visually impaired people. Five of them found that the belt was easy to use and could interpret the feedback while conversing with another person. Meanwhile, the other three found its use difficult. Gonzalez-Canete et al. have proposed Tactons, whereby they identified sixteen applications with different vibration signals so that they could be distinguished from one another. The authors found that musical techniques for haptic icons are more recognizable and can be further distinguished. In addition, adding complicated vibrotactile sensations to smartphones is a significant benefit for users with any kind of sensory disability. The authors measured the recognition rate of the VI users and non-VI users. They found that non-VI users scored higher rates, especially with identification applications that they were familiar with, but when they used the reinforcement learning stage, in which some feedback is provided to the users, the recognition rate of the VI users increased [84].

2.3.2. Acoustic

Masking auditory signals by binaurally re-displaying environmental information through headphones or earbuds blocks vital environmental signals on which many visually impaired people rely for secure navigation [22]. Currently, bone-transmitting helmets enable the user to obtain 3D auditory feedback, leaving the ear canal open and enabling the user to operate with free eyes, hands, and mind. The algorithm reduces sound production that does not indicate a change in order to further minimize the auditory output. Thus, the audible output sound is only produced when the user is confronted with an impediment, limiting possibly irritating sounds to a minimum [85].

2.4. Functions and Applications

Here, we review the necessary functions and applications that the visually impaired use to solve difficult matters. These applications are obstacle detection, navigation, facial recognition, color and texture recognition, micro-interactions, text recognition, informing services, and braille displayers and printers. Next, a detailed review is presented.

2.4.1. Obstacle Detection

A great deal of research and many studies have focused on obstacle detection due to its significance for the visually impaired, as it is considered to be a major challenge for them. An ETA using a microwave radar to detect obstacles up to the head level through the vertical beam of the sensor was presented in [37]. To overcome the issue of power consumption, the authors switched off the transmitter during the listening time of the echo. Moreover, the pulsed chirp scheme was adapted to manage the spatial resolution. To improve the precision of the indoor blind guide robot's obstacle recognition, Du et al. [86] presented a sensor data fusion approach based on the DS evidence theory of the genetic algorithm. The system uses ultrasonic sensors, infrared sensors, and LiDAR to collect data from the surroundings. The optimized weight is replaced in DS evidence theory by data fusion for the purpose of determining the weight range of various sensors using the genetic algorithm. In practice, weighing and fusing evidence requires the determination of the weight of the evidence. Their technique has an accuracy of 0.94 for indoor obstacle identification. Bleau et al. [87] presented EyeCane, which can identify four kinds of obstacles: cubes, doors, posts, and steps. However, its bottom sensor failed to properly identify objects on the ground, making downwards navigation more dangerous.

2.4.2. Navigation

Navigation can be divided into two main categories, including internal and external navigation, because the set of techniques used in each one is different from the other. For example, the global positioning system (GPS) is not suitable for indoor localization due to the power of satellite signals, which become weak and cannot determine whether the user is close to a building or a wall [88]. However, some studies have developed techniques that may apply to both.

AL-Madani et al. [88] adopted a fingerprinting localization algorithm with fuzzy logic type-2 to navigate indoors in rooms with six BLE (Bluetooth low energy) beacons. The algorithm calculation was performed on the smartphone. The algorithm achieved an accuracy of 98.2% in indoor navigation precision and an accuracy of 0.5 m on average. Jafri et al. [89] used Google Tango Project to serve the visually impaired. The Unity engine's built-in functions in the Tango SDK were used to build a 3D reconstruction of the local area, and then a Unity collider component was provided to the user, who used it for obstacle detection by determining its relationship with the reconstructed mesh. A method of indoor navigation assistance using an optical head-mounted screen that directs the visually impaired is presented in [66]. The program creates indoor maps by monitoring a sighted person's activities inside of the facility, develops and prints QR code location markers for locations of interest, and then gives blind users vocal direction. Pare et al. [90] investigated a smartphone-based sensory replacement device that provides navigation direction based on strictly spatial signals in the form of horizontally spatialized sounds. The system employs numerous sensors to identify impediments in front of the user at a distance or to generate a 3D map of the environment and provide audio feedback to the user. A navigation system based on binaural bone-conducted sound was proposed by the authors of [91]. The system performs the following functions to correctly direct the user to their desired point. Initially, the best bone conduction device for use is described, as well as the best contact circumstances between the device and the human skull. Secondly, using the head-related transfer functions (HRTFs) acquired in the airborne sound field, the fundamental performance of the sound localization replicated by the chosen bone conduction device with binaural sounds is validated. A panned sound approach was also approved here, which may accentuate the sound's location.

iMove Around and Seeing Assistant [92,93] enable users to know their current location, including the street address, receive immediate area details (Open Street Map), manage their points, manage paths, create automated paths, navigate to the selected point or path, and exchange newly generated data with other users. The app can use voice commands to facilitate the control of the program. Seeing Assistant Move has an exploration mode that utilizes a magnetic compass to measure the direction correctly, which transmits this knowledge using clock hours. It also has a light source detector that allows the user to interact with devices that use a diode as an information tool. For people who are fully blind, the light source detector is extremely useful. When leaving the house or planning to sleep, the blind consumer can avoid leaving a lamp turned on. In order to be able to accommodate signaling devices (such as diodes and control lights), the program has a feature that helps a user to detect a blinking light. This can be used to indicate whether or not a device is turned on, or whether or not a battery level is low or high. On the other hand, a great deal of speech, correlated with the user position, is registered by the iMove around app. A note of speech is played any time when the user is near the position where it was captured. BlindExplorer [94] utilizes 3D sounds as auditory stimuli, which offers the app a type of feedback that helps a consumer to travel to the route or destination or in the right direction without needing to visualize the screen and without moving their eyes away from the ground. Right Hear [95] is a virtual access assistant that helps users to easily navigate new environments. It has two modes, including indoor and outdoor. It locates the visually impaired user's current location and nearby points in indoor and outdoor environments. However, indoor location is limited by supported locations.

Ariadne GPS [96] has a feature that makes the app suitable for the visually impaired. Namely, it uses VoiceOver in the app to inform the user about the street names and numbers that are around them, activated by touch. By simply placing the finger on the device's screen, the user can be told about the streets while viewing the map and moving it. The user location is in the middle of the screen, and everything in front of the user is in the top half of the screen, while the bottom half of the screen shows that wis behind the user. This app reports on the user's position at all times. It has a monitor function that works during activation, and it informs the user about their location continuously. BlindSquare [97] is a self-voicing software combined with third-party navigation applications that provides detailed points of interest and intersections for navigating both outside and inside and is designed especially for the visually impaired, blind, deafblind, and partially sighted. To determine what data are most important, BlindSquare has some special algorithms and talks to the user through high-quality speech synthesis. The app can be controlled by voice commands. The Voice Command feature is a paid service that requires credits to be purchased for its continuous use. Voice Command credits are available on the App Store as an in-app purchase.

2.4.3. Facial and Emotion Recognition

Morrison et al. [98] investigated the technological needs of the visually impaired through tactile ideation. Their findings were critical for people with visual disabilities and pointed to the need for social information, such as facial recognition and emotion recognition. In addition, social engagement and the ability to watch what others are doing, as well as the simulation of a variety of visual skills such as object recognition or text recognition, were important abilities. The importance of knowledge of people's emotions was discussed in [62]. The authors used computer vision technology to solve the problem. Their system uses facial recognition applications to capture six basic emotions. Then, it conveys this emotion to the user by means of a vibration belt. The proposal faced several challenges, involving lighting conditions and the movements of the person opposite the user who was facing the camera directly while the pictures were captured. Therefore, the recognition accuracy was affected. The authors did not present any numerical accuracy information in their paper.

2.4.4. Color and Texture Recognition

Medeiros et al. [65] present a finger-mounted wearable that can recognize colors and visual textures. They developed a wearable device that included a tiny camera and a co-located light source (an LED) that was placed on the finger. This technology allows users to obtain color and texture information by touch, allowing them to glide their fingers across a piece of clothing and combine their understanding of physical materials with automated aural input regarding the visual appeal. The authors used a special camera for this purpose. To identify visual textures, they used a machine learning approach, while color identification was performed through super-pixel segmentation to help with the higher cognition of clothing appearance.

Color Inspector [99] was developed to help blind and other visually disabled people to distinguish color by analyzing live footage in order to explain the color in view and to recognize complicated colors. It supports VoiceOver, which can audibly read out the color. Color Reader [100] allows for the real-time identification of colors solely by pointing the camera at the object. The app has a feature for reading the colors in the Arabic language.

ColoredEye [101] provides different color categories with different color descriptions, such as BASIC, with 16 fundamental colors, CRAYOLA, with 134 fun pigments, and DETAILED, with 134 descriptive colors.

2.4.5. Micro-Interactions

Micro-interaction refers to the animations and configuration adjustments that occur when a user interacts with something [102]. There are four stages involved in setting micro-interactions: First, a micro-interaction is started when a trigger is engaged. Second, triggers can be initiated by the user or by the system. Then, a user-initiated trigger requires the user to take action. In a system-initiated trigger, the software recognizes the presence of particular criteria and takes action. When a micro-interaction is initiated, rules dictate what occurs next. Finally, feedback informs individuals about what is happening. Feedback is defined as whatever a user sees, hears, or feels during a micro-interaction. The meta-rules of the micro-interaction are determined by Loops and Modes [103]. A device mounted on a finger was presented by Oh et al. [58] that uses physical gestures to facilitate the micro-interactions of some common and daily applications, such as setting the alarm and finding and opening an app. The authors carried out a survey to study the efficiency of their methods.

2.4.6. Text Recognition

The important questions that we must consider are how to use this feature and how to convey the important information only, rather than all of the information detected in the surrounding area [104]. The blob is a collection of pixels whose intensity is different from the other nearby pixels. Although the MSER (maximally stable external region) can detect the blob faster, the SWT (stroke width transform) algorithm can detect characters in an image with no separate learning process [104]. Shilkrot et al. [105] developed FingerReader, a reading support system for visually impaired people to assist impaired persons in reading printed texts with a real-time response. This gadget is a close-up scanning device that can be worn on the index finger. As a result, the gadget reads the printed text one line at a time and then provides haptic feedback and audible feedback. The Text Extraction Algorithm, which is integrated with Flite Text-To-Speech, was utilized by the authors of [106]. The proposed technique uses a close-up camera to retrieve the printed text. The trimmed curves are then matched with the lines. The 2D histogram ignores the repeated words. The program then defines the words from the characters and transmit them to ORC. As the user continues to scan, the identified words are recorded in a template. As a result, the system keeps a note of those terms in the event of a match. However, when the user deviates from the current line, they receive audible and tactile feedback. Moreover, if the device does not discover any more printed text blocks, the visually impaired receive signals via tactile feedback, informing them of the line's ending. SeeNSpeak [107] supports VoiceOver and can audibly interpret text from the photographs of books, newspapers, posters, bottles, or any item with text. The app can also use a wide range of target languages to translate the detected text.

2.4.7. Information Services

In the context of location-based services, a beacon is a tiny hardware device that allows data to be transmitted to mobile devices within a certain range. Most apps require that receivers have Bluetooth switched on and that they download the corresponding mobile app and have location services turned on. Moreover, they require that the receiver accepts the sender's messages. Beacons are frequently mounted on walls or other surfaces in the area as small standalone devices. Beacons can be basic, sending a signal to nearby devices, but they can also be Wi-Fi- and cloud-connected, with memory and processing resources. Some are equipped with temperature and motion sensors [108]. Perakovic et al. [60] used the beacon technology to inform the visually impaired of the required information, such as notifications about possible obstacles, location of an object, and information about a facility or discounts, and to provide navigation in indoor environments.

2.4.8. Braille Display and Printer

Braille technology is an assistive technology that helps blind or visually impaired individuals to perform basic tasks, such as writing, internet searches, braille typing, text processing, chatting, file downloading, recording, electronic mail, song burning, and reading [109]. Major challenges include the high cost of some current braille technologies and the large and heavy format of some braille documents or books with embossed paper. Braille is a representation of the alphabet, numbers, marks of punctuation, and symbols, composed of cells of dots. In a cell, there are 6–8 possible dots, and a single letter, number, or punctuation mark is created by one cell.

Displayer is an electromechanical mechanism for viewing braille characters that commonly utilizes round-tipped pins lifted through holes on a flat surface, which is often called a refreshable braille display or braille terminal. The visually impaired use it instead of a monitor. Via the braille display, they can insert commands and text, and it conveys text and images on the screen to them by modifying or refreshing the braille characters on the keyboard. They can browse the internet, draft documents, and use a computer in general. Up to 80 characters on the screen can be shown on a braille display, which can be updated when the user moves the cursor across the monitor using the command keys, cursor routing keys, or Windows and screen reader controls. Braille displays of 40, 70, or 80 characters are typically available. For most occupations, a 40-character display is appropriate and sufficient.

There are other important devices that can be used by the visually impaired, especially in the case of learning or employment in the office, such as the Refreshable Braille Display and Braille Embosser. A braille printer or a braille embosser is a device that uses solenoids to regulate embossing pins. It extracts data from computing devices and embosses the information on paper in braille. It produces tactile dots on hard paper, rendering written documents that are clear to the blind. Depending on the number of characters depicted, the cost of braille displays varies from USD 3500 to USD 15,000. In 2012, sixty-three studies aimed at finding new ways to developed refreshable braille were conducted by the Transforming Braille Group. Orbit Reader 20 is the result of these studies, with a 20-cell refreshable braille display [110] that currently costs USD 599. However, Orbit Reader 20 is the basic version and has limited braille characters.

Special braille paper is needed for braille embossers, which is heavier and more expensive than standard printer paper. More pages with the same volume of data are required for braille printing compared to a standard printer. They are also sluggish and noisier. Some braille printers are capable of printing single- or double-sided pages. Although embossers are relatively simple to use, they can be messy and can somewhat differ from one device to another in regard to the quality of the finished product [111]. Braille displays range in price from USD 3500 to USD 15,000, depending on the number of characters depicted. The cost of a braille printer is relative to the volume of braille it generates. Low-volume braille printers cost between USD 1800 and USD 5000, whereas high-volume printers cost between USD 10,000 and USD 80,000 [112].

2.4.9. Money Identifier App

Blind people have difficulty determining the value of the banknotes in their possession, and they usually ask and rely on others to discover the value of the banknotes, which exposes them to the risk of theft or fraud. Money identifier apps are a type of software that can recognize the denomination of banknotes used as currencies. The "Cash Reader: Bill Identifier" app [113] can identify a wide range of currencies. However, it requires a monthly payment subscription. MCT Money Reader [114] is another app that can recognize currencies, including Saudi Riyal, with a cost of SR 58.99 for lifetime use.

2.4.10. Light Detection

The Light Detector [115] translates any natural or artificial light source it encounters into sound. The software locates the light by aiming the mobile camera in its direction. Depending on the strength of the light, the user can notice a greater or lower tone. It is useful for helping the visually impaired to determine if the lights at home are on or off or to if the blinds are drawn by moving the device upwards and downwards.

2.4.11. External Assistance App

Be My Eyes [116] is an app with which blind or visually impaired users can request help from sighted volunteers. Once the first sighted user accepts the request, a live audio and video call is established between the two parties. With a back-camera video call function, the sighted aide can assist the blind or visually impaired. However, the app depends on an insufficient number of volunteers and has privacy issues because the users share videos and personal information during the connection.

2.4.12. Multifunctional App

Sullivan+ (blind, visually impaired, low vision) [117] depends on the camera shots of the mobile and analyzes them using the AI mode, which automatically seeks the top results that match the pictures taken. The app supports the following functions: text recognition, facial recognition, image description, color recognition, light brightness, and a magnifier. However, its capacities for facial and object recognition need further improvements. Visualize-Vision AI [118] makes use of different neural networks and AI to identify pictures and texts. The software is intended to provide visual assistance to the visually disabled, while still providing a developer mode that enables various AIs to be explored. Another app that can identify an object using artificial neural networks was explored in [119].

A visually impaired person can find an object by calling the name of the object, and the app finds it. Seeing Assistance Home [120] allows users to use an electronic lens for partially visually impaired persons, providing color identification, light source detection, and the ability to scan and produce barcodes and QR codes. VocalEyes AI [121] can assist the visually impaired in the following functions: object recognition; reading text; describing environments, label brands, and logos; facial recognition; emotion classification; age recognition; and currency recognition. LetSee [122] has three functions, including money recognition, which supports several currencies but does not support the Saudi Riyal; plastic card recognition; and light measurement tools to help users to locate sources of light, such as spotlights, cameras, or windows. The higher the brightness of the light is, the louder the sound the user hears is. TapTapSee [123] supports object recognition, barcode and QR code reading, auto-focus notification, and Flash toggle. Aipoly Vision [124] provides its service, including full object recognition features, for monthly fees of USD 4.99. The software also has the functions of currency recognition, text reading, color recognition, and light detection. However, the software is only supported by designated iPhone and iPad devices.

2.4.13. VoiceOver

For visually impaired people, voice over is one of the most useful functions. VoiceOver is a gesture-based screen reader. The user can utilize a mobile device even if they cannot see the screen. VoiceOver provides auditory explanations of what is on the screen.

On an iPhone, for example, when the user touches the screen or drags his finger over it, VoiceOver reads out the name of whatever the user touches, including icons and text. To interact with an item, such as a button or link, or to go to another item, the user can use VoiceOver gestures. VoiceOver creates a sound when the user moves to a new screen and then picks and speaks the name of the first item on the screen (typically in the top-left corner). VoiceOver tells the user when the display changes to the landscape or portrait orientation, the screen dims or locks, and what is active on the lock screen when the user turns on their iPhone [125].

The VoiceOver on iOS communicates with the user through a variety of "gestures", or motions made with one or more fingers on the screen. Many gestures are location sensitive. Sliding one's finger over the screen, for example, reveals the screen's visual contents as the finger passes over them. This allows visually impaired users to explore an application's actual on-screen layout. A person can activate a selected element by double-tapping, similar to double-clicking a mouse, in the same way a sighted user would. VoiceOver can also switch off the display while keeping the touch screen responsive, conserving battery life. This function is called the "Screen Curtain" by Apple [126].

2.4.14. Virtual Assistant Apps (Voice Commands)

Virtual assistants can help the visually impaired through their ability to control their mobile with voice commands. Here, we investigate the three most popular and recent virtual assistants: Siri, Google Assistant, and Bixby. One of the main concerns that we explore is privacy.

Siri

Siri is an assistant that uses voice queries and a natural language user interface to respond to queries, make recommendations, and take action by delegating requests to a set of internet services [127]. Siri assists in a series of tasks, such as phone calls; messaging, setting alarms, timers, and reminders; handling device settings; getting directions; scheduling events and reminders; previewing the calendar; running smart homes; making payments; playing music; checking facts; making calculations; and/or translating a phrase into another language [128]. However, some of the functions need to be visualized in order to be completed, because the assistant is designed for sighted people, not for the visually impaired. It needs improvements in order satisfy their needs. Siri provides various languages, including Arabic.

Apple notes that Siri searches and requests are paired with a specific identifier and not an Apple ID; thus, personal information is not stored for sale to advertisers or organizations. Apple declares that users can reset the identifier by turning Siri off and back on at any point, essentially restarting their interaction with Siri, which would erase user data associated with the Siri identifier. The terms state that personal information can only be used by Apple in order to provide or enhance third-party applications for their products, services, and ads. Private data are not exchanged with third parties for marketing purposes of their own. However, for whatever reason, Apple can use, pass, and reveal non-personal information. This means that Apple's sites, internet platforms, mobile software, email messages, and third-party product ads use monitoring tools to help Apple to better identify customer behavior, inform the business about the areas of its website accessed by the users, and promote and evaluate the efficacy of advertising and searches. Users may see advertisements dependent on other details in third-party applications; however, the Apple ID of a child also receives non-targeted advertisements on such platforms. The terms of Apple state that the protection of all children is a significant priority of Apple. Apple provides parents with the information they need to determine what would be best for their child.

Google Assistant

Google Assistant is a virtual assistant that is powered by artificial intelligence created by Google and is mostly available on smartphones and smart home platforms. Google Assistant can be accessed through its website and can be downloaded from the iOS App Store and the Google Play Store. For the sake of privacy, Google Account is designed with on/off data controls, allowing users to choose the privacy settings that suit them. In addition, as technology advances, Google's terms note that its privacy policies often change, meaning that privacy is still a user-determined individual option. The terms state that Google can use the personal information of users to provide ads to third parties but report that it does not sell the personal information of users to third parties. Moreover, the terms state that Google can show targeted ads to users, but that users can alter their preferences, choose whether their personal information is used to make ads more applicable to them, and turn such advertising services on or off. The terms also state that Google enables particular collaborators to use their cookies or related technology to retrieve information from a user's account or computer for advertisement and measurement purposes. Google's rules, however, note that a child's customized advertising is not displayed, meaning that advertisements are not focused on the information received from a child's account.

The terms of Google note that all of its services allow users to connect with other trustworthy and untrusted users and exchange information with other people, such as others with whom a user may chat or share content. If a user has a Google Account, their profile name, profile photo, and activities that they carry out on Google, or on third-party applications that are linked to their Google Account, can appear. In addition, information about a child, including their name, photo, email address, and transactions on Google Play, can be exchanged with members of a family using Google Family Link. These terms and conditions note that Google does not gather or use data for advertising purposes in the Google Cloud or G Suite services and that there are no commercials in the G Suite or Google Cloud Platform. Finally, the terms of Google note that it does not send users personalized advertisements on the basis of specific categories, such as religion, race, health, or sexual orientation.

Bixby

For Samsung devices, Bixby [129] is an intelligent digital assistant that hears and records according to the user's desire and works with their favorite applications. Bixby Vision's scene description function describes what is displayed on the screen. Personalized Bixby allows the assistant to learn the user's preferences based on their usage in order to facilitate its utility in the future. In addition, Bixby can handle smart devices with voice commands while attaching the apps to SmartThings. The user can change the TV channel or turn the lights on/off.

There is a privacy notice that describes what is recorded and saved. The information that Bixby requires—and it will not work until the user gives their permission—include the username, birthdate, phone number, email, device identifiers, voice commands, health information, and any information that has been provided through the application, such as the user's interaction with the app. In addition, some of the information can be sent to an external third party to convert the voice command into text. Some of Samsung's services allow users to communicate with others, and those other users may view information stored or displayed in the user's account on the social networking service that they are connecting to. In addition, Samsung can use third-party monitoring technology for a range of purposes, such as evaluating the usage of its services and (in combination with cookies) delivering user-relevant content and advertising. Some third parties may serve to advertise or keep track of which advertisements users see, how frequently they see those advertisements, and what users do in response. The terms note that only restricted representatives of Samsung's Bixby voice service team can access and otherwise process personal data in accordance with their job or contractual duties. Nevertheless, the terms do not reveal how common security measures in the sector, such as encryption, are used to secure sensitive details in transit or at rest [130]. The word Bixby is not easy to pronounce in Arabic. Currently, Bixby does not support the Arabic language.

Alexa

Alexa began as a smart speaker equipped with Alexa software, capable of listening to user questions and answering with replies. Over time, more household gadgets were interconnected through Alexa, and they can be operated by smartphones from anywhere. Amazon first built the Alexa platform to work as a digital assistant and entertainment device, but its application and use grew to encompass IoT, online searching, smart office, and smart home features, substantially improving the way that the average person interacts with technology [131]. To make it easier to interact with Alexa, the developers provide a set of tools, APIs, reference solutions, and documentation [132].

2.5. Research Gap

We provide a comparison of LidSonic V2.0 with the related works in Table 2. In Column 2, the technologies used in the particular works are mentioned in their respective rows. In Column 3, we discuss the work settings (i.e., whether they were indoor or outdoor). After that, the studies are examined in regard to the capacity for detecting transparent object features. We verify whether the gadget is handsfree. It is critical to know whether the device can operate at night, which is documented in Column 7. We also note whether or not machine learning techniques were used in the research. We also examine the different forms of feedback that they provided and whether they used verbal feedback. In addition, we examine the processing speed, because the solution requires real-time and quick data processing. We also discuss whether the gadget has a low energy consumption. We also explore the device's cost effectiveness and whether it is inexpensive, and also whether or not the solutions given require low memory, as well as their weights. The various studies relate to, and satisfied the requirements of, some of the system's essential features. All of these aspects of system design are addressed in our work. To ensure maturity and robustness, further system optimization and assessments are required. A detailed comparison of LidSonic V1.0, which also applies to LidSonic V2.0, is provided in [39].

We noted earlier that, despite the fact that several devices and systems for the visually impaired have been developed in academic and commercial settings, the current devices and systems lack maturity and do not completely fulfil user requirements and satisfaction. We created a low-cost, miniature green device that can be built into or mounted on any pair of glasses or even a wheelchair to assist the visually impaired. Our method allows for faster inference and decision-making while using relatively little energy and smaller data sets. The focus of this paper is the facilitation of the mobility of the visually impaired for the reason that this is one of the most basic and important tasks required for the visually impaired to be self-reliant, as explained in Section 1. The broader literature review was provided in this section to make the reader aware of other requirements of, and solutions for, the visually impaired, to break research barriers, and enable collaboration between different solution providers, leading to the integration of different solutions to create holistic solutions for the visually impaired. Increased and collaborative research activity in this field will encourage the development, commercialization, and widespread acceptance of devices for the visually impaired.

Table 2. System Aspects and a Comparison with Related Works.

Research	Technology	Environment		Transparent Object Detection	Handsfree	Functioning in Dark	ML/DL	Vocal Feed-back	High-Speed Processing	Low Energy Consump-tion	Low Cost	Low Memory Usage	Lightweight
		Indoor	Outdoor										
[133]	Solid-state LiDAR Sensor, RealSense L515 (LiDAR Depth Camera), Laptop	✓	✗	✗	✓	-	✓	✓	✗	✗	✗	✗	✓
[47]	LiDARs, Vibrotactile Units	✓	✓	✗	✓	✗	✓	✓	✗	✗	✓	✗	✓
[134]	Ultrasonic, PIR Motion Sensor, Accelerometer, Smartphone	✓	✓	✓	✓	✓	✗	✓	✓	✓	✓	✓	✓
This Work	TF-mini LiDAR, Ultrasonic	✓	✓	✓	✓	✓	✓	✓	✓	✓	✓	✓	✓

3. A High-Level View

In Sections 3.1–3.3, we present a high-level view of the LidSonic V2.0 system, the user view, the developer view, and the system view. A detailed description of the system design is provided in Section 4.

3.1. User View

Figure 2 shows the user view. The user puts on the LidSonic V2.0 gadget, which is fixed in a glass frame. The user installs the LidSonic V2.0 smartphone app after downloading it. Bluetooth connection between the LidSonic V2.0 mobile app and the LidSonic V2.0 device is used. LidSonic V2.0 is intensively trained in both indoor and outdoor settings. The user wanders around in both indoor and outdoor surroundings, allowing the LidSonic V2.0 gadget to be further trained and validated. Furthermore, a visually impaired person's family member or a volunteer may move around and retrain and check the gadget as needed. The gadget has a warning system in case the user encounters any impediments. When the user encounters an obstacle, a buzzer is activated. Additionally, the system may provide vocal input, such as "Ascending Stairs", to warn the user of an impending challenge. By pressing the prediction mode screen, the user may also hear the result. A user or his/her assistant can also use voice commands to label or relabel an obstacle class and create a dataset. This enables the validation and refining of the machine learning model, such as the revision of an object's label in the case that it was incorrectly categorized.

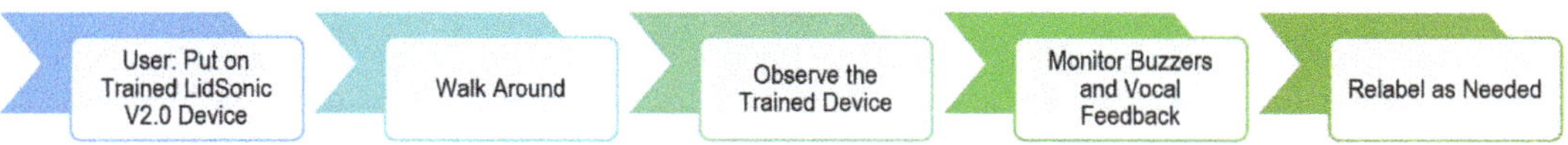

Figure 2. LidSonic: A User's View.

3.2. Developer View

The development of modules, as seen in Figure 3, starts with the construction of the LidSonic V2.0 device. A LiDAR sensor, ultrasonic sensor, servo, buzzer, laser, and Bluetooth are all connected to an Arduino Uno CPU used to build the LidSonic V2.0 gadget. Then, using an Arduino sketch, we combined and handled the different components (sensors and actuators), as well as their communication. The LidSonic V2.0 smartphone app was created with Android Studio (LidSonic V2.0). We created the dataset module to help with the dataset generation. Then, the chosen machine or deep learning module was used to construct and train the models. We utilized the Weka library for the machine learning and the TensorFlow framework for the deep learning models. Bluetooth is used to create a connection between the LidSonic V2.0 device and the mobile app, which is also used to send data between the device and the app. The Google speech-to-text and text-to-speech APIs were used to develop the speech module.

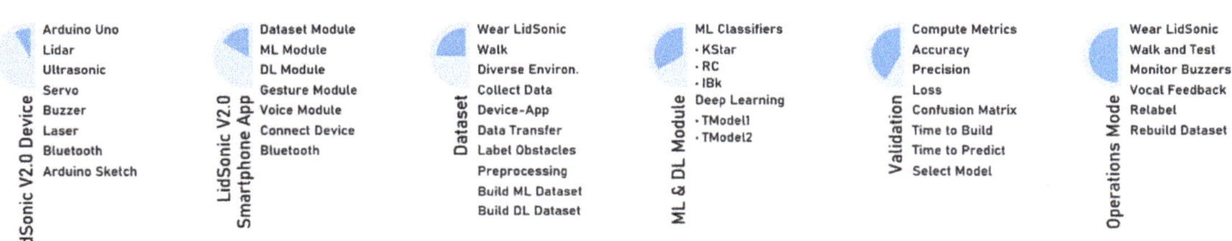

Figure 3. LidSonic: A Developer's View.

The developer wears the LidSonic V2.0 device and walks around to create the dataset. The LidSonic V2.0 device provides sensor data to the smartphone app, which classifies obstacle data and generates the dataset. To verify our findings, we used standard machine and deep learning performance metrics. The developer wore the trained LidSonic V2.0 gadget and went for a walk to test it in the operational mode. The developer observed the system's buzzer and vocal feedback. The dataset can be expanded and recreated by the developers, users, or their assistants to increase the device's accuracy and precision.

3.3. System View

LidSonic V2.0 detects hazards in the environment using various sensors, analyzes the data using multiple channels, and issues buzzer warnings and vocal information. With the use of an edge device and an app that collects data for recognition, we propose a technique for detecting and recognizing obstacles. Figure 4 presents a high-level functional overview of the system. When the Bluetooth connection is established, the data is collected from the LiDAR and ultrasonic sensors. An obstacle dataset should be established if the system does not already have one. The dataset is created using LiDAR data only. Two distinct channels or procedures are used to process the data. First, simple logic is used by the Arduino unit. The sensors operated by the Arduino Uno controller unit offer the essential data required for visually impaired people to perceive the obstacles surrounding them. It processes the ultrasonic and basic LiDAR data for rapid processing and feedback through a buzzer. The second channel is the use of deep learning or machine learning techniques to analyze the LiDAR data via a smartphone app and produce vocal feedback. These two channels are unrelated to one another. The recognition process employs deep learning and machine learning approaches and is examined and evaluated in the sections below.

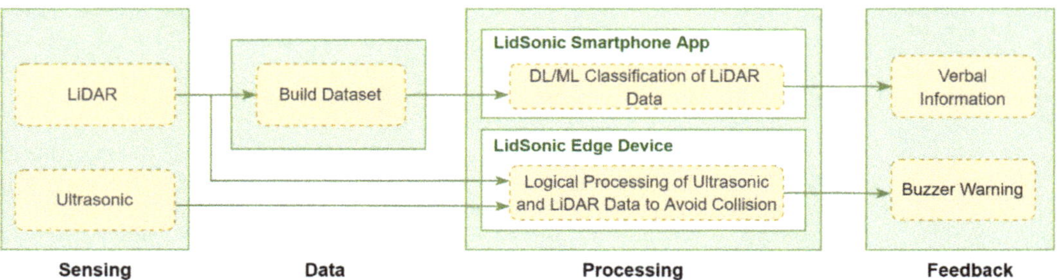

Figure 4. LidSonic V2.0 Overview (Functional).

Figure 5 depicts a high-level architectural overview of the system. The hardware, machine learning and deep learning models, software, datasets, validation, and the platform are all part of the system. The hardware includes all of the components required by the LidSonic V2.0 gadget. We created several models using ML and DL techniques that are explained further in Section 4. The system makes use of two types of software: one for controlling the sensors and performing the obstacle detection tasks with an Arduino skitch device, and another for the recognition tasks using the smartphone app. The accuracy, precision, loss, time to train a model, time to predict an object, and confusion matrix were employed as validation metrics in this work. Depending on the type and performance of the classifier, the system can be used on a variety of platforms, including edge and cloud. In the next section, we expand this system perspective with comprehensive diagrams and methods.

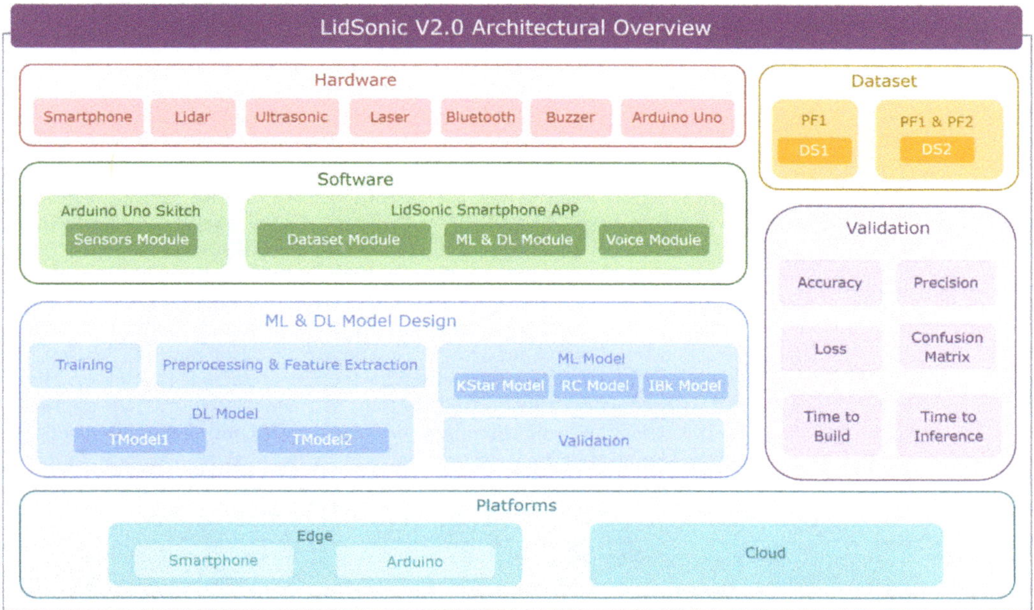

Figure 5. LidSonic V2.0 Overview (Architectural).

4. Design and Implementation

This section explains the LidSonic V2.0 System's design in detail. The hardware components and design are described in Section 4.1. Section 4.2 provides an overview of the system's software design. The sensor module is illustrated in Section 4.3 and the dataset and machine and deep learning modules are explained in Sections 4.4 and 4.5, respectively.

4.1. System Hardware

The system incorporates the following hardware components, as shown in Figure 6: TFmini Plus LiDAR, an ultrasonic sensor, Bluetooth, Arduino Uno, and the user's smartphone. A servo, buzzer, and power bank are used to operate the device.

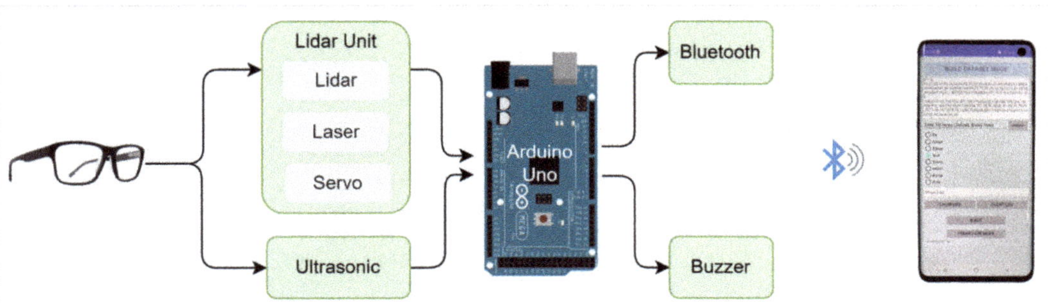

Figure 6. System Hardware.

Figure 7 displays a photo of the LidSonic V2.0 gadget, which includes smart glasses with sensors and an Arduino Uno board. The LidSonic V2.0 device's central nervous system is the Arduino Uno microcontroller unit, which is used to integrate and manage the sensors and actuators and to transfer sensor data to the smartphone app through Bluetooth. It is configured so as to control how the servo motions, sensors, and other components interact. The LiDAR Unit contains the TFmini Plus LiDAR sensor [135] that is connected with a servo and laser as a unit. The laser beam is installed above the TFmini Plus LiDAR and helps to indicate where the LiDAR is pointed so that one may scan and categorize various things in order to build a valid dataset. The data collected by the TFmini Plus LiDAR from its spatial environment is transferred to the Arduino unit. Some of this information is used by Arduino to detect obstacles and activate the buzzers as needed, while other information is relayed through Bluetooth to the smartphone app. The servo that controls the movement of the two devices comprises both the TFmini Plus LiDAR and the laser. The ultrasonic sensor is capable of detecting a wide range of obstructions. It is also utilized to compensate for the TFmini's LiDAR's inadequacies by recognizing transparent obstructions on the route of visually impaired people. The ultrasonic sensor detects objects at 30 degrees and has a detection range of 0.02 m–4.5 m [136]. The Arduino unit analyzes the data from the ultrasonic sensor, and if an item is detected, the buzzer is actuated. The buzzer sounds with distinct tones to notify visually impaired persons of different sorts of objects detected by the sensors. The buzzer and sound frequencies, or tones, are controlled by the Arduino CPU based on the identified items. A microphone for user instructions, Bluetooth for interfacing with the LidSonic V2.0 device, and speakers for vocal feedback regarding the detected objects are all included in the smartphone app's hardware.

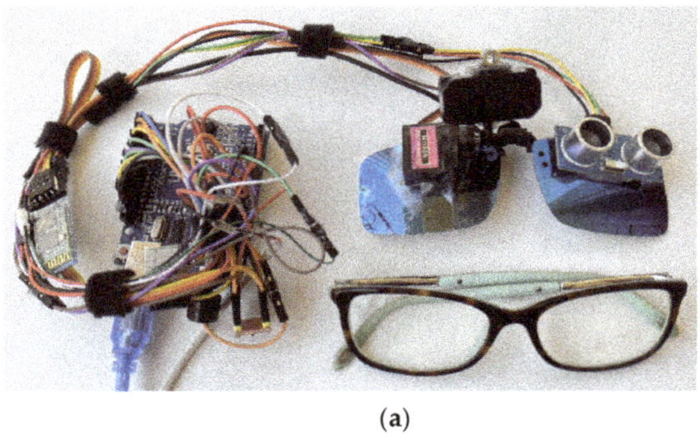

(a)

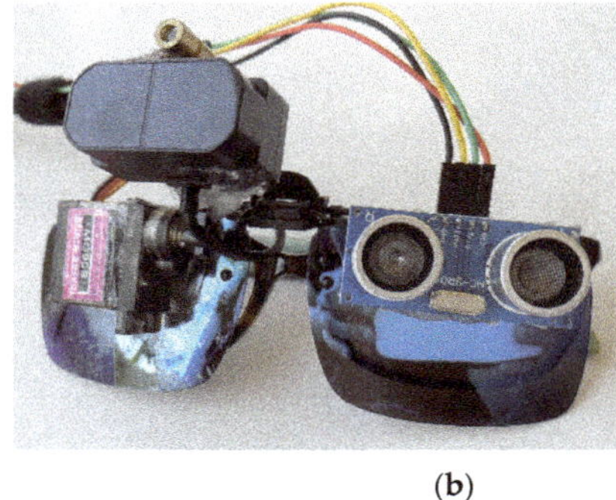

(b)

Figure 7. (**a**) LidSonic V2.0 device and a glass (**b**) LidSonic V2.0 device mounted into a glass frame.

4.1.1. TFmini-S LiDAR

A laser diode emits a light pulse, which is used in a LiDAR. Light strikes and is reflected by an item. A sensor detects the reflected light and determines the time of flight (ToF). The TF-mini-S device is based on the OPT3101 and is a high-performance, single-point, short-range LiDAR sensor manufactured by Benewake [137]. It is based on long-range proximity and distance sensor analog front end (AFE) technology based on ToF [137]. A TFmini-S LiDAR operates on the networking protocol UART (TTL)/I2C, can be powered by a conventional 5 V supply, and has a total power consumption of 0.6 w.

The TFmini-S LiDAR has a refresh rate of 1000 Hz and a size range of 10 cm to 12 m. It provides a ±6 cm accuracy between 0.1 m and 6 m, and a 1 percent accuracy between 6 m and 12 m. The operational temperature range is from around 0 °C to 60 °C. The range of the angles is 3.5° [138]. Data from the TFmini-S LiDAR may be collected quickly and precisely. There are no geometric distortions in the LiDAR, and it may be utilized at any time of day or night [138]. When no item is identified within a 12 m range, the sensor sends a value of 65,535.

The TFmini-S has the advantages of being inexpensive in cost, having a small volume, low energy consumption, and many interfaces in order to satisfy various requirements, but it has the disadvantage of not being able to detect transparent objects, such as glass doors (we used an ultrasonic sensor to compensate for this). It improves the outdoor efficiency and accuracy with various degrees of reflectivity by detecting stable, accurate, sensitive, and high-frequency ranges. Few studies have been conducted on the utilization of LiDAR to assist the visually impaired and identify their needs. The gadgets that aid the visually impaired make use of a very expensive Linux-based LiDAR [139].

4.1.2. Ultrasonic Sensor

An ultrasonic sensor is one of the best tools for detecting barriers because of its cheap price, low energy consumption, sensitivity to practically all types of artifacts [40], and the fact that the ultrasonic waves may be transmitted up to a distance from 2 cm to 300 cm. Furthermore, ultrasonic sensors can detect items in the dark, dust, smoke, in cases of electromagnetic interference, and tough atmospheres [140].

A transducer, in an ultrasonic sensor, transmits and receives ultrasonic pulses, which carry information about the distance between an item and the sensor. It sends and receives signals using a single ultrasonic unit [41]. The HC SR04 ultrasonic sensor has a <15° effective angle, a resolution of 0.3 cm, a frequency of operation of 40 kHz, and a measurement angle of 30°. The range limit of ultrasonic sensors is reduced when they are reflected off smooth surfaces, when they have a low incidence beam and when they open narrowly. Optical sensors, on the other hand, are unaffected by these issues. Nonetheless, the optical sensors' shortcomings include the fact that they are sensitive to natural ambient light and rely on the optical properties of the object [37]. Sensors are often employed in industrial systems to calculate object distance and flow velocity. ToF is the time required for an ultrasonic wave to travel from the transmitter to the receiver after being reflected by an object. Equation (1) can be used to calculate the distance from the transmitter, where c is the velocity of the sound [141]:

$$d = [c \times (ToF)]/2$$

Infrared sensors and lasers are outperformed by ultrasonic sensors. Infrared sensors cannot work in the dark and produce incorrect findings when there is no light. However, there are inherent drawbacks that restrict the application of ultrasonic instruments to mapping or other jobs requiring great accuracy in enclosed spaces. Due to sonar cross-talk, they are less reliable and have a reduced range, large beam coverage, latency, and update rates [12]. The receiver detects an undetectable small volume of the reflected energy if the obstacle surface is inclined (i.e., surfaces formed of triangles or with rough edges), which causes the ultrasonic range estimations to fail [142].

4.2. System Software

The LidSonic V2.0 system consists of the LidSonic V2.0 device and the LidSonic V2.0 Smartphone App (see Figure 8). The LidSonic V2.0 device's sensor module contains software that controls and manages the sensors (LiDAR and ultrasonic sensors) and actuators (the servo and laser beam). This module also carries out the basic logical processing of sensor data in order to generate buzzer alerts regarding discovered items.

The smartphone app's dataset module collects data from the LidSonic V2.0 device and appropriately stores the dataset, including the labels. The machine and deep learning module is located in the smartphone app and allows the models to be trained, inferred, and evaluated. Two Google APIs are used by the voice module. The text-to-speech API is used to provide audio feedback from the smartphone app, such as spoken input regarding adjacent objects identified by the sensors, using the mobile speakers. The Google speech-to-text API is used to transform user voice instructions and evaluate them so that the app can take relevant actions, such as labeling and relabeling data objects.

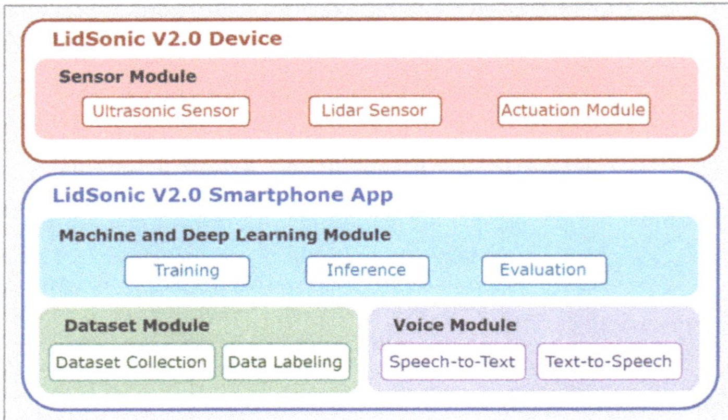

Figure 8. LidSonic V2.0 Software Modules.

The master algorithm is given in Algorithm 1. The array VoiceCommands (various commands sent to the LidSonic V2.0 System) and AIType are the master algorithm's inputs. AItype indicates the type of classification approach to be used, either machine Learning (ML) or deep learning (DL). Label, Relabel, VoiceOff, VoiceOn, and Classify are the VoiceCommands. The user gives the system the Label and Relabel voice commands to Label or Relabel an object observed by the system. The commands VoiceOff and VoiceOn are used to switch voice commands on and off if the user simply wants to hear the buzzer sound that alerts them when an object is close rather than hearing the names of all the things being recognized in the surroundings. When the user wants to identify a specific obstacle, they can use the voice command Classify. This command can be used even if the vocal feedback is turned off. The master algorithm produces three outputs: LFalert, HFalert, and VoiceFeedback, which are used to notify the user about various items through a buzzer or voice instruction.

Algorithm 1: The Master algorithm: LidSonic V2.0

Input: VoiceCommands (Label, Relabel, VoiceOff, VoiceOn, Classify),
AIType (ML, DL)
Output: LFalert, HFalert, VoiceFeedback

1.   ServoSensorsModuleSubSystem (Angle, Position)
2.   LaserSensorsModuleSubSystem ( )
3.   UD2O ← UltrasonicSensorsModuleSubSystem ( )
4.   [LD2O, LDO] ← LiDARSensorsModuleSubSystem ( )
5.   FeedbackType ← ObsDetWarnSensorsModuleSubSystem ( )
6.   **switch** (AIType) **do**
7.   **case:** ML
8.   [MLDataset] ← MLDatasetModule (LDO, Label, Relabel)
9.   [MOL, VoiceCommands] ← MLModule (MLDataset, VoiceCommands)
10.   **case:** DL
11.   [DLDataset] ← DLDatasetModule (LDO, Label, Relabel)
12.   [DOL, VoiceCommands] ← DLModule (DLDataset, VoiceCommands)
13.   **End switch**
14.   VoiceModule (VoiceCommands, VoiceFeedback)

The LidSonic V2.0 system operates different modules and subsystems for numerous purposes, as shown by the master algorithm. The ServoSensorsModuleSubSystem, a subsystem of the Sensors module, uses the angle and position as inputs to determine the servo starting position and motion and control the position of the LiDAR sensor. The LaserSensorsModuleSubSystem displays the direction in which the LiDAR is pointing (this is only for development purposes and assists the developer in identifying the object being scanned by the LiDAR). The UltrasonicSensorsModuleSubSystem returns the data output from the ultrasonic sensor, "UD2O" (the user's distance from the object computed based on the data from the ultrasonic sensor). The LiDARSensorsModuleSubSystem returns two outputs from the LiDAR sensor, "LD2O" (the user's distance from the object computed based on the data from the LiDAR sensor) and "LDO" (the user's distance from the object (LiDAR data object that contains detailed data about the objects). The ObsDetWarnSensorsModule-SubSystem returns "FeedbackType", detects objects, and informs the user about them via buzzers and voice feedback. The MLDatasetModule provides the Weka datasets, labeled "MLDataset", after receiving the inputs "LDO", "Label", and "Relabel". The DLDataset-Module provides CSV files, the "DLDataset". The MLModule returns "MOL" (the object level below or above the floor) and "VoiceCommands". The DLModule returns "DOL" (the object level below or above the floor) and "VoiceCommands". The VoiceModule transforms speech to text and vice versa using VoiceCommands and VoiceFeedback as inputs. In the next sections, more algorithms, pictures, and text are used to describe the four modules, as well as the inputs and outputs.

4.3. Sensor Module

Figure 9 illustrates how the LidSonic V2.0 pair of glasses use ultrasonic and LiDAR sensors to observe the world. The ultrasonic sound pulse is directed in front of the user, as seen by the dotted green line, to detect any objects in front of the user. It can also detect obstacles that are transparent, such as glass doors or walls, which LiDAR may miss. The LiDAR sensor range is represented by dotted blue lines. The LiDAR sensor has a range of 10 cm to 12 m. We covered a 60-degree region in front of the user with a servo motor that moves the LiDAR sensor, which equates to an area of "m" meters on the floor. This floor area "m", covered by the LiDAR sensor for a user of 1.7 m in height, would be around 3.5 m. Note that we ignored the fact that the glasses are at eye level rather than head level. The figure also displays the floor area "n", which is the closest floor area to the user that is not covered by the LiDAR sensor, since we deactivated it in order to minimize false alarms triggered by the user's knee while walking. This floor space closest to the user would be around 0.15 m for a person of 1.7 m in height. Within this "m" floor space, the LidSonic V2.0 system identifies any obstacles, including descending the stairs, using the LiDAR sensor.

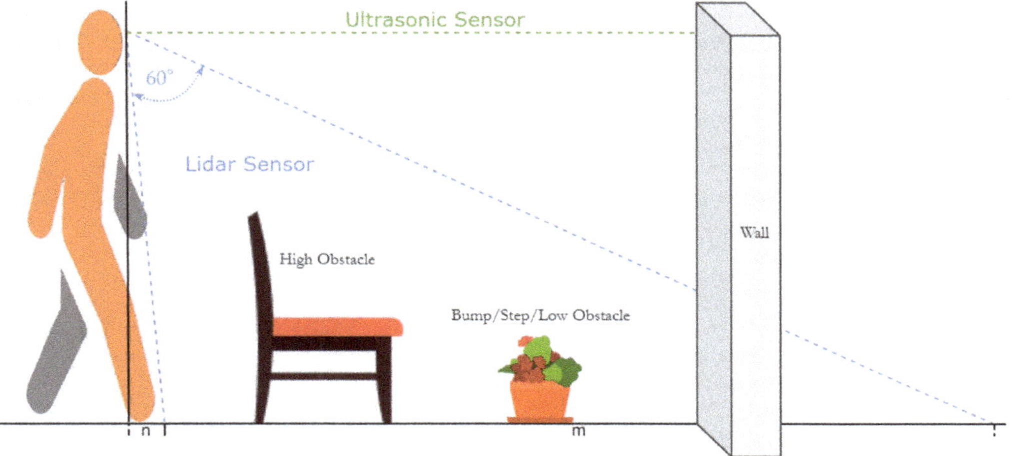

Figure 9. Sensor Coverage.

The flow diagram of the obstacle detection and warning subsystem is shown in Figure 10. The LidSonic V2.0 device uses data from an ultrasonic sensor (UD2O stands for distance to object detected by the ultrasonic sensor) and activates a buzzer with a low-frequency warning, providing an LFalert if it falls below a threshold of 0.5. The LidSonic V2.0 device additionally checks the closest point read by the LiDAR sensor (LD2O is the D2O detected by the LiDAR sensor), and if it is higher than the floor, the LFalert is triggered, signaling that the obstacle might be a high obstacle, a bump, and/or ascending stairs, etc. If it is below the floor, this means that there are obstructions such as descending stairs and/or holes, and a high-frequency alarm HFalert buzzer tone is triggered. The ML module provides the user with voice input depending on the anticipated obstacle type. MOL is converted from the predicted value (object level detected by the ML module). The buzzer is actuated with the LFalert if the predicted value type is an object above the floor; otherwise, the HFalert is triggered. The predicted value is converted to DOL (object level detected by the DL module). If the predicted value type is an object above the floor, the buzzer is activated with the LFalert; otherwise, the HFalert is activated. The figure only shows "MOL < Floor"; however, the values of MOL or DOL are used based on the algorithm used.

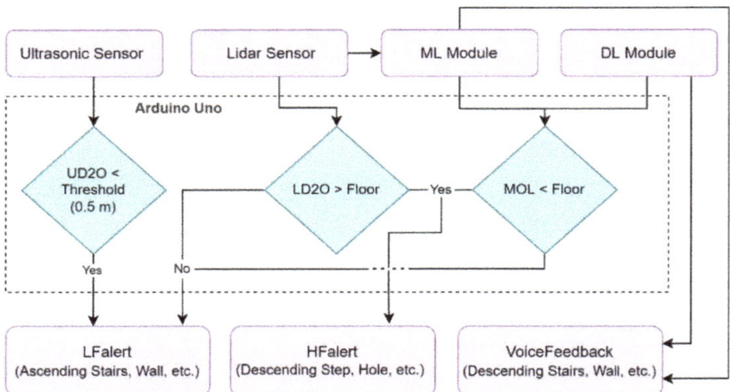

Figure 10. Detection and Warning System.

Algorithm 2 depicts the algorithm for the obstacle detection, warning, and feedback subsystem. It takes the ultrasonic data (UD2O), nearby LiDAR distance readings (LD2O), the object level calculated by the machine learning module (MOL), and the object level calculated by the deep learning module (DOL) as inputs. The ObsDetWarnSensorsModuleSub-System function analyzes data for detection and generates audio alarms. High-frequency buzzer tones (HFalert), low-frequency buzzer tones (LFalert), and VoiceFeedback are the output alerts. The subsystem has a logical function that accepts the inputs UD2O, LD2O, and MOL and returns the determination of the kind of obstacle (whether the obstacle is an object above floor level, etc.). No action is required if the output is a floor. If the obstacle returns as HighObs, however, it is either a wall or a high obstacle, and so on.

An LFalert instruction is delivered to the buzzer to start the low-frequency tone buzzer. The buzzer parameter HFalert is used to activate the high-frequency tone buzzer if the obstacle is of the LowObs type. The goal of selecting a high-frequency tone for low obstacle outputs is to ensure that low obstacle outputs are possibly more hazardous and destructive than high obstacle outputs. The high-frequency tone may be more noticeable than the low-frequency tone.

Algorithm 2: Obstacle Detection, Warning, and Feedback
Input: UD2O, LD2O, MOL, DOL
Output: FeedbackType (HFalert, LFalert, VoiceFeedback)
1.   **Function** ObsDetWarnSensorsModuleSubSystem ( )
2.   Obstacle ← Check (UD2O, LD2O, MOL, DOL)
3.   **switch** (Obstacle) **do**
4.   **case:** Floor
5.   skip
6.   **case:** HighObs
7.   Buzzer (LFalert)
8.   VoiceModule (VoiceCommands, VoiceFeedback)
9.   **case:** LowObs
10.   Buzzer (HFalert)
11.   VoiceModule (VoiceCommands, VoiceFeedback)
12.   **End switch**

4.4. Dataset Module

Machine learning is strongly reliant on data. It is the most important factor that enables algorithm training to become possible and to obtain accurate results from the trained models. Our dataset includes a 680-example training set and is formatted using ARFF files for Weka and CSV files for TensorFlow deep learning. The collection provides distance data obtained from the LiDAR equipment by LidSonic V2.0.

Table 3 shows the eight classes in the dataset, the kinds of obstacles that we accounted for, as well as the number of examples/instances of each. Note that the images in the table include both indoor and outdoor conditions. Deep Hole, Ascending Stairs, Descending Stairs, and Wall were trained in indoor environments, while the rest of the objects were trained in outdoor environments. For our smart glasses, we were not especially concerned with the exact specifications of the objects but rather with the broader types of the objects. For example, differentiating between Descending Stairs and a Deep Hole is important, because the former is an object that blind people might wish to use (go down), while they would avoid the Deep Hole. The type of Deep Hole is not important, because they would aim to avoid a deep hole. We trained the machine and deep learning algorithms with the data generated by the LiDAR sensor for the objects and did not program the software with the specification of the objects. Hence, the objects are not defined.

Table 4 shows the preprocessing and feature extraction approaches. PF1 requires that the LidSonic V2.0 device's upward line scan is set to the same angle index as the downward scan and ends with the class label. PF2 must complete PF1 and then extract just eleven angle readings by dividing the 60 readings by ten along with the last angle distance data (in essence, skipping every five readings, assuming that an object does not exist in this gap and considering that the user is moving, so that this gap is moving too). It also computes the height of the angle nearest to the user, which is the LidSonic V2.0 device's starting point, as well as the middle angle of the LidSonic V2.0 device's scan. We need to calculate the distance from the user to the obstacle d2 and the distance from the user to the ground d1 (y-axis), the points for both angles, once we have the two height calculations, which are the x-axis points. The slope between h1 and h2 can then be calculated. The two heights and the slope are added to the 11-angle distance readings to create the 14-feature dataset, DS2. The 60-angle distance readings are the dataset features of DS1.

Table 3. Obstacle Dataset.

Image Class	No. of Instances	Example	Image Class	No. of Instances	Example
Floor	111		Deep Hole	20	
Ascending Stairs	108		High Obstacle	149	
Descending Stairs	89		Ascending-Step/Bump	62	
Wall	87		Descending-Step/Hole	34	

Table 4. Preprocessing and Feature Extraction.

Name	Preprocessing and Feature Extraction
PF1	Adjust the upwards lines of the readings to be the same (angle index) with the downwards readings
PF2	PF1 Extract 11 angle reading by dividing the 60 readings by 10 + angle no. 60 Add three features: Calculate the height of the obstacle of the starting angle of the LidSonic V2.0 device (angle closest to the user h1) Calculate the height of the obstacle of the middle angle of the LidSonic V2.0 device scan (h2) Calculate the slope between h1 and h2

Figure 11 depicts the model used for calculating the obstacle height. The distance between the LidSonic V2.0 device and the ground is represented by g. The larger triangle's hypotenuse, which is colored blue, is represented by g. The LidSonic V2.0 device's LiDAR distance from an obstacle is c. We can compute the height of the object h using the similar triangle law and the value of c. Two triangles that have the same ratio of their comparable sides and an identical pair of corresponding angles are called similar triangles.

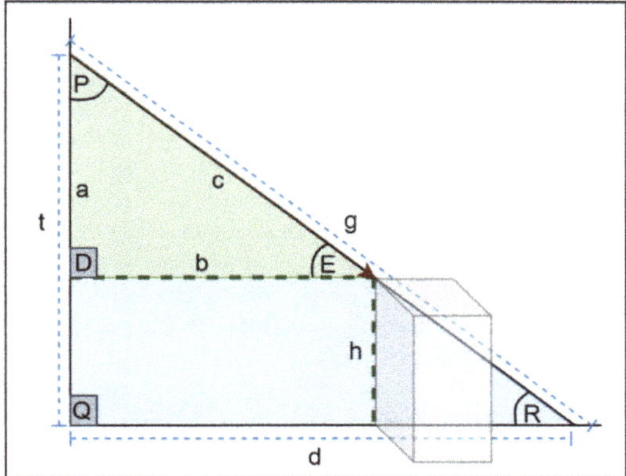

Figure 11. Obstacle Height.

In ΔPQR and PDE, ∠DPE is common and ∠PDE = ∠PQR (corresponding angles). (1)

$$\Rightarrow \Delta PQR \sim \Delta PDE \text{ (PP criterion for similar triengles)} \quad (2)$$

Hence, from (1) & (2):

$$\Rightarrow PR/PE = PQ/PD \quad (3)$$

From Equation (3), calculate r as:

$$r = c/g \quad (4)$$

then,

$$a = t * r \quad (5)$$

The height of the obstacle is:

$$h = t - a \quad (6)$$

The horizontal distance from the user to the obstacle is calculated by Equation (7):

$$b = d * r \quad (7)$$

There is a $\cong$ ±3 cm error when computing the height of an object, which we consider insignificant in our case, because we do not require the exact height but, rather, the nature of it (high, low, etc.). The height is calculated and used as a feature in the dataset. We added two height features, h1 and h2, for the purpose of this computation.

Another crucial parameter that we included in our dataset is the slope in Figure 12, between h2 and h1. Since the value of the slope fluctuates depending on the slope of the ground level or the kind of obstacle, especially in the case of stairs, it is a significant factor. Equation (8) calculates the slope as follows:

$$s = \frac{h2 - h1}{b2 - b1} \quad (8)$$

We created two types of datasets: one that collects 60 values of the features using the LiDAR distance of 60 angles, which we called DS1, and the second dataset, DS2, which extracts 11 features from DS1. Three more features were added: two obstacle heights from two different angles, as well as the slope between these two positions, giving a total of 14 features. We constructed six training models to be examined, evaluated, and analyzed for optimal utilization. Two different approaches were investigated: Weka-based machine learning and TensorFlow-based neural networks. We used K* (KStar), Random Committee, and IBk as classifiers in Weka in order to train six machine learning models. For our system, these were the most successful classification methods [39]. The second is a TensorFlow-based deep learning technique that utilizes the two datasets. We constructed two deep learning models for each. Machine and deep learning algorithms are discussed in the next subsection. The DS1 labels range from a01–a60 and end with the obstacle class to give a total of 61 features. DS2 has 11 angle labels in addition to h1, h2, s, and the obstacle class. The DS1 labels range from a01–a60 and end with the obstacle class to give a total of 61 features. DS2 has 11 angle labels in addition to h1, h2, s, and the obstacle class.

Tables 5 and 6 list the two types of datasets that we used in our research, along with a sample of collected data for each obstacle class. The DS1 labels range from a01–a60 and end with the obstacle class to give a total of 61 features. DS2 has 11 angle labels in addition to h1, h2, s, and the obstacle class.

Table 5. Dataset 1 (D1) Sample.

Obstacle Type	Data
DS1 Labels	a01, a02, a03, a04, a05, a06, a07, a08, a09, a10, a11, a12, a13, a14, a15, a16, a17, a18, a19, a20, a21, a22, a23, a24, a25, a26, a27, a28, a29, a30, a31, a32, a33, a34, a35, a36, a37, a38, a39, a40, a41, a42, a43, a44, a45, a46, a47, a48, a49, a50, a51, a52, a53, a54, a55, a56, a57, a58, a59, a60, Obstacle_class
Floor	309, 305, 301, 295, 288, 285, 274, 266, 264, 260, 259, 253, 249, 243, 240, 229, 227, 211, 215, 214, 211, 208, 208, 205, 205, 197, 193, 191, 187, 186, 180, 176, 172, 169, 167, 172, 173, 172, 171, 169, 168, 167, 166, 165, 164, 163, 161, 161, 159, 158, 33, 25, 24, 23, 23, 26, 29, 120, 154, 157, 0
Ascending Stairs	141, 143, 143, 142, 143, 127, 145, 145, 142, 143, 144, 146, 147, 145, 141, 129, 133, 134, 127, 130, 134, 135, 135, 134, 134, 134, 134, 133, 131, 130, 128, 127, 124, 123, 122, 121, 119, 118, 119, 121, 130, 130, 129, 129, 128, 128, 124, 123, 124, 123, 123, 123, 122, 122, 132, 133, 132, 131, 133, 133, 1
Descending Stairs	696, 740, 841, 834, 575, 372, 380, 396, 608, 659, 654, 614, 546, 353, 343, 386, 389, 385, 382, 374, 359, 356, 355, 354, 350, 342, 336, 321, 308, 305, 303, 299, 296, 268, 264, 264, 262, 264, 263, 261, 254, 190, 189, 201, 215, 217, 211, 211, 209, 207, 208, 208, 206, 200, 154, 186, 186, 186, 185, 183, 2
Wall	49, 52, 46, 46, 46, 50, 51, 50, 52, 49, 53, 52, 50, 50, 55, 54, 57, 55, 55, 57, 58, 56, 57, 58, 62, 62, 62, 63, 65, 64, 66, 66, 69, 70, 78, 80, 83, 83, 84, 86, 88, 93, 95, 96, 102, 111, 123, 126, 118, 120, 120, 120, 122, 125, 129, 141, 152, 152, 152, 152, 3
Deep Hole	638, 643, 646, 650, 654, 659, 654, 661, 669, 663, 666, 668, 669, 672, 638, 631, 628, 630, 592, 588, 589, 592, 577, 554, 555, 532, 531, 531, 530, 528, 523, 520, 495, 494, 490, 487, 485, 482, 480, 476, 472, 450, 441, 443, 446, 438, 434, 434, 434, 436, 436, 430, 429, 427, 426, 423, 422, 422, 422, 421, 4
High Obstacle	87, 85, 85, 85, 85, 84, 84, 85, 89, 88, 91, 93, 93, 94, 99, 99, 100, 100, 98, 98, 98, 100, 100, 102, 103, 103, 103, 103, 102, 101, 99, 98, 96, 96, 95, 94, 93, 92, 91, 91, 91, 90, 88, 85, 86, 86, 84, 80, 78, 77, 77, 81, 81, 81, 81, 81, 81, 79, 78, 75, 5
Bump	237, 225, 217, 231, 239, 242, 227, 219, 219, 220, 228, 205, 205, 207, 210, 212, 212, 200, 196, 196, 196, 195, 194, 183, 184, 176, 174, 173, 170, 168, 167, 166, 164, 163, 160, 159, 158, 157, 153, 151, 151, 147, 144, 144, 144, 148, 148, 148, 147, 144, 143, 143, 143, 143, 143, 143, 142, 141, 140, 139, 6
Hole	275, 275, 298, 298, 273, 276, 284, 294, 296, 263, 254, 260, 262, 262, 254, 252, 252, 251, 249, 248, 245, 243, 231, 227, 228, 215, 214, 212, 212, 210, 209, 206, 203, 201, 201, 201, 200, 196, 189, 191, 191, 186, 184, 184, 185, 180, 181, 177, 178, 177, 178, 175, 173, 174, 174, 175, 174, 172, 171, 171, 7

Table 6. Dataset 2 (D2) Sample.

Obstacle Type	Data
DS2 Labels	a01, a06, a12, a18, a24, a30, a36, a42, a48, a54, a60, h1, h2, s, Obstacle_class
Floor	309, 285, 253, 211, 205, 186, 172, 167, 161, 23, 157, 0.96, 2.42, 0.02, 0
Ascending Stairs	141, 127, 146, 134, 134, 130, 121, 130, 123, 122, 133, 24.21, 47.76, 0.54, 1
Descending Stairs	696, 372, 614, 385, 354, 305, 264, 190, 211, 200, 183, −24.20, −93.90, −0.55, 2
Wall	49, 50, 52, 55, 58, 64, 80, 93, 126, 125, 152, 5.81, 101.19, 19.06, 3
Deep Hole	638, 659, 668, 630, 554, 528, 487, 450, 434, 427, 421, −254.67, −274.42, −0.09, 4
High Obstacle	87, 84, 93, 100, 102, 101, 94, 90, 80, 81, 75, 80.37, 71.23, −0.23, 5
Bump	236, 264, 221, 209, 182, 173, 162, 152, 148, 141, 139, 18.39, 12.95, −0.08, 6
Hole	275, 276, 260, 251, 227, 210, 201, 186, 177, 174, 171, −12.58, −17.0, −0.05, 7

Algorithm 3 outlines how our system's dataset is created. It takes the CSV Header, LDO, and Features as inputs. Using the Building Dataset function, the CSV header file from CSVHeader is first placed in the new dataset, CSVFile. Data are collected from LDO and saved in a LogFile using the DataCollection method. LDO represents the LiDAR distance readings, while the loop records the data in the proper format, saving the LiDAR downwards data in the original order and reversing the order of the LiDAR upward data.

Algorithm 3: DatasetModule: Building Dataset Algorithm

Input: Header, LDO
Output: Dataset

1.	**Function:** BuildingDataset ( )
2.	Insert Header into the File
3.	LogFile ← LDO
4.	**While** (not end of LogFile) //Bluetooth incoming data stored in LogFile
5.	strLine ← BufferLine //BufferLine is a line taken from LogFile
6.	mutualFlag ← true
7.	**While** (strLine ! = 0)
8.	**If** (mutualFlag)
9.	myData ← strLine + Obstacle class
10.	Write myData in the File
11.	Clear myData
12.	mutu-alFlag ← false
13.	**Else**
14.	Store numbers of strLine into an array called strarr
15.	**For** (x = (strarr.length)—1; x ≥ 0; x–)
16.	reverseStr ← reverseStr + strarr[x] + ","
17.	**End For**
18.	myData ← myData + reverseStr + Obstacle class
19.	Write myData in the File
20.	Clear my-Data and reverseStr
21.	**End If**
22.	**End While**
23.	**End While**

Figure 13 depicts the LidSonic V2.0 Smartphone app's user interface, which is used for building the dataset. The LiDAR sensor sends data to the mobile app through Bluetooth, which it saves in a file named LogFile. On the left-hand side is the prediction mode in which the user hears the verbal feedback of the recognized hazard. In addition, it shows some of the evaluation measurements that are conducted for the three classifiers. For example, KstarT-Elapsed Time (ms) shows the time in milliseconds that is required to build the Kstar classifier (we provide more details regarding the classifiers in the next subsection) for the DS1 dataset in the white box and the DS2 dataset in the blue box. KstarI-Elapsed Time (ms) shows the inference time required to predict an object for datasets DS1 and DS2, respectively. When the D1 INFERENCE button is pressed, the evaluation measurement of DS1 is presented in the white boxes for each classifier, while the D2 INFERENCE button shows the results obtained for DS2.

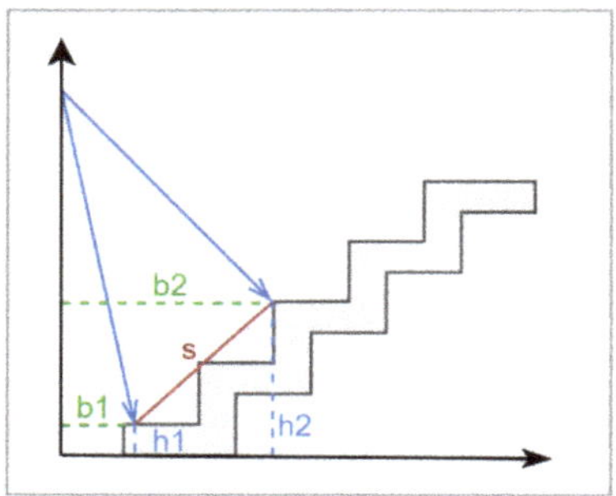

Figure 12. Slope.

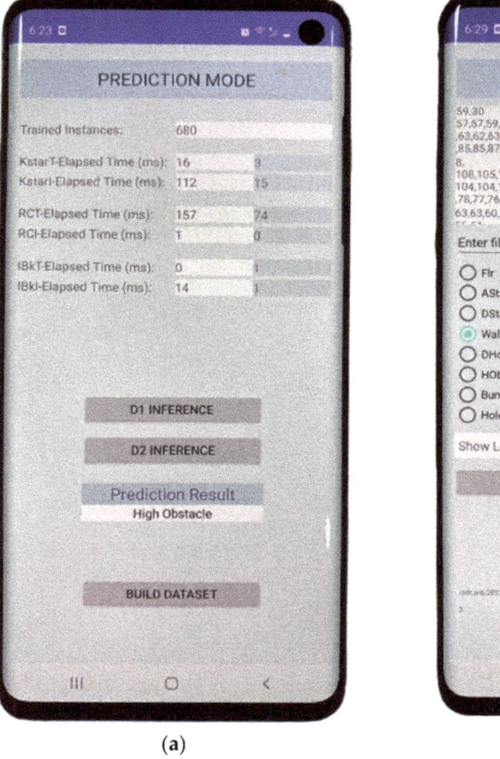

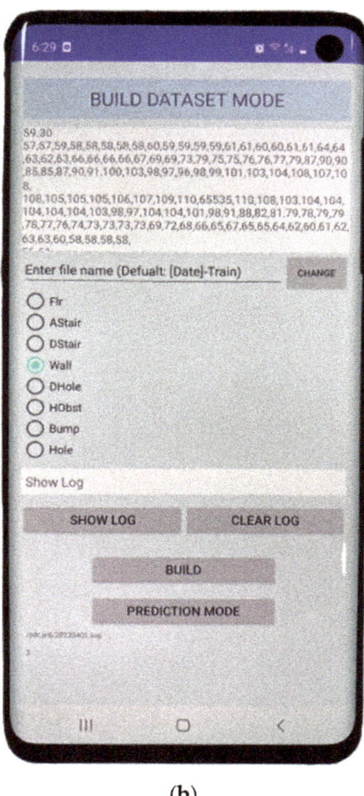

Figure 13. LidSonic V2.0 App. (**a**) Prediction Mode. (**b**) Build Dataset Mode.

The LogFile data are displayed in the figure's right-hand mobile app view. The real number ultrasonic sensor measurements is shown in the first line. We will explore this further in future work to determine whether it is worthwhile to include it as a feature. The ultrasonic measurements were not included in the dataset for this study. The LiDAR sensor's downward and upward 60-degree readings are presented in the first two lines, which include 60 comma-separated numbers. The LiDAR sensor is linked to a servo that rotates in one-degree increments downwards and upwards, capturing the distance from the device to the object at each degree position. The 60-degree downward and upward measurements are acquired in this manner. Each line of data basically provides 60 measurements of the distance from the user's eye to the object, each with a distinct line of sight angle. Every two lines of the 60-degree downward and upward measures are followed by a real number, and so on.

4.5. Machine and Deep Learning Module

We tested multiple types of models on different types of datasets to determine which one performed the best. Algorithm 4 shows the method used to preprocess the data and extract features. The inputs are Dataset, LIndex1, and LIndex2. This is the method we use to preprocess and extract features from the dataset and the variation between the ARFF files (for WEKA) and CSV files (for TensorFlow) based on the file format. To begin, we proceed to the locations where the dataset's data begin. The SelectedFeatures function extracts 11 data values from 60 values for each line. The CalculateHeight function takes the two pointing locations of the Lidar sensor, LIndex1 and LIndex2, to acquire their heights, yielding Height [1] and Height [2]. The two numbers are then sent to the CalculateSlope function to determine the slope, finally incorporating the findings into the file.

Algorithm 4: PreProFXModuleDL: Preprocessing and Feature Extraction

---

Input: Dataset, LIndex1, LIndex2
Output: Dataset

1.   Go to the first instance
2.   **While** (not end of File)
3.   strLine ← BufferLine //BufferLine is a line taken from the Dataset
4.   **While** (strLine ! = 0)
5.   SData [1, . . . , 11] ← SelectedFeatures()
6.   Height [1], Height [2] ← CalculateHeight (LIndex1,LIndex2)//Equations (4)–(6)
7.   Slope ← CalculateSlope(Height [1], Height [2])//Equations (7) and (8)
8.   DataLine ← AddFeatures(SData[], Height[], Slope)
9.   Write DataLine in the File
10.   Clear DataLine
11.   **EndWhile**
12.   **EndWhile**

Algorithms 5 and 6 provide high-level algorithms for the machine and deep learning modules. A detailed explanation is presented in Sections 4.5.1 and 4.5.2.

Algorithm 5: Machine learning Module

Input: Dataset, PrFx
Output: MOL, VoiceCommands

1.   **If** (PrFx)
2.   Dataset ← PreProFXModuleML(Dataset)
3.   **End If**
4.   MLModel ← Train (Dataset)
5.   [MOL, VoiceCommands] ← Inference (MLModel)

Algorithm 6: Deep Learning Module
Input: Dataset, PrFx
Output: DOL, VoiceCommands
1.   **If** (PrFx)
2.   Dataset ← PreProFXModuleDL(DLDataset)
3.   **End If**
4.   DLModel ← Train (Dataset)
5.   [DOL, VoiceCommands] ← Inference (DLModel)

The prediction mode can be used to aid visually impaired users in three distinct ways: the prediction button, throw gesture, or voice instruction. To utilize the voice instruction, the user must double-tap the screen so as to access the speech-to-text API and then speak "Prediction Mode" on the command line. The prediction mode is accessed by flinging the screen.

4.5.1. Machine Learning Models (WEKA)

WEKA, a java-based open source program that contains a collection of machine learning algorithms for data mining applications [143], was employed to train the dataset with three classifiers that were carefully selected from detailed experiments carried out in our previous works, which provided the best results, including KStar, IBk, and Random Committee [39].

KStar Algorithm

KStar is an instance-based classifier, which means that the class of a test instance is decided by the class of related training examples, as defined by a certain similarity function. It utilizes an entropy-based distance function, which sets it apart from other instance-based learners. Instance-based learners use a dataset of pre-classified examples to categorize an instance. The essential hypothesis is that comparable instances are classified similarly. The issue is that we must determine how to define the terms "similar instance" and "similar class". The distance function, which defines how similar two examples are, and the classification function, which describes how the instance similarity creates a final classification for the new instance, are the related components of an instance-based learner. The KStar algorithm employs an entropic measure, which is based on the chance of random selection from among all the conceivable transformations, turning one instance into another. It is particularly helpful to use entropy as a metric for the instance distance, and information theory aids in the determining the distance between the instances [144]. The distance between instances determines the complexity of a transition from one instance to another. This is accomplished in two stages. To begin, a limited set of transformations are created that transfer one instance to another. Then, using the program, we can convert one instance from x to y in a limited sequence of transformations that begins with x and ends with y.

Instance-Based Learner (IBk) Algorithm

An ideal description is found in the principal output of IBk algorithms (or concept). This is a function that maps instances to create categories. It returns a classification for an instance chosen from the instance space, which is the anticipated value for the instance's category attribute. A collection of stored examples and, possibly, some information about their historical performances during classifying are included in an instance-based concept description (e.g., their number of correct and incorrect classification predictions). After each training instance is handled, this list of instances may vary. IBk algorithms, on the other hand, do not generate extensive idea descriptions. Instead, the IBk algorithm's chosen similarity and classification functions determine the concept descriptions based on the current collection of stored instances. The framework that describes all IBk algorithms comprises three sections: the Similarity Function, Classification Function, and Concept

Description Updater. These are explained as follows. (1) The Similarity Function determines how similar a training instance x is to the concept description's examples. Similarities are given as numerical values. (2) The Classification Function takes the results of the similarity function and the classification performance records of the instances in the concept description and uses them to classify them. This leads to an x classification. (3) The Updater for Concept Descriptions is a program that keeps track of the results of the classification and decides which instances should be included in the concept description. Include '*I*' inputs, the similarity outcomes, the classifying results, and the current concept description are all inputs. This results in an updated concept description.

Unlike most other supervised learning approaches, IBk algorithms do not create explicit abstractions, such as decision trees or rules. When cases are provided, most learning methods produce generalizations from these cases and utilize simple matching processes to classify subsequent instances. At the time of presentation, this includes the objective of the generalizations. Because IBk algorithms do not store explicit generalizations, they perform less work at the presentation time. However, when they are supplied with more cases for classification, their workload increases, as they compute the similarities of their previously saved instances with the newly presented instance. This eliminates the need for IBk algorithms to keep rigid generalizations in concept descriptions, which may incur significant costs for their updating in order to account for prediction errors [145].

Random Committee Algorithm

The Random Committee algorithm is an ensemble of randomizable base classifiers that may be built using this class. A distinct random number seed is used to build each base classifier (but each is based on the same data). The final prediction is an arithmetic mean of the predictions made by each of the base classifiers [146].

Figure 14 shows the model procedure using the Weka and TensorFlow frameworks. The data are obtained from the LidSonic V2.0 gadget via its sensors and labeled by the user. Next, using the preprocessing and extraction module, we can produce two datasets, DS1 and DS2. Then, these datasets are trained using three machine-learning methods, IBk, Random Committee, and Kstar, in the machine learning obstacle recognition module. In the evaluation and visualization module, six Weka models were evaluated using three classifiers and two datasets. We used a 10-fold cross-validation to evaluate the training datasets. Weka runs the learning algorithm eleven times in the 10-fold cross-validation, once for each fold of the cross-validation and once more for the complete dataset. Each fit is performed using a training set made up of 90% of the entire training set, chosen at random, with the remaining 10% utilized as a hold-out set for validation. The deployment may be performed in a variety of ways, and we chose the optimal deployment method on the basis of the performance and analysis of each classifier.

The training model building time was evaluated on a Samsung Galaxy S8 mobile (see Figure 15). The mobile has 4 GB RAM; Exynos 8895 (10 nm), EMEA shipset; and Octa-core (4 × 2.3 GHz Mongoose M2 & 4 × 1.7 GHz Cortex-A53), EMEA CPU. In the results section, the results are fully clarified.

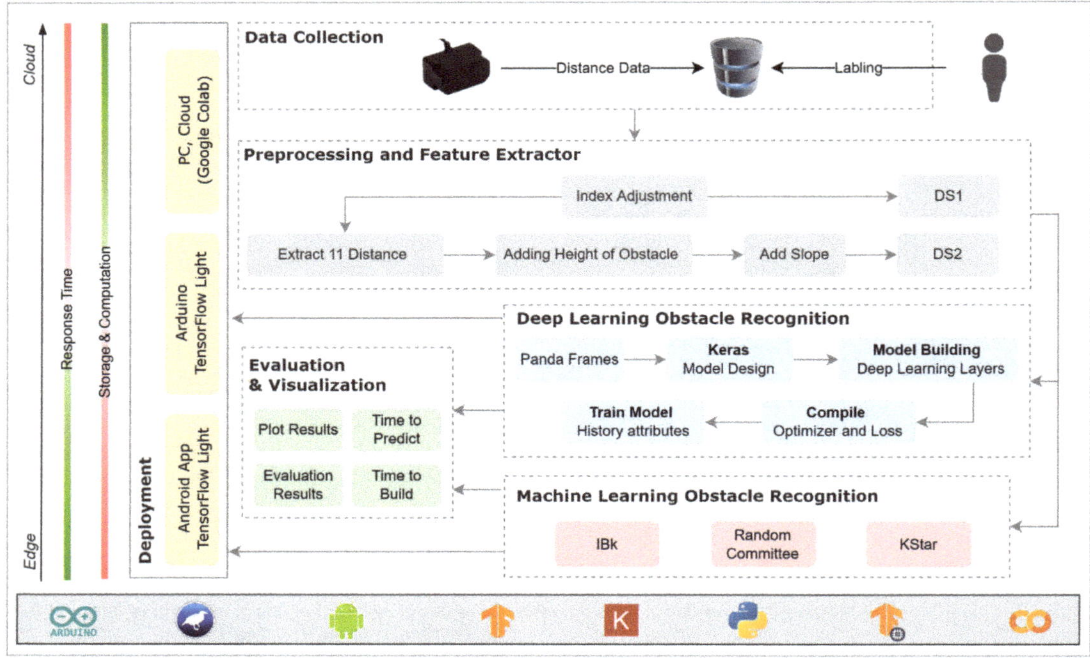

Figure 14. Machine and Deep Learning Modules.

Model	Galaxy S8
Platform	Android
Memory	4GB RAM
Shipset	Exynos 8895 (10 nm) – EMEA
CPU	Octa-core (4x2.3 GHz Mongoose M2 & 4x1.7 GHz Cortex-A53) – EMEA

Figure 15. Samsung Galaxy S8 Specification.

4.5.2. Deep Learning Models: TensorFlow

The model procedure of the TensorFlow framework is depicted in Figure 14. The user labels the data that is acquired from the LidSonic V2.0 device via its sensors. After that step, we created two datasets, DS1, and DS2 (these are the same datasets that are used for the machine learning models). We built two models, TModel1 and TModel2, to evaluate the two datasets we constructed. The deep models used are convolutional neural networks (CNNs). In the evaluation and visualization module, the models were evaluated, and the results were plotted and analyzed.

The datasets were divided into three sections: training, validation, and testing (see Table 7). The validation set was used to evaluate the loss and any metrics during the model fitting; however, the model was not fitted using this data. In the deployment phase, we put the module into production so that users would be able to make predictions with it. TensorFlow has great capabilities and offers a variety of choices regarding the models to be deployed, including TensorFlow Serving, TensorFlow Light (TinyML), and more. TensorFlow Serving is a TensorFlow library that enables models to be served through HTTP/REST or gRPC/Protocol Buffers. TensorFlow Serving is a model deployment strategy used for machine learning and deep learning models that are flexible and have a high performance. TensorFlow Serving makes it simple to deploy models. TensorFlow Lite is a lightweight TensorFlow solution for mobile and embedded devices that focuses on running machine learning (mostly deep learning) algorithms directly on edge devices, such as Android and iOS, as well as embedded systems, such as Arduino Uno. Tiny machine learning refers to a branch of machine learning microcontrollers and mobile phones. Because most of these devices are low powered, the algorithms must be carefully tuned so as to operate on them. TinyML has become one of the fastest developing subjects in deep learning due to the ability to perform machine learning directly on edge devices and the ease that comes with it. The smartphone, or Arduino Uno microprocessor, in our scenario, is an edge device that employs the final output of machine learning algorithms. Many operators run machine learning models on more capable devices and then send the results to edge devices. This method is starting to change because of the emergence of TinyML.

Table 7. TensorFlow Training Model Shapes.

Dataset	TModel1	TModel2
Training feature shape	(435, 60)	(435, 14)
Validation feature shape	(109, 60)	(109, 14)
Test feature shape	(136, 60)	(136, 14)

The datasets are divided as shown in Table 7. During the training phase, the test set is ignored, and it is only utilized at the end in order to assess how well the model generalizes to new data. This is especially essential in the case of unbalanced datasets, when the absence of training data poses a considerable risk of overfitting.

ReLU

The model was fine-tuned with layers to increase its accuracy and precision. We employed three layers for the Deep Neural Network in addition to the input and output layers, as shown in Table 8 for DS1 and Table 9 for DS2, and applied the Rectified Linear Unit activation function (ReLU).

Table 8. TModel1 Summary.

Model: "Sequential_16"		
Layer (Type)	**Output Shape**	**Parameters**
flatten_16 (Flatten)	(None, 60)	0
dense_74 (Dense)	(None, 60)	3660
dense_75 (Dense)	(None, 120)	7320
dense_76 (Dense)	(None, 60)	7260
dense_77 (Dense)	(None, 60)	3660
dense_78 (Dense)	(None, 30)	1830
dense_79 (Dense)	(None, 8)	248

Total params: 23,978
Trainable params: 23,978
Non-trainable params: 0

Table 9. TModel2 Summary.

Model: "Sequential_2"		
Layer (Type)	**Output Shape**	**Parameters**
flatten_2 (Flatten)	(None, 14)	0
dense_8 (Dense)	(None, 140)	2100
dense_9 (Dense)	(None, 64)	9024
dense_10 (Dense)	(None, 30)	1950
dense_11 (Dense)	(None, 8)	248

Total params: 13,322
Trainable params: 13,322
Non-trainable params: 0

Softmax Regression

Since we had a multi-class dataset, we employed softmax regression. Softmax regression (also known as multinomial logistic regression) is a generalization of logistic regression used for dealing with several classes. As a result, the softmax function performs two tasks: First, it converts all of the scores into probabilities. Then, the total probability equals 1. The sigmoid function is used for the same problem in the binary logistic classifier to classify two classes. The softmax function is little more than a generalized sigmoid function.

Cost Function

We must create a cost function, with which the softmax probability and one-hot encoded target vector must be compared to determine the similarity. For this purpose, we employed the conception of cross-entropy. Cross-entropy is a distance computation function that uses the softmax function's estimated probability and the one-hot-encoding matrix to determine the distance. The distance values for the correct target classes are lower, while the distance values for the incorrect target classes are greater. Passing an input through the model and comparing the predictions to ground-truth labels are the means by which a neural network is trained. A loss function is used to make this comparison. Categorical cross-entropy loss is the loss function of choice for multiclass classification issues. It does, however, necessitate the one-hot encoding of the labels. Sparse categorical cross-entropy loss may be a useful option in this instance. The loss function offers the same type of loss as categorical cross-entropy loss but on integer targets rather than one-hot encoded targets. This eliminates the categorical step that is so prevalent in TensorFlow/Keras models. In artificial neural networks, the softmax function is employed in a variety of multiclass classification algorithms. The outcome of K's unique linear functions is used as the input for the multinomial logistic regression and linear discriminant analysis, and the predicted probability is calculated by Equation (9):

$$P(y = j|x) = \frac{e^{x^T w_j}}{\sum_{k=1}^{K} e^{x^T w_k}} \tag{9}$$

with the jth class given as the x sample vector and w weighting vector.

Adam Optimizer

Next, we employed Adam as an optimizer. Adam describes the phrase "adaptive moment estimation". Adam is an optimization algorithm that may be used to update network weights iteratively based on training data instead of the traditional stochastic gradient descent procedure. We can note the following advantages of employing Adam for non-convex optimization problems. Its implementation is simple. It is effective in terms of computation. There are not that many memory demands. The gradients are invariant in regard to diagonal rescaling. It is ideally suited to issues involving a large number of data and/or parameters. It is a great option for non-stationary objectives. It is suitable for gradients that are exceedingly noisy or sparse. Finally, the hyper-parameters are easy to read and usually do not require much adjustment. Adam is a stochastic gradient descent extension that combines the benefits of two earlier extensions, the Adaptive Gradient Algorithm (AdaGrad) and Root Mean Square Propagation (RMSProp). Adam is a popular deep learning method, since it produces good results swiftly. The results were plotted and reviewed during the evaluation and visualization phase. The model configuration of the experiments is shown in Table 10.

Table 10. Deep Learning Model Configurations.

Specifications	Value
Number of hidden layers—Tmodel1	5
Number of hidden layers—Tmodel2	3
Activation function—hidden layers	Relu
Activation function—output layer	Softmax
Loss function categorical	Sparse_categorical_crossentropy
Optimizer	Adam
Epochs	200

4.6. Voice Module

A multiplicity of application programming interfaces (APIs) are now accessible for a variety of activities that formerly required significant programming effort on behalf of developers. When working with audio file data, the job becomes more challenging. As a result, we relied on Google's speech-to-text engine [147], which can transcribe any audio while maintaining the context and language. The API supports up to 120 languages. Other functions include voice command and control, call center audio transcription, real-time streaming, pre-recorded audio processing, and others. The Google speech-to-text tool can successfully translate written text into grammatically and contextually relevant speech using a range of natural voices. The Google text-to-speech API enables developers to interact with customers through speech user interfaces in devices and applications and customize the communication depending on voice and language preferences.

The Voice Module, for example, allows the user to generate the dataset and transition between different development and operation phases using voice commands. To begin the process of producing a dataset, the user types the command "Train". The system then asks the user "what is the obstacle class?" in order to classify the incoming data. The user specifies the obstacle, such as "Wall". The system then requests that the user to "Specify the dataset file name". Finally, the user enters the file name verbally.

5. Performance Evaluation

We now analyze the performance of the LidSonic V2.0 system: Section 5.1 discusses the performance using the machine learning models and Section 5.2 discusses the system performance using the deep learning models.

5.1. Machine-Learning-Based Performance

There are several metrics defined in the Weka software that can be computed by the model and are useful for measuring the performance. Accuracy is defined as the percentage of properly classified instances. Precision is defined as the percentage of expected positives that were correctly classified. Table 11 displays the accuracy and precision of the six machine learning models, adopting three classifiers to construct models from the two datasets, DS1 and DS2.

Table 11. Evaluation of the Machine Learning Models.

Trained Model	Dataset	No. of Features	Classifier	Precision	Accuracy
IBk-DS1	DS1	60	IBk	94.0	93.68
RC-DS1	DS1	60	Random Committee	94.7	94.56
KStar-DS1	DS1	60	KStar	95.6	95.44
IBk-DS2	DS2	14	IBk	95.2	95
RC-DS2	DS2	14	Random Committee	95.2	95.15
KStar-DS2	DS2	14	KStar	94.1	93.82

The results are depicted in Figure 16. The results indicate that using DS2 with Random Committee and IBk classifiers increases accuracy to (95%) and (95.15%), respectively, with equal precision results of (95.2%). The KStar classifier, on the other hand, has greater accuracy (95.44%) and precision (95.6%) when utilizing DS1.

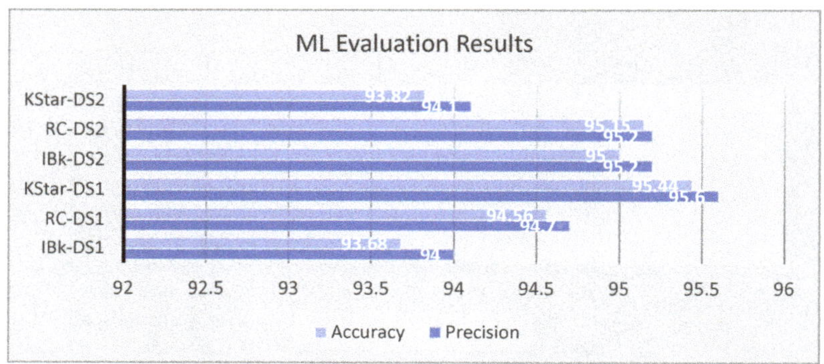

Figure 16. Weka Classifier Evaluation Results.

Figure 17 plots the model training times required for the top three classifiers. The longest time was spent building the classification model using DS1, compared with the time spent building the classifiers using DS2. Using DS1, RC required the longest time for building (299 ms), followed by KStar (42 ms) and IBk (1 ms). For DS2, RC required the longest time for building (128 ms), followed by KStar (11 ms) and then IBk (1 ms). While RC classifier requires the longest time to build its model, it requires the shortest time to predict an object (see Figure 18).

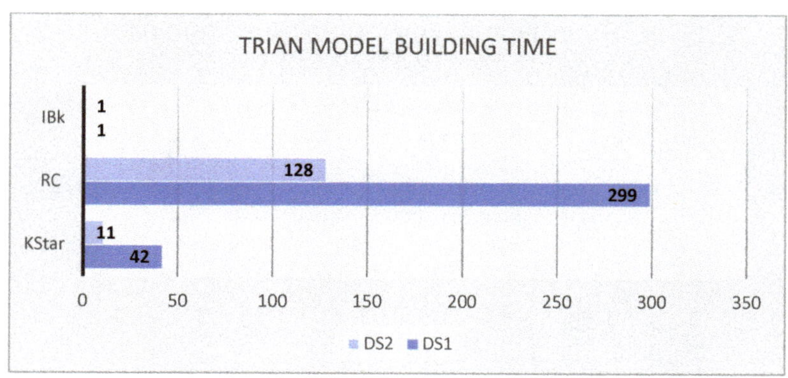

Figure 17. Time Required to Build a Model for each Classifier.

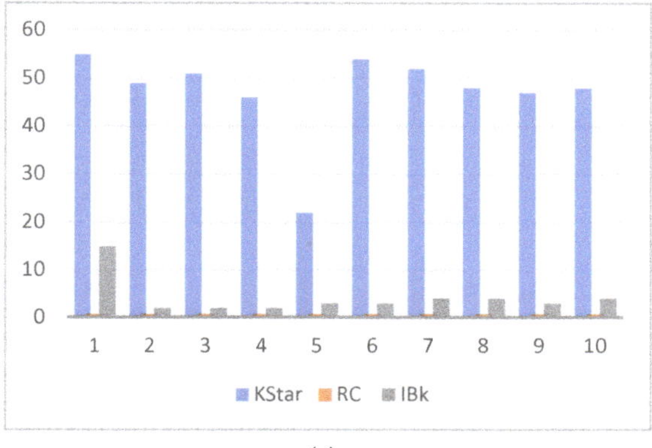
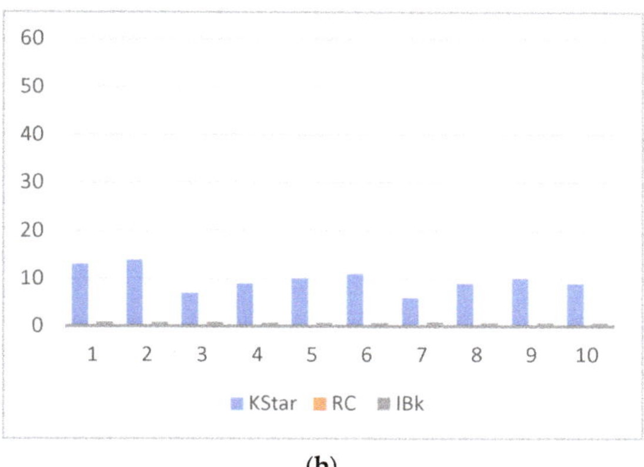

(a)

(b)

Figure 18. Classifiers' Inference Elapsed Time in Milliseconds for (**a**) DS1 Test Samples and (**b**) DS2 Test Samples.

Figure 18 plots the model inference times for the top three classifiers. For both the DS1 and DS2 test samples, we constructed 10 test predictions and timed each classifier in order to produce the outcome. The time required for KStar to forecast an object varied between 22 and 55 milliseconds when using the DS1 test samples and between 9 and 13 milliseconds when using the DS2 test samples. The IBk classifier, on the other hand, showed faster times than KStar, with an average of 4–5 milliseconds for the DS1 test samples and 0–1 milliseconds for the DS2 test samples. Random Committee required a substantial amount of time to develop its training model, and it predicted the test samples in less than 1 millisecond for both the DS1 and DS2 test samples.

It is worth mentioning that the IBk and KStar algorithms trained and generated models faster than the Random Committee algorithm (see Figure 17). Random Committee required 299 milliseconds to create its trained model using DS1, whereas for DS2 it required 128 milliseconds. KStar additionally required a long time to construct its training model, taking 42 ms for DS1 and 11 ms for DS2. On the other hand, for both DS1 and DS2,the IBk classifier built the trained model in 1 ms. As a consequence, we suggest that the IBk classifier is preferable over the Random Committee and KStar classifiers for the purpose of mobile adaption, embedded microprocessors, and/or large datasets. Although there is a modest accuracy trade-off, the IBk classifier delivers a significant decrease in the mobile computing and battery usage.

When approaching the design of a system architecture that takes advantage of fog or cloud in the training phase, we suggest the KStar classifier, especially in the case of a larger training dataset that will be trained in the layers of fog or cloud, because it can exploit its computation processing powers and return results for the edge level. In the case of the RC classifier, it can be trained in the higher layers (cloud or fog) and the constructed model can then transfer to the edge, because its prediction time is the shortest among the three classifiers.

Figure 19 depicts the confusion matrix with the best training model score, which was obtained using the KStar classification method with the D1 dataset. Figure 20 shows the confusion matrix of the Random Committee classifiers trained on DS2. Figure 21 shows the confusion matrix for the IBk classifier. Because these were the top three highest performing classifiers identified in our previous study, we chose to exhibit the confusion matrices of these three classifiers. The abbreviations used in the figures are listed in Table 12.

Table 12. Class Abbreviation.

Class	Abbreviation	Class	Abbreviation
Floor	F	Deep Hole	DH
Ascending Stairs	AS	High Obstacle	HO
Descending Stairs	DS	Ascending Step	Ast
Wall	W	Descending Step	DSt

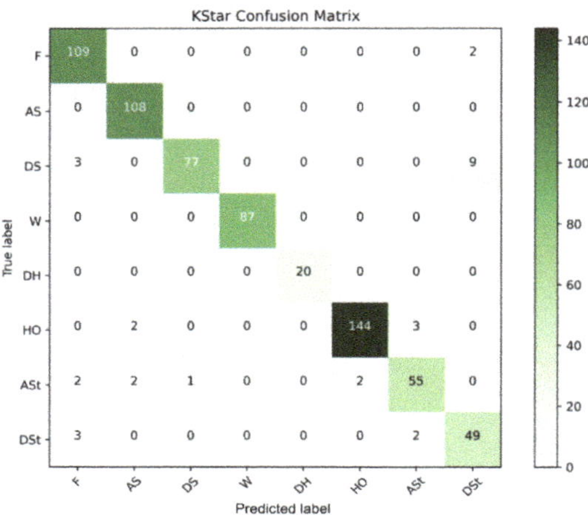

Figure 19. KStar Confusion Matrix using DS1.

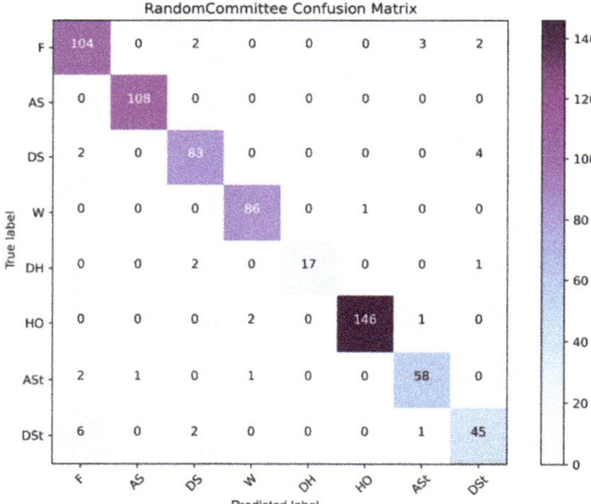

Figure 20. Random Committee Confusion Matrix Using DS2.

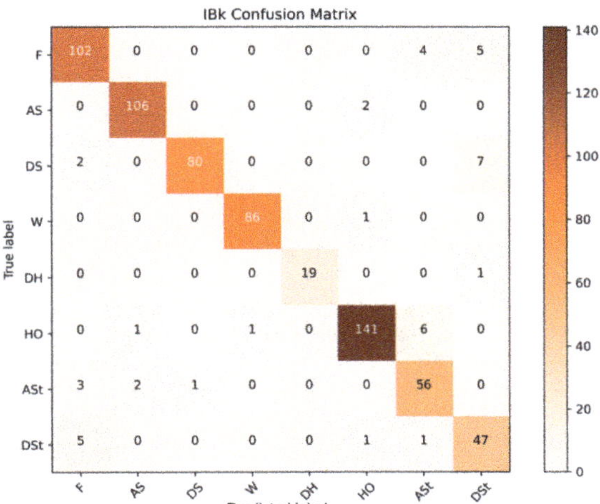

Figure 21. IBk Confusion Matrix Using DS2.

Figure 19 plots the confusion matrix of the KStar classifier. The highest number of true positives was obtained for the Floor class (109), followed by Ascending Stairs (108), Descending Stairs (77), Wall (87), Deep Hole (20), High Obstacle (144), Ascending Step (55), and Descending Step (49). The total number of wrong predictions for the classes are Floor (2), Ascending Stairs (0), Descending Stairs (3), Wall (0), Deep Hole (0), High Obstacle (5), Ascending Step (7), and Descending Step (5). Obviously, the number of wrong predictions should be considered relative to the total number of instances. It is possible that the higher number of wrong predictions for some classes is due to the low number of instances of the data objects for those classes. Note also that the Floor was misclassified two times as Descending Step. Descending Stairs were misclassified 3 times as Floor and 9 times as Descending Step. The High Obstacle class was misclassified 2 times as Ascending Stairs and 3 times as Ascending Step. Ascending Step was misclassified 2 times as Floor, 2 times as Ascending Stairs, 1 time as Descending Stairs, and 2 times as High Obstacle. Descending Step was misclassified 3 times as Floor and 2 times as Ascending Step. On the other hand, Ascending Stairs, Wall, and Deep Hole classes had no misclassified results.

Figure 20 plots the confusion matrix for the Random Committee (RC) classifier. The highest number of true positives was obtained for the High Obstacle (146) and Ascending Stairs (108) classes, followed by Floor (104), Wall (86), Descending Stairs (83), Ascending Step (58), Descending Step (45), and Deep Hole (17). The total number of wrong predictions for the classes are Floor (7), Ascending Stairs (0), Descending Stairs (6), Wall (1), Deep Hole (3), High Obstacle (3), Ascending Step (4), and Descending Step (9). Note that Floor was misclassified 2 times as Descending Stairs, 3 times as Ascending Step, and 2 times as Descending Step. Descending Stairs were misclassified 2 times as Floor and 4 times as Descending Step. The Wall class was misclassified 1 time as High Obstacle. Deep Hole was misclassified 2 times as Descending Stairs and 1 time as Descending Step. The High Obstacle class was misclassified 2 times as Wall and 1 time as Ascending Step. Ascending Step was misclassified 2 times as Floor, 1 time as Ascending Stairs, and 1 time as Wall. Descending Step was misclassified 6 times as Floor, 2 times as Descending Stairs, and 1 time as Ascending Step. Ascending Stairs had no misclassifications.

Figure 21 plots the confusion matrix for the IBk classifier. The highest number of true positives was obtained for the High Obstacle (141) and Ascending Stairs (106) classes, followed by Floor (102), Wall (86), Descending Stairs (80), Ascending Step (56), Descending Step (47), and Deep Hole (19). The total number of misclassifications for the classes were Floor (9), Ascending Stairs (2), Descending Stairs (9), Wall (1), Deep Hole (1), High Obstacle

(8), Ascending Step (6), and Descending Step (7). The greater numbers of incorrect predictions for some classes are likely due to the low number of instances of the data objects for those classes, as we saw with the KStar classifier. Note that Floor was misclassified 4 times as Ascending Step and 5 times as Descending Step. Ascending Stairs were misclassified 2 times as High Obstacles. The Descending Stairs class was misclassified 2 times as Floor and 7 times as Descending Step. The Wall class was misclassified 1 time as High Object and zero times as any other class. Deep Hole was misclassified 1 time as Descending Step and zero times as any other class. The High Obstacle class was misclassified 1 time as Ascending Stairs, 1 time as Wall, and 6 times Ascending Step. Ascending Step was misclassified 3 times as Floor, 2 times as Ascending stairs, and 1 time as Descending Stairs. Descending Step was misclassified 5 times as Floor, 1 time as High Obstacle, and 1 time as Ascending Step. Wall and Deep Hole had the least number of misclassifications.

5.2. Deep-Learning-Based Performance

Observing the performance of neural networks and deep learning models over time during training helps to provide researchers with knowledge about them. Keras is a Python framework that encapsulates the more technical TensorFlow backends and provides a clear interface for generating deep learning models. We used Keras in Python to evaluate and display the performance of deep learning models over time during training so as to measure their accuracy and loss. Note that, here, the deep learning models were trained and executed on a laptop device. Future work will attempt to implement deep learning models on mobile phones and other edge devices using TFLite, as in our other strands of research [148]. Table 13 summarizes the findings.

Table 13. TensorFlow Model Evaluation.

Trained Model	Dataset	Features	Model Accuracy	Test Accuracy	Model Loss	Test Loss
TModel1	D1	60	88.05	76.32	0.3374	0.7190
TModel2	D2	14	98.01	96.49	0.0883	0.3672

The TModel1 and TModel2 accuracy and loss results were plotted throughout each period and are presented in Figures 22 and 23, respectively. Using different datasets, we can see the significant differences in performance. The TModel2 trained model, which is used for DS2, has a 98.01 percent training accuracy and a model loss of 0. 0883 percent. The model accuracy for the test dataset was 96.49 and the test model loss was 0.3672. The model accuracy of the TModel1 is 88.05 and it has a loss of 0.3374. The test model accuracy is 76.32 and it has a loss of 0.7190.

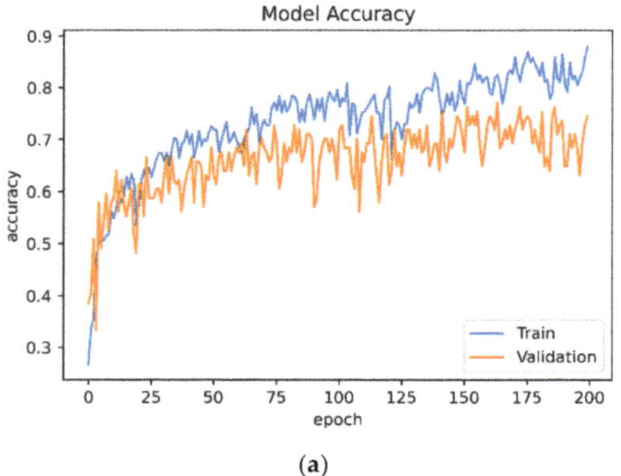

(a)

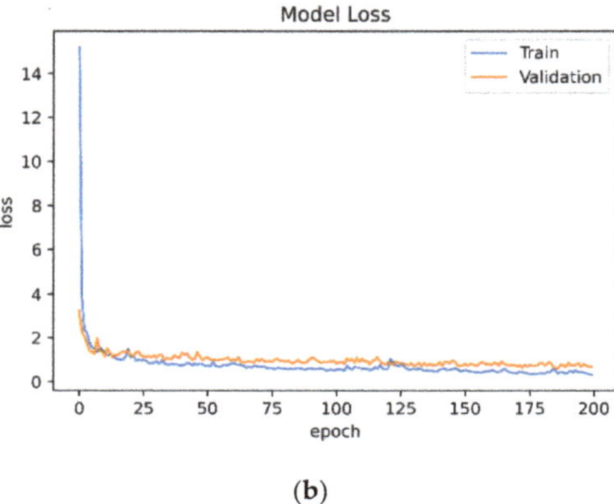

(b)

Figure 22. Deep Learning Evaluation Using DS1: (**a**) TModel1 Accuracy and (**b**) TModel1 Loss.

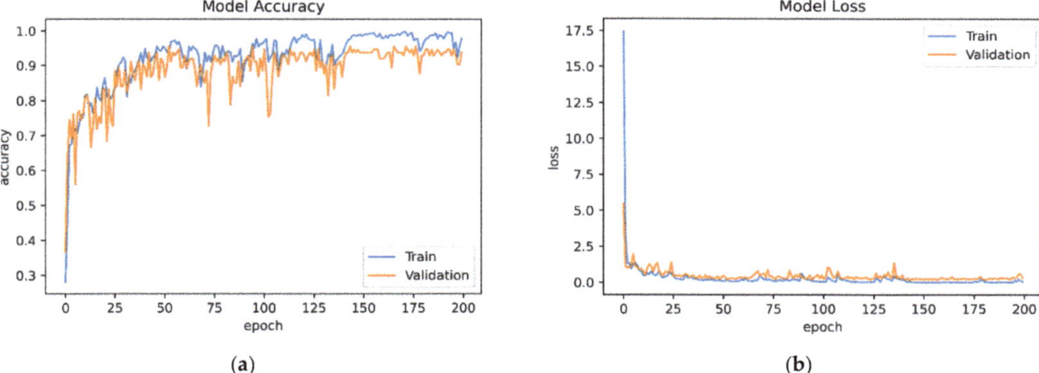

(**a**)

(**b**)

Figure 23. Deep Learning Evaluation Using DS2: (**a**) TModel2 Accuracy and (**b**) TModel2 Loss.

Figure 24 plots the confusion matrices of the TModel1 and TModel2 deep learning models used on a test dataset that the trained model was not exposed to. For TModel1, the highest number of true positives were obtained for the Floor (20) and Ascending Stairs (18), followed by Wall and High Obstacle (15), Descending Stairs (9), Descending Step (5), Ascending Step (4), and Deep Hole (1). The total number of wrong predictions for the classes are Floor (13), Ascending Stairs (2), Descending Stairs (3), Deep Hole (1), High Obstacle (2), Ascending Step (3), and Descending Step (4). Wall was classified correctly. The Floor was misclassified 1 time as Descending Stairs, 7 times as Ascending Step, and 4 times as Descending Step. Ascending Step was misclassified 2 times as High Obstacle. Descending Stairs has been misclassified 1 time as High Obstacle and 2 times as Ascending Step. Deep Hole had one misclassification as Descending Step. High Obstacle was misclassified 1 time as Floor and 1 time as Ascending Stairs. Ascending Step was misclassified 1 time as Floor, 1 time as Ascending Stairs, and 1 time as High Obstacle. Descending Step was misclassified 2 times as Floor, 1 time as Descending Stairs, and 1 time as Ascending Step.

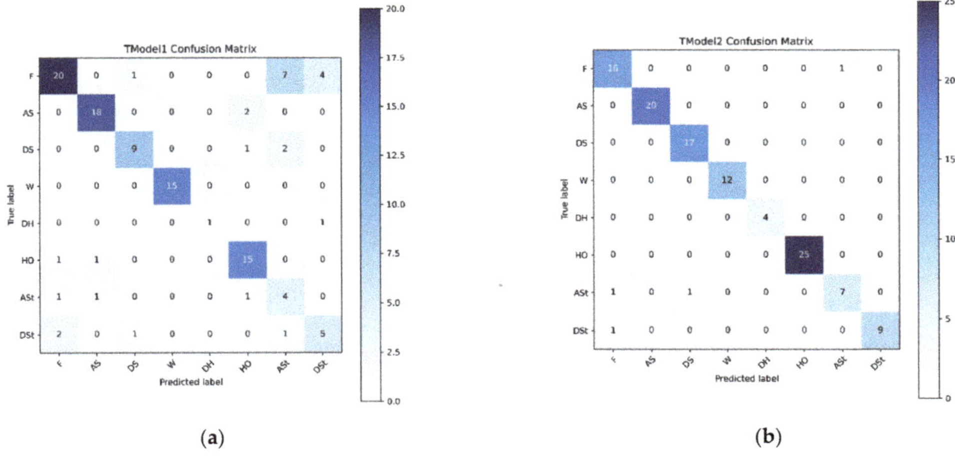

(**a**)

(**b**)

Figure 24. Deep Learning Confusion Matrix on the Test Dataset: (**a**) TModel1 and (**b**) TModel2.

For TModel2, the highest number of true positives was obtained for the High Obstacle (25) and Ascending Stairs (20) classes, followed by Descending Stairs (17), Wall (12), Descending Step (9), Ascending Step (7), and Deep Hole (4). The total number of wrong predictions for the classes are Floor (1), Ascending Step (2), and Descending Step (1). Note that the Floor was misclassified 1 time as Ascending Stairs. Ascending Step was misclassified 1 time as Floor and 1 time as Descending Stairs. Descending Step was misclassified 1 time as Floor. Ascending Stairs, Descending Stairs, Wall, Deep Hole, and High Obstacle had no misclassifications.

6. Conclusions

In this paper, we developed the LidSonic V2.0 system by leveraging a comprehensive understanding of the state-of-the-art requirements and solutions involving assistive technologies for the visually impaired through a detailed literature review and a survey. We explained in Section 1 that it is difficult for visually impaired people to orient themselves and move in an unfamiliar environment without assistance and, hence, the focus regarding the LidSonic system in this paper is placed on the mobility tasks of the visually impaired. The system is based on a novel approach using a combination of a LiDAR with a servo motor and an ultrasonic sensor to collect data and predict objects using machine and deep learning for environment perception and navigation. The deep learning model TModel2, for the DS2 dataset, provided the overall best accuracy results at 96.49%. The second-best accuracy was provided by the KStar classifier at 95.44%, with a precision of 95.6%. The IBk and RC classifiers provided the same precision at 95.2% and similar accuracy results at 95.15% and 95%, respectively, using the DS2 dataset. Note that the IBk classifier was seen to be relatively non-dependent on the size of the datasets. This could be because both the datasets are numeric, and the difference between their sizes is small. It took 1 ms to train both the D2 and D1 datasets, and these were the fastest training times overall. The IBk classifier also provided the second fastest prediction time at 0.8 ms and, hence, we recommend using it at the edge for training and prediction. As for the KStar classifier, the training time is influenced by the size of the training dataset. It took 11 ms to train KStar with the DS2 dataset and 42 ms to train it with the larger DS1 dataset. Moreover, the KStar classifier required much longer times for the prediction, ranging between 22 ms and 55 ms, compared to the other classifiers in our experiments. Hence, we proposed using the KStar classifier at the fog or cloud layers.

We evaluated the proposed system from multiple perspectives. For instance, we proposed, based on the results, using the Random Committee classifier at the edge for prediction due to its faster prediction time, although it needs to be trained at the fog or cloud layers because it requires larger resources. In this respect, we plan to extend and integrate this work with other strands of our work on big data analytics and edge, fog, and cloud computing [148–151]. For example, we plan to experiment with different machine learning and deep learning methods at the edge, fog, and cloud layers, assessing their performance and the applicability of the use of edge, fog, and cloud computing for smart glasses, and considering new applications for the integration of smart glasses with cloud, fog, and edge layers. Another direction of our research is green and explainable AI [152,153], and we will also explore the expandability of the LidSonic system.

We created the second prototype of our LidSonic system in this work. The team constructed and tested the prototype. We also benefitted from the assistance of four other people aged 18 to 47 (who were not visually impaired) who helped to test and evaluate the LidSonic system. The time required to explain the device's operation to the selected users was only a few minutes, but it varied depending on the user's age and digital affinity. The tests were carried both indoors and outdoors on the campus of King Abdulaziz University. The purpose of this paper was to put the system's machine learning and other technical capabilities to the test. Future work will involve testing the device with blind and visually impaired users so as to provide more details about the LidSonic system's usability, human training, and testing aspects.

We conclude this paper with the remark that the technologies developed in this study show a high potential and are expected to open new directions for the design of smart glasses and other solutions for the visually impaired using open software tools and off-the-shelf hardware.

Author Contributions: Conceptualization, S.B. and R.M.; methodology, S.B. and R.M.; software, S.B.; validation, S.B. and R.M.; formal analysis, S.B., R.M., I.K., A.A., T.Y. and J.M.C.; investigation, S.B., R.M., I.K., A.A., T.Y. and J.M.C.; resources, R.M., I.K. and A.A.; data curation, S.B.; writing—original draft preparation, S.B. and R.M.; writing—review and editing, R.M., I.K., A.A., T.Y. and J.M.C.; visualization, S.B.; supervision, R.M. and I.K.; project administration, R.M., I.K. and A.A.; funding acquisition, R.M., I.K. and A.A. All authors have read and agreed to the published version of the manuscript.

Funding: The authors acknowledge, with thanks, the technical and financial support from the Deanship of Scientific Research (DSR) at King Abdulaziz University (KAU), Jeddah, Saudi Arabia, under the grant No. RG-11-611-38. The experiments reported in this paper were performed on the Aziz supercomputer at KAU.

Institutional Review Board Statement: Not applicable.

Informed Consent Statement: Not applicable.

Data Availability Statement: The dataset developed in this work can be provided on request.

Acknowledgments: The work carried out in this paper was supported by the HPC Center of King Abdulaziz University. The training and software development work reported in this paper was carried out on the Aziz supercomputer.

Conflicts of Interest: The authors declare no conflict of interest.

Abbreviations

The following abbreviations are used in this paper.

ETA	Electronic Travel Aids
ToF	Time-of-flight
RNA	Robotic Navigation Aid
ALVU	Array of LiDARs and Vibrotactile Units
BLE	Bluetooth Low Energy
SDK	Software Development Kit
RGB-D	RGB-Depth
CNN	Convolutional Neural Networks
RFID	Radio Frequency Identification Reader
HRTFs	Head-Related Transfer Functions
GPS	Global Positioning System
MSER	Maximally Stable External Region
SWT	Stroke Width Transform

References

1. World Health Organisation (WHO). Disability and Health (24 November 2021). Available online: https://www.who.int/news-room/fact-sheets/detail/disability-and-health (accessed on 19 July 2022).
2. World Health Organisation (WHO). Disability. Available online: https://www.who.int/health-topics/disability#tab=tab_1 (accessed on 19 July 2022).
3. DISABLED | Meaning in the Cambridge English Dictionary. Available online: https://dictionary.cambridge.org/dictionary/english/disabled (accessed on 7 February 2020).
4. Physical Disability—Wikipedia. Available online: https://en.wikipedia.org/wiki/Physical_disability (accessed on 30 July 2022).
5. Disability and Health Overview | CDC. Available online: https://www.cdc.gov/ncbddd/disabilityandhealth/disability.html (accessed on 30 July 2022).
6. Disability Facts and Figures | Disability Charity Scope UK. Available online: https://www.scope.org.uk/media/disability-facts-figures/ (accessed on 30 July 2022).
7. Disability among People in the U.S. 2008–2019 | Statista. Available online: https://www.statista.com/statistics/792697/disability-in-the-us-population-share/ (accessed on 30 July 2022).
8. Disabled People in the World: Facts and Figures. Available online: https://www.inclusivecitymaker.com/disabled-people-in-the-world-in-2021-facts-and-figures/ (accessed on 30 July 2022).

9. Questions and Answers about Blindness and Vision Impairments in the Workplace and the Americans with Disabilities Act | U.S. Equal Employment Opportunity Commission. Available online: https://www.eeoc.gov/fact-sheet/questions-and-answers-about-blindness-and-vision-impairments-workplace-and-americans (accessed on 8 June 2020).

10. Blind vs. Visually Impaired: What's the Difference? | IBVI | Blog. Available online: https://ibvi.org/blog/blind-vs-visually-impaired-whats-the-difference/ (accessed on 8 June 2020).

11. What Is Visual Impairment? Available online: https://www.news-medical.net/health/What-is-visual-impairment.aspx (accessed on 8 June 2020).

12. Katzschmann, R.K.; Araki, B.; Rus, D. Safe local navigation for visually impaired users with a time-of-flight and haptic feedback device. *IEEE Trans. Neural Syst. Rehabil. Eng.* **2018**, *26*, 583–593. [CrossRef] [PubMed]

13. Alotaibi, S.; Mehmood, R.; Katib, I.; Rana, O.; Albeshri, A. Sehaa: A Big Data Analytics Tool for Healthcare Symptoms and Diseases Detection Using Twitter, Apache Spark, and Machine Learning. *Appl. Sci.* **2020**, *10*, 1398. [CrossRef]

14. Praveen Kumar, M.; Poornima; Mamidala, E.; Al-Ghanim, K.; Al-Misned, F.; Ahmed, Z.; Mahboob, S. Effects of D-Limonene on aldose reductase and protein glycation in diabetic rats. *J. King Saud Univ.—Sci.* **2020**, *32*, 1953–1958. [CrossRef]

15. Ekstrom, A.D. Why vision is important to how we navigate. *Hippocampus* **2015**, *25*, 731–735. [CrossRef]

16. Deverell, L.; Bentley, S.A.; Ayton, L.N.; Delany, C.; Keeffe, J.E. Effective mobility framework: A tool for designing comprehensive O&M outcomes research. *IJOM* **2015**, *7*, 74–86.

17. Andò, B. Electronic sensory systems for the visually impaired. *IEEE Instrum. Meas. Mag.* **2003**, *6*, 62–67. [CrossRef]

18. Ranaweera, P.S.; Madhuranga, S.H.R.; Fonseka, H.F.A.S.; Karunathilaka, D.M.L.D. Electronic travel aid system for visually impaired people. In Proceedings of the 2017 5th International Conference on Information and Communication Technology (ICoIC7), Melaka, Malaysia, 17–19 May 2017; IEEE: Piscataway, NJ, USA, 2017. [CrossRef]

19. Bindawas, S.M.; Vennu, V. The National and Regional Prevalence Rates of Disability, Type, of Disability and Severity in Saudi Arabia-Analysis of 2016 Demographic Survey Data. *Int. J. Environ. Res. Public Health* **2018**, *15*, 419. [CrossRef]

20. General Authority for Statistics. GaStat: (2.9%) of Saudi Population Have Disability with (Extreme) Difficulty. Available online: https://www.stats.gov.sa/en/news/230 (accessed on 30 July 2022).

21. Patel, S.; Kumar, A.; Yadav, P.; Desai, J.; Patil, D. Smartphone-based obstacle detection for visually impaired people. In Proceedings of the 2017 International Conference on Innovations in Information, Embedded and Communication Systems (ICIIECS), Coimbatore, India, 17–18 March 2017; IEEE: Piscataway, NJ, USA, 2018. [CrossRef]

22. Rizzo, J.R.; Pan, Y.; Hudson, T.; Wong, E.K.; Fang, Y. Sensor fusion for ecologically valid obstacle identification: Building a comprehensive assistive technology platform for the visually impaired. In Proceedings of the 2017 7th International Conference on Modeling, Simulation, and Applied Optimization (ICMSAO), Sharjah, United Arab Emirates, 4–6 April 2017; Institute of Electrical and Electronics Engineers Inc.: Piscataway, NJ, USA, 2017.

23. Meshram, V.V.; Patil, K.; Meshram, V.A.; Shu, F.C. An Astute Assistive Device for Mobility and Object Recognition for Visually Impaired People. *IEEE Trans. Hum.-Mach. Syst.* **2019**, *49*, 449–460. [CrossRef]

24. Omoregbee, H.O.; Olanipekun, M.U.; Kalesanwo, A.; Muraina, O.A. Design and Construction of A Smart Ultrasonic Walking Stick for the Visually Impaired. In Proceedings of the 2021 Southern African Universities Power Engineering Conference/Robotics and Mechatronics/Pattern Recognition Association of South Africa (SAUPEC/RobMech/PRASA), Potchefstroom, South Africa, 27–29 January 2021. [CrossRef]

25. Mehmood, R.; See, S.; Katib, I.; Chlamtac, I. *Smart Infrastructure and Applications: Foundations for Smarter Cities and Societies*; Springer International Publishing, Springer Nature: Cham, Switzerland, 2020; ISBN 9783030137045.

26. Yigitcanlar, T.; Butler, L.; Windle, E.; Desouza, K.C.; Mehmood, R.; Corchado, J.M. Can Building "Artificially Intelligent Cities" Safeguard Humanity from Natural Disasters, Pandemics, and Other Catastrophes? An Urban Scholar's Perspective. *Sensors* **2020**, *20*, 2988. [CrossRef]

27. Mehmood, R.; Bhaduri, B.; Katib, I.; Chlamtac, I. Smart Societies, Infrastructure, Technologies and Applications. In Proceedings of the Smart Societies, Infrastructure, Technologies and Applications, Lecture Notes of the Institute for Computer Sciences, Social Informatics and Telecommunications Engineering (LNICST), Jeddah, Saudi Arabia, 27–29 November 2017; Springer International Publishing: Cham, Switzerland, 2018; Volume 224.

28. Mehmood, R.; Sheikh, A.; Catlett, C.; Chlamtac, I. Editorial: Smart Societies, Infrastructure, Systems, Technologies, and Applications. *Mob. Netw. Appl.* **2022**, *1*, 1–5. [CrossRef]

29. Electronic Travel Aids for the Blind. Available online: https://www.tsbvi.edu/orientation-and-mobility-items/1974-electronic-travel-aids-for-the-blind (accessed on 5 February 2020).

30. INTRODUCTION—Electronic Travel AIDS: New Directions for Research—NCBI Bookshelf. Available online: https://www.ncbi.nlm.nih.gov/books/NBK218018/ (accessed on 5 February 2020).

31. Bai, J.; Lian, S.; Liu, Z.; Wang, K.; Liu, D. Smart guiding glasses for visually impaired people in indoor environment. *IEEE Trans. Consum. Electron.* **2017**, *63*, 258–266. [CrossRef]

32. Islam, M.M.; Sadi, M.S.; Zamli, K.Z.; Ahmed, M.M. Developing Walking Assistants for Visually Impaired People: A Review. *IEEE Sens. J.* **2019**, *19*, 2814–2828. [CrossRef]

33. Siddesh, G.M.; Srinivasa, K.G. IoT Solution for Enhancing the Quality of Life of Visually Impaired People. *Int. J. Grid High Perform. Comput.* **2021**, *13*, 1–23. [CrossRef]

34. Mallikarjuna, G.C.P.; Hajare, R.; Pavan, P.S.S. Cognitive IoT System for visually impaired: Machine Learning Approach. *Mater. Today Proc.* **2021**, *49*, 529–535. [CrossRef]

35. Stearns, L.; DeSouza, V.; Yin, J.; Findlater, L.; Froehlich, J.E. Augmented reality magnification for low vision users with the microsoft hololens and a finger-worn camera. In Proceedings of the 19th International ACM SIGACCESS Conference on Computers and Accessibility, Baltimore, MD, USA, 29 October–1 November 2017; pp. 361–362. [CrossRef]

36. Busaeed, S.; Mehmood, R.; Katib, I. Requirements, Challenges, and Use of Digital Devices and Apps for Blind and Visually Impaired. *Preprints* **2022**, 2022070068. [CrossRef]

37. Cardillo, E.; Di Mattia, V.; Manfredi, G.; Russo, P.; De Leo, A.; Caddemi, A.; Cerri, G. An Electromagnetic Sensor Prototype to Assist Visually Impaired and Blind People in Autonomous Walking. *IEEE Sens. J.* **2018**, *18*, 2568–2576. [CrossRef]

38. O'Keeffe, R.; Gnecchi, S.; Buckley, S.; O'Murchu, C.; Mathewson, A.; Lesecq, S.; Foucault, J. Long Range LiDAR Characterisation for Obstacle Detection for use by the Visually Impaired and Blind. In Proceedings of the 2018 IEEE 68th Electronic Components and Technology Conference (ECTC), San Diego, CA, USA, 29 May–1 June 2018; Institute of Electrical and Electronics Engineers Inc.: Piscataway, NJ, USA, 2018; pp. 533–538.

39. Busaeed, S.; Mehmood, R.; Katib, I.; Corchado, J.M. LidSonic for Visually Impaired: Green Machine Learning-Based Assistive Smart Glasses with Smart App and Arduino. *Electronics* **2022**, *11*, 1076. [CrossRef]

40. Magori, V. Ultrasonic sensors in air. In Proceedings of the IEEE Ultrasonics Symposium, Cannes, France, 31 October–3 November 1994; IEEE: Piscataway, NJ, USA, 1994; Volume 1, pp. 471–481.

41. Different Types of Sensors, Applications. Available online: https://www.electronicshub.org/different-types-sensors/ (accessed on 8 March 2020).

42. Fauzul, M.A.H.; Salleh, N.D.H.M. Navigation for the Vision Impaired with Spatial Audio and Ultrasonic Obstacle Sensors. In Proceedings of the International Conference on Computational Intelligence in Information System, Bandar Seri Begawan, Brunei Darussalam, 25–27 January 2021; Springer: Cham, Switzerland, 2021; pp. 43–53. [CrossRef]

43. Gearhart, C.; Herold, A.; Self, B.; Birdsong, C.; Slivovsky, L. Use of Ultrasonic sensors in the development of an Electronic Travel Aid. In Proceedings of the SAS 2009—IEEE Sensors Applications Symposium Proceedings, New Orleans, LA, USA, 17–19 February 2009; pp. 275–280.

44. Tudor, D.; Dobrescu, L.; Dobrescu, D. Ultrasonic electronic system for blind people navigation. In Proceedings of the 2015 E-Health and Bioengineering Conference, EHB, Lasi, Romania, 19–21 November 2015; Institute of Electrical and Electronics Engineers Inc.: Piscataway, NJ, USA, 2016.

45. Khan, A.A.; Khan, A.A.; Waleed, M. Wearable navigation assistance system for the blind and visually impaired. In Proceedings of the 2018 International Conference on Innovation and Intelligence for Informatics, Computing, and Technologies, 3ICT, Sakhier, Bahrain, 18–20 November 2018; Institute of Electrical and Electronics Engineers Inc.: Piscataway, NJ, USA, 2018.

46. Noman, A.T.; Chowdhury, M.A.M.; Rashid, H.; Faisal, S.M.S.R.; Ahmed, I.U.; Reza, S.M.T. Design and implementation of microcontroller based assistive robot for person with blind autism and visual impairment. In Proceedings of the 20th International Conference of Computer and Information Technology, ICCIT, Dhaka, Bangladesh, 22–24 December 2017; Institute of Electrical and Electronics Engineers Inc.: Piscataway, NJ, USA, 2018; pp. 1–5.

47. Chitra, P.; Balamurugan, V.; Sumathi, M.; Mathan, N.; Srilatha, K.; Narmadha, R. Voice Navigation Based Guiding Device for Visually Impaired People. In Proceedings of the 2021 International Conference on Artificial Intelligence and Smart Systems (ICAIS), Coimbatore, India, 25–27 March 2021; pp. 911–915. [CrossRef]

48. What Is an IR Sensor? | FierceElectronics. Available online: https://www.fierceelectronics.com/sensors/what-ir-sensor (accessed on 16 March 2020).

49. Nada, A.A.; Fakhr, M.A.; Seddik, A.F. Assistive infrared sensor based smart stick for blind people. In Proceedings of the Proceedings of the 2015 Science and Information Conference, SAI, London, UK, 28–30 July 2015; Institute of Electrical and Electronics Engineers Inc.: Piscataway, NJ, USA, 2015; pp. 1149–1154.

50. Elmannai, W.; Elleithy, K. Sensor-Based Assistive Devices for Visually-Impaired People: Current Status, Challenges, and Future Directions. *Sensors* **2017**, *17*, 565. [CrossRef]

51. Chaitrali, K.S.; Yogita, D.A.; Snehal, K.K.; Swati, D.D.; Aarti, D.V. An Intelligent Walking Stick for the Blind. *Int. J. Eng. Res. Gen. Sci.* **2015**, *3*, 1057–1062.

52. Cardillo, E.; Li, C.; Caddemi, A. Millimeter-Wave Radar Cane: A Blind People Aid with Moving Human Recognition Capabilities. *IEEE J. Electromagn. RF Microw. Med. Biol.* **2021**, *6*, 204–211. [CrossRef]

53. Ikbal, M.A.; Rahman, F.; Hasnat Kabir, M. Microcontroller based smart walking stick for visually impaired people. In Proceedings of the 4th International Conference on Electrical Engineering and Information and Communication Technology, iCEEiCT, Dhaka, Bangladesh, 13–15 September 2018; Institute of Electrical and Electronics Engineers Inc.: Piscataway, NJ, USA, 2019; pp. 255–259.

54. Gbenga, D.E.; Shani, A.I.; Adekunle, A.L. Smart Walking Stick for Visually Impaired People Using Ultrasonic Sensors and Arduino. *Int. J. Eng. Technol.* **2017**, *9*, 3435–3447. [CrossRef]

55. Yang, K.; Wang, K.; Cheng, R.; Hu, W.; Huang, X.; Bai, J. Detecting Traversable Area and Water Hazards for the Visually Impaired with a pRGB-D Sensor. *Sensors* **2017**, *17*, 1890. [CrossRef]

56. Ye, C.; Qian, X. 3-D Object Recognition of a Robotic Navigation Aid for the Visually Impaired. *IEEE Trans. Neural Syst. Rehabil. Eng.* **2018**, *26*, 441–450. [CrossRef]

57. Cornacchia, M.; Kakillioglu, B.; Zheng, Y.; Velipasalar, S. Deep Learning-Based Obstacle Detection and Classification with Portable Uncalibrated Patterned Light. *IEEE Sens. J.* **2018**, *18*, 8416–8425. [CrossRef]

58. Oh, U.; Stearns, L.; Pradhan, A.; Froehlich, J.E.; Indlater, L.F. Investigating microinteractions for people with visual impairments and the potential role of on-body interaction. In Proceedings of the 19th International ACM SIGACCESS Conference on Computers and Accessibility, Baltimore, MD, USA, 29 October–1 November 2017; pp. 22–31. [CrossRef]

59. Zhu, J.; Hu, J.; Zhang, M.; Chen, Y.; Bi, S. A fog computing model for implementing motion guide to visually impaired. *Simul. Model. Pract. Theory* **2020**, *101*, 102015. [CrossRef]

60. Peraković, D.; Periša, M.; Cvitić, I.; Brletić, L. Innovative services for informing visually impaired persons in indoor environments. *EAI Endorsed Trans. Internet Things* **2018**, *4*, 156720. [CrossRef]

61. Terven, J.R.; Salas, J.; Raducanu, B. New Opportunities for computer vision-based assistive technology systems for the visually impaired. Computer (Long. Beach. Calif). *Computer* **2014**, *47*, 52–58. [CrossRef]

62. Buimer, H.; Van Der Geest, T.; Nemri, A.; Schellens, R.; Van Wezel, R.; Zhao, Y. Making facial expressions of emotions accessible for visually impaired persons. In Proceedings of the 19th International ACM SIGACCESS Conference on Computers and Accessibility, Baltimore, MD, USA, 29 October–1 November 2017; Volume 46, pp. 331–332. [CrossRef]

63. Gutierrez-Gomez, D.; Guerrero, J.J. True scaled 6 DoF egocentric localisation with monocular wearable systems. *Image Vis. Comput.* **2016**, *52*, 178–194. [CrossRef]

64. Lee, Y.H.; Medioni, G. RGB-D camera based wearable navigation system for the visually impaired. *Comput. Vis. Image Underst.* **2016**, *149*, 3–20. [CrossRef]

65. Medeiros, A.J.; Stearns, L.; Findlater, L.; Chen, C.; Froehlich, J.E. Recognizing clothing colors and visual T extures using a finger-mounted camera: An initial investigation. In Proceedings of the 19th International ACM SIGACCESS Conference on Computers and Accessibility, Baltimore, MD, USA, 29 October–1 November 2017; pp. 393–394. [CrossRef]

66. Al-Khalifa, S.; Al-Razgan, M. Ebsar: Indoor guidance for the visually impaired. *Comput. Electr. Eng.* **2016**, *54*, 26–39. [CrossRef]

67. Alomari, E.; Katib, I.; Albeshri, A.; Mehmood, R. COVID-19: Detecting government pandemic measures and public concerns from twitter arabic data using distributed machine learning. *Int. J. Environ. Res. Public Health* **2021**, *18*, 282. [CrossRef] [PubMed]

68. Alahmari, N.; Alswedani, S.; Alzahrani, A.; Katib, I.; Albeshri, A.; Mehmood, R.; Sa, A.A. Musawah: A Data-Driven AI Approach and Tool to Co-Create Healthcare Services with a Case Study on Cancer Disease in Saudi Arabia. *Sustainability* **2022**, *14*, 3313. [CrossRef]

69. Alotaibi, H.; Alsolami, F.; Abozinadah, E.; Mehmood, R. TAWSEEM: A Deep-Learning-Based Tool for Estimating the Number of Unknown Contributors in DNA Profiling. *Electronics* **2022**, *11*, 548. [CrossRef]

70. Alomari, E.; Katib, I.; Albeshri, A.; Yigitcanlar, T.; Mehmood, R. Iktishaf+: A Big Data Tool with Automatic Labeling for Road Traffic Social Sensing and Event Detection Using Distributed Machine Learning. *Sensors* **2021**, *21*, 2993. [CrossRef]

71. Alomari, E.; Katib, I.; Mehmood, R. Iktishaf: A Big Data Road-Traffic Event Detection Tool Using Twitter and Spark Machine Learning. *Mob. Netw. Appl.* **2020**, 1–6. [CrossRef]

72. Ahmad, I.; Alqurashi, F.; Abozinadah, E.; Mehmood, R. Deep Journalism and DeepJournal V1.0: A Data-Driven Deep Learning Approach to Discover Parameters for Transportation. *Sustainability* **2022**, *14*, 5711. [CrossRef]

73. Yigitcanlar, T.; Regona, M.; Kankanamge, N.; Mehmood, R.; D'Costa, J.; Lindsay, S.; Nelson, S.; Brhane, A. Detecting Natural Hazard-Related Disaster Impacts with Social Media Analytics: The Case of Australian States and Territories. *Sustainability* **2022**, *14*, 810. [CrossRef]

74. Aqib, M.; Mehmood, R.; Albeshri, A.; Alzahrani, A. Disaster management in smart cities by forecasting traffic plan using deep learning and GPUs. In *Smart Societies, Infrastructure, Technologies and Applications*; SCITA 2017. Lecture Notes of the Institute for Computer Sciences, Social Informatics and Telecommunications Engineering; Springer: Cham, Switzerland, 2018; Volume 224, pp. 139–154.

75. Alswedani, S.; Katib, I.; Abozinadah, E.; Mehmood, R. Discovering Urban Governance Parameters for Online Learning in Saudi Arabia During COVID-19 Using Topic Modeling of Twitter Data. *Front. Sustain. Cities* **2022**, *4*, 1–24. [CrossRef]

76. Mehmood, R.; Alam, F.; Albogami, N.N.; Katib, I.; Albeshri, A.; Altowaijri, S.M. UTiLearn: A Personalised Ubiquitous Teaching and Learning System for Smart Societies. *IEEE Access* **2017**, *5*, 2615–2635. [CrossRef]

77. Alswedani, S.; Mehmood, R.; Katib, I. Sustainable Participatory Governance: Data-Driven Discovery of Parameters for Planning Online and In-Class Education in Saudi Arabia during COVID-19. *Front. Sustain. Cities* **2022**, *4*, 871171. [CrossRef]

78. Mohammed, T.; Albeshri, A.; Katib, I.; Mehmood, R. DIESEL: A Novel Deep Learning based Tool for SpMV Computations and Solving Sparse Linear Equation Systems. *J. Supercomput.* **2020**, *77*, 6313–6355. [CrossRef]

79. Hong, J.; Pradhan, A.; Froehlich, J.E.; Findlater, L. Evaluating wrist-based haptic feedback for non-visual target finding and path tracing on a 2D surface. In Proceedings of the 19th International ACM SIGACCESS Conference on Computers and Accessibility, Baltimore, MD, USA, 29 October–1 November 2017; pp. 210–219. [CrossRef]

80. Chun, A.C.B.; Theng, L.B.; WeiYen, A.C.; Deverell, L.; Mahmud, A.A.L.; McCarthy, C. An autonomous LiDAR based ground plane hazards detector for the visually impaired. In Proceedings of the 2018 IEEE EMBS Conference on Biomedical Engineering and Sciences, IECBES 2018—Proceedings, Sarawak, Malaysia, 3–6 December 2018; Institute of Electrical and Electronics Engineers Inc.: Piscataway, NJ, USA, 2019; pp. 346–351.

81. Gurumoorthy, S.; Padmavathy, T.; Jayasree, L.; Radhika, G. Design and implementation assertive structure aimed at visually impaired people using artificial intelligence techniques. *Mater. Today Proc.* **2021**, *in press*. [CrossRef]

82. Rao, S.; Singh, V.M. Computer Vision and Iot Based Smart System for Visually Impaired People; Computer Vision and Iot Based Smart System for Visually Impaired People. In Proceedings of the 2021 11th International Conference on Cloud Computing, Data Science & Engineering (Confluence), Noida, India, 28–29 January 2021. [CrossRef]

83. D'angiulli, A.M.; Waraich, P. Enhanced tactile encoding and memory recognition in congenital blindness. *Int. J. Rehabil. Res.* **2002**, *25*, 143–145. [CrossRef] [PubMed]

84. González-Cañete, F.J.; López Rodríguez, J.L.; Galdón, P.M.; Díaz-Estrella, A. Improvements in the learnability of smartphone haptic interfaces for visually impaired users. *PLoS ONE* **2019**, *14*, e0225053. [CrossRef] [PubMed]

85. Neugebauer, A.; Rifai, K.; Getzlaff, M.; Wahl, S. Navigation aid for blind persons by visual-to-auditory sensory substitution: A pilot study. *PLoS ONE* **2020**, *15*, e0237344. [CrossRef]

86. Du, D.; Xu, J.; Wang, Y. Obstacle recognition of indoor blind guide robot based on improved D-S evidence theory. *J. Phys. Conf. Ser.* **2021**, *1820*, 012053. [CrossRef]

87. Bleau, M.; Paré, S.; Djerourou, I.; Chebat, D.R.; Kupers, R.; Ptito, M. Blindness and the Reliability of Downwards Sensors to Avoid Obstacles: A Study with the EyeCane. *Sensors* **2021**, *21*, 2700. [CrossRef]

88. Al-Madani, B.; Orujov, F.; Maskeliūnas, R.; Damaševičius, R.; Venčkauskas, A. Fuzzy Logic Type-2 Based Wireless Indoor Localization System for Navigation of Visually Impaired People in Buildings. *Sensors* **2019**, *19*, 2114. [CrossRef]

89. Jafri, R.; Campos, R.L.; Ali, S.A.; Arabnia, H.R. Visual and Infrared Sensor Data-Based Obstacle Detection for the Visually Impaired Using the Google Project Tango Tablet Development Kit and the Unity Engine. *IEEE Access* **2017**, *6*, 443–454. [CrossRef]

90. Pare, S.; Bleau, M.; Djerourou, I.; Malotaux, V.; Kupers, R.; Ptito, M. Spatial navigation with horizontally spatialized sounds in early and late blind individuals. *PLoS ONE* **2021**, *16*, e0247448. [CrossRef]

91. Asakura, T. Bone Conduction Auditory Navigation Device for Blind People. *Appl. Sci.* **2021**, *11*, 3356. [CrossRef]

92. iMove around on the App Store. Available online: https://apps.apple.com/us/app/imove-around/id593874954 (accessed on 13 July 2020).

93. Seeing Assistant—Move—Tutorial. Available online: http://seeingassistant.tt.com.pl/en/move/tutorial/#basics (accessed on 13 July 2020).

94. BlindExplorer on the App Store. Available online: https://apps.apple.com/us/app/blindexplorer/id1345905790 (accessed on 13 July 2020).

95. RightHear—Blind Assistant on the App Store. Available online: https://apps.apple.com/us/app/righthear-blind-assistant/id1061791840 (accessed on 13 July 2020).

96. Ariadne GPS on the App Store. Available online: https://apps.apple.com/us/app/ariadne-gps/id441063072 (accessed on 19 September 2020).

97. BlindSquare on the App Store. Available online: https://apps.apple.com/us/app/blindsquare/id500557255 (accessed on 19 September 2020).

98. Morrison, C.; Cutrell, E.; Dhareshwar, A.; Doherty, K.; Thieme, A.; Taylor, A. Imagining artificial intelligence applications with people with visual disabilities using tactile ideation. In Proceedings of the 19th International ACM SIGACCESS Conference on Computers and Accessibility, Baltimore, MD, USA, 29 October–1 November 2017; pp. 81–90. [CrossRef]

99. Color Inspector on the App Store. Available online: https://apps.apple.com/us/app/color-inspector/id645516384 (accessed on 13 July 2020).

100. Color Reader on the App Store. Available online: https://iphony.net/details-nWhpnA.html (accessed on 13 July 2020).

101. ColoredEye on the App Store. Available online: https://apps.apple.com/us/app/coloredeye/id388886679 (accessed on 14 July 2020).

102. Wolf, K.; Naumann, A.; Rohs, M.; Müller, J. LNCS 6946—A Taxonomy of Microinteractions: Defining Microgestures Based on Ergonomic and Scenario-Dependent Requirements. In Proceedings of the IFIP Conference on Human-Computer Interaction, Lisbon, Portugal, 5–9 September 2011; Springer: Berlin/Heidelberg, Germany, 2011; Volume 6946.

103. Micro-Interactions: Why, When and How to Use Them to Improve the User Experience | by Vamsi Batchu | UX Collective. Available online: https://uxdesign.cc/micro-interactions-why-when-and-how-to-use-them-to-boost-the-ux-17094b3baaa0 (accessed on 3 April 2022).

104. Kim, J. Application on character recognition system on road sign for visually impaired: Case study approach and future. *Int. J. Electr. Comput. Eng.* **2020**, *10*, 778–785. [CrossRef]

105. Shilkrot, R.; Maes, P.; Huber, J.; Nanayakkara, S.C.; Liu, C.K. FingerReader: A wearable device to support text reading on the go. In Proceedings of the CHI'14 Extended Abstract on Human Factors in Computing Systems, Toronto, ON, Canada, 26 April–1 May 2014; pp. 2359–2364. [CrossRef]

106. Black, A. Flite: A small fast run-time synthesis engine. In Proceedings of the 4th ISCA Worskop on Speech Synthesis, Pitlochry, Scotland, 29 August–1 September 2001.

107. SeeNSpeak on the App Store. Available online: https://apps.apple.com/us/app/seenspeak/id1217183447 (accessed on 13 July 2020).

108. What Is Beacon (Proximity Beacon)?—Definition from WhatIs.com. Available online: https://whatis.techtarget.com/definition/beacon-proximity-beacon (accessed on 3 April 2022).

109. Braille Technology—Wikipedia. Available online: https://en.wikipedia.org/wiki/Braille_technology (accessed on 28 September 2020).

110. Refreshable Braille Displays | American Foundation for the Blind. Available online: https://www.afb.org/node/16207/refreshable-braille-displays (accessed on 27 September 2020).
111. 5 Best Braille Printers and Embossers—Everyday Sight. Available online: https://www.everydaysight.com/braille-printers-embossers/ (accessed on 27 September 2020).
112. Home | American Foundation for the Blind. Available online: https://www.afb.org/ (accessed on 28 September 2020).
113. Cash Reader: Bill Identifier on the App Store. Available online: https://apps.apple.com/us/app/cash-reader-bill-identifier/id1344802905 (accessed on 7 July 2020).
114. MCT Money Reader—Google Play. Available online: https://play.google.com/store/apps/details?id=com.mctdata.ParaTanima (accessed on 7 July 2020).
115. Light Detector on the App Store. Available online: https://apps.apple.com/us/app/light-detector/id420929143 (accessed on 13 July 2020).
116. Be My Eyes—on the App Sotre—Google Play. Available online: https://play.google.com/store/apps/details?id=com.bemyeyes.bemyeyes (accessed on 7 July 2020).
117. Sullivan+ (Blind, Visually Impaired, Low Vision)—on the App Store Google Play. Available online: https://play.google.com/store/apps/details?id=tuat.kr.sullivan&showAllReviews=true (accessed on 7 July 2020).
118. Visualize—Vision AI on the App Store. Available online: https://apps.apple.com/us/app/visualize-vision-ai/id1329324101 (accessed on 13 July 2020).
119. iCanSee world on the App Store. Available online: https://apps.apple.com/us/app/icansee-world/id1302090656 (accessed on 13 July 2020).
120. Seeing Assistant Home on the App Store. Available online: https://apps.apple.com/us/app/seeing-assistant-home/id625146680 (accessed on 13 July 2020).
121. VocalEyes AI on the App Store. Available online: https://apps.apple.com/us/app/vocaleyes-ai/id1260344127 (accessed on 13 July 2020).
122. LetSeeApp on the App Store. Available online: https://apps.apple.com/us/app/letseeapp/id1170643143 (accessed on 13 July 2020).
123. TapTapSee on the App Store. Available online: https://apps.apple.com/us/app/taptapsee/id567635020 (accessed on 13 July 2020).
124. Aipoly Vision: Sight for Blind & Visually Impaired on the App Store. Available online: https://apps.apple.com/us/app/aipoly-vision-sight-for-blind-visually-impaired/id1069166437 (accessed on 13 July 2020).
125. Turn on and Practice VoiceOver on iPhone—Apple Support (SA). Available online: https://support.apple.com/en-sa/guide/iphone/iph3e2e415f/ios (accessed on 3 January 2022).
126. VoiceOver—Wikipedia. Available online: https://en.wikipedia.org/wiki/VoiceOver (accessed on 3 January 2022).
127. Siri—Wikipedia. Available online: https://en.wikipedia.org/wiki/Siri (accessed on 10 August 2020).
128. Siri—Apple. Available online: https://www.apple.com/siri/ (accessed on 10 August 2020).
129. Samsung Bixby: Your Personal Voice Assistant | Samsung US. Available online: https://www.samsung.com/us/explore/bixby/ (accessed on 17 August 2020).
130. Samsung Bixby—Privacy Evaluation. Available online: https://privacy.commonsense.org/evaluation/Samsung-Bixby (accessed on 17 August 2020).
131. Alexa vs Siri | Top 14 Differences You Should Know. Available online: https://www.educba.com/alexa-vs-siri/ (accessed on 17 July 2022).
132. Amazon Alexa Voice AI | Alexa Developer Official Site. Available online: https://developer.amazon.com/en-US/alexa (accessed on 18 July 2022).
133. Liu, H.; Liu, R.; Yang, K.; Zhang, J.; Peng, K.; Stiefelhagen, R. HIDA: Towards Holistic Indoor Understanding for the Visually Impaired via Semantic Instance Segmentation with a Wearable Solid-State LiDAR Sensor. In Proceedings of the IEEE/CVF International Conference on Computer Vision (ICCV) Workshops, Virtual, 11–17 October 2021; pp. 1780–1790. [CrossRef]
134. Rahman, M.M.; Islam, M.M.; Ahmmed, S.; Khan, S.A. Obstacle and Fall Detection to Guide the Visually Impaired People with Real Time Monitoring. *SN Comput. Sci.* **2020**, *1*, 1–10. [CrossRef]
135. 12m IP65 Distance Sensor. Available online: http://en.benewake.com/product/detail/5c345cd0e5b3a844c472329b.html (accessed on 31 March 2022).
136. HC-SR04 Ultrasonic Sensor Working, Pinout, Features & Datasheet. Available online: https://components101.com/sensors/ultrasonic-sensor-working-pinout-datasheet (accessed on 31 March 2022).
137. Benewake TF Mini Series LiDAR Module (Short-Range Distance Sensor): Benewake—BW-3P-TFMINI-S—Third Party Tool Folder. Available online: https://www.ti.com/tool/BW-3P-TFMINI-S (accessed on 12 October 2020).
138. Co Ltd, B. SJ-PM-TFmini-S A00 Specified Product Manufacturer Product Certification. Available online: https://www.gotronic.fr/pj2-sj-pm-tfmini-s-a00-product-mannual-en-2155.pdf (accessed on 8 March 2022).
139. Krishnan, N. A LiDAR based proximity sensing system for the visually impaired spectrum. In Proceedings of the Midwest Symposium on Circuits and Systems, Dallas, TX, USA, 4–7 August 2019; Institute of Electrical and Electronics Engineers Inc.: Piscataway, NJ, USA, 2019; pp. 1211–1214.

140. Agarwal, R.; Ladha, N.; Agarwal, M.; Majee, K.K.; Das, A.; Kumar, S.; Rai, S.K.; Singh, A.K.; Nayak, S.; Dey, S.; et al. Low cost ultrasonic smart glasses for blind. In Proceedings of the 2017 8th IEEE Annual Information Technology, Electronics and Mobile Communication Conference, IEMCON 2017, Vancouver, BC, Canada, 3–5 October 2017; Institute of Electrical and Electronics Engineers Inc.: Piscataway, NJ, USA, 2017; pp. 210–213.

141. Marioli, D.; Narduzzi, C.; Offelli, C.; Petri, D.; Sardini, E.; Taroni, A. Digital Time-of-Flight Measurement for Ultrasonic Sensors. *IEEE Trans. Instrum. Meas.* **1992**, *41*, 93–97. [CrossRef]

142. Borenstein, J.; Koren, Y. Obstacle Avoidance with Ultrasonic Sensors. *IEEE J. Robot. Autom.* **1988**, *4*, 213–218. [CrossRef]

143. Bosaeed, S.; Katib, I.; Mehmood, R. A Fog-Augmented Machine Learning based SMS Spam Detection and Classification System. In Proceedings of the 2020 Fifth International Conference on Fog and Mobile Edge Computing (FMEC), Paris, France, 20–23 April 2020; Volume 2020, pp. 325–330. [CrossRef]

144. Cleary, J.G.; Trigg, L.E. K*: An Instance-based Learner Using an Entropic Distance Measure. *Mach. Learn. Proc.* **1995**, *1995*, 108–114. [CrossRef]

145. Aha, D.W.; Kibler, D.; Albert, M.K.; Quinian, J.R. Instance-based learning algorithms. *Mach. Learn.* **1991**, *6*, 37–66. [CrossRef]

146. RandomCommittee. Available online: https://weka.sourceforge.io/doc.dev/weka/classifiers/meta/RandomCommittee.html (accessed on 7 February 2022).

147. Speech-to-Text Basics | Cloud Speech-to-Text Documentation | Google Cloud. Available online: https://cloud.google.com/speech-to-text/docs/basics (accessed on 20 February 2022).

148. Janbi, N.; Mehmood, R.; Katib, I.; Albeshri, A.; Corchado, J.M.; Yigitcanlar, T.; Sa, A.A. Imtidad: A Reference Architecture and a Case Study on Developing Distributed AI Services for Skin Disease Diagnosis over Cloud, Fog and Edge. *Sensors* **2022**, *22*, 1854. [CrossRef]

149. Janbi, N.; Katib, I.; Albeshri, A.; Mehmood, R. Distributed Artificial Intelligence-as-a-Service (DAIaaS) for Smarter IoE and 6G Environments. *Sensors* **2020**, *20*, 5796. [CrossRef]

150. Mohammed, T.; Albeshri, A.; Katib, I.; Mehmood, R. UbiPriSEQ—Deep reinforcement learning to manage privacy, security, energy, and QoS in 5G IoT hetnets. *Appl. Sci.* **2020**, *10*, 7120. [CrossRef]

151. Usman, S.; Mehmood, R.; Katib, I. Big data and hpc convergence for smart infrastructures: A review and proposed architecture. In *Smart Infrastructure and Applications Foundations for Smarter Cities and Societies*; Springer: Cham, Switzerland, 2020; pp. 561–586.

152. Yigitcanlar, T.; Corchado, J.M.; Mehmood, R.; Li, R.Y.M.; Mossberger, K.; Desouza, K. Responsible Urban Innovation with Local Government Artificial Intelligence (AI): A Conceptual Framework and Research Agenda. *J. Open Innov. Technol. Mark. Complex.* **2021**, *7*, 71. [CrossRef]

153. Yigitcanlar, T.; Mehmood, R.; Corchado, J.M. Green Artificial Intelligence: Towards an Efficient, Sustainable and Equitable Technology for Smart Cities and Futures. *Sustainability* **2021**, *13*, 8952. [CrossRef]

Article

Cloud-Based Monitoring of Thermal Anomalies in Industrial Environments Using AI and the Internet of Robotic Things

Mohammed Ghazal [1,*], **Tasnim Basmaji** [1,†], **Maha Yaghi** [1,†], **Mohammad Alkhedher** [2], **Mohamed Mahmoud** [3] **and Ayman S. El-Baz** [4]

[1]   Electrical and Computer Engineering Department, College of Engineering, Abu Dhabi University, Abu Dhabi 59911, UAE; tasnim.basmaji@adu.ac.ae (T.B.); maha.yaghi@adu.ac.ae (M.Y.)

[2]   Mechanical Engineering Department, College of Engineering, Abu Dhabi University, Abu Dhabi 59911, UAE; mohammad.alkhedher@adu.ac.ae

[3]   Emirates Global Aluminium, Technology Development and Transfer Midstream, Abu Dhabi 109111, UAE; mmahmoud@ega.ae

[4]   Bioengineering Department, University of Louisville, Louisville, KY 40292, USA; ayman.elbaz@louisville.edu

*   Correspondence: mohammed.ghazal@adu.ac.ae

†   These authors contributed equally to this work.

Received: 21 September 2020; Accepted: 3 November 2020; Published: 7 November 2020

Abstract: Recent advancements in cloud computing, artificial intelligence, and the internet of things (IoT) create new opportunities for autonomous industrial environments monitoring. Nevertheless, detecting anomalies in harsh industrial settings remains challenging. This paper proposes an edge-fog-cloud architecture with mobile IoT edge nodes carried on autonomous robots for thermal anomalies detection in aluminum factories. We use companion drones as fog nodes to deliver first response services and a cloud back-end for thermal anomalies analysis. We also propose a self-driving deep learning architecture and a thermal anomalies detection and visualization algorithm. Our results show our robot surveyors are low-cost, deliver reduced response time, and more accurately detect anomalies compared to human surveyors or fixed IoT nodes monitoring the same industrial area. Our self-driving architecture has a root mean square error of 0.19 comparable to VGG-19 with a significantly reduced complexity and three times the frame rate at 60 frames per second. Our thermal to visual registration algorithm maximizes mutual information in the image-gradient domain while adapting to different resolutions and camera frame rates.

Keywords: edge-fog-cloud computing; Internet of Things; robotics; artificial intelligence; autonomous driving; image registration

---

1. Introduction

Autonomous robots are smart machines capable of performing complex and repetitive tasks with a high level of independence from human intervention [1]. According to the International Federation of Robotics, industrial robots represent the largest percentage of all robots used and are expected to continue to be the leading robotic category until 2025 [2]. Recent advancements in AI and computer vision paved the way for robots to navigate challenging industrial environments for condition monitoring and assessment. They require less control and observation and can work efficiently and accurately [3]. Some of the reasons behind the challenging nature of industrial environments include the size of the monitored area, sensors cost, process effect on electronics and communications networks, obstacles, accessibility of monitored areas, high temperature, and low-lighting conditions. An example of challenging environments is industrial aluminum factories. In these factories, steel pot shell surfaces' temperature needs to be regularly measured to analyze sick pots, amperage increase, and design changes in pot lining. The pots' conditions must be continuously monitored and assessed, typically done manually by humans despite being unsafe and carrying a risk of exposure to harm. An alternative approach is to use the Internet of Things (IoT) architectures and install sensors-equipped nodes to cover the monitored area. Using IoT nodes provides an early response to anomalies and reduces the risk of injury. There are two main disadvantages to this approach when applied to some industrial settings including: (1) industrial processes may produce conditions which affect the electronics and communication capabilities of the IoT nodes (e.g., high temperatures or magnetic fields); (2) the cost of covering a large industrial area reduces deployment, especially as the cost of the sensors onboard increase.

To reduce or eliminate the need for human involvement in monitoring challenging industrial environments and reduce the cost of large-scale IoT nodes deployment, we propose mobile IoT nodes carried onboard autonomous robots. These robots must be designed to simultaneously address the application's requirements and the challenges posed by the environment. In our example case of aluminum factories, the application requirements are to thermally monitor a sizeable industrial area without a large scale deployment of thermal-camera-equipped nodes. Instead, we are to use low-cost AI-powered robots to navigate and monitor the environment. The robot detects, localizes, and visualizes thermal anomalies within a short response time. The environment's challenges include a strong magnetic field interfering with electronic circuits and virtually non-existent wired or wireless communication channels. The large scale installation of custom-designed communication solutions is also not feasible. The temperature and humidity are high, and the surface on which the robot will move is rough, with many obstacles to avoid. The system is to report the visual and thermal videos to a centralized repository for archiving and further short- and long-term automated analyses. Reports are to be available on mobile devices to facilitate access to them on-site. While we have used the aluminum pot-lines as a basis for our design and validation, our proposed solution can be used in other automated condition monitoring in challenging industrial environments after an initial setup stage which includes: (1) additional training for the self-driving sub-system; (2) thermal cameras and detectors re-calibration; and (3) placement of localization QR codes.

To address the requirements and constraints, we propose an edge-fog-cloud architecture for a low response time, larger covered area, and overall reduced deployment costs. The mobile IoT nodes at the edge layer comprise of multiple anomalies detecting autonomous robots. The robots initially use real-time global thresholding of thermal images for anomalies detection but cannot communicate the alarm to maintenance staff due to the lack of communication channels. The robot dispatches its drone with the thermal and visual images and coarse location data to leave the dead communication zone and deliver the alarm information as soon as it regains connectivity, thus maintaining a fast response time. We place the drones in the fog layer of our architecture, which, along with the robots, form an Internet of Robotic Things. The drones can also be used to initiate automated early response actions. When a robot run (i.e., a full area scan) is complete, their ability to communicate is restored by leaving the dead communication zone. The thermal and visual videos and location information are uploaded to a cloud server for registration, fine localization, and further analysis of the anomalies (e.g., anomaly classification or thermal distribution analysis).

The robot integrates high- and low-level sub-systems to perform its function, including obstacle avoidance, self-driving, localization, remote control, and anomaly detection and reporting. We propose a three-layered architecture for these sub-systems to govern their communication and coordination. In addition to the above contributions, we also propose a deep learning convolutional neural network for regressing the steering angle of differential robots that can deliver a root mean square error comparable to well-known architecture but with a faster frame rate and less computational complexity.

Moreover, since the frame rates and image resolutions of the thermal and visual imagers are different, our cloud server needs to align and overlay the thermal to the visual video to help the end-users localize and analyze the detected thermal anomalies. To this end, we also propose a thermal to visual frame-rate and resolution adaptive multi-modal image registration technique based on maximizing mutual information in the image gradient domain.

The main contribution of our work is an end to end system for the autonomous monitoring of thermal anomalies in industrial settings. To achieve this goal, we also propose: (1) an autonomous robot surveyor operating system architecture for mobile IoT nodes; (2) an edge-fog-cloud communication architecture that reduces monitoring cost and anomalies detection time; (3) a real-time and accurate self-driving deep learning network; (4) a localization system based deep object detection of visual beacons; (5) a frame-rate and resolution adaptive thermal to visual registration algorithm based on maximizing mutual derivative information.

2. Literature Review

Several approaches in the literature propose autonomous robots for industrial applications. According to the authors in ref. [4], the modern industry is devoting efforts to the construction of unmanned factories, autonomous industrial robots, and the development of the digital twin techniques to eliminate potential human mistakes. The authors presented an industrial cyber-physical system monitoring and a control framework for large-scale hierarchical systems. The main purpose is to define the tasks to be performed at each stage of operation in modern large-scale industrial systems and the techniques that need to be applied to carry out these tasks. An autonomous vehicle used in smart factories was presented in ref. [5]. The system is based on HD mapping technology. The methodology involves real-time object detection and localization. Recognized objects are then compared with stored ones in the database to check their validity in the HD map and update them accordingly. The ATV can then use the updated map. Another use of such robots is for gas leakage detection, as proposed in ref. [6]. The robot introduced has two modes of operation: line following and obstacle avoidance. The robot is equipped with ultrasonic sensors for obstacle avoidance and an infrared sensor for leakage detection. Real-time leakage detection is based on the navigation through a dangerous environment and transmission of the intensity through a wireless signal to the receiving module for gas leakage monitoring. In ref. [7], the authors proposed the ARCO robot, which aims to work with laborers to collect material, facilitate transportation, and manage schedules in a factory. Similar to some of the previously discussed systems, the ARCO robot needs a map for localization purposes. However, the ARCO robot's localization uses a previously generated map and computes a fixed map during operation. The robot has two modes of operation: straight-line and omnidirectional and includes an obstacle avoidance system for safety purposes.

A recent study in ref. [8] provides a comprehensive review of the main challenges proposed in the literature for autonomous robots in various environments. The challenges mainly include navigation and localization. Nowadays, some navigation systems involve lidar-based techniques for localization and mapping; however, such a technique does not perform well in smooth environments due to the lack of details and irregularities. On the other hand, Visual SLAM techniques are also used for localization in some systems. These techniques depend on features that describe the environment in terms of color and, therefore, need proper illumination, which becomes a problem in some environments like tunnels. To ensure better performance, multiple other systems rely on QR codes for navigation. The work in ref. [9] is a warehouse management robot that aims to automate tasks in an unknown environment. The robot is equipped with a robotic arm for handling items and a QR scanner for navigation and localization. It also involves a route development technique based on QR codes' placement in strategic locations around the environment. The robot can extract location information for less computation and navigate to the required location from these codes. Recent systems developed in ref. [10,11] also use a similar indoor navigation system based on QR codes. A system developed in ref. [12] for assisting the elderly uses the Unscented Kalman Filter and QR codes for localization. The system provides the user with the ability to choose the desired location on the map and navigates autonomously to the chosen location.

Several other papers focused on autonomous robots in industrial environments. Authors in ref. [13] developed an autonomous robot for mapping unfamiliar environments and identifying target objects. The methodology involves extracting and collecting a set of features during the setup stage. Accordingly, the robot will search for the features that match the stored features after developing a map using ultrasonic sensors and an RGB stream. In ref. [14], the authors proposed an autonomous robot used to monitor oil and gas locations. The robot operates in 4 modes: autonomous, manual, rail, and unsafe modes depending on the task. The rail mode enables the user to manually control the forward speed for manual inspections, where complicated operations are performed automatically. In comparison, the unsafe manual mode disables all safety features to operate in an unknown environment. The system also includes obstacle detection and lidar-based localization algorithms to drive the robot autonomously. Various approaches were proposed in the literature to solve the autonomous robot trajectory. The work in ref. [15] is an autonomous surveillance robot used to detect anomalies, including temperature instabilities and unauthorized personnel. The robot is equipped with multiple sensors, a near-infrared camera, and a Wi-Fi adapter. It navigates autonomously by collecting Lidar data and transmitting them to a central server for processing and map generation. The robot has an obstacle avoidance system to avoid any collisions with obstacles. The user needs to drive the robot manually in the environment to complete the mapping process in a semi-autonomous mode. Authors in ref. [16,17] used neural networks and artificial neural networks mainly for self-driving purposes. Whereas Deep and Convolutional Neural Networks were explicitly used in ref. [18].

Anomalies detection and recognition is typically one of the challenging requirements in industrial systems. An activity recognition system is proposed in ref. [19] designed for industrial surveillance systems. The framework proposed starts by using CNN-based human saliency features to select shots captured from surveillance videos. Temporal features are then extracted from the shots using the layers of the FlowNet2 CNN model to represent an activity. Accordingly, activities are recognized using a multi-layer long short-term memory. The proposed method achieved the highest detection accuracy results compared to previous methods for the Youtube activities dataset in terms of accuracy. The time complexity of the system is almost real-time making it suitable for industrial surveillance environments. Another framework is proposed in ref. [20] for similar environments. The framework aims to detect real-world anomalies in surveillance environments. The pipeline starts with extracting CNN features from frames and then inputting them to a multi-layer BD-LSTM for class detection. Compared to other anomaly detection methods, the proposed framework achieved higher accuracy and a low false alarm rate. Authors in ref. [21] presented a data-driven fault detection approach based on the integration of the modified principal component analysis (MPCA) into a locally weighted projection regression framework and compared it to several other approaches. The approach aims to monitor processes based on sensing measurements in complex nonlinear systems using the normalized weighted mean of test statistics. Depending on the modeling phase, different models are used to approximate the nonlinear systems. Then, the fault detection approach is applied for fault diagnosis. Compared to other approaches, the MPCA approach is less complicated, has lower computational complexity, a fast learning capacity, enhanced robustness, and a high fault detection rate.

The remainder of the paper is organized as follows: Section 3 introduces the materials and methods. Section 4 discusses the testing and validation results of our proposed system and sub-systems. Finally, conclusions are drawn in Section 5.

3. Materials and Methods

3.1. Environment and Requirements

Automatic and on-time detection of machine anomalies in harsh industrial environments is essential for establishing reliable preventive or predictive maintenance schemes. The current manual diagnosis methods are limited to operator training capabilities, machine accessibility, and worker ability to endure the harsh industrial conditions [22].

Industrial environmental conditions are very harsh; this has restrained the use of electronics and communication systems in anomalies detection due to high temperatures, high humidity, dust, and magnetic fields. To improve normal practices in monitoring operating conditions of equipment and machines and overcome the harsh operating conditions, digitization of inspection processes is ramping up due to the availability of several sensing devices, data processing and storage capabilities using cloud computing, and affordable machine learning technologies. This has allowed maintenance protocols to utilize real-time sensing data to forecast possible anomalies and failures in machines due to several operating conditions and causes. Big data analysis is needed to study failure history and system performance that could require performing predictive maintenance. This analysis is powered by machine learning techniques to optimize available solutions.

Researchers have proposed using sensor data anomalies with statistical machine learning methods to predict machines' unfavorable temperature in industrial manufacturing environments [23,24]. To improve the performance of anomaly detection systems in industrial facilities, machine learning has been implemented in predictive maintenance of industrial plants. These models are used to define the condition of the machines [25,26].

In aluminum production's industrial environment, regular thermal measurements of the electrolytes take place in the basement of the potrooms to maintain heat balance and prolong the pot shell life cycle. The concept of the heat balance in the aluminum reduction pots is quite simple; almost half of the fed energy is actually used to produce aluminum. The remaining part must be dissipated as heat losses by the cell for it to maintain its thermal equilibrium. To maintain consistent pot performance, the thermal balance is regularly assessed during the whole pot life. These measurements are performed manually using thermocouples and thermometers at an ambient temperature that could reach 60 °C and humidity of 100% [27]. Moreover, sending human operators poses health and safety risks due to aluminum tap out and exposure to a high magnetic field of 300–500 Gauss.

Based on potline operation procedures, if the surface temperature of the potshell exceeds approximately 500 °C, the system operators consider that hotspot as an abnormal measurement. This detected thermal anomaly has to be monitored for any future excessive temperatures which could be caused by faults associated with the aluminum production process [28]. Regular potshell surface temperature measurements are performed at different locations to help in studying and analyzing potshell thermal balance. Detecting thermal anomalies will help operators to activate a cooling system immediately to maintain the current efficiency of the produced aluminum, identify damaged pots, and improve potline design [27].

Pot shell temperature varies between 120 °C to 500 °C at standards conditions. The measurements can vary based on location and process parameters. In this work, the robot surveyor should monitor any excessive change above 500 °C. A regular inspection of potshell surface measurement is carried out every two-three weeks. Rapid detection of thermal anomalies can help operators and process engineers to timely respond to problems that occur in the process and avoid damage in the potshell. Figure 1 shows a depiction of the operating environment. Due to the characteristics of the working environment, the system is inaccessible to readers.

Figure 1. System's working environment characterised by a high magnetic field up to 300–500 Gauss, ambient temperature up to 60 °C, high pot shell temperature varying between 120 °C and 500 °C, high relative humidity up to 100%, and concrete debris on the floor.

3.2. Proposed Robot Surveyor Operating System Architecture

The central unit of the overall system is an IoT node carried on by our autonomous robot surveyor. The IoT node integrates all sensors needed for inspection. We use modular sub-systems organized in the layered architecture shown in Figure 2 to design our robot surveyor operating system. Communication between all sub-systems, regardless of their level, uses a publish-subscribe pattern. Low-level sub-systems include ones for motor control, remote control, line-following, and obstacle avoidance. Motor control receives command signals from mid-level sub-systems such as self-driving or low-level ones such as remote control and obstacle avoidance, with conflicts resolved through a priority setting. Motor control also publishes robot speed feedback to subscribed sub-systems (e.g., for localization and navigation). Mid-level sub-systems handle localization and behavior cloning and are thus responsible for autonomous driving. High-level sub-systems are application layer systems customizable to the inspection requirements and the industrial environment challenges. For example, since our proposed architecture and robots are tested and validated on the problem of thermal anomalies detection in aluminum factories, we deploy thermal and visual imagers as part of the IoT node sensors. Also, since our testing showed communication networks are inaccessible in the environment, we add to the application layer a messenger drone for early response services (e.g., early alerts). Long-term analyses of anomalies are done by offloading to the cloud. In our application area, this includes the frame rate and resolution adaptive multi-modal (i.e., light and temperature) image registration for the detection and visualization of thermal anomalies. Other analyses can be tailed to the application needs.

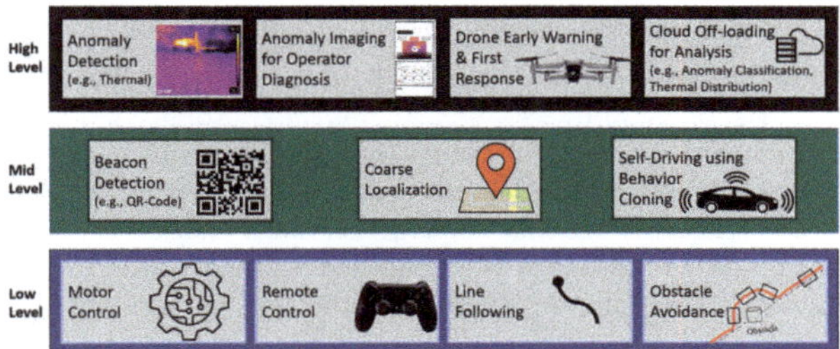

Figure 2. Robot surveyor operating system architecture with modular sub-systems. The architecture is general and adaptable to different applications through the high-level layer.

3.3. Proposed Edge-Fog-Cloud Architecture for Industrial Monitoring

While the IoT nodes carried on the robot surveyors are the central system, the companion drones and the cloud back-end are critical for early response and detailed reporting and analysis. We propose an edge-fog-cloud architecture for autonomous industrial monitoring communication needs to increase response time and reduce surveying costs. We depict the architecture in Figure 3. In our architecture, robot surveyors are mobile edge computing nodes providing surveying and early anomaly detection services, companion drones are fog computing nodes providing early warning and first response services and the back-end is powered by cloud computing nodes providing analysis, storage, and tracking services. What distinguishes our architecture is the AI-powered autonomous nature, or smartness, of our edge and fog nodes, reducing the number of nodes needed to cover the monitored area. As discussed in Section 1, placing fixed IoT nodes to monitor a large industrial area has practicality challenges. Limiting computing to the cloud increases the response latency. Our edge-fog-cloud computing architecture distributes computing closer to the source of the industrial monitoring data reducing latency and improving network utilization.

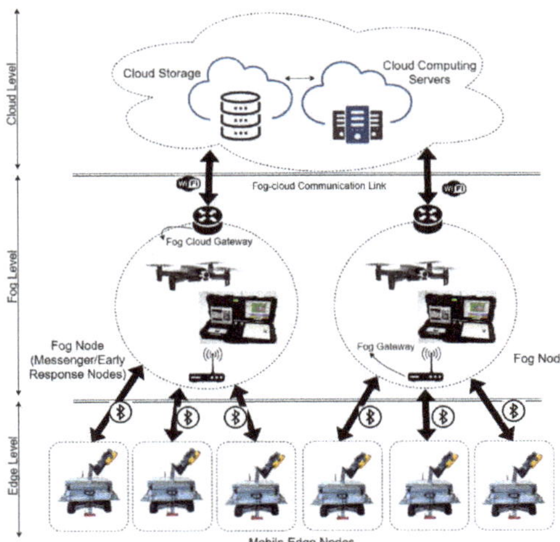

Figure 3. Edge-fog-cloud architecture for the communication needs of autonomous industrial monitoring.

3.4. Thermal Anomalies Detection in Aluminium Factories

In Sections 3.2 and 3.3, we present our proposed architectures designed to address industrial monitoring needs and challenges. In this section, we implement these architectures as we detail our proposed system for thermal anomalies detection in aluminum factories. The overall system diagram is depicted in Figure 4 with numbers indicating the sequence of operations. Our system uses a mobile application to deliver early alerts from the edge and fog layers and detailed reports from the cloud layer to the end-users. The drone is used to reduce the alert time and to initiate automated early response actions if needed (e.g., activate shutdown or cooling routines).

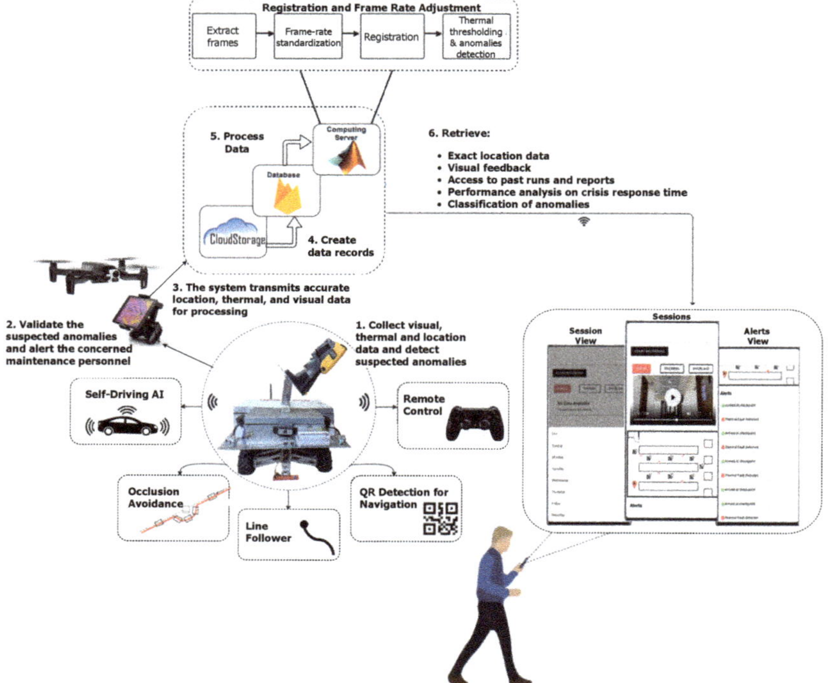

Figure 4. Proposed system architecture.

In Section 3.5, we present the robot's mechanical and electronic design, including shielding design needed to protect the carried IoT node from the harsh environment (e.g., the magnetic field). In Sections 3.6–3.8, we discuss our proposed self-driving, QR detection navigation, and remote control and robot-preservation sub-systems, respectively. We use a line following sub-system in parts of the environment where visual navigation is not possible (e.g., due to low lighting conditions).

Our robot uses a high frame rate global thermal thresholding to detect anomalies while on survey run. While the robot cannot communicate with the cloud server due to the effects of the environment, it can establish a short-range communication link over Bluetooth to a companion drone as long as a line of sight is available. Once a thermal anomaly is detected, the robot dispatches the drone to the nearest location with access to communication networks to send the thermal and visual images and coarse location data to the user as an early alarm system. The drone can find and join a communication network faster by immediately navigating out of the dead-zone. When the drone leaves the robot to deliver the alert, the connection is lost but is also no longer needed. After the completion of a survey run, recorded thermal and visual videos are uploaded to the cloud. Uploaded videos are processed by a computing server that applies our proposed multi-modal thermal to visual registration to detect and localize thermal anomalies. Analysis reports are delivered to the end user's mobile application, as explained in Section 3.9.

3.5. Robot Body and Magnetic Shielding

Our AI-powered autonomous robot surveyor is a six-wheeled all-terrain vehicular robot, which consists of an 18" long boxed channel chassis connected to six 5.4" off-road tires. The tires are driven by six 313 rpm ball-bearing precision planetary gear-motors, which allow the robot to navigate tough terrain with ease. The suspension incorporates 4.62" aluminum beams and 130mm, oil-filled, aluminum-bodied shocks. We shield the computing engine, controllers, and sensors from the magnetic field as illustrated in Figure 5. We also added an extension to safely position the line follower sensor array closer to the ground for improved and reliable performance. The robot mechanical and electronic design is shown in Figure 5.

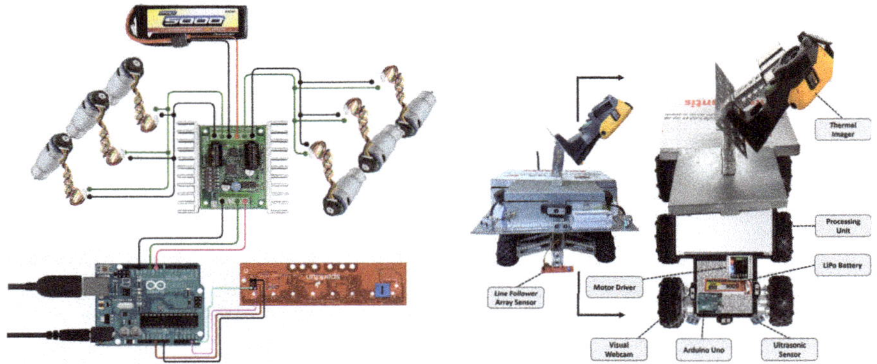

Figure 5. Mechanical and electronic design.

3.6. Proposed Self-Driving Deep Architecture for Steering Angle and Speed Regression

We use a behavior-cloning approach to self-driving. An operator remotely controls the robot to complete several runs using real-time video streams as visual feedback. We collect training data by capturing both the speed, the steering angle, and the video signal from all three cameras mounted facing forward, right, and left with a small degree of visual overlap. We then train a deep convolutional neural network to predict the robot's speed and steering angle. After training, we use the network to autonomously control the robot using real-time images from the three cameras. We cascade the images captured from the camera into a single image.

Our network, detailed in Figure 6, consists of five convolutional layers with an exponential linear unit (ELU) activation followed by three fully-connected layers. The output layer produces predictions for the right and left DC motor speeds controlled by pulse-width modulation (PWM) signals. Controlling the speed of the motors simultaneously achieves speed and steering angle control. To generalize better and reduce over-fitting, we add dropout layers after every convolutional layer, reduce temporal redundancy by artificially lowering the frame rate, use wide-angle cameras to increase visual details, and capture our videos under varying lighting conditions.

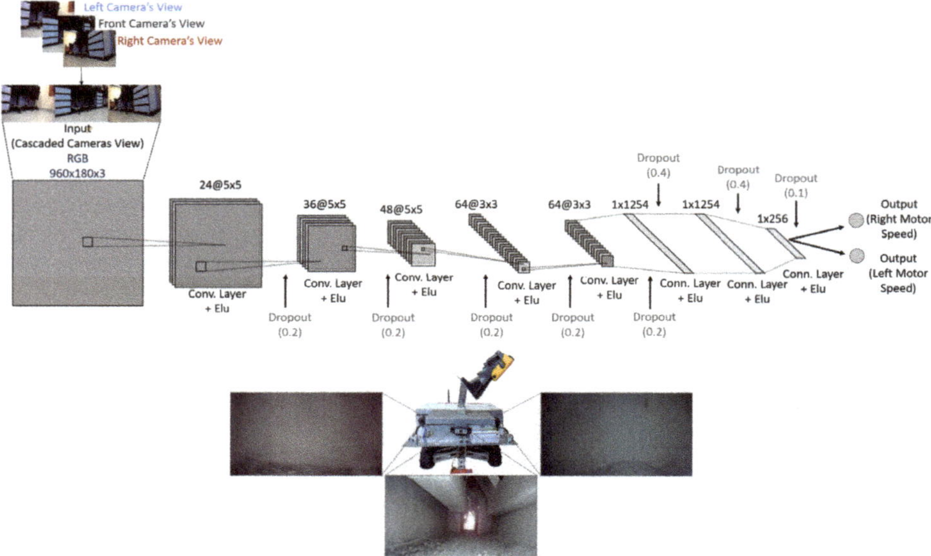

Figure 6. Top: Proposed self-driving deep architecture for real-time speed and steering angle control. Bottom: mounted wide-angle cameras and corresponding views on-site.

Both the architecture design and the computational power of the robot's computing unit influence the frame rate and energy needs of the robot. For example, a high frame rate reduces the motor controller sample time and improves stability. We aim to minimize the root mean square error for the PWM levels prediction while simultaneously maximizing the frame rate and battery life. We tune the hyper-parameters using a grid search approach using the lab environment videos and use the tuned parameter for the site environment operations. During training, we optimize the network weights using stochastic gradient descent with adaptive moment estimation.

3.7. Robot Surveyor Localization and Navigation Using QR Code Detection

We use a modified localization system to the one in ref. [29], as shown in Figure 7. Since the use of radio beacons is impractical due to environmental limitations, we replace them in [29] with visual Quick Response (QR) beacons encoding location data. The QR codes are placed along the surveyed path. The three wide-angle cameras feeding video to the self-driving sub-system are also monitored for QR codes. We use the You Only Look Once (YOLO) object detector in Figure 7 to detect the QR code in the image and decode the absolute location information embedded within. We also replace steps counting in ref. [29] used for dead-reckoning with rotations counting from the DC motors encoders. We consider the output of localization as noisy measurements with Gaussian noise and use a Kalman filter to improve location accuracy. We finally utilize location information to navigate within the environment using a waypoint navigation system.

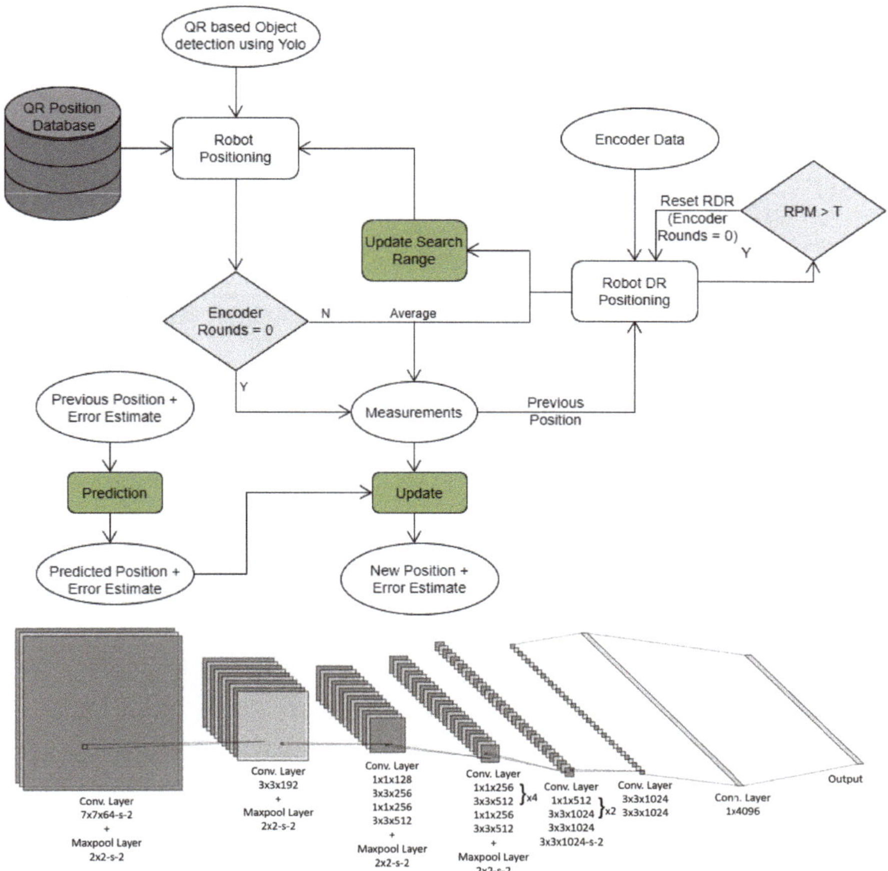

Figure 7. Top: Flow chart of QR-based robot positioning [29] with QR codes instead of radio beacons, round counting instead of steps counting, and the addition of a Kalman filter. Bottom: Used YOLOv2 QR code detector.

3.8. Remote Control and Collision Avoidance

We use a PlayStation controller to transmit control commands to the robot over Bluetooth. We map the controller's two analog joysticks to the PWM levels of the right and left motor groups. The PWM levels are the ones recorded during training and regressed while self-driving. We use six ultrasonic sensors mounted on the robot to detect obstacles and avoid collisions. The controller unified modeling language (UML) state chart is shown in Figure 8.

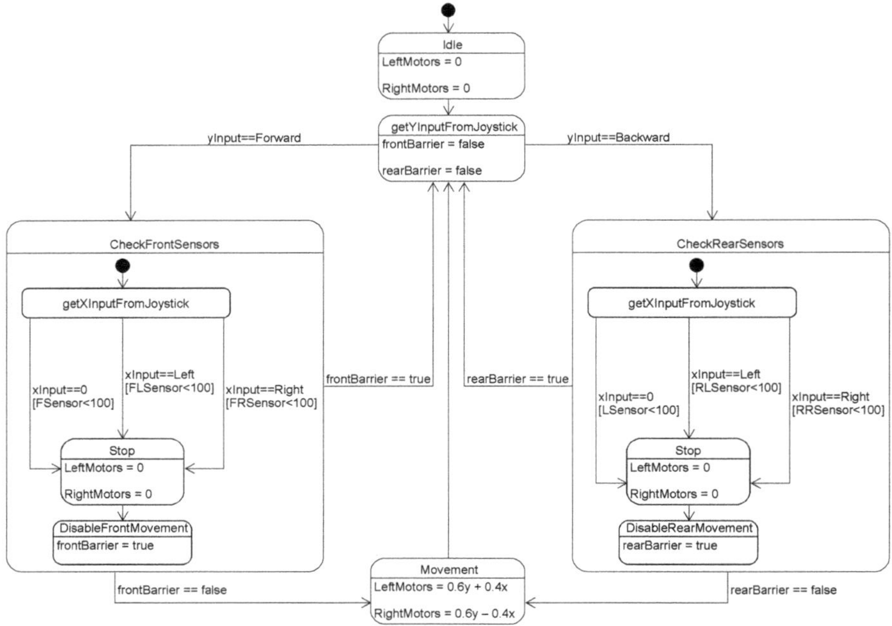

Figure 8. Remote control with collision avoidance UML statechart.

3.9. Proposed Anomaly Localization Using Thermal to Visual Registration

After a robot surveyor run is complete, thermal and visual videos are offloaded to our cloud server for storage and processing. The thermal imager is sensitive to temperature ranging from $-20\,°\text{C}$ to $550\,°\text{C}$ and captures 9 VGA frames per second. The visual camera capture 29 HD frames per second. The first step of our proposed algorithm is to align the two frames using the capture timestamp by spatiotemporally down-sampling the visible video to match the frame rate and (approximate) resolution of the thermal imager.

We discard color information while registering the two frames and convert the visual frame $V(t)$ from the RGB color-space to gray-scale using the sRGB standard luma coefficients. We represent each pixel in the false-color thermal frame $N(t)$ by its hue angle in the CIE L^*C^*h color-space with $180° \leq h < 540°$. This particular range of hue angles preserves relative temperature information, i.e., $h_1 < h_2$ iff $T_1 < T_2$, for thermal images using the Iron color-map. We high-pass filter $V(t)$ and $N(t)$ using Sobel and Gaussian ($\sigma^2 = 1$ pixel) kernels to compute the gradient magnitude images highlighting object boundaries, $V'(t)$ and $N'(t)$, derived from edges in the visible light and thermal infrared images, respectively.

We estimate a sequence of affine transformations $A(t)$ that aligns $N'(t)$ with $V'(t)$ frame by frame by maximizing mutual information (MI) using

$$\text{MI}(A * N', V') = \sum_{l=0}^{255} \sum_{m=0}^{255} p(l,m) \log_2 \left(\frac{p(l,m)}{p_{N'}(l)\,p_{V'}(m)} \right), \tag{1}$$

where $A * N'$ is the warped thermal gradient magnitude, l is the temperature level (remapped to 0–255), m is the intensity level, $p(l, m)$ is the joint probability distribution, and $p_{N'}(l)$ and $p_{V'}(m)$ are the marginal probability distributions of the thermal and visual frames, respectively. We initialize $A(t)$ for each frame with a pure translation aligning the frames' geometric centers and find the optimal transformation argmax_A MI using the anisotropic $(1 + 1)$-ES algorithm [30].

We smooth the transformations to reduce jitter in the registered videos. Let each frame-specific affine transformation matrix $A(t)$ be the parameterized six-element vector

$$a(t) = \left(\Delta_x, \Delta_y, h, s_x, s_y, \sin \theta \right), \tag{2}$$

where (Δ_x, Δ_y) are translations, h, s_x, and s_y together determine shear and non-uniform scaling, and θ is the angle of rotation. We smooth $a(t)$ elementwise using local cubic polynomial fitting and generate the smoothed affine transformations $A_s(t)$. The corresponding transformed frames $N_V(t) = A_s(t) * N(t)$ then smoothly track the down-sampled $V(t)$, and are interpolated in t and resized to match the original video resolution and frame rate. We detect and mark anomalies by thermal thresholding of the blended video and generate a report with the robot location and blended video temporally aligned. Our cloud server delivers the report directly to the end user's mobile application.

4. Validation and Results

In this section, we present the testing results for each of the robot's sub-systems. We also validate our proposed architectures by discussing the end to end results of deploying our system on-site at an aluminum factory in the United Arab Emirates.

4.1. Objective Results for Proposed Self-Driving Network

To validate our self-driving quantitatively, we trained our proposed, VGG-19, and Resnet-18 networks on the same training data set. We first trained on 5000 images and tested on an equal number in the lab environment collected from 10 complete runs. We then used transfer learning and further optimized all networks using an additional 5000 images collected from site runs. We also added 5000 site images to our testing set. We optimized the hyper-parameters using a grid search approach for all networks. The VGG-19 regression network delivered the lowest validation root mean squared error (RMSE) at 0.17, took approximately 989 min to train, and delivered a maximum frame rate of 22 frames per second while in operation. The results for the VGG-19 networks are shown in Figure 9. We terminated the Resenet-18 training early due to repeated signs of over-fitting, which can be observed from Figure 10. Our proposed network performance is shown in Figure 11. It delivered an RMSE close to that of VGG-19 at 0.19 in 224 min of training time. Our network had a higher frame rate of 65 frames per second. All training and testing were carried out on the same single GPU computer as the central processing unit for the robot.

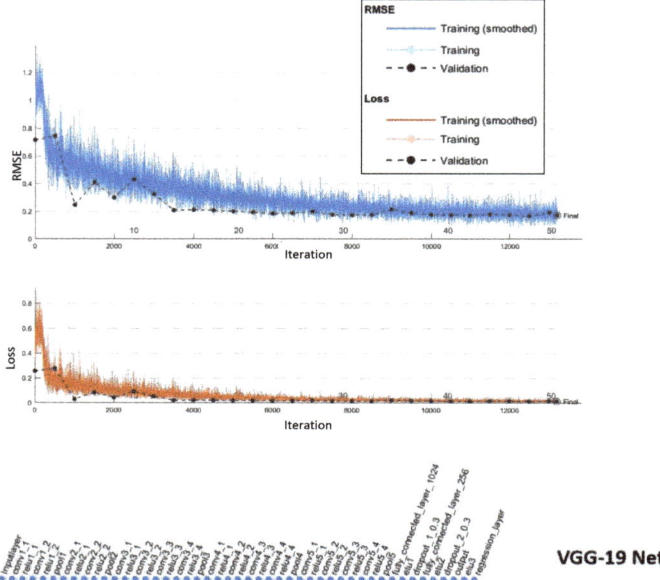

Figure 9. VGG-19: configured with a learning rate of 10^{-1}, piece-wise scheduling, and 50 Epochs. Trained in 989 min and a final validation RMSE of 0.17.

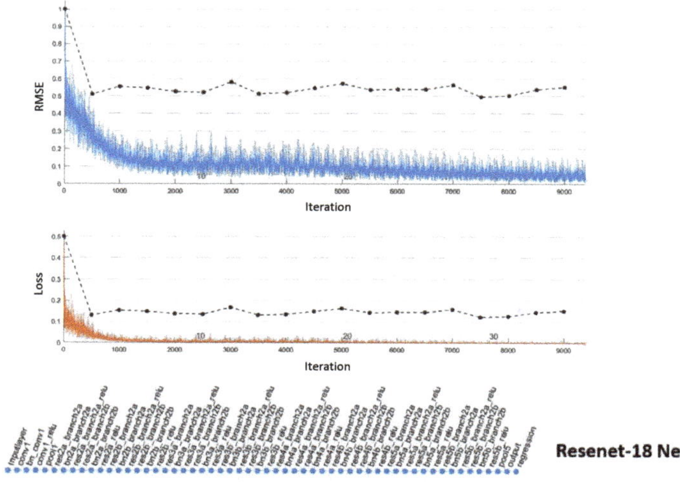

Figure 10. Resnet-18: configured with a learning rate of 10^{-1}, piece-wise scheduling, and 50 Epochs. Terminated early in 77 min and a final RMSE of 0.55 with clear signs of over-fitting.

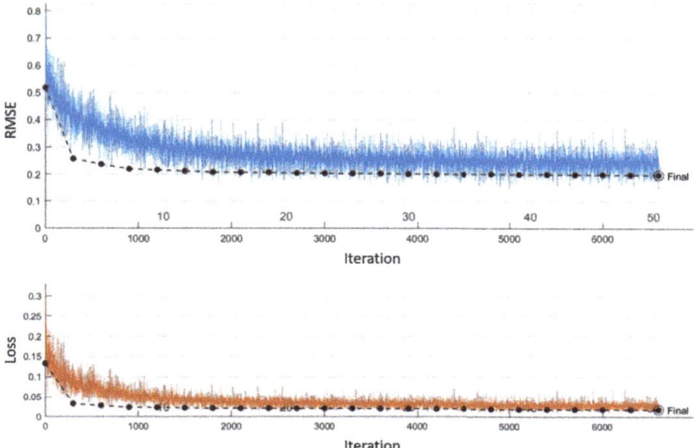

Figure 11. Proposed network: configured with a learning rate of 10^{-1}, piece-wise scheduling, and 50 Epochs. Completed 50 Epochs in 23% of the VGG-19's time and final RMSE of 0.19.

4.2. Subjective Results for Proposed Self-Driving Network

We tested our robot surveyor in both the lab and site environments and it was able to complete all 20 runs without collisions. Figure 12 shows the model of the potline's basement and depicts the path conditions and obstacles configuration. The potline consists of several parallel pots with narrow spacing and paths for operators to carry regular manual measurements. The robot path usually suffers debris from the aluminum production process, bricks, and sandstones. We visually compare the true and regressed steering angles over time including both the robot and observer viewpoints in Figures 13 and 14 for two test runs in the lab environment and Figures 15 and 16 for two test runs in the site environment, respectively. Our robot cloned the remote operator behavior and faithfully reproduced the same control signals.

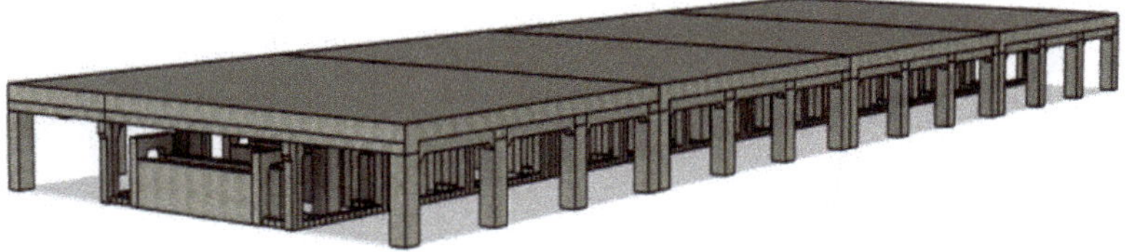

Figure 12. Model of the basement of the potline.

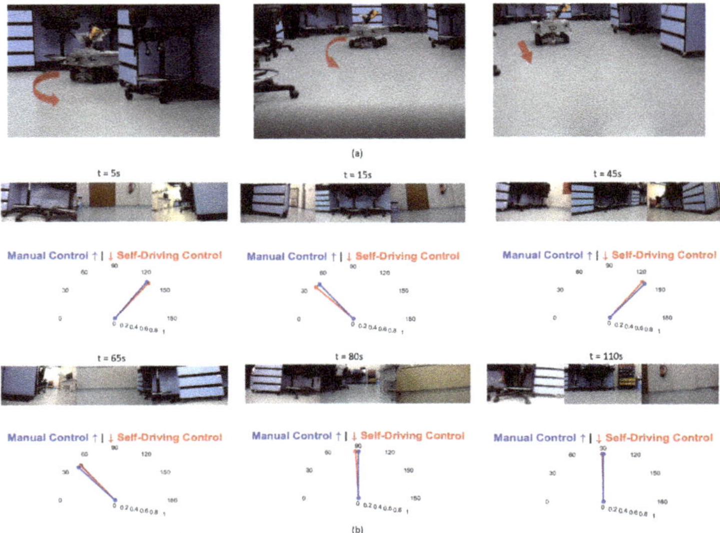

Figure 13. [Lab test run 1] Self-Driving in the lab environment results: Figures in (**a**) show the robot turning around the center bench autonomously then moving forward after turning. Figures in (**b**) represent the robot's view and the true vs. regressed steering angles.

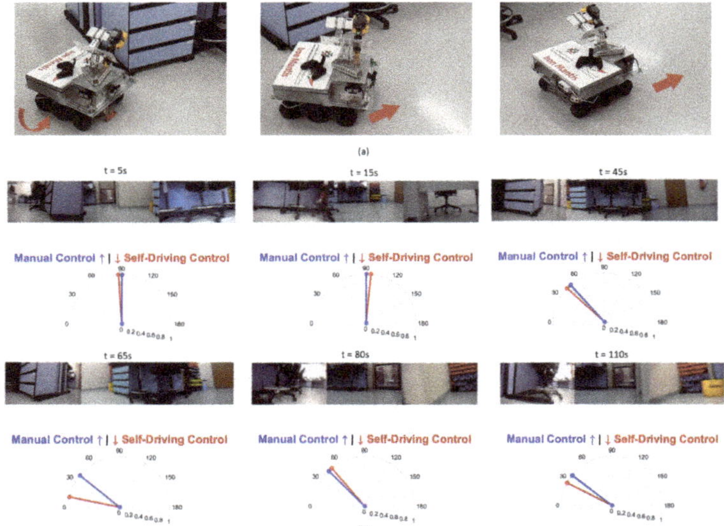

Figure 14. [Lab test run 2] Self-Driving in the lab environment results: Figures in (**a**) show the robot turning around the side bench autonomously then moving forward after turning. Figures in (**b**) represent the robot's view and the true vs. regressed steering angles.

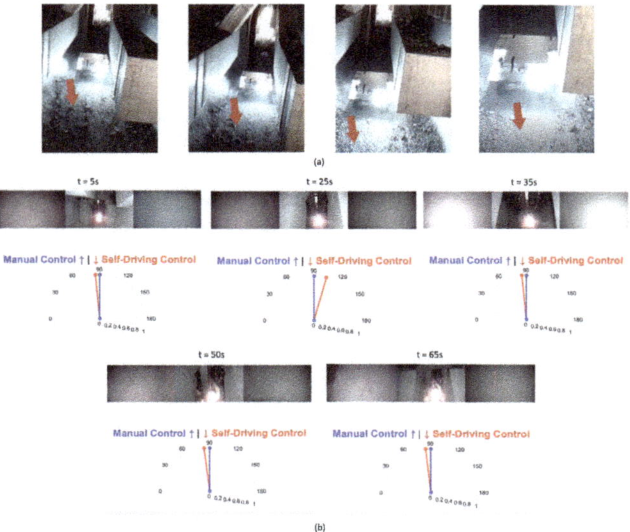

Figure 15. [Site test run 1] Self-Driving on-site results: Figures in (**a**) show the robot going into the environment autonomously. Figures in (**b**) represent the robot's view and the true vs. regressed steering angles.

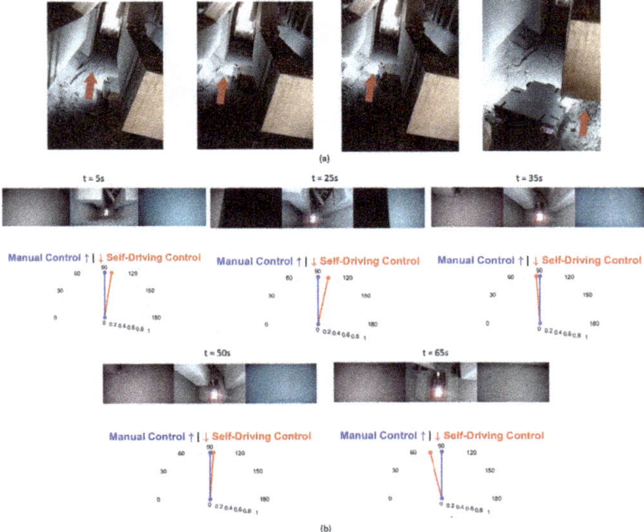

Figure 16. [Site test run 2] Self-Driving on-site results: Figures in (**a**) show the robot going out of the environment autonomously. Figures in (**b**) represent the robot's view and the true vs. regressed steering angles.

4.3. QR Detection Localization Results

We generated 5000 QR images with random location data to train our YOLO QR object detector and used 500 images to test the detector's performance using an intersection over union (IoU) threshold of 0.5. In Figure 17a, we show the histogram of confidence scores for the 500 test images with a clear bias to high confidence scores. The detector was able to recognize 90% of the QR codes in the testing set with an average precision of 0.96 and the precision vs. recall curve shown in Figure 17b. We observed that QR detection is more precise on-site than in the lab environment, which is attributed to the site's higher visual homogeneity. Figure 17c shows a sequence of QR codes detections, decoding, and use in navigation.

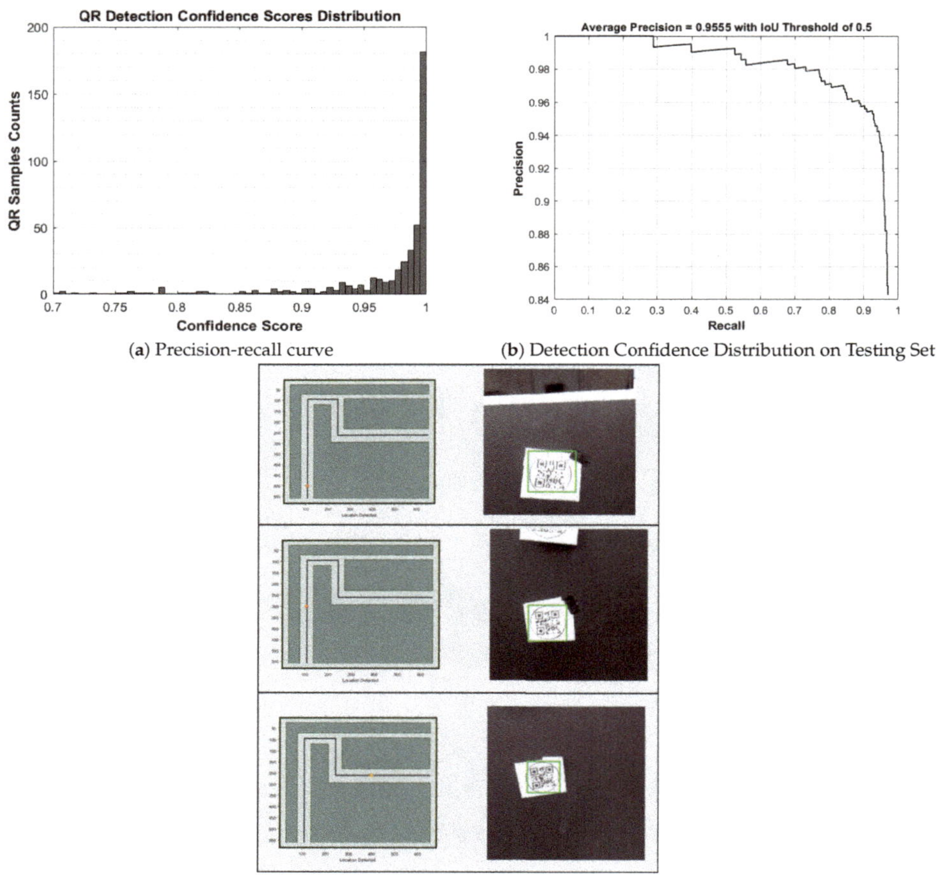

(**a**) Precision-recall curve   (**b**) Detection Confidence Distribution on Testing Set

(**c**) Localization and navigation

Figure 17. Detection performance for the Quick Response (QR) robot localization and navigation.

4.4. Obstacle Avoidance Response Time

To test the self-preservation sub-system, we designed seven testing scenarios in the lab environment. In each scenario, we measured the detection response time. Obstacles can be detected from the front middle sensor, front right sensor, front left sensor, rear middle sensor, rear left sensor, rear right sensor, and front and rear sensors at the same time. We detail three testing scenarios below:

- Scenario 1: An obstacle is detected from the front middle sensor. When the distance retrieved from the ultrasonic sensor centered at the front is small, indicating an obstacle exists in front of the robot, the robot moves to the stop state. After that, all motion is blocked except backward, as illustrated in Figure 18a.
- Scenario 2: An obstacle is detected from the front right sensor. When an obstacle exists on the front right side, the robot stops, and motion is restricted to backward Figure 18b.
- Scenario 3: Obstacles are detected from the front and rear sensors. As depicted in Figure 18c, when the front and backward sensors detect obstacles in rare cases, the robot stops, all motion commands are blocked.

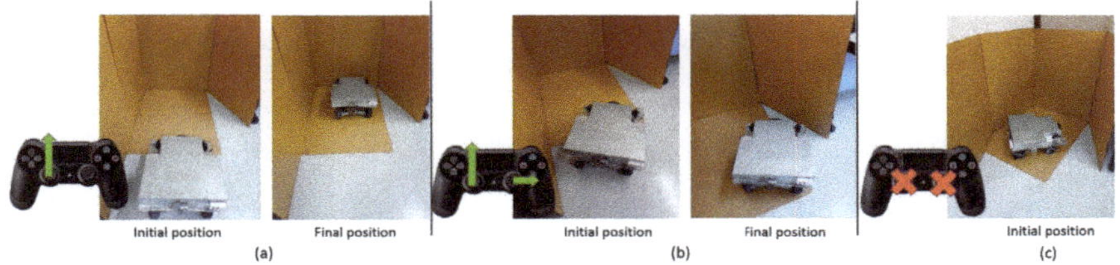

Figure 18. Obstacle avoidance algorithm testing results shows the robot stopping as an obstacle is detected by the (**a**) front middle sensor, (**b**) front right sensor, and (**c**) front and rear sensors.

The detection and response times of each scenario are tabled in Table 1. Our system can respond within 0.3 s of being close to an obstacle by 60 cm from the front and rear sides and 30 cm from the left and right sides as measured in the lab.

Table 1. Obstacles detection and response times.

Scenario	Obstacle Location	Obstacle Detection Time (s)	Obstacle Response Time (s)
Scenario 1	front center	0.00350	0.29350
Scenario 2	front right	0.00146	0.23146
Scenario 3	front left	0.00146	0.22146
Scenario 4	rear center	0.00350	0.21350
Scenario 5	rear right	0.00146	0.24146
Scenario 6	rear left	0.00146	0.26146
Scenario 7	front and rear	0.00350	0.20350

Our robot reverts to line-following when lighting conditions are too low for self-driving as demonstrated in Figure 19.

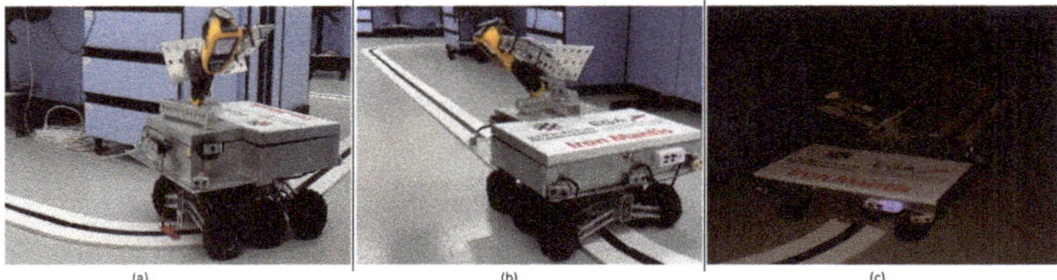

Figure 19. The Line Follower algorithm results in the lab environment: (**a**) shows the robot following the track and turning around the center bench while (**b**) shows the robot while it continues after turning and moves forward. (**c**) shows the results of testing the line follower algorithm under different lighting conditions.

4.5. Visual Results of Thermal Anomaly Segmentation and Visualization

We tested our thermal to visual registration and anomaly detection algorithm in three different lab environments and five runs in the site environment and concluded that it can accurately warp the thermal frame and align it to the visual frame despite being captured at different resolutions and frame rates. Figures 20–22 illustrate the results of testing in the labs, and Figures 23–27 on-site. In the lab setting, we heat objects to a target temperature guided by real-time measurements. We then configure our thermal detectors to detect this target temperature range. These surfaces may not be as hot as the anomaly range since it is difficult to replicate conditions leading to an anomaly, but they are hot enough to test our system. On-site, we manually measure the temperature of hot surfaces using handheld thermal cameras. In all cases, we detected and segmented the thermal anomalies. To produce sufficient testing video sequences, we lowered the thermal threshold for testing and re-calibrated the color-map to create more testing scenarios. During operations, the sensors and thresholds can be re-calibrated to meet the specific needs of the process.

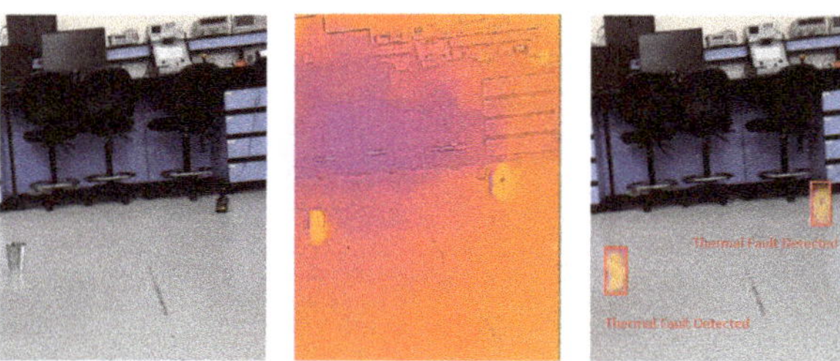

Figure 20. Thermal anomalies detection in lab environment 1.

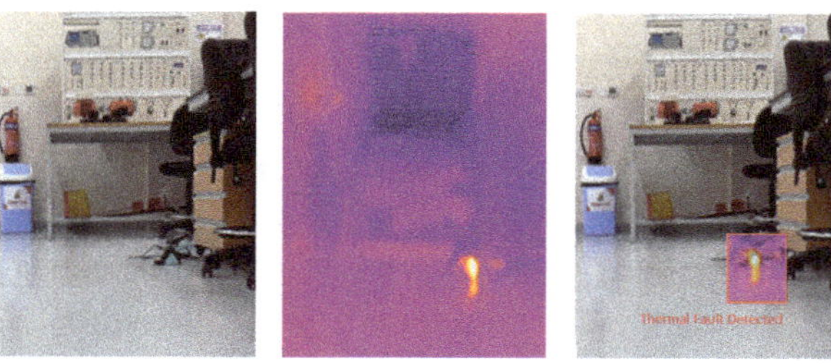

Figure 21. Thermal anomalies detection in lab environment 2.

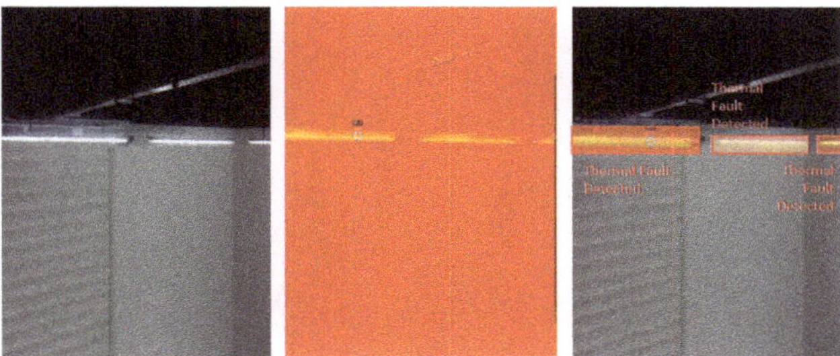

Figure 22. Thermal anomalies detection in lab environment 3.

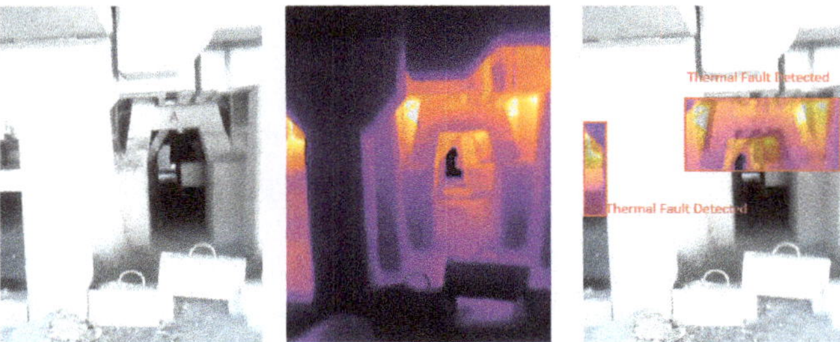

Figure 23. Thermal anomalies detection on-site for test run 1.

Figure 24. Thermal anomalies detection on-site for test run 2.

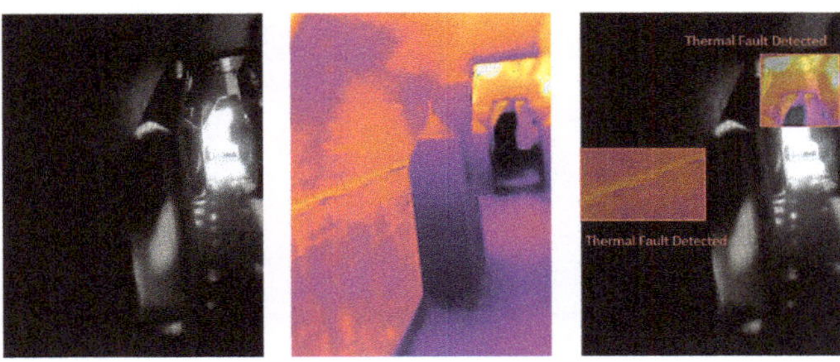

Figure 25. Thermal anomalies detection on-site for test run 3.

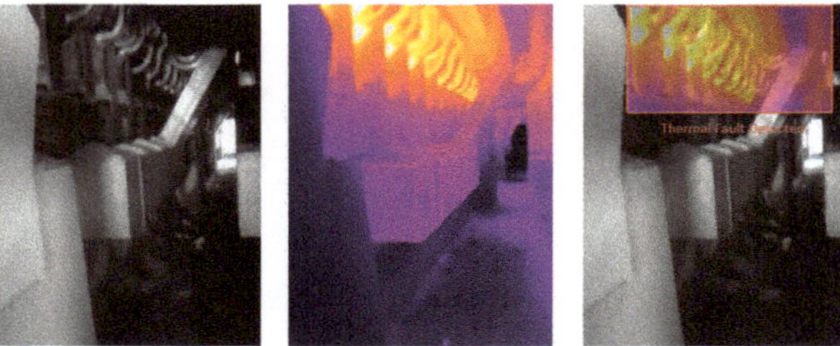

Figure 26. Thermal anomalies detection on-site for test run 4.

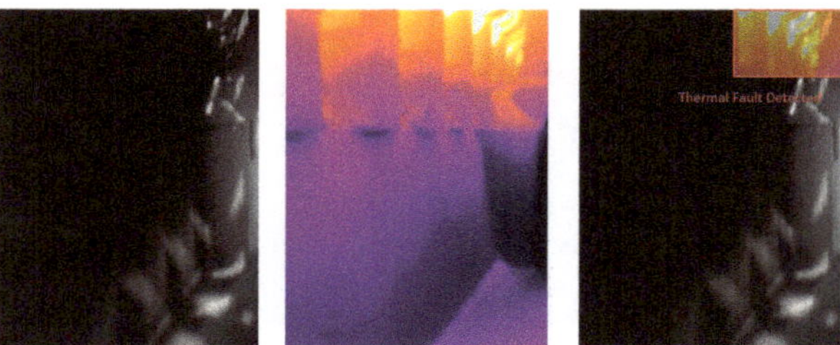

Figure 27. Thermal anomalies detection on-site for test run 5.

4.6. End to End Integration Testing and Validation

We performed end to end integration testing by placing hot objects in five different locations in the lab. Our robot surveyor completed its lab run then uploaded the thermal and visual videos and location data to the cloud server. All five hot objects were detected and localized on the lab environment map, as shown in Figure 28. Early warnings from the drone were delivered in a short response time (less than 1 s) along with location data as can be seen in Figure 28d.

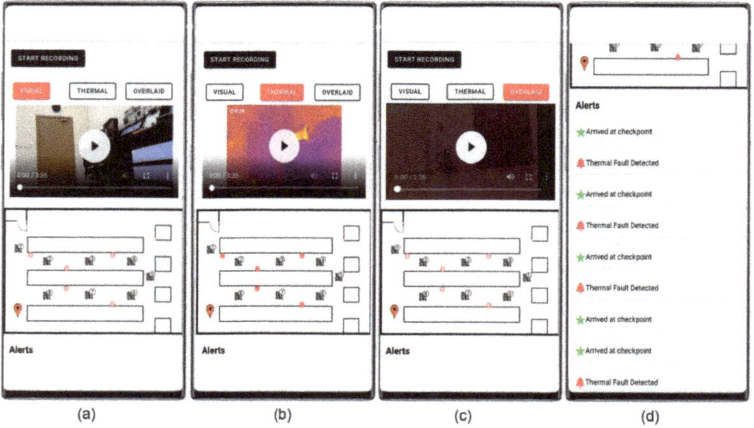

Figure 28. Mobile app displaying alerts and showing the localization and the ability to switch between the different views to be able to analyze thermal anomalies: (**a**) shows the visual view, (**b**) shows the thermal view while (**c**) shows the registered view. (**d**) shows the alerts received.

The user can switch between the visual, thermal, or blended views to analyze the thermal anomaly.

4.7. Computational Complexity and Response Time Testing

The most computationally demanding sub-system during a survey run is self-driving. We compare the number of layers, iterations until convergence, training time, validation RMSE, and frame rate for our self-driving network to VGG-19 and Resnet-19 in Table 2.

Table 2. Comparison of the different network architectures.

Network Architecture	Number of Layers	Iterations	Training Time (mins)	Validation RMSE	Frame Rate
VGG-19	47	13,200	989	0.17	22
Resenet-18	65	-	77	0.55	-
Proposed architecture	28	6600	228	0.19	65

Our proposed network outperforms VGG-19 with respect to training and testing complexities but slightly lags in terms of validation RMSE. Nevertheless, our self-driving tests confirm our proposed network's ability to regress the control signals for the robot accurately to autonomously complete all runs. Thermal to visual registration on the cloud happens offline at a frame rate of 3 frames per second.

We define alert time to be the time between the robot surveyor detecting a thermal anomaly while on a run until the initial alert with images and location information is delivered to the end user's mobile application. This performance parameter is mostly affected by the length of the run and the location of the robot at the anomaly detection time since it influences the time the messenger drone needs to find a communication network. If the communication network is available during a run, the alert time is practically insignificant for our application needs. Our average alert time in the lab environment is measured at 6.7 s. We define response time to be the time between the robot detecting an anomaly until the full run report with complete thermal anomaly visualization and location information becomes available for the end-user. Response time is affected by the length of the run and the location of the robot. Our average response time in the lab environment is measured to be 16.03 min. Both these measures are far less than the time it takes a human surveyor to detect and report the anomaly.

4.8. Initial Setup Stage for New Environments

In this Section, we discuss the initial setup needed to use our system in a different environment. We design the mechanical and electronic systems to operate under challenging conditions. The robot has an A-arm style suspension and nearly 5″ of independent wheel travel. It also has 313 rpm ball-bearing precision planetary gearmotors giving enough torque to turn 5.4″ off-road tires allowing it to overcome obstacles. The suspension incorporates 4.62″ aluminum beams and our 130 mm, oil-filled, aluminum bodied shocks. The motor driver supplies the DC motors with 25A for each group with a peak current of 50A per group for a few seconds. The shielding protects from high magnetic fields and operating temperatures. The initial setup stage requires additional training of the self-driving sub-system for improved performance. This can be accomplished by putting the robot in remote control mode and driving it in the environment for 5–10 runs. The robot automatically records the videos and controls and uploads them to the cloud. The self-driving architecture is trained on the newly collected data. The robot downloads the trained model and is ready to go. This additional training does not mean the robot trains from scratch. We still use the trained network from our lab and environment videos. The additional training serves to optimize the network to the new environment. We also re-calibrate the thermal camera and set the detectors thresholds to target the expected anomalies range, much like tuning a bandpass filter.

Finally, QR codes need to be fixed around the environment for localization. The same initial setup stage was used in the three lab environments and the on-site environment.

5. Conclusions

In this paper, we propose an edge-fog-cloud architecture for autonomous industrial monitoring of thermal anomalies. At the edge layer, robot surveyors carrying IoT nodes scan the environment's temperature in real-time for thermal anomalies in the absence of communication networks to report alerts. The robots use our proposed deep network running at 65 frames per second with a root mean squared error of 0.19, 0.02 less than the deeper VGG-19 network, and three times faster. If anomalies are detected, a mobile fog node in the form of a companion drone is dispatched looking for connectivity to deliver an early alert in 6.7 s on average. When a run is complete, the robot uploads thermal and visual videos and location information to a cloud back-end server. We use a proposed thermal to visual registration algorithm to maximize mutual derivative information and spatio-temporally align and localize thermal anomalies. End-users receive detailed reports, including the aligned video with thermal anomalies localized in 16.03 min, far sooner than the human survey time. We tested our proposed architecture in both the lab and onsite and concluded it efficiently monitors a sizeable industrial area despite its challenging characteristics. Our system's limitations are survey path segments with low lighting conditions affecting self-driving and the lack of connectivity in the environment, which we have addressed by shifting autonomous driving to rely on line-following in these segments and adding a companion drone to deliver alerts in a short time. Further research is needed to address these limitations better.

Author Contributions: Funding acquisition, M.M.; Investigation, M.G., T.B., M.Y., M.A., and A.S.E.-B.; methodology, M.G., T.B., M.Y., M.A., and A.S.E.-B.; project administration, M.G.; software, M.G., A.S.E.-B., T.B., and M.Y.; supervision, M.G., M.A., and A.S.E.-B.; Validation, M.G., T.B., M.Y., M.A., and A.S.E.-B.; writing—original draft, T.B., M.Y., and M.G.; writing—review and editing, M.G., M.A., A.S.E.-B., and M.M. All authors have read and agreed to the published version of the manuscript.

Funding: This research was funded by Emirates Global Aluminium, UAE.

Acknowledgments: This work is supported by Emirates Global Aluminum. Moreover, the authors thank Abdalla Rashed and Razib Sarker for roles in the implementation and testing.

Conflicts of Interest: The authors declare no conflict of interest.

References

1. Bekey, G. *Autonomous Robots*; Bradford Books: Cambridge, UK, 2017.
2. Murphy, A. Industrial: Robotics Outlook 2025. *Loup Ventures.* 6 July 2018. Available online: https://loupventures.com/industrial-robotics-outlook-2025/ (accessed on 7 September 2020).
3. Rembold, U.; Lueth, T.; Ogasawara, T. From autonomous assembly robots to service robots for factories. In Proceedings of the IEEE/RSJ International Conference on Intelligent Robots and Systems (IROS'94), Munich, Germany, 12–16 September 1994; Volume 3, pp. 2160–2167. [CrossRef]
4. Yin, S.; Rodriguez-Andina, J.J.; Jiang, Y. Real-Time Monitoring and Control of Industrial Cyberphysical Systems: With Integrated Plant-Wide Monitoring and Control Framework. *IEEE Ind. Electron. Mag.* **2019**, *13*, 38–47. [CrossRef]
5. Tas, M.O.; Yavuz, H.S.; Yazici, A. Updating HD-Maps for Autonomous Transfer Vehicles in Smart Factories. In Proceedings of the 6th International Conference on Control Engineering & Information Technology (CEIT), Istanbul, Turkey, 25–27 October 2018; pp. 1–5. [CrossRef]
6. Saeed, M.S.; Alim, N. Design and Implementation of a Dual Mode Autonomous Gas Leakage Detecting Robot. In Proceedings of the International Conference on Robotics, Electrical and Signal Processing Techniques (ICREST), Dhaka, Bangladesh, 10–12 January 2019; pp. 79–84. [CrossRef]

7. Rey, R.; Corzetto, M.; Cobano, J.A.; Merino, L.; Caballero, F. Human-robot co-working system for warehouse automation. In Proceedings of the 24th IEEE International Conference on Emerging Technologies and Factory Automation (ETFA), Zaragoza, Spain, 10–13 September 2019; pp. 578–585. [CrossRef]

8. Montano, L. Robots in challenging environments. In Proceedings of the 24th IEEE International Conference on Emerging Technologies and Factory Automation (ETFA), Zaragoza, Spain, 10–13 September 2019; pp. 23–26. [CrossRef]

9. Teja, P.R.; Kumaar, A.A.N. QR Code based Path Planning for Warehouse Management Robot. In Proceedings of the International Conference on Advances in Computing, Communications and Informatics (ICACCI), Bangalore, India, 19–22 September 2018; pp. 1239–1244. [CrossRef]

10. Chie, L.C.; Juin, Y.W. Artificial Landmark-based Indoor Navigation System for an Autonomous Unmanned Aerial Vehicle. In Proceedings of the IEEE 7th International Conference on Industrial Engineering and Applications (ICIEA), Bangkok, Thailand, 16–21 April 2020; pp. 756–760. [CrossRef]

11. Limeira, M.A.; Piardi, L.; Kalempa, V.C.; de Oliveira, A.S.; Leitão, P. WsBot: A Tiny, Low-Cost Swarm Robot for Experimentation on Industry 4.0. In Proceedings of the 2019 Latin American Robotics Symposium (LARS), 2019 Brazilian Symposium on Robotics (SBR) and 2019 Workshop on Robotics in Education (WRE), Rio Grande, Brazil, 23–25 October 2019; pp. 293–298. [CrossRef]

12. Ciuccarelli, L.; Freddi, A.; Longhi, S.; Monteriu, A.; Ortenzi, D.; Pagnotta, D.P. Cooperative Robots Architecture for an Assistive Scenario. In Proceedings of the Zooming Innovation in Consumer Technologies Conference (ZINC), Novi Sad, Serbia, 30–31 May 2018; pp. 128–129. [CrossRef]

13. Avola, D.; Foresti, G.L.; Cinque, L.; Massaroni, C.; Vitale, G.; Lombardi, L. A multipurpose autonomous robot for target recognition in unknown environments. In Proceedings of the IEEE 14th International Conference on Industrial Informatics (INDIN), Poitiers, France, 19–21 May 2016; pp. 766–771. [CrossRef]

14. Merriaux, P.; Rossie, R.; Boutteau, R.; Vauchey, V.; Qin, L.; Chanuc, P.; Rigaud, F.; Roger, F.; Decoux, B.; Savatier, X. The VIKINGS Autonomous Inspection Robot: Competing in the ARGOS Challenge. *IEEE Robot. Autom. Mag.* **2019**, *26*, 21–34. [CrossRef]

15. Dharmasena, T.; Abeygunawardhana, P. Design and Implementation of an Autonomous Indoor Surveillance Robot based on Raspberry Pi. In Proceedings of the International Conference on Advancements in Computing (ICAC), Malabe, Sri Lanka, 5–7 December 2019; pp. 244–248. [CrossRef]

16. Mogaveera, A.; Giri, R.; Mahadik, M.; Patil, A. Self Driving Robot using Neural Network. In Proceedings of the International Conference on Information, Communication, Engineering and Technology (ICICET), Pune, India, 29–31 August 2018; pp. 1–6. [CrossRef]

17. Omrane, H.; Masmoudi, M.S.; Masmoudi, M. Neural controller of autonomous driving mobile robot by an embedded camera. In Proceedings of the 4th International Conference on Advanced Technologies for Signal and Image Processing (ATSIP), Sousse, Tunisia, 21–24 March 2018; pp. 1–5. [CrossRef]

18. Ebuchi, T.; Yamamoto, H. Vehicle/Pedestrian Localization System Using Multiple Radio Beacons and Machine Learning for Smart Parking. In Proceedings of the International Conference on Artificial Intelligence in Information and Communication (ICAIIC), Okinawa, Japan, 11–13 February 2019; pp. 86–91. [CrossRef]

19. Ullah, A.; Muhammad, K.; del Ser, J.; Baik, S.W.; de Albuquerque, V.H.C. Activity Recognition Using Temporal Optical Flow Convolutional Features and Multilayer LSTM. *IEEE Trans. Ind. Electron.* **2019**, *66*, 9692–9702. [CrossRef]

20. Ullah, W.; Ullah, A.; Haq, I.U.; Muhammad, K.; Sajjad, M.; Baik, S.W. CNN features with bi-directional LSTM for real-time anomaly detection in surveillance networks. *Multimed. Tools Appl.* **2020**. [CrossRef]

21. Yin, S.; Yang, C.; Zhang, J.; Jiang, Y. A Data-Driven Learning Approach for Nonlinear Process Monitoring Based on Available Sensing Measurements. *IEEE Trans. Ind. Electron.* **2017**, *64*, 643–653. [CrossRef]

22. Zhou, D.; Wei, M.; Si, X. A Survey on Anomaly Detection, Life Prediction and Maintenance Decision for Industrial Processes. *Acta Autom. Sin.* **2014**, *39*, 711–722. [CrossRef]

23. Kammerer, K.; Hoppenstedt, B.; Pryss, R.; Stökler, S.; Allgaier, J.; Reichert, M. Anomaly Detections for Manufacturing Systems Based on Sensor Data—Insights into Two Challenging Real-World Production Settings. *Sensors* **2019**, *19*, 5370. [CrossRef] [PubMed]

24. Pittino, F.; Puggl, M.; Moldaschl, T.; Hirschl, C. Automatic Anomaly Detection on In-Production Manufacturing Machines Using Statistical Learning Methods. *Sensors* **2020**, *20*, 2344. [CrossRef] [PubMed]
25. Kroll, B.; Schaffranek, D.; Schriegel, S.; Niggemann, O. System modeling based on machine learning for anomaly detection and predictive maintenance in industrial plants. In Proceedings of the IEEE Emerging Technology and Factory Automation (ETFA), Barcelona, Spain, 16–19 September 2014; pp. 1–7. [CrossRef]
26. Kubota, T.; Yamamoto, W. Anomaly Detection from Online Monitoring of System Operations Using Recurrent Neural Network. *Procedia Manuf.* **2019**, *30*, 83–89. [CrossRef]
27. Arkhipov, A.; Zarouni, A.; Baggash, M.J.I.; Akhmetov, S.; Reverdy, M.; Potocnik, V. *Cell Electrical Preheating Practices at Dubal*; Hyland, M., Ed.; Light Metals 2015; Springer: Cham, Switzerland, 2016; pp. 445–449. [CrossRef]
28. Majid, N.A.; Taylor, M.; Chen, J.J.; Young, B.R. Aluminium Process Fault Detection and Diagnosis. *Adv. Mater. Sci. Eng.* **2015**, *2015*, 682786. [CrossRef]
29. Gang, H.; Pyun, J. A Smartphone Indoor Positioning System Using Hybrid Localization Technology. *Energies* **2019**, *12*, 3702. [CrossRef]
30. Styner, M.; Brechbuhler, C.; Szckely, G.; Gerig, G. Parametric estimate of intensity inhomogeneities applied to MRI. *IEEE Trans. Med. Imaging* **2000**, *19*, 153–165. [CrossRef] [PubMed]

Publisher's Note: MDPI stays neutral with regard to jurisdictional claims in published maps and institutional affiliations.

Article

Smart Helmet 5.0 for Industrial Internet of Things Using Artificial Intelligence

Israel Campero-Jurado [1,†], Sergio Márquez-Sánchez [2,*,†], Juan Quintanar-Gómez [3], Sara Rodríguez [2] and Juan M. Corchado [2,4,5,6]

[1] Laboratoire de l'Informatique du Parallélisme 46 allée d'Italie, 69007 Lyon, France; israel.campero_jurado@ens-lyon.fr
[2] BISITE Research Group, University of Salamanca, Calle Espejo s/n. Edificio Multiusos I+D+i, 37007 Salamanca, Spain; srg@usal.es (S.R.); corchado@usal.es (J.M.C.)
[3] Graduate School in Information Technology and Communications Research Department, Universidad Politécnica de Pachuca, Zempoala Hidalgo 43830, Mexico; juanquintanargomez@micorreo.upp.edu.mx
[4] Air Institute, IoT Digital Innovation Hub (Spain), 37188 Salamanca, Spain
[5] Department of Electronics, Information and Communication, Faculty of Engineering, Osaka Institute of Technology, Osaka 535-8585, Japan
[6] Faculty of Creative Technology & Heritage, Universiti Malaysia Kelantan, Locked Bag 01, 16300 Bachok, Malaysia
* Correspondence: smarquez@usal.es; Tel.: +34-685-043-554
† These authors contributed equally to this work.

Received: 10 September 2020; Accepted: 27 October 2020; Published: 1 November 2020

Abstract: Information and communication technologies (ICTs) have contributed to advances in Occupational Health and Safety, improving the security of workers. The use of Personal Protective Equipment (PPE) based on ICTs reduces the risk of accidents in the workplace, thanks to the capacity of the equipment to make decisions on the basis of environmental factors. Paradigms such as the Industrial Internet of Things (IIoT) and Artificial Intelligence (AI) make it possible to generate PPE models feasibly and create devices with more advanced characteristics such as monitoring, sensing the environment and risk detection between others. The working environment is monitored continuously by these models and they notify the employees and their supervisors of any anomalies and threats. This paper presents a smart helmet prototype that monitors the conditions in the workers' environment and performs a near real-time evaluation of risks. The data collected by sensors is sent to an AI-driven platform for analysis. The training dataset consisted of 11,755 samples and 12 different scenarios. As part of this research, a comparative study of the state-of-the-art models of supervised learning is carried out. Moreover, the use of a Deep Convolutional Neural Network (ConvNet/CNN) is proposed for the detection of possible occupational risks. The data are processed to make them suitable for the CNN and the results are compared against a Static Neural Network (NN), Naive Bayes Classifier (NB) and Support Vector Machine (SVM), where the CNN had an accuracy of 92.05% in cross-validation.

Keywords: PPE; OHS; risk detection; naive Bayes; support vector machine; convolutional neural network; deep learning; microcontroller

---

1. Introduction

Industrial security is achieved when adequate measures and procedures are applied to obtain access to, handle or generate classified information during the execution of a classified contract or program. Industrial safety is the set of rules and activities aimed at preventing and limiting the potential risks associated with an industry, including both transient and permanent risks [1,2].

Many safety protocols have been proposed to improve the quality of life of workers using different techniques [3,4]. Several studies have examined how the availability of artificial intelligence (AI) techniques could affect the industrial organization of both AI service providers and industries adopting AI technology [5]. Above all, the impact of AI on industry 4.0 and its possible applications in other fields have been studied in depth [6].

In recent years, research has also been conducted on the applications of AI in the manufacturing industry [7–11]. The system architecture described in the article integrates technology together with communication systems and permits analyzing intelligent manufacturing. The provided information shows an overview of the possible applications of AI in all industrial areas.

AI allows to maximize decision making in simple or very complex situations. The AI boom that has taken place in the last decades has led to the development of countless AI applications in numerous areas. At present, increasingly better solutions are available to protect the lives of workers when they are exposed to high-risk conditions. That is why, in industry, AI is combined with security measures in order to create an environment that offers better conditions for industrial development.

The objective of the proposed device is to improve occupational health and safety (OHS); increasing employee performance by reducing the probability of illness, injury, absence or death [12]. Another objective is to contribute to the third wave, as proposed by Niu et al. [13,14], through the implementation of intelligent systems for early risk detection in the working environment.

Different studies have been conducted in creation of devices for occupational safety and health (OSH), which indicate the need to implement increasingly innovative solutions for workers in high-risk areas. For example, in 2014 [15] a study was conducted among 209 welders in India and it was found that all of them had more than 2 injuries and 44% (92) of them had more than 10 injuries. Furthermore, in 2020 [16] an analysis of workplace-related injuries in major industries such as agriculture, construction, manufacturing and health care has been carried out. The data for this analysis have been obtained from a Bureau of Labor Statistics and it was found that from 1992 to 2018, the number amounted to 4,471,340 injuries in the upper extremities, 3,296,547 in the lower extremities, and 5,889,940 in the trunk ($p < 0.05$). Therefore, the motivation behind this research is to propose an innovative helmet with different sensors such as temperature, humidity and atmospheric pressure, the force exerted between the helmet and the head of the user, the variations in axes, air quality and luminosity, through specialized IoT modules being able to have a faster reaction time to an accident in a work team. All the research papers that address the problem of occupational safety and health (OSH) are summarised in Table 1 with the purpose of comparing the improvements and advantages of similar research.

The information coming from the sensors is analyzed through a platform known as ThingsBoard. Independent alarms are configured using this information. Likewise, the data coming from the sensors are adapted to classify them in a Convolutional Neural Network, whose accuracy is of 92.05% in cross-validation compared to 3 other supervised learning models.

The remaining part of this work is organized as follows: Section 2 gives an overview of the related literature. Section 3 describes the system design. A multisensory helmet with communication in IIoT and AI-based information analysis is presented in Section 4. Finally, the last section describes future lines of research.

Table 1. OSH-related proposals.

Bibliography	Keywords	Novelty of the Proposal
Vaughn Jr, Rayford B. et al. (2002)	Security-engineering, Risk assessment	The state of security-engineering practices by three information security practitioners with different perspectives.
Choudhry, R. M., and Fang, D. (2008)		This work discusses empirical research aimed at why construction workers engage in unsafe behavior.
Niu, Yuhan, et al. (2019)		This research seeks to develop a smart construction object enabled OHS management system.
Champoux, D., and Brun, J. P. (2003)	Occupational health and safety (OHS), Construction safety, Artificial Inteligence (AI)	This exploratory study based on telephone interviews with the owner-managers of small manufacturing enterprises gives an overview of the most characteristic OHS representations and practices in small firms.
Podgorski, Daniel, et al. (2017)		A proposed framework based on a new paradigm of OSH risk management consisting of real-time risk assessment and risk level detection of every worker individually.
Barata, Joao et al. (2019)		Viable System Model (VSM) to design smart products that adhere to the organization strategy in disruptive transformations
Sun, Shengjing, et al. (2020)		A unified architecture to support the integration of different enabling technologies
Hasle, P., and Limborg, H. J. (2006)	Occupational health and safety, Accident Prevention	The scientific literature regarding preventive occupational health and safety activities in small enterprise.
Hasle, P., et al. (2011)	Occupational health and safety, Accident prevention	The investigation applied qualitative methods and theoretical approaches to CSR, small and medium-sized enterprises (SMEs), and occupational health and safety.
Abdelhamid, T. S., Everett, J. G. (2000)	Occupational safety, Construction safety, Accidents prevention	Accident root causes tracing model (ARCTM) tailored to the needs of the construction industry.
Chi, S., Han, S. (2013)		This study incorporates the systems theory into Heinrich's domino theory to explore the interrelationships of risks and break the chain of accident causation.
Cambraia, F. B., et al. (2010)	Incident reporting systems, Safety management	Guidelines for identifying, analyzing and disseminating information on near misses in construction sites.
Chevalier, Yannick, et al. (2004)	Network security, Cryptographic protocols	High level protocol specification language for the modelling of security-sensitive cryptographic protocols.

2. Related Works

Protective equipment is of obligatory use in cases where the safety of the worker is at risk. However, detecting hazardous situations in a timely manner is not always possible, leading to the occurrence of accidents. Such events call the worker's health and safety into question; the confidence of the worker in the company for which they work decreases [17–19]. For effective prevention of injuries or fatal accidents in the working environment, the integration of electronic components is crucial given their ability for early risk detection. The research of Henley, E.J. and Kumamoto, H [20] proposed a quantitative approach for the optimal design of safety systems which focused on information links (human and computer), sensors, and control systems. In 2003, Condition Monitoring (CM) was addressed in the research of Y. Han and Y. H. Song [21] including a review of popular CM methods, as well as the research status of CM transformer, generator, and induction motor, respectively. In December 2001, the factor structure of a safety climate within a road construction organization was determined by A.I Glendon and D.K Litherland [22]; a modified version of the safety climate questionnaire (SCQ). They also investigated the relationship between safety climate and safety performance. In March 2011, Intelligent Internet of Things for Equipment Maintenance (IITEM) was presented by Xu Xiaoli et al. [23]. The static and dynamic information on electrical and mechanical equipment is collected by IITEM from all kinds of sensors, and the different types of information are standardized, facilitating Internet of Things information transmission [24,25]. The investigations that address motion monitoring and sensor networks have been compiled in Table 2.

Table 2. Proposals related to sensor networks.

Bibliography	Keywords	Novelty of the Proposal
Zhou, Yinghui, et al. (2012)	Internet of Things, Wearable Computing, Robot sensing systems, Acceleration Feature analysis Human-computer interaction	Wearable device based on a tri-axis accelerometer, which can detect acceleration change of human body based on the position of the device being set.
Zhu, C., and Sheng, W. (2009)		A human daily activity recognition method by fusing the data from two wearable inertial sensors attached on one foot and the waist of the subject.
Lindeman, Robert W., et al. (2006)		A development history of a wearable, scalable vibrotactile stimulus delivery system.
Kim, Sung Hun, et al. (2018)		Experiments were performed in which the sensing data were classified whether the safety helmet was being worn properly, not worn, or worn improperly during construction workers' activities.
Nithya, T., et al. (2018)	Head motion recognition, Hazardous gas, Temperature measurement, Sensor System, IMU, Electroencephealography (EEG)	Smart helmet able to detect hazardous events in the mining industry and design a mine safety system using wireless sensor networks.
Li, Ping, et al. (2014)		Smart Safety Helmet (SSH) in order to tack the head gestures and the brain activity of the worker to recognize anomalous behavior.
Fang, Y., et al. (2016)	Crane safety, Human errorReal-time, Crane motion capturing	A prototype system was developed based on the framework and deployed on a mobile crane.
Cao, Teng, et al. (2014)	Steady-state visual evoked potential (SSCEP), Brain-computer interfaces (BCIs)	Propose a method for the real-time evaluation of fatigue in SSVEP-based BCIs.

Moreover, an Accident Root Causes Tracing Model (ARCTM), tailored to the needs of the construction industry, has been presented by Tariq S. Abdelhamid and John G. Everett [26]. In January 2010, guidelines for identifying, analyzing and disseminating information on near misses at construction sites were defined by Fabricio BorgesCambraia et al. [27]. In September 2013, three case studies were presented by Tao Cheng and Jochen Teizer [28] which employed methods for recording data and visualizing information on construction activities at a (1) simulated virtual construction site, (2) outdoor construction setting, and (3) worker training environment. Furthermore, systems theory has been incorporated in Heinrich's domino theory by Seokho Chia and Sangwon Han [29] to explore the interrelationships between risks and to break the chain of accident causation. In April 2008, the reasons for which construction workers engage in unsafe behavior were discussed in the empirical research of Rafiq M. Choudhry and Dongping Fang [30]. Interviews were conducted in Hong Kong with workers who had been accident victims. In addition, Daniel Fitton et al. [31] applied augmented technology with sensing and communication technologies which can measure use in order to enable new pay-per-use payment models for equipment hire. The areas in which it is necessary to create a safer working environment are listed in Table 3. This can be achieved through the use of sensors for the monitoring environmental parameters and capturing motion.

In December 2008, the underlying biomechanical elements required to understand and study human movement were identified by A. Godfrey et al. [32]. A method for investigating the kinematics and dynamics of locomotion without any laboratory-related limitations has been developed by Yasuaki Ohtaki et al. [33]. In April 2012, the usage of the Unscented Kalman Filter (UKF) as the integration algorithm for the inertial measurements was proposed by Francisco Zampella et al. [34]. Furthermore, in 2012, a micro wearable device based on a tri-axis accelerometer was introduced by Yinghui Zhou et al. [35]. It can detect change in the acceleration of the human body on the basis of the position of the device. In 2009, a method for the recognition of daily human activities was developed by Chun Zhu and Weihua Sheng [36]. This method involved fusing the data from two wearable inertial sensors attached to the foot and the waist of the subject. In October 2012, Martin J.-D. Otis and Bob-Antoine J. Menelas [37] reported an ongoing project whose objective was to create intelligent clothes for fall prevention in the work environment. In 2007, a signal transform method, called Common Spatial

Pattern, was introduced by Hong Yu et al. [38] for Electroencephalographic (EEG) data processing. In March 2006, the development history of a wearable, called the scalable vibrotactile stimulus delivery system, was presented by Robert W. Lindeman et al. [39]. In 2014, an objective and real-time approach based on EEG spectral analysis for the evaluation of fatigue in SSVEP-based BCIs was proposed by Teng Cao et al. [40].

Table 3. Proposals related to safety environment and motion recognition.

Bibliography	Keywords	Novelty of the Proposal
Fernández-Muñiz, B., et al. (2012)	Safety climate, Employee perceptions, Safety performance	The current work aims to analyse the safety climate in diverse sectors, identify its dimensions, and propose to test a structural equation model that will help determine the antecedents and consequences of employees' safety behaviour.
Glendon, A. I., Litherland, D. K. (2001)		A behavioral observation measure of safety performance and a road construction organization using a modified version of the safety climate questionnaire (SCQ).
Han, Y., and Song, Y. H. (2003)	IMU, Magnetometers, Gyroscopes, Accelerometer, Human motion	After introducing the concepts and functions of CM, this paper describes the popular monitoring methods and research status of CM on transformer, generator, and induction motor, respectively.
Godfrey, A. C. R. M. D. O. G., et al. (2008)		The underlying biomechanical elements necessary to understand and study human movement.
Ohtaki, Y., et al. (2001)		A new method is proposed to investigate kinematics and dynamics of locomotion without any limitation of laboratory conditions.
Zampella, Francisco, et al. (2012)		The usage of the Unscented Kalman Filter (UKF) as the integration algorithm for the inertial measurements.
Cheng, T., and Teizer, J. (2013)	Body sensor network (BSN), Vision Algorithms, Augmented reality (AR) Virtual Reality (VR), Location tracking,	A novel framework is presented that explains the method of streaming data from real-time positioning sensors to a real-time data visualization platform.
Bleser, Gabriele, et al. (2015)		Assistance system based in the last advances in hardware, software and system level.
Fitton, Daniel, et al. (2008)		Investigation into how physical objects augmented with sensing and communication technologies can measure use in order to enable new pay-per-use payment models for equipment hire.
Yu, H., et al. (2007)	Measuring vigilance, Sensor network, Intelligent sensors	Signal transform method, Common Spatial Pattern, to process the EEG data.
Qiang, Cheng, et al. (2009)		A cost effective ZigBee-based wireless mine supervising system

Thanks to the implementation of communication technologies, it is possible to notify both the managing staff and the workers about the hazards encountered in a particular working area. A helmet that implements Zigbee transmission technologies for the analysis of variables such as humidity, temperature and methane in mines has been developed by Qiang et al. (2009) [41]. This helmet helps decrease the risk of suffering an accident during the coal extraction process. An intelligent helmet for the detection of anomalies in mining environments was also proposed by Nithya et al. (2018) [42]. This research points to the possibility of integrating components in the PPE that would alert the worker of the presence of danger. Moreover, the vital signs of the worker are monitored by their helmet, making it possible to monitor their state of health. An emergency button on the helmet is used for the transmission of alerts via Zigbee technologies to the personnel nearest to the working environment. Accelerometers have been integrated in safety helmets by Kim et al. (2018) [43], with the purpose of detecting if the safety helmet is being worn properly, improperly or not worn at all while the worker performs their tasks. In December 2016, a framework for real-time pro-active safety assistance was developed by Yihai Fang et al. [44] for mobile crane lifting operations.

Ensuring the physical well-being of workers is the responsibility of employers. Better protection is offered to today's workers thanks to PPE helmets by protecting the worker from blows to the head. However, monitoring other aspects for the worker's security is important in some cases. Li et al. (2014) [45] developed a helmet which, by means of sensors, measures the impact of blows to the worker's head. Sensors for brain activity detection are also implemented in the helmet. In terms of movement, identifying the position of the worker is essential in order to detect falls that result in physical injury or fatal accidents.

In 2019, Machine Learning (ML) algorithms for the prediction and classification of motorcycle crash severity were employed in a research by Wahab, L., and Jiang, H. [46]. Machine-learning-based techniques are non-parametric models without any presumption of the relationships between endogenous and exogenous variables. Another objective of this paper was to evaluate and compare different approaches to modeling motorcycle crash severity as well as investigating the risk factors involved and the effects of motorcycle crashes. In 2015, a scalable concept and an integrated system demonstrator was designed by Bleser, G. et al. [47]. The basic idea is to learn workflows from observing multiple expert operators and then transferring the learned workflow models to demonstrate the severity of motorcycle crashes. In 2019, an intelligent video surveillance system which detected motorcycles automatically was developed by Yogameena, B., Menaka, K., and Perumaal, S. S. [48]. Its purpose was to identify whether motorcyclists were wearing safety helmets or not. If the motorcyclists were found without the helmet, their License Plate (LP) number was recognised and legal action was taken against them by the traffic police and the legal authority, such as assigning penalty points on the motorcyclists' vehicle license and Aadhar Number (Applicable to Indian Scenario). In 2017, a comparison of four statistical and ML methods was presented by Iranitalab, A., and Khattak [49], including Multinomial Logit (MNL), Nearest Neighbor Classification (NNC), Support Vector Machines (SVM) and Random Forests (RF), in relation to their ability to predict traffic crash severity. A crash costs-based approach was developed to compare crash severity prediction methods, and to investigate the effects of data clustering methods—K-means Clustering (KC) and Latent Class Clustering (LCC)—on the performance of crash severity prediction models. These novel proposals are compiled in Table 4. They employ artificial intelligence and machine learning, and suppose a significant improvement in different scenarios.

In 2005, the results obtained with the random forest classifier were presented in the research of M. Pal [50] and its performance was compared with that of the support vector machines (SVMs) in terms of classification accuracy, training time and user defined parameters. In January 2012, the performance of the RF classifier for land cover classification of a complex area was explored by V. F. Rodriguez-Galiano et al. [51]; the evaluation was based on several criteria: mapping accuracy, sensitivity to data set size and noise. Furthermore, in February 2014, a random forest classifier (RF) approach was proposed by Ahmad Taher Aza et al. [52] for the diagnosis of lymph diseases. In April 2016, the use of the RF classifier in remote sensing was reviewed by Mariana Belgiua and Lucian Drăguţ [53]. Besides, in 2015, machine learning approaches including k-nearest neighbor (k-NN), a rules-based classifier (JRip), and random forest, were investigated by Esrafil Jedari et al. [54] to estimate the indoor location of a user or an object using RSSI based fingerprinting method. Finally, in July 2011, a method utilizing Healthcare Cost and Utilization Project (HCUP) dataset was presented by Mohammed Khalilia et al. [55] for predicting disease risk in individuals on the basis of their medical history.

With regard to CNN in 2020, an automated system for the identification of motorcyclists without helmets from real-time traffic surveillance videos was presented by Shine L. and Jiji, C. V. [56]. A two-stage sorter was used to detect motorcycles in surveillance videos. The detected motorcycles were fed in a helmet identification stage based on a CNN. Moreover, in July 2019, the same approach to detecting the absence of helmets on motorcyclists with or without helmets was presented by Yogameena B. et al. [48]; it was different in that it combined a CNN with a Gaussian Mixture Model (GMM) [57]. Furthermore, in 2020, a system that uses image processing and CNN networks was developed by Raj K. C. et al. [58] for the identification of the motorcyclists who violate helmet laws.

The system includes motorcycle detection, helmet vs. helmetless classification and motorcycle license plate recognition. As can be observed, CNNs have been used mainly for real-time image processing. However, the use of CNN for linear data evaluation is proposed in this paper. Here, CNN is integrated (input–output) in a rules model for the classification of different problems in working environments. The presented papers are examples and inspired the given research as a support for this paper. A diagram of the most represented technologies in the state of the art is given in Figure 1. These technologies are the main basis of the proposal.

Table 4. Proposals related to Smart manufacturing and Machine Learning.

Bibliography	Keywords	Novelty of the Proposal
Lee, Jay, et al. (2018)		State of AI technologies and the eco-system required to harness the power of AI in industrial applications.
Henley, E. J., and Kumamoto, H. (1985)	Artificial Inteligent Smart manufacturing, Failt diagnosis	Provides a quantitative treatment of the optimal design of safety systems focusing on information links (human and computer), sensors, and control systems.
Li, Bo-hu, et al. (2017)		Based on research into the applications of artificial intelligence (AI) technology in the manufacturing industry in recent years.
Xiaoli, X. et al. (2011)		A presentation of Intelligent internet of things for equipment maintenance (IITEM) which we can make intelligent processing of device information.
Varian, Hal. (2018)		Summary of some of the forces at work and to describe some possible areas for future research.
Wahab, L., and Jiang, H. (2019)		Traffic crash analysis using machine learning techniques.
Azar, A. T., et al. (2014)		A random forest classifier (RFC) approach is proposed to diagnose lymph diseases.
Belgiu, M., and Drăguţ, L. (2016)		This review has revealed that RF classifier can successfully handle high data dimensionality and multicolinearity, being both fast and insensitive to overfitting.
Khalilia, M., et al.	Machine Learning (ML), Decision Tree Classifier (DTC), Random Forest (RF), Multinomial logic model (MNLM), Support vector machine (SVMs), Receiver operating characteristic (ROS)	Method for predicting disease risk of individuals using random forest.
Jedari, E., et al. (2015)		Machine learning approaches including k-nearest neighbor (k-NN), a rules-based classifier (JRip), and random forest have been investigated to estimate the indoor location of a user or an object using RSSI based fingerprinting method.
Iranitalab, A., and Khattak, A. (2017)		This paper had three main objectives: comparison of the performance of four statistical and machine learning methods including Multinomial Logit (MNL), Nearest Neighbor Classification (NNC), Support Vector Machines (SVM) and Random Forests (RF), in predicting traffic crash severity.
Pal, M. (2005)		To present the results obtained with the random forest classifier and to compare its performance with the support vector machines (SVMs) in terms of classification accuracy, training time and user defined parameters.
Rodriguez-Galiano, V. F., et al. (2012)		The performance of the RF classifier for land cover classification of a complex area is explored.
Yogameena, B., et al. (2019)	Complex software system, Mixture models, Convolutional neural networks	Intelligent video surveillance system for automatically detecting the motorcyclists with and without safety helmets.
Cockburn, D. (1996)		The benefit of taking a holistic perspective to developing complex software systems.

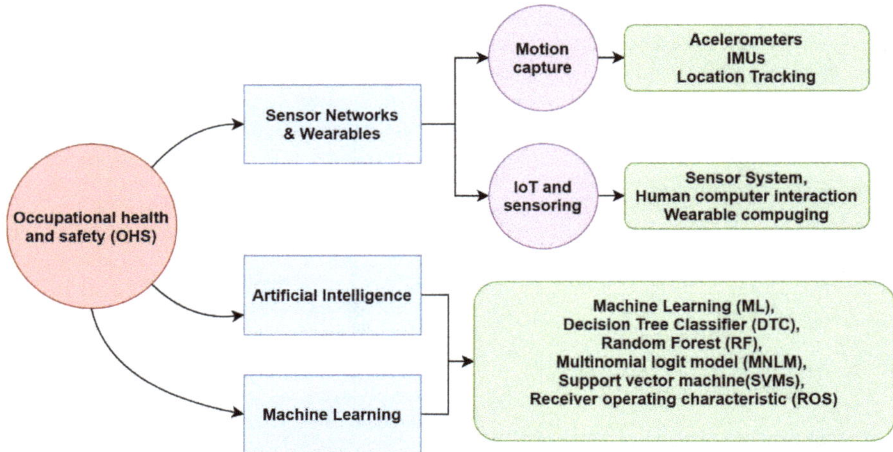

Figure 1. A block diagram of the devices.

3. Smart Helmet 5.0 Platform

There are different methodologies for carrying out research on electronics and system design. Thus, in this section, a description of the hardware and software used for the development of the fifth version of the smart helmet will be presented, and the procedure followed for its subsequent validation through the AI model will be detailed. The four previous helmets included less sensorisation and conectivity, which is why we developed a new version with all the improvements.

3.1. Hardware Platform

The structure followed in the development of the proposed helmet are the steps involved in the prototype development methodology, identifying the parameters to be monitored in the environment. A Job Safety Analysis (JSA) was performed, identifying the risk factors that lead to injuries and accidents in the worker [59]. The deficiencies that have been observed are presented in Table 5 for work places such us mines, construction places and electrical work areas. They are related to aspects such as lighting, detection of blows to the worker's helmet (PPE detection), dangerous temperature levels for human activityand poor air quality in the environment. Other parameters that could be interesting such as, noise, rate pulse and body temperature are implemented in other devices for better ergonomics.

Table 5. Identification of common risk situations in the worker's environment.

Risk Factors	Associated Hazards
Lack of Adequate Lighting	- The inability of the worker to see their environment clearly leads to accidental hits, slips, trips and falls. - The worker is unaware of the events occurring in their environment.
Temperature	- Extreme temperature changes leading to a heat stroke
Air Quality	- Harmful air in the environment
Operator Movement	-Slips, trips and falls -Blows to the worker's head

Given the above, a series of specialized sensors are proposed to counteract the difficulties that usually occur in a high-risk work environment [1], see Table 6. As seen in the literature review, agriculture and industrial activities involve high risk, among others.

Table 6. Identification of electronic components for the prevention of risks in the worker's environment.

Risk Factors	Solution
Lack of Adequate Lighting	- Implementation of a brightness sensor in the helmet - Inclusion of torches as one of the tools of worker
Temperature	- Implementation of temperature sensors in the devices of the worker or environment
Air Quality	-Moisture and gas sensors.
Operator Movement	- The use of wearable devices with accelerometers capable of detecting falls. - Integration of sensitive force resistors in the helmet of the operator.

In terms of the transmission of information from sensors, the use of Wi-Fi technologies has been selected due to their ability to transmit the information in Local Area Networks (LAN) to a web server responsible for collecting, processing and transmitting anomaly warnings to the worker or administrative personnel. The following describes the system design and the interaction of the components.

The elements used in the smart helmet and the risks it seeks to prevent or detect are detailed below. The operation of the Smart PPE and the distribution of the circuits will also be discussed. In addition, the architecture and technologies are explained, as well as the operating rules of the different sensors and actuators that make up the system. Finally, their communication system is considered, as well as the technology used for both the management of the data and for its visualization and treatment once obtained.

The aim of this Smart PPE is to protect the operator from possible impacts, while monitoring variables in their environment such as the amount of light, humidity, temperature, atmospheric pressure, presence of gases and air quality. At the same time, the Smart PPE is to be bright enough to be seen by other workers, and the light source will provide extra vision to the operator. All these alerts will be transmitted to the operator by means of sound beeps. The sensors described below were selected as part of the set of electronic devices to be implemented:

- Temperature, gas and pressure sensor;
- Brightness sensor;
- Shock sensor;
- Accelerometer.

In the process of the visualization of environmental data, a LED strip is deployed on the helmet as a means of notifying the worker of anomalies through color codes presented in the environment. The block diagram shown in Figure 2 is a representation of the electronic system integrated in the helmet.

The specifications of the sensors and the microcontroller used to monitor the environment are defined as follows:

The component used to supervise the parameters of gas, pressure, temperature and humidity is the low power environmental sensor DFRobot BME680. It is a MEMS (Micro-Electromechanical System) multifunctional 4 in 1 environmental sensor that integrates a VOC (Volatile Organic Compounds) sensor, temperature sensor, humidity sensor and barometer. The environmental pressure is subject to many short-term changes caused by external disturbances. To suppress disturbances in the output data without causing additional interface traffic and processor work load, the BME680 features an internal IIR filter. The output of the subsequent measurement step is filtered using the following Equation (1):

$$x_{filt}[n] = \frac{x_{filt}[n-1] * (c-1) + x_{ADC}}{c} \tag{1}$$

where $x_{filt}[n-1]$ is the data coming from the current filter memory, and x_{ADC} the data coming from current ADC acquisition and where $x_{filt}[n]$ denotes the new value of filter memory and the value that will be sent to the output registers.

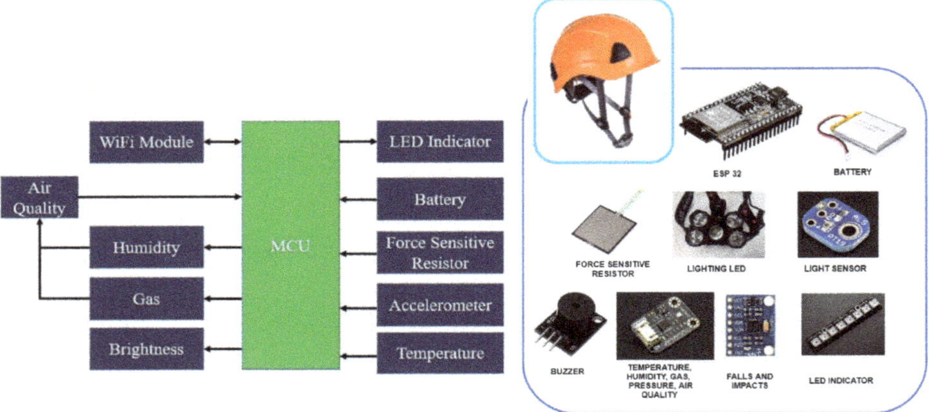

Figure 2. A block diagram of the devices.

The sensor implemented for the monitoring of the level of brightness is the ALS-PT19 ambient light sensor. Due to the high rejection ratio of infrared radiation, the spectral response of the ambient light sensor resembles that of the human eyes.

The sensor implemented for shock detection is a sensitive force resistor, the sensor emits shock alerts if the readings obtained in the environment exceed a threshold value.

The sensor responsible for the detection of falls suffered by the worker is the MPU6050 module, it is an electronic component that has six axes (three corresponding to the gyroscope system and three to the accelerometer) making it possible to obtain the values of positioning in the X, Y and Z axes.

The light source integrated in the helmet is a NeoPixel Adafruit LED strip, the component integrates a multicolor LED in each section of the strip. The algorithm implemented in the microcontroller is configured in such a way that it is possible to control the color of the LED strip.

The microcontroller used for processing, transmitting and displaying the information transmitted to the web platform is the dual-core ESP-WROOM-32 module of the DFRobot FireBeetle series, which supports communication through Wi-Fi and Bluetooth. The main controller supports two power methods: USB and 3.7 V external lithium battery.

The components are integrated in the microcontroller, which obtains and processes the information coming from the sensors. This information is then transmitted to the implemented web server by means of the Wi-Fi module. The designed electronic system is located in the backside of the helmet, as shown in Figure 3. It also integrates a lamp which is activated automatically if the brightness value of the sensor is below the threshold value established in the programming of the microcontroller. The information transmitted by the helmet can be viewed on a web platform.

This section describes the developed software and the interaction that takes place between the different components.

Figure 3. The electronic system of the helmet.

3.2. Intelligence Module

Firstly, the communication between the active sensors is enabled by Thingsboard. ThingsBoard is an open source IoT platform for data collection, processing, visualization and IoT device management. It is free for both personal and commercial use and can be implemented anywhere.

It enables device connectivity through industry-standard IoT protocols (MQTT, CoAP and HTTP) and supports both cloud and on-premise deployments. ThingsBoard combines scalability, fault tolerance and performance, ensuring that the users' data are never lost.

ESP32 is a series of low-power, low-consumption system-on-a-chip microcontrollers with integrated Wi-Fi and dual-mode Bluetooth, as mentioned in the previous section. The device is responsible for transmitting the information to the ThingsBoard platform and its subsequent processing by the intelligent model see Figure 4, to interact with the helmet.

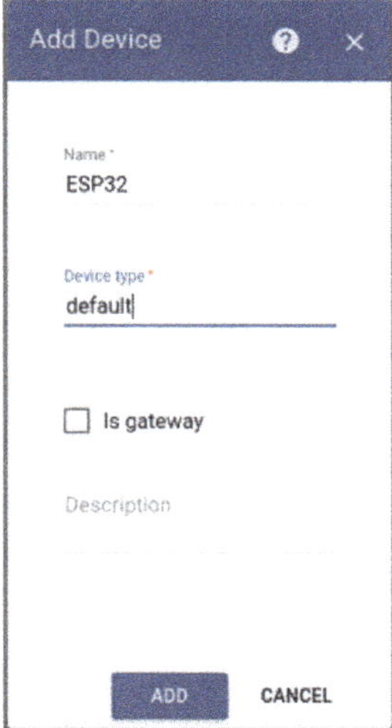

Figure 4. Setting up the ThingsBoard platform to operate according to the information received from ESP32, IoT module added to ThingsBoard and Multi-sensorial configuration.

Simple steps are required to link the devices to the platform:

- The automatically generated access token is copied from the Access token field.
- Go to Devices, locate ESP32 device, open the device details, and select the Latest telemetry tab.
- It is now possible to view the data regarding an asset.

The data obtained through ThingsBoard is later processed by an intelligent model, the model confirms or denies the existence of a real emergency. This is the reason why configuring the platform is very important.

An association must be created between the different values of the sensors and the corresponding response. Once these associations are created, it is possible to modify any value depending on the values to be tested empirically or in the alarms. Alarms are configured in the device settings so that the respective notifications appear on the panel. A rule chain must be added.

A selection of the attributes placed on the server and on the device's threshold panel must be carried out. The names of the attributes on the server must correspond with those on the panel so that when the data are dynamically configured, they will be recognized correctly and will appear on the diagram generated by the platform, Figure 5.

Subsequently, in the script block, it is verified that the information coming from the device does not exceed the established threshold value. If the script is positive, an alarm is configured and the information to be displayed is defined.

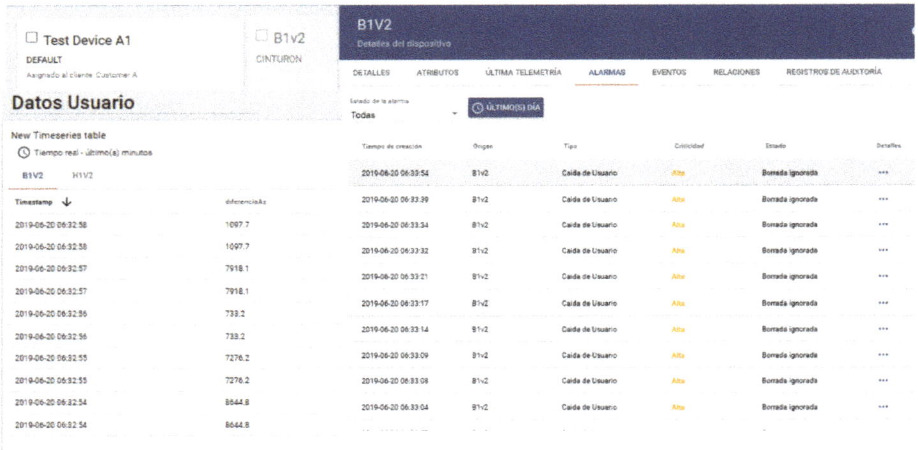

Figure 5. Alarm configuration on ThingsBoard, Block alarm creation method and Connecting alarms with sensors.

Moreover, the root string, which is in charge of obtaining and processing the information coming from the devices, has been modified. In this case, an originator type section has been added, where the devices that transmit the information are identified. Likewise, code strings have been generated to implement the customized code blocks in the panels. Finally, the information on the data panel may be visualized.

In cases where it is not necessary to perform this procedure, it is possible to view the notifications generated by the different devices. To this end, it is necessary to enter the Device section. Select one of the devices for which an alarm has been configured and go to the alarms tab, see Figure 6, where the notifications generated by that device are displayed.

Once the alarms have been configured on the platform, validation is carried out through the explanation of the AI [60].

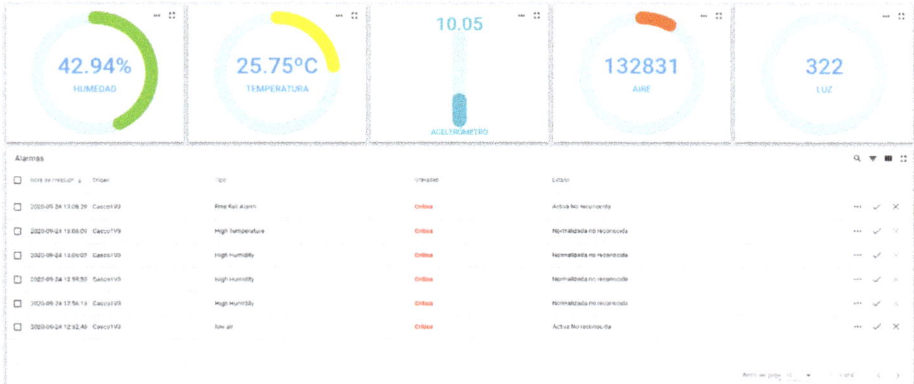

Figure 6. Final configuration of ThingsBoard platform to be validated through an intelligent algorithm.

4. Platform Evaluation

This section compares the different algorithms used in the state of the art to solve problems similar or related to the one being addressed here [61–64], these models have been accepted for real world problems due to their dataset results with data unbalance and saturation issues, this comparison will be performed with the same amount of data and on an objective quantitative basis. Furthermore, the present proposal is described in detail.

4.1. Data Model

In this study, samples of data from a real environment have been obtained, where a subject was subjected to various scenarios in simulated environments, considering the different risks that could arise. The five analyzed parameters are shown in Table 7. The acquired dataset consists of a total of 11,755 samples, where five descriptive variants are proposed with respect to the target of the study.

Table 7. Parameters for which data are collected.

1. Brightness
2. Variation in X, Y and Z axis
3. Force Sensitive Resistor
4. Temperature, Humidity, Pressure
5. Air quality

This research tackles a multi-class type of problem, for this reason there is a set of labels that have a different meaning. When the programming of the microcontroller was carried out, the different parameter values that could trigger an alarm signal were investigated, for example, if the air quality falls below the threshold (measured by the Air Quality Index, AQI) it is possible to associate this situation with the values for other parameters measured by neighboring sensors. The 12 labels proposed in this work are described below, where research was carried out on the most common problems in industrial areas and from there the type of sensors in the helmet were included [65,66]:

0. Good for health air (AQI from 0 to 50) with sufficient illumination in the working environment.
1. Moderate air quality (AQI of 51 to 100) with slight variation in temperature and humidity.
2. Harmful air to health for sensitive groups (AQI 101-150) with moderate variation in temperature and humidity.
3. Harmful air to health (AQI 151 to 200) with considerable variation in temperature and humidity
4. Very harmful air to health (AQI 201 to 300) with high variation in temperature and humidity.
5. Hazardous air (AQI greater than 300) with atypical variation in temperature and humidity.

6.   Lack of illumination and variation equivalent to a fall in axes.
7.   Lack of illumination and variation equivalent to a fall in axes and considerable force exerted on the helmet.
8.   Atypical variation on the detected axes and moderate force detected on the FSR.
9.   Illumination problems, air quality and sudden variation in axes.
10.   Very high force exerted on the FSR.
11.   Variation in axes with illumination problems.
12.   Outliers on the 5 sensors.

Once the information has been understood, it is cleaned. As proposed by [67], the data were cleaned due to common problems such as missing values solved with the clamp transformation, see Equation (2).

$$a_i = \begin{cases} lower & if\, a_i < lower \\ upper & if\, a_i > upper \\ a_i & Otherside \end{cases} \tag{2}$$

where a_i represents the $i-$th sample of the data set, lower and upper thresholds respectively.

The upper and lower thresholds can be calculated from the data. A common way of calculating thresholds for the clamp transformation is to establish:

- The lower threshold value $= Q_1 - 1.5IQR$;
- The upper threshold value $= Q_3 + 1.5IQR$.

Where Q_1 is the first quartile, Q_3 is the third quartile and IQR is the interquartile range ($IQR = Q_3 - Q1$ the interquartile range). Any value outside these thresholds would become the threshold values. This research takes into account the fact that the variation in a data set may be different on either side of a central trend. Each sample that had missing data was eliminated so as not to bias the model. However, the search for outliers was only used to find erroneous data generated by the electronic acquisition system since outliers usually provide a large source of information for the analysis of a dataset.

4.2. Intelligent Models Evaluation

The comparison part describes each of the models used for the current project, detailing the Support Vector Machine, Naïve Bayes classifier, Static Neural Network and a Convolutional Neural Network. Each model used the dataset of 11,755 where 80% was used for modeling and 20% for evaluation, in other words 9404 in training and 2351 in evaluation. The following confusion matrices reports the result of the validation and after that, we include a figure for each model in order to present the information clearly. It is worth mentioning that all models were trained and validated with the same data division in relation 80-20, it is also notable above an imbalance of classes, given the imbalance some models had unfavorable behavior in cross validation. To handle the imbalance it is possible to opt for techniques such as oversampling or undersampling but it is not desired to change the quality of the data, that is why the model with the best performance will be chosen and evaluated with 10 folds for validation.

Support Vector Machine

SVMs belong to the categories of linear classifiers, since they introduce linear separators better known as hyperplanes, regularly made within a space transformed to the original space.

The first implementation used for the multi-class classification of SVM is probably the one against all method (one-against-all). The SVM is trained with all the examples of the $m-$th class with positive labels, and all the other examples with negative labels. Therefore, given the data of the $(x_1, y_1), ..., (x_l, y_l)$ where $x_i \in R^n, i = 1, ..., l$ y $y_i \in \{1, ..., k\}$ is the class of x_i, the $m-$th SVM, and solve the problem in Equation (3) [68], which involves finding a hyper plane so that points of the same kind are on the same side of the hyperplane, this is finding a b and w such:

$$y_i(w'x_i + b) > 0, \; i = 1, \ldots, N \tag{3}$$

Equation (4) looks for a hyper plane to ensure that the data are linearly separable.

$$\min_{1 \leq i \leq N} \; y_i(w'x_i + b) \geq 1 \tag{4}$$

where $w \in \mathbb{R}^d$, $b \in \mathbb{R}$ and the training dataset x_i is mapped to a higher dimensional space. Thus, it is possible to search among the various hyperplanes that exist for the one whose distance to the nearest point is the maximum, in other words, the optimum hyperplane [68], see Equation (5).

$$\min_{w,b} \frac{1}{2} w^T w \tag{5}$$
$$individual \; a \; y_i(w^T, x_i + b) \geq 1, \forall_i$$

Given the above, we are looking for a plane with the maximum distance between the samples of different classes on a higher dimension. As mentioned above, the SVM was of the type one against all in the mathematical description since it is a multi-class problem. Furthermore, the type of kernel was linear. The modeling was performed and the confusion matrix was obtained with 20% of data for evaluation. The accuracy of each class in comparison to the others can be observed in Table 8. The SVM was the model with the worst performance out of the four evaluated according to the recommendation in the literature where the overall accuracy was 68.51%.

Table 8. Confusion matrix SVM.

Predicted Class (Vertical)/ True Class (Horizontal)	Class 0	Class 1	Class 2	Class 3	Class 4	Class 5	Class 6	Class 7	Class 8	Class 9	Class 10	Class 11
Class 0	153	10	31	0	0	0	0	0	0	1	1	6
Class 1	0	122	0	0	1	6	0	12	0	2	1	5
Class 2	20	0	192	0	0	0	0	0	0	120	0	0
Class 3	20	0	0	158	3	0	13	25	2	101	0	7
Class 4	1	0	0	0	12	0	0	0	3	5	9	0
Class 5	0	25	23	0	0	135	0	0	0	2	8	7
Class 6	0	0	0	0	0	0	110	30	0	1	0	0
Class 7	15	5	30	0	0	57	0	159	0	5	0	0
Class 8	1	9	0	30	0	0	20	0	11	1	0	9
Class 9	13	0	5	40	0	0	10	0	0	432	0	0
Class 10	0	8	0	4	0	0	0	6	0	3	53	0
Class 11	0	0	0	0	0	0	8	0	0	2	5	72

Naive Bayes Classifier

A Gaussian NB classifier is proposed that is capable of predicting when an accident has occurred in a work environment through different descriptive characteristics, which is based on Bayes' theorem. Bayes' theorem establishes the following relationship, given the class variable y and the vector of the dependent characteristic x_1 through x_n [69,70], Equation (6).

$$P(y \mid x_1, \ldots, x_n) = \frac{P(y)P(x_1, \ldots x_n \mid y)}{P(x_1, \ldots, x_n)} \tag{6}$$

where $\forall i$, the relationship can be simplified as shown in Equation (7).

$$P(y \mid x_1, \ldots, x_n) = \frac{P(y) \prod_{i=1}^{n} P(x_i \mid y)}{P(x_1, \ldots, x_n)} \tag{7}$$

where $P(x_1, \ldots, x_n)$ is constant based on the input; the classification rule presented in Equation (8) can also be used.

$$P(y \mid x_1, \ldots, x_n) \propto P(y) \prod_{i=1}^{n} P(x_i \mid y)$$

$$\Downarrow \tag{8}$$

$$\hat{y} = \arg\max_{y} P(y) \prod_{i=1}^{n} P(x_i \mid y),$$

The difference in the distributions of each class in the dataset means that each distribution can be independently estimated as a one-dimensional distribution. This in turn helps reduce the problems associated with high dimensionality. For a Gaussian NB classifier the probability of the characteristics is assumed to be Gaussian, see Equation (9).

$$P(x_i \mid y) = \frac{1}{\sqrt{2\pi\sigma_y^2}} \exp\left(-\frac{(x_i - \mu_y)^2}{2\sigma_y^2}\right) \tag{9}$$

In other words, in order to use the NB classifier in the grouping of the different work circumstances that put the worker at risk, it is assumed that the presence or absence of a particular characteristic is not related to the presence or absence of any other characteristic, given the variable class. The confusion matrix of the NB is shown in Table 9, where on average the accuracy was of 78.26%.

Table 9. Confusion matrix NB.

Predicted Class (Vertical)/ True Class (Horizontal)	Class 0	Class 1	Class 2	Class 3	Class 4	Class 5	Class 6	Class 7	Class 8	Class 9	Class 10	Class 11
Class 0	174	10	31	0	1	0	0	0	0	5	1	0
Class 1	0	140	0	0	1	0	0	12	0	2	1	9
Class 2	10	0	220	0	0	0	0	0	0	70	0	0
Class 3	15	0	0	181	0	0	13	5	1	62	0	7
Class 4	1	0	0	0	13	0	0	7	0	1	5	0
Class 5	0	25	13	0	0	155	0	9	0	11	4	1
Class 6	0	0	0	0	0	0	126	1	0	0	0	0
Class 7	9	2	15	0	0	37	0	180	0	0	0	0
Class 8	1	1	0	30	0	0	20	0	13	0	2	4
Class 9	3	0	1	20	1	6	2	18	1	520	0	3
Class 10	0	1	1	1	0	0	0	0	1	0	60	0
Class 11	0	0	0	0	0	0	0	0	0	3	4	82

Static Neural Network

Neural networks are simple models of the functioning of the nervous system. The basic units are the neurons, which are usually organized in layers. The processing units are also organized in layers. A neural network normally consists of three parts [71]:

1. An input layer, with units representing the input in the dataset.
2. One or more hidden layers.
3. An output layer, with a unit or units representing the target field or fields.

The units are regularly connected with varying connection forces (or weights). Input data is presented in the first layer, and values are propagated from one neuron to another in the next layer. At the end, a result is sent from the output layer. All the weights assigned to each layer are random in the first instance of the training. However, there are a series of methods that can be employed to optimize this phase. Furthermore, the responses that result from the network are offline. The network learns through training [71]. Data for which the result is known are continuously presented to the network, and the responses it provides are compared with the known results.

The use of a static NN is proposed in this research. The performance of the classic model Adam has been compared with the performance of a CNN. The architecture of the NN is shown in Figure 7, which is a three-layer static model, where the first layer contains five neurons that correspond to each of the five data being obtained from the multisensory case, the hidden layer has 32 neurons with the ReLU activation function and finally the output layer has 12 neurons representing the situations a worker may find themselves in. They range from safe to risky situations. The last layer has a SoftMax activation function because it is a multi-class problem. The learning step was 0.05 and the model was trained with 500 epochs. In which the approach for the proposed structure is based on "trial and error", since as it is well known establishing a neural network is more an art than a science. That is why the number of neurons on the second layer was modified, which obtained a better result than adding other layers on the network. However, CNN showed better results than the rest of the models with a predetermined structure (12 neurons in the hidden layer).

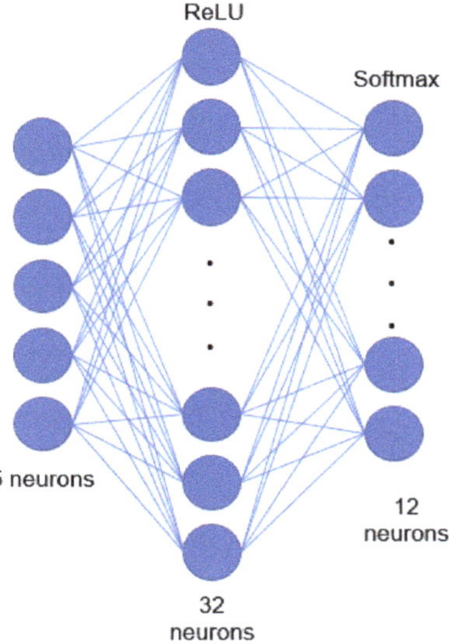

Figure 7. Proposed architecture, static neural network.

The result of the static NN are given in Table 10. It is possible to observe its performance was not very different from the NB classifier, where an average accuracy of 78.56% was obtained.

Table 10. Confusion matrix NN.

Predicted Class (Vertical)/ True Class (Horizontal)	Class 0	Class 1	Class 2	Class 3	Class 4	Class 5	Class 6	Class 7	Class 8	Class 9	Class 10	Class 11
Class 0	175	10	30	0	0	0	0	0	0	5	1	0
Class 1	0	141	0	0	0	0	0	7	0	2	1	9
Class 2	9	0	221	0	0	0	0	0	0	60	0	0
Class 3	13	0	0	181	0	0	13	5	0	50	0	7
Class 4	1	0	0	0	15	0	0	7	0	0	25	0
Class 5	0	23	12	0	0	150	0	9	0	9	9	1
Class 6	0	0	0	0	0	0	134	1	0	2	0	0
Class 7	9	3	16	0	0	32	0	185	0	15	0	0
Class 8	1	1	0	30	0	0	12	0	14	0	2	4
Class 9	3	0	1	20	1	16	2	18	1	526	0	2
Class 10	0	1	1	1	0	0	0	0	1	0	30	0
Class 11	0	0	0	0	0	0	0	0	0	5	9	83

4.3. Convolutional Neural Network

A Convolutional Neural Network (CNN) was the selected model, it is a deep learning algorithm mainly used to work with images in which it is possible to use an input image (instead of a single vector as in static NNs), assign importance, weights, learnable biases to various aspects/objects of the image and be able to differentiate one from another [72]. The advantage of NNs is their ability to learn these filters/characteristics. Given the abovel we propose the use of a CNN to classify the data coming from the multisensorial helmet.

The proposed CNN's operation is illustrated in Figure 8. The CNN consists of segmenting groups of pixels close to the input image and mathematically operating against a small matrix called a kernel. However, the part of the image is replaced with the input vector of size 5, where a re-shape is made to obtain a vector of 5×1. Therefore, the kernel proposed in the current CNN is of size 1, and moves from 1×1 pixel, in our case it would be different dimensions. With that size it manages to visualize all the input neurons and thus it can generate a new output matrix; a matrix that will be our new layer of hidden neurons.

A CNN can contain the spatial and temporal dependency characteristics in an image by applying relevant filters, the same applies to a data set that has been re-organized. The proposed architecture is an input layer for the transformed vector with size $5 \times 1 \times 1$ with two hidden convolutional layers for two-dimensional data (Conv2D) and ReLU activation functions with a total of 64 and 32 neurons respectively. Finally a layer with 12 output neurons with SoftMax activation function for multiclass classification.

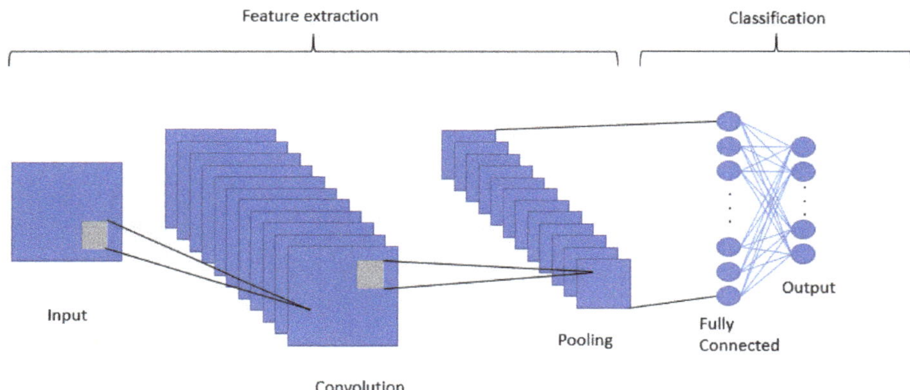

Figure 8. Deep convolutional neural network operation.

A classical model of Adam was proposed and trained with 500 epochs, the parameters were the same for the static NN and CNN to have an objective margin with respect to their evaluation. The following are the results on the AI models used for their implementation in the multisensorial helmet. Table 11 shows the evaluation for CNN where an overall accuracy of 92.05% was achieved.

Table 11. Confusion matrix CNN.

Predicted Class (Vertical)/ True Class (Horizontal)	Class 0	Class 1	Class 2	Class 3	Class 4	Class 5	Class 6	Class 7	Class 8	Class 9	Class 10	Class 11
Class 0	190	0	8	0	0	0	0	0	0	0	0	0
Class 1	8	169	0	0	0	12	0	3	0	0	0	0
Class 2	18	0	267	0	0	3	0	2	0	3	1	0
Class 3	4	0	0	209	0	1	0	0	0	30	0	0
Class 4	1	0	1	0	15	0	0	0	0	0	1	0
Class 5	1	8	0	0	0	179	0	4	0	0	0	0
Class 6	0	2	0	4	0	0	150	0	4	29	0	0
Class 7	1	0	0	0	0	0	0	115	0	0	0	0
Class 8	0	0	0	3	0	0	0	0	67	1	0	0
Class 9	0	0	4	16	0	2	11	1	6	612	0	0
Class 10	0	0	1	0	1	0	0	0	0	1	73	0
Class 11	0	0	0	0	0	1	0	0	0	0	2	106

4.4. Results

As mentioned above, each model was evaluated with 20% of data for cross validation. The SVM presented a general accuracy of 68.51 % which was the model with the lowest performance in cross-validation. Its behavior is compared with that of the rest of the analyzed classes in Figure 9. Therefore, the use of this model in the multisensory helmet has been discarded.

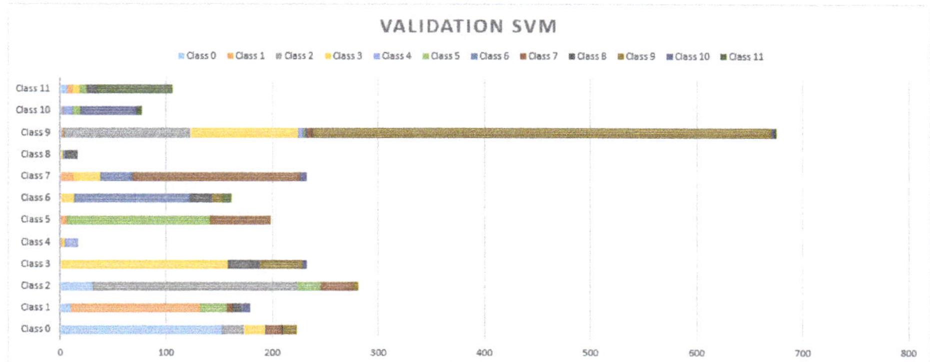

Figure 9. Cross-validation results with 20% for the SVM.

An average accuracy of 78.26% has been achieved by NB in all the classes, as shown in Figure 10. Its performance has been better in class 5 and class 11. Despite having a better result than SVM it has been discarded since there were models that had better performance.

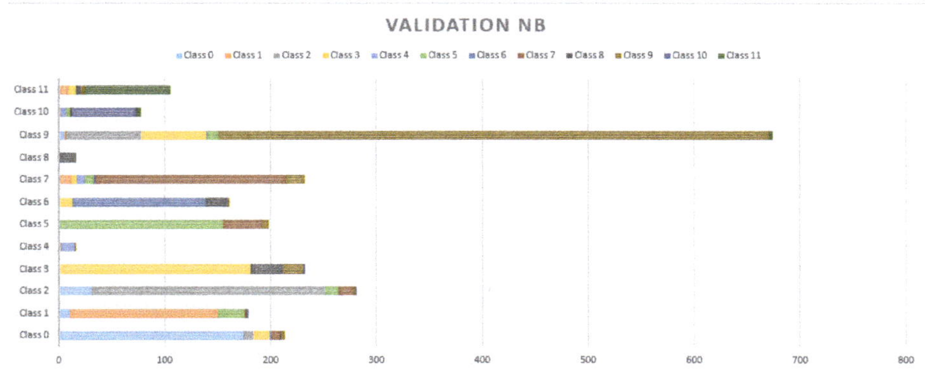

Figure 10. Cross-validation results with 20% for the NB.

Figures 11 and 12 show the performance of static NN and CNN respectively. In Figure 8 it can be observed that there is not a significant difference in the performance of NB, which had an accuracy of 78.56%. On the contrary, CNN, which allows for the implicit extraction of characteristics and for maintaining the relationships between the information regarding the dataset, had a considerably better result, with an accuracy of 92.05%. Our innovation comes on the proposed implementation of a CNN in a safety helmet as a proposal to reduce accidents and fractures in work areas, also through the use of technologies such as IoT for rapid synchronization of alarms that are sent to supervisors to take immediate action.

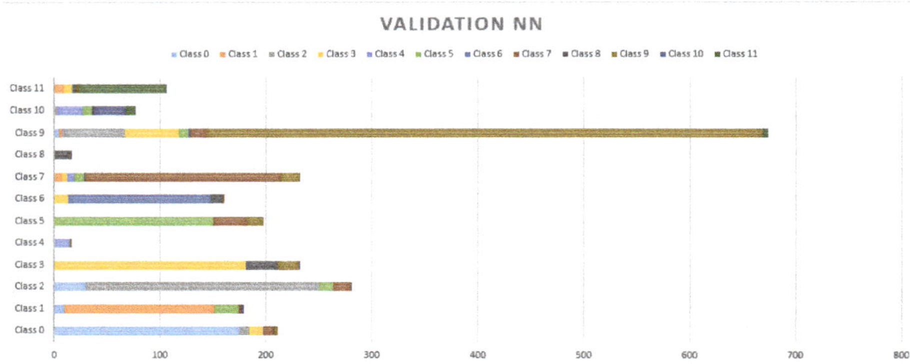

Figure 11. Cross-validation results with 20% for the NN.

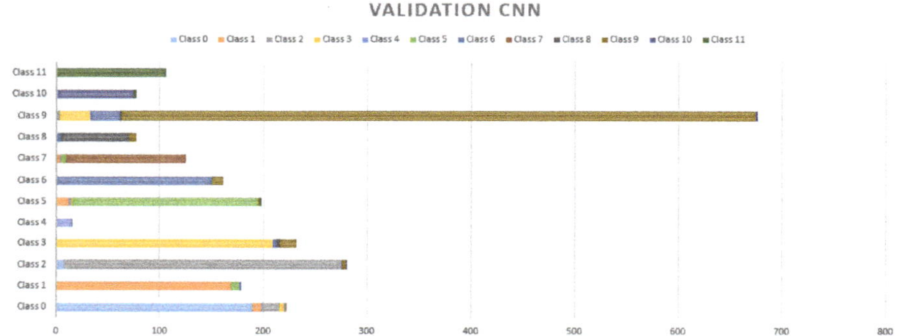

Figure 12. Cross-validation results with 20% for the CNN.

Given the above, CNN is the model that has been implemented in conjunction with the ThingsBoard platform. ThingsBoard and CNN work independently, creting an alarm system in a simulated environment that can serve as an higher security approach to a work environment. CNN is in charge of validating the information obtained from the platform, see Figure 13.

Previously it was mentioned that the creation of the deep models was through the "trial and error" approach, but the possible problem of overfiting should not be left aside, that is why Table 12 shows the results for the CNN in 10-Folds that shows the average performance from an objective point of view of the models.

Table 12. Ten-fold validation for CNN.

Ten Fold Cross-Validation Test Sets	Accuracy (%) (Automated Risk Situations Develop in This Research)
1	93.18
2	93.09
3	90.73
4	94.12
5	91.27
6	92.75
7	92.61
8	92.59
9	92.76
10	91.99
Average Accuracy	92.509

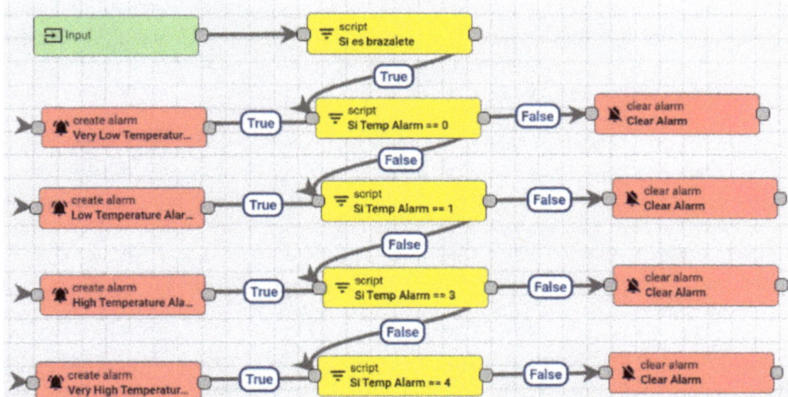

Figure 13. System of alarm rules established in ThingsBoard.

In the next section, the conclusions drawn from the conducted research are described, and the contributions of this work to the state of the art are highlighted.

5. Conclusions and Discussion

Our work has a history of electronic development in which the use of a multisensory helmet was established. Through a conditional model of input–output rules, we tried to detect the different situations to which a worker was subjected. However, the input–output techniques presented false positives and false negatives with 60% accuracy in the best of cases, which is why after several stages, it was decided to implement AI in the helmet. The 60% that was described a moment ago is due to the combination of different circumstances, that is to say, the correlation that exists on the independent characteristics, is for that reason, that through techniques that find linear and nonlinear relations we decided to innovate in the present work. Since it is necessary to find the patterns that determine a particular action, for them there are the techniques of deep learning as our work presents

The comparison between different models of AI has been made in this research. Our innovation comes from the moment of using a CNN that in the literature has been used to analyze images or videos in intelligent helmets with the aim of saving lives. However, we proposed a multisensory approach to real-time feature analysis. Through the transmission of data through specialized IoT devices, a smart helmet has been designed to monitor the conditions in a working environment. The application areas of this proposal are industrial and agricultural sectors and any other sector that involves risk for the workers. Thanks to the helmet, different injuries can be avoided, and in case an accident occurs the damage caused to the worker is lessened through prompt attention or detection.

It is possible to observe in Figures 9–11 that the MSA presented many false positives on majority classes in sample size, and even false positives of repeated classes (class 6) on more than five different classifications. NB and NN had a better performance in minority classes, however, there are three different classifications in false positives in classes such as 11, 9, 7, 2, 1 and 0. The NN has a strong resolution where the classes mentioned above still present false positives but with a decrease to 2 wrong classes in almost all cases.

5.1. Limitations

The work has different limitations. It is well known that artificial intelligence has the ability to find patterns that can hardly be found in linear analysis models. However, as stated in [73] risk analyses are not yet common in project-oriented industries. A problem with current risk analysis procedures is that procedures that are simple enough to be used by normal project staff are too simplistic to capture the subtlety of risk situations. Those that are complex enough to capture the essence and subtlety of risk situations are so complex that they require an expert to operate them. That is why the combination of possible risk situations can be counterproductive in the industrial area, an area that should be analyzed in more detail, with the following consequences:

- False positives would result in economic losses that would eventually affect the services and production areas involved, since the medical service and will be attending to situations that were not risky, the industry part will have to make production stops every time a false positive is found.
- On the other hand, false negatives are even more dangerous because the misinterpretation of data due to the complexity that can cause the unbalance of classes with less data set would result in losses not only economic but also of human personnel due to situations that were not attended to in the indicated time.

Our system has limitations on the amount of data that can be processed due to the microcontroller and the data that the model supports through the ESP32 module. That is why other techniques can be adopted, as will be seen in the next part of future work.

5.2. Future Work Opportunities

The use of paradigms such as edge computing or fog computing for the processing of many data as would be the integration of images or video would be the viable option to allow a transmission of information in real time, avoiding problems of saturation by the microcontroller. Several state-of-the-art researches have proposed smart helmets, among them is the US6798392B2 patent [74], which integrates a global location system, an environmental interaction sensor, a mobile communications network device, a small display panel, a microphone and a speaker. The helmet knows the location of the user and their interaction with the environment. The helmet can provide data to a user, monitor the actions of the user and conditions. This work is quite interesting since it offers device–user interaction. On the contrary, the advantage of our proposal is that it strives towards the autonomy of the system where decisions are made by the convolutional method.

Furthermore, the US9389677B2 patent [75], which is a smart helmet that includes a camera, a communications subsystem and a control subsystem. The control subsystem processes the video data from the camera, and the communications subsystem transmits this video data from the smart helmet to a target device. This work can be taken as a reference for a future sensor integration, since in our proposal it would be possible to integrate a camera that can process data through Deep CNN, for example thermal radiation data or even data regarding those who are infected with COVID-19.

Furthermore, the US registered patent, US20150130945A1 [76], in which a smart helmet is proposed that includes a helmet shell, a visor and a projector mounted on the helmet shell. The projector is configured so that content can be selected for display on the visor. The visor is rotatably attached to the helmet shell, and is configured to expose or cover the passage. The hull of the helmet defines an internal cavity and a passage that communicates with the internal cavity. The internal cavity is designed to receive the head of a user. This proposal's focus is directed at the ergonomic part for the user, in addition to having navigation systems and control modules. This research is comparable to our proposal.

Moreover, in 2013 a helmet was proposed by Rasli Mohd Khairul Afiq Mohd et al. [77] for the prevention of accidents in which an FSR and a BLDC fan were used to detect the head of the driver and the speed of the motorcycle, respectively. A 315 MHz radio frequency module was used as a wireless link for communication between the transmitter circuit and the receiver circuit. PIC16F84a is a microcontroller for the control of the different components of the system. The motorcyclist could start the engine only when they had fastened their helmet. In comparison, our proposal communication takes place through IIoT for optimized decision making in case of accidents.

With reference to smart helmets connected to IoT, in 2016 [78], Sreenithy Chandran et al. presented a design whose objective is to provide a channel and a device for monitoring and reporting accidents.

Sensors **2020**, *20*, 6241

Sensors, a Wi-Fi enabled processor, and cloud computing infrastructures were used to build the system. The accident detection system communicates the accelerometer values to the processor that continuously monitors erratic variations. When an accident occurs, details about the accident are sent to emergency contacts using a cloud-based service. The location of the vehicle is obtained using the global positioning system. This work has a close relationship with the one proposed by us where there is optimized communication to reduce the consequences of accidents, the approach is different since we propose it for a work environment that can later be adapted for a case focused on vehicle safety, mainly on motorcycles.

Author Contributions: I.C.-J. data cleaning, analysis and modeling; S.M.-S. in electronic design; J.Q.-G. firmware programming and IIoT communications management, S.R. in data interpretation analysis, J.M.C. in supervision and direction of the entire project. All authors have read and agreed to the published version of the manuscript.

Funding: This research was funded by Research group in Bioinformatics, Intelligent Computer Systems and Educational Technology (BISITE), R&B building, Calle Espejo s/n, 37008, Salamanca, Spain.

Acknowledgments: Thanks are due to the entire BISITE research group for their invaluable work from electronic design to management for information analysis. Likewise, for providing the economic resources to make this work possible.

Conflicts of Interest: The authors declare no conflict of interest. Each of the researchers and scientists who have developed the current work presents no conflict of interest with respect to the design of the study; in the collection, analyses or interpretation of data; in the writing of the manuscript, or in the decision to publish the results

Abbreviations

The following abbreviations are used in this manuscript sorted alphabetically:

AI	Artificial Intelligence
AQI	Air Quality Index
ARCTM	Accident Root Causes Tracing Model
CM	Condition Monitoring
CNN/ConvNet	Convolutional Neural Network
EEG	Electroencephalographic
GMM	Gaussian Mixture Model
ICTs	Information and communication technologies
IIoT	Industrial Internet of Things
IITEM	Internet of Things for Equipment Maintenance
JSA	Job Safety Analysis
KC	K-means Clustering
LAN	Local Area Networks
LCC	Latent Class Clustering
LP	License Plate
MEMS	Micro-Electromechanical System
ML	Machine Learning
MNL	Multinomial Logit
NB	Naive Bayes Classifier
NN	Neural Netowork
NNC	Nearest Neighbor Classification
OHS	Occupational Health and Safety
PPE	Personal Protective Equipment
RF	Random Forests
SVM	Support Vector Machine
UKF	Unscented Kalman Filter
VOC	Volatile Organic Compounds

References

1. Formación Superior en Prevención de Riesgos Laborales. Parte Obligatoria y Común. Available online: http://www.paraprofesionales.com/indices/ind50276.pdf (accessed on 1 November 2020).
2. Seguridad e Higiene Industrial. Available online: http://168.121.45.184/bitstream/handle/20.500.11818/599/Seguridad%20e%20Higiene%20Industrial-1-79.pdf?sequence=1&isAllowed=y (accessed on 1 November 2020).
3. Chevalier, Y.; Compagna, L.; Cuellar, J.; Drielsma, P.H.; Mantovani, J.; Mödersheim, S.; Vigneron, L. A High Level Protocol Specification Language for Industrial Security-Sensitive Protocols. Available online: https://hal.inria.fr/inria-00099882/document (accessed on 1 November 2020).
4. Vaughn, R.B., Jr.; Henning, R.; Fox, K. An empirical study of industrial security-engineering practices. *J. Syst. Softw.* **2002**, *61*, 225–232. [CrossRef]
5. Varian, H. *Artificial Intelligence, Economics, and Industrial Organization*; Technical Report; National Bureau of Economic Research: Cambridge, MA, USA, 2018.
6. Lee, J.; Davari, H.; Singh, J.; Pandhare, V. Industrial Artificial Intelligence for industry 4.0-based manufacturing systems. *Manuf. Lett.* **2018**, *18*, 20–23. [CrossRef]
7. Li, B.H.; Hou, B.C.; Yu, W.T.; Lu, X.B.; Yang, C.W. Applications of artificial intelligence in intelligent manufacturing: A review. *Front. Inf. Technol. Electron. Eng.* **2017**, *18*, 86–96. [CrossRef]
8. Cockburn, D. Artificial Intelligence System for Industrial Applications. *Found. Distrib. Artif. Intell.* **1996**, *9*, 319.
9. Chamoso, P.; González-Briones, A.; Rodríguez, S.; Corchado, J.M. Tendencies of technologies and platforms in smart cities: A state-of-the-art review. *Wirel. Commun. Mob. Comput.* **2018**, *2018*, 3086854. [CrossRef]
10. Sun, S.; Zheng, X.; Gong, B.; García Paredes, J.; Ordieres-Meré, J. Healthy Operator 4.0: A Human Cyber–Physical System Architecture for Smart Workplaces. *Sensors* **2020**, *20*, 2011. [CrossRef]
11. Podgorski, D.; Majchrzycka, K.; Dąbrowska, A.; Gralewicz, G.; Okrasa, M. Towards a conceptual framework of OSH risk management in smart working environments based on smart PPE, ambient intelligence and the Internet of Things technologies. *Int. J. Occup. Saf. Ergon.* **2017**, *23*, 1–20. [CrossRef]
12. Fernández-Muñiz, B.; Montes-Peón, J.M.; Vázquez-Ordás, C.J. Safety climate in OHSAS 18001-certified organisations: Antecedents and consequences of safety behaviour. *Accid. Anal. Prev.* **2012**, *45*, 745–758. [CrossRef]
13. Niu, Y.; Lu, W.; Xue, F.; Liu, D.; Chen, K.; Fang, D.; Anumba, C. Towards the "third wave": An SCO-enabled occupational health and safety management system for construction. *Saf. Sci.* **2019**, *111*, 213–223. [CrossRef]
14. Barata, J.; da Cunha, P.R. The Viable Smart Product Model: Designing Products that Undergo Disruptive Transformations. *Cybern. Syst.* **2019**, *50*, 629–655. [CrossRef]
15. Kumar, S.G.; Dharanipriya, A. Prevalence and pattern of occupational injuries at workplace among welders in coastal south India. *Indian J. Occup. Environ. Med.* **2014**, *18*, 135. [CrossRef]
16. Yedulla, N.R.; Koolmees, D.S.; Battista, E.B.; Raza, S.S.; Montgomery, Z.A.; Day, C.S. *Upper-Extremity Injuries Are the 2nd Most Common Workplace Injuries from 1992 to 2018*; Wayne State University: Detroit, MI, USA, 2020.
17. Champoux, D.; Brun, J.P. Occupational health and safety management in small size enterprises: an overview of the situation and avenues for intervention and research. *Saf. Sci.* **2003**, *41*, 301–318. [CrossRef]
18. Hasle, P.; Limborg, H.J. A review of the literature on preventive occupational health and safety activities in small enterprises. *Ind. Health* **2006**, *44*, 6–12. [CrossRef] [PubMed]
19. Hasle, P.; Limborg, H.J.; Granerud, L. Social responsibility as an intermediary for health and safety in small firms. *Int. J. Workplace Health Manag.* **2011**, *4*, 109–122
20. Henley, E.J.; Kumamoto, H. *Designing for Reliability and Safety Control*; Prentice Hall: Upper Saddle River, NJ, USA, 1985.
21. Han, Y.; Song, Y. Condition monitoring techniques for electrical equipment-a literature survey. *IEEE Trans. Power Deliv.* **2003**, *18*, 4–13. [CrossRef]
22. Glendon, A.I.; Litherland, D.K. Safety climate factors, group differences and safety behaviour in road construction. *Saf. Sci.* **2001**, *39*, 157–188. [CrossRef]
23. Xiaoli, X.; Yunbo, Z.; Guoxin, W. Design of intelligent internet of things for equipment maintenance. In Proceedings of the 2011 Fourth International Conference on Intelligent Computation Technology and Automation, Shenzhen, China, 28–29 March 2011; Volume 2, pp. 509–511.

24. Casado-Vara, R.; Chamoso, P.; De la Prieta, F.; Prieto, J.; Corchado, J.M. Non-linear adaptive closed-loop control system for improved efficiency in IoT-blockchain management. *Inf. Fusion* **2019**, *49*, 227–239. [CrossRef]

25. Casado-Vara, R.; González-Briones, A.; Prieto, J.; Corchado, J.M. Smart contract for monitoring and control of logistics activities: pharmaceutical utilities case study. In *The 13th International Conference on Soft Computing Models in Industrial and Environmental Applications*; Springer: Berlin/Heidelberg, Germany, 2018; pp. 509–517.

26. Abdelhamid, T.S.; Everett, J.G. Identifying root causes of construction accidents. *J. Constr. Eng. Manag.* **2000**, *126*, 52–60. [CrossRef]

27. Cambraia, F.B.; Saurin, T.A.; Formoso, C.T. Identification, analysis and dissemination of information on near misses: A case study in the construction industry. *Saf. Sci.* **2010**, *48*, 91–99. [CrossRef]

28. Cheng, T.; Teizer, J. Real-time resource location data collection and visualization technology for construction safety and activity monitoring applications. *Autom. Constr.* **2013**, *34*, 3–15. [CrossRef]

29. Chi, S.; Han, S. Analyses of systems theory for construction accident prevention with specific reference to OSHA accident reports. *Int. J. Proj. Manag.* **2013**, *31*, 1027–1041. [CrossRef]

30. Choudhry, R.M.; Fang, D. Why operatives engage in unsafe work behavior: Investigating factors on construction sites. *Saf. Sci.* **2008**, *46*, 566–584. [CrossRef]

31. Fitton, D.; Sundramoorthy, V.; Kortuem, G.; Brown, J.; Efstratiou, C.; Finney, J.; Davies, N. Exploring the design of pay-per-use objects in the construction domain. In *European Conference on Smart Sensing and Context*; Springer: Berlin/Heidelberg, Germany, 2008; pp. 192–205.

32. Godfrey, A.; Conway, R.; Meagher, D.; ÓLaighin, G. Direct measurement of human movement by accelerometry. *Med Eng. Phys.* **2008**, *30*, 1364–1386. [CrossRef] [PubMed]

33. Ohtaki, Y.; Sagawa, K.; Inooka, H. A method for gait analysis in a daily living environment by body-mounted instruments. *JSME Int. J. Ser. C Mech. Syst. Mach. Elem. Manuf.* **2001**, *44*, 1125–1132. [CrossRef]

34. Zampella, F.; Khider, M.; Robertson, P.; Jiménez, A. Unscented kalman filter and magnetic angular rate update (maru) for an improved pedestrian dead-reckoning. In Proceedings of the 2012 IEEE/ION Position, Location and Navigation Symposium, Myrtle Beach, SC, USA, 23–26 April 2012; pp. 129–139.

35. Zhou, Y.; Jing, L.; Wang, J.; Cheng, Z. Analysis and selection of features for gesture recognition based on a micro wearable device. *Int. J. Adv. Comput. Sci. Appl.* **2012**, *3*, 15–24. [CrossRef]

36. Zhu, C.; Sheng, W. Human daily activity recognition in robot-assisted living using multi-sensor fusion. In Proceedings of the 2009 IEEE International Conference on Robotics and Automation, Kobe, Japan, 12–17 May 2009; pp. 2154–2159.

37. Otis, M.J.D.; Menelas, B.A.J. Toward an augmented shoe for preventing falls related to physical conditions of the soil. In Proceedings of the 2012 IEEE International Conference on Systems, Man, and Cybernetics (SMC), Seoul, Korea, 14–17 October 2012; pp. 3281–3285.

38. Yu, H.; Shi, L.C.; Lu, B.L. Vigilance estimation based on EEG signals. In Proceedings of the IEEE/ICME International Conference on Complex Medical Engineering (CME2007), Beijing, China, 23–27 May 2007.

39. Lindeman, R.W.; Yanagida, Y.; Noma, H.; Hosaka, K. Wearable vibrotactile systems for virtual contact and information display. *Virtual Real.* **2006**, *9*, 203–213. [CrossRef]

40. Cao, T.; Wan, F.; Wong, C.M.; da Cruz, J.N.; Hu, Y. Objective evaluation of fatigue by EEG spectral analysis in steady-state visual evoked potential-based brain-computer interfaces. *Biomed. Eng. Online* **2014**, *13*, 28. [CrossRef]

41. Qiang, C.; Ji-Ping, S.; Zhe, Z.; Fan, Z. ZigBee based intelligent helmet for coal miners. In Proceedings of the 2009 WRI World Congress on Computer Science and Information Engineering, Los Angeles, CA, USA, 31 March–2 April 2009; Volume 3, pp. 433–435.

42. Nithya, T.; Ezak, M.M.; Kumar, K.R.; Vignesh, V.; Vimala, D. Rescue and protection system for underground mine workers based on Zigbee. *Int. J. Recent Res. Asp.* **2018**, *4*, 194–197.

43. Kim, S.H.; Wang, C.; Min, S.D.; Lee, S.H. Safety Helmet Wearing Management System for Construction Workers Using Three-Axis Accelerometer Sensor. *Appl. Sci.* **2018**, *8*, 2400. [CrossRef]

44. Fang, Y.; Cho, Y.K.; Chen, J. A framework for real-time pro-active safety assistance for mobile crane lifting operations. *Autom. Constr.* **2016**, *72*, 367–379. [CrossRef]

45. Li, P.; Meziane, R.; Otis, M.J.D.; Ezzaidi, H.; Cardou, P. A Smart Safety Helmet using IMU and EEG sensors for worker fatigue detection. In Proceedings of the 2014 IEEE International Symposium on Robotic and Sensors Environments (ROSE) Proceedings, Timisoara, Romania, 16–18 October 2014; pp. 55–60.

46. Wahab, L.; Jiang, H. A comparative study on machine learning based algorithms for prediction of motorcycle crash severity. *PLoS ONE* **2019**, *14*, e0214966. [CrossRef]

47. Bleser, G.; Damen, D.; Behera, A.; Hendeby, G.; Mura, K.; Miezal, M.; Gee, A.; Petersen, N.; Maçães, G.; Domingues, H.; et al. Cognitive learning, monitoring and assistance of industrial workflows using egocentric sensor networks. *PLoS ONE* **2015**, *10*, e0127769. [CrossRef] [PubMed]

48. Yogameena, B.; Menaka, K.; Perumaal, S.S. Deep learning-based helmet wear analysis of a motorcycle rider for intelligent surveillance system. *IET Intell. Transp. Syst.* **2019**, *13*, 1190–1198. [CrossRef]

49. Iranitalab, A.; Khattak, A. Comparison of four statistical and machine learning methods for crash severity prediction. *Accid. Anal. Prev.* **2017**, *108*, 27–36. [CrossRef] [PubMed]

50. Pal, M. Random forest classifier for remote sensing classification. *Int. J. Remote Sens.* **2005**, *26*, 217–222. [CrossRef]

51. Rodriguez-Galiano, V.F.; Ghimire, B.; Rogan, J.; Chica-Olmo, M.; Rigol-Sanchez, J.P. An assessment of the effectiveness of a random forest classifier for land-cover classification. *ISPRS J. Photogramm. Remote Sens.* **2012**, *67*, 93–104. [CrossRef]

52. Azar, A.T.; Elshazly, H.I.; Hassanien, A.E.; Elkorany, A.M. A random forest classifier for lymph diseases. *Comput. Methods Programs Biomed.* **2014**, *113*, 465–473. [CrossRef]

53. Belgiu, M.; Drăguţ, L. Random forest in remote sensing: A review of applications and future directions. *ISPRS J. Photogramm. Remote Sens.* **2016**, *114*, 24–31. [CrossRef]

54. Jedari, E.; Wu, Z.; Rashidzadeh, R.; Saif, M. Wi-Fi based indoor location positioning employing random forest classifier. In Proceedings of the 2015 International Conference on Indoor Positioning and Indoor Navigation (IPIN), Banff, AB, Canada, 13–16 October 2015; pp. 1–5.

55. Khalilia, M.; Chakraborty, S.; Popescu, M. Predicting disease risks from highly imbalanced data using random forest. *BMC Med. Inform. Decis. Mak.* **2011**, *11*, 51. [CrossRef]

56. Shine, L.; Jiji, C.V. Automated detection of helmet on motorcyclists from traffic surveillance videos: A comparative analysis using hand-crafted features and CNN. *Multimed. Tools Appl.* **2020**, *79*, 14179–14199. [CrossRef]

57. Li, T.C.; Su, J.Y.; Liu, W.; Corchado, J.M. Approximate Gaussian conjugacy: Parametric recursive filtering under nonlinearity, multimodality, uncertainty, and constraint, and beyond. *Front. Inf. Technol. Electron. Eng.* **2017**, *18*, 1913–1939. [CrossRef]

58. Raj, K.D.; Chairat, A.; Timtong, V.; Dailey, M.N.; Ekpanyapong, M. Helmet violation processing using deep learning. In Proceedings of the 2018 International Workshop on Advanced Image Technology (IWAIT), Chiang Mai, Thailand, 7–9 January 2018; pp. 1–4.

59. Albrechtsen, E.; Solberg, I.; Svensli, E. The application and benefits of job safety analysis. *Saf. Sci.* **2019**, *113*, 425–437. [CrossRef]

60. González-Briones, A.; Chamoso, P.; Yoe, H.; Corchado, J.M. GreenVMAS: Virtual organization based platform for heating greenhouses using waste energy from power plants. *Sensors* **2018**, *18*, 861. [CrossRef] [PubMed]

61. Ali, M.S. A., Helmet deduction using image processing. *Indones. J. Electr. Eng. Comput. Sci* **2018**, *9*, 342–344. [CrossRef]

62. Preetham, D.A.; Rohit, M.S.; Ghontale, A.G.; Priyadarsini, M.J.P. Safety helmet with alcohol detection and theft control for bikers. In Proceedings of the 2017 International Conference on Intelligent Sustainable Systems (ICISS), Palladam, India, 7–8 December 2017; pp. 668–673.

63. Bisio, I.; Fedeli, A.; Lavagetto, F.; Pastorino, M.; Randazzo, A.; Sciarrone, A.; Tavanti, E. Mobile smart helmet for brain stroke early detection through neural network-based signals analysis. In Proceedings of the GLOBECOM 2017–2017 IEEE Global Communications Conference, Singapore, 4–8 December 2017; pp. 1–6.

64. Dasgupta, M.; Bandyopadhyay, O.; Chatterji, S. Automated Helmet Detection for Multiple Motorcycle Riders using CNN. In Proceedings of the 2019 IEEE Conference on Information and Communication Technology, Allahabad, India, 6–8 December 2019; pp. 1–4.

65. Cauvin, S.; Cordier, M.O.; Dousson, C.; Laborie, P.; Lévy, F.; Montmain, J.; Porcheron, M.; Servet, I.; Travé-Massuyès, L. Monitoring and alarm interpretation in industrial environments. *AI Commun.* **1998**, *11*, 139–173.

66. Gryllias, K.C.; Antoniadis, I.A. A Support Vector Machine approach based on physical model training for rolling element bearing fault detection in industrial environments. *Eng. Appl. Artif. Intell.* **2012**, *25*, 326–344. [CrossRef]
67. Kelleher, J.D.; Mac Namee, B.; D'arcy, A. *Fundamentals of Machine Learning for Predictive Data Analytics: Algorithms, Worked Examples, and Case Studies*; MIT Press: Cambridge, MA, USA, 2015.
68. Suykens, J.A.; Vandewalle, J. Least squares support vector machine classifiers. *Neural Process. Lett.* **1999**, *9*, 293–300. [CrossRef]
69. An empirical study of the naive Bayes classifier. In *IJCAI 2001 Workshop on Empirical Methods in Artificial Intelligence*. Available online: https://www.semanticscholar.org/paper/An-empirical-study-of-the-naive-Bayes-classifier-Watson/2825733f97124013e8841b3f8a0f5bd4ee4af88a (accessed on 1 November 2020).
70. D'Agostini, G. *A Multidimensional Unfolding Method Based on Bayes' Theorem*; Technical Report, P00024378; INFN: Roma, Italy, 1994.
71. Hassoun, M.H. *Fundamentals of Artificial Neural Networks*; MIT Press: Cambridge, MA, USA, 1995.
72. Kalchbrenner, N.; Grefenstette, E.; Blunsom, P. A convolutional neural network for modelling sentences. *arXiv* **2014**, arXiv:1404.2188.
73. Diekmann, J.E. Risk analysis: Lessons from artificial intelligence. *Int. J. Proj. Manag.* **1992**, *10*, 75–80. [CrossRef]
74. Hartwell, P.G.; Brug, J.A. Smart Helmet. U.S. Patent 6,798,392, 28 September 2004.
75. Hobby, K.C.; Gowing, B.; Matt, D.P. Smart Helmet. U.S. Patent 9,389,677, 12 July 2016
76. Yu, C.C.; Chu, B.H.; Chien, H.W. Smart Helmet. U.S. Patent Application 14/539,040, 14 May 2015.
77. Rasli, M.K.A.M.; Madzhi, N.K.; Johari, J. Smart helmet with sensors for accident prevention. In Proceedings of the 2013 International Conference on Electrical, Electronics and System Engineering (ICEESE), Kuala Lumpur, Malaysia, 4–5 December 2013; pp. 21–26.
78. Chandran, S.; Chandrasekar, S.; Elizabeth, N.E. Konnect: An Internet of Things (IoT) based smart helmet for accident detection and notification. In Proceedings of the 2016 IEEE Annual India Conference (INDICON), Bangalore, India, 16–18 December 2016; pp. 1–4.

Publisher's Note: MDPI stays neutral with regard to jurisdictional claims in published maps and institutional affiliations.

sensors

MDPI

Article

Imtidad: A Reference Architecture and a Case Study on Developing Distributed AI Services for Skin Disease Diagnosis over Cloud, Fog and Edge

Nourah Janbi [1], Rashid Mehmood [2,*], Iyad Katib [1], Aiiad Albeshri [1], Juan M. Corchado [3,4,5] and Tan Yigitcanlar [6]

1 Department of Computer Science, Faculty of Computing and Information Technology, King Abdulaziz University, Jeddah 21589, Saudi Arabia; njanbi0006@stu.kau.edu.sa (N.J.); iakatib@kau.edu.sa (I.K.); aaalbeshri@kau.edu.sa (A.A.)
2 High Performance Computing Center, King Abdulaziz University, Jeddah 21589, Saudi Arabia
3 Bisite Research Group, University of Salamanca, 37007 Salamanca, Spain; corchado@usal.es
4 Air Institute, IoT Digital Innovation Hub, 37188 Salamanca, Spain
5 Department of Electronics, Information and Communication, Faculty of Engineering, Osaka Institute of Technology, Osaka 535-8585, Japan
6 School of Architecture and Built Environment, Queensland University of Technology, 2 George Street, Brisbane, QLD 4000, Australia; tan.yigitcanlar@qut.edu.au
* Correspondence: rmehmood@kau.edu.sa

Citation: Janbi, N.; Mehmood, R.; Katib, I.; Albeshri, A.; Corchado, J.M.; Yigitcanlar, T. Imtidad: A Reference Architecture and a Case Study on Developing Distributed AI Services for Skin Disease Diagnosis over Cloud, Fog and Edge. *Sensors* **2022**, *22*, 1854. https://doi.org/10.3390/s22051854

Academic Editor: James (Jong Hyuk) Park

Received: 5 January 2022
Accepted: 21 February 2022
Published: 26 February 2022

Publisher's Note: MDPI stays neutral with regard to jurisdictional claims in published maps and institutional affiliations.

Abstract: Several factors are motivating the development of preventive, personalized, connected, virtual, and ubiquitous healthcare services. These factors include declining public health, increase in chronic diseases, an ageing population, rising healthcare costs, the need to bring intelligence near the user for privacy, security, performance, and costs reasons, as well as COVID-19. Motivated by these drivers, this paper proposes, implements, and evaluates a reference architecture called Imtidad that provides Distributed Artificial Intelligence (AI) as a Service (DAIaaS) over cloud, fog, and edge using a service catalog case study containing 22 AI skin disease diagnosis services. These services belong to four service classes that are distinguished based on software platforms (containerized gRPC, gRPC, Android, and Android Nearby) and are executed on a range of hardware platforms (Google Cloud, HP Pavilion Laptop, NVIDIA Jetson nano, Raspberry Pi Model B, Samsung Galaxy S9, and Samsung Galaxy Note 4) and four network types (Fiber, Cellular, Wi-Fi, and Bluetooth). The AI models for the diagnosis include two standard Deep Neural Networks and two Tiny AI deep models to enable their execution at the edge, trained and tested using 10,015 real-life dermatoscopic images. The services are evaluated using several benchmarks including model service value, response time, energy consumption, and network transfer time. A DL service on a local smartphone provides the best service in terms of both energy and speed, followed by a Raspberry Pi edge device and a laptop in fog. The services are designed to enable different use cases, such as patient diagnosis at home or sending diagnosis requests to travelling medical professionals through a fog device or cloud. This is the pioneering work that provides a reference architecture and such a detailed implementation and treatment of DAIaaS services, and is also expected to have an extensive impact on developing smart distributed service infrastructures for healthcare and other sectors.

Keywords: tiny AI; tiny ML; distributed AI as a service (DAIaaS); fog computing; edge computing; cloud computing; skin disease diagnosis; healthcare; smart societies; smart cities; smart healthcare; reference architecture; TensorFlow

1. Introduction

Smart cities and societies are at the vanguard of driving digital transformation [1–5]. The digital transformation process involves developing digital services and systems that allow us to sense, analyze, and act on our environment with the optimality of our objectives [6,7]. Various industrial sectors are undergoing this transformation and healthcare is among the most critical sectors in need of this [8]. Several drivers are motivating the need to transform healthcare and develop preventive, personalized, connected, virtual, and everywhere healthcare services and systems [9–12]. These drivers include, among others, declining public health (due to processed food, lifestyles, etc.), increase in chronic diseases (e.g., hypertension, diabetes, heart disease), ageing population, decreasing quality of healthcare, and rising healthcare costs for the public and governments [6]. Due to the restrictions placed because of COVID-19, the difficulty of accessing to public healthcare has aggravated and amplified the need for virtual and everywhere healthcare [4].

The technology-related drivers to provide distributed services include the need to bring intelligence near the user (at the fog and edge layers) for reasons such as privacy, security, performance, and costs [13–18]. These drivers are not specific to healthcare alone and are driving all of the sectors in which data is generated at the edge and/or in which decisions need to be made instantaneously and intelligently by the user at the edge [19–23].

Motivated by these drivers, this paper proposes, implements, and evaluates a reference (software) architecture called Imtidad that provides distributed Artificial Intelligence (AI) as a Service (DAIaaS) over the cloud, fog, and edge layers using a case study of a service catalog with 22 Deep Learning-based skin disease diagnosis services. These services belong to four service classes that are distinguished by software platforms (containerized gRPC, gRPC, Android, and Android Nearby) and are executed on a range of hardware platforms (Google Cloud, HP Pavilion Laptop, NVIDIA Jetson nano, Raspberry Pi Model B, Samsung Galaxy S9, and Samsung Galaxy Note 4) and four network types (Fiber, Cellular, Wi-Fi, and Bluetooth). A selection of four AI models are provided for the diagnosis; two of these are standard Deep Neural Networks, and the other two are Tiny AI versions to enable their execution on smaller devices at the edge. The models have been trained and tested on the HAM10000 dataset containing 10,015 dermatoscopic images.

The services have been evaluated against several benchmark criteria, including model service value, processing time, response time, data transfer rate, energy consumption, and network transfer time. The service values have been computed and compared in terms of their speed and energy consumption. A Deep Learning (DL) service on a local smartphone provides the best service in terms of energy, followed by a Raspberry Pi edge device. A DL service on a local smartphone provides the best service (also in terms of speed), followed by a laptop device in the fog layer.

Imtidad is an Arabic word indicating the "extending" or "extension" (to the cloud, fog, and edge) nature of our reference architecture. The services are being extended in both directions, from cloud to fog and edge, and from edge to fog and cloud.

To help the reader conceptualize the proposed work, Figure 1 provides a high-level view of the Imtidad reference architecture. The reference architecture is described at length in this paper. The three perspectives of the reference architecture are: the service development and deployment perspective, the user view perspective, and the validation perspective. The service development and deployment perspective provides guidelines on developing and operationalizing the services: an application such as skin lesion diagnosis is selected for the provision of related distributed services followed by acquiring the necessary data, AI model designs, service use cases, service design, composing these into a service catalog, porting these to the execution platforms and networks, operationalizing the services, evaluating and validating them against benchmark criteria, medical professionals, and other sources of knowledge. The user view perspective includes selecting and requesting a service from the service catalog, receiving the diagnosis, and validating it. The validation perspective is shared with both the user view and the service designers and providers view because it is meant to allow all of them, as well as third parties, such as auditors, to validate. A more detailed view and discussion of the reference architecture is provided in Sections 3 and 4.

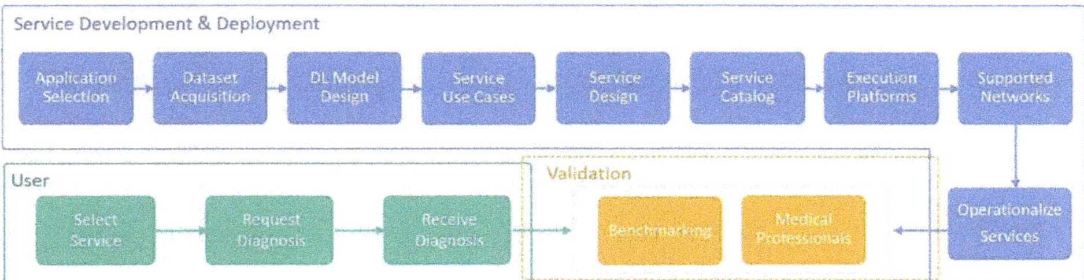

Figure 1. The Imtidad reference architecture (a high-level view).

The contributions of this paper can be outlined as follows:

- This is the first paper in which a reference architecture for distributed AI-as-a-service is proposed and implemented; a healthcare application (skin lesion diagnosis) is developed and studied in great detail, with a catalog containing several AI and Tiny AI services supported on multiple software, hardware, and networking platforms; and several use cases are evaluated using multiple benchmarks.

- The services are designed considering innovative use cases, such as a patient at home taking images of their skin lesion and performing the diagnosis by themselves with the help of a service or a travelling medical professional requesting a diagnosis from a fog device or cloud. The users of the services provided by this architecture can be patients, medical professionals, the patients' family members, or any other stakeholder. Similarly, the services can be used by someone who has the disease diagnosis model, or the image, or both, since the resource (image or model) may be requested from other providers.

- The proposed work is highly novel and is expected to produce high impact due to the developed reference architecture; the service catalog offering a large number of services; the potential for the implementation of innovative use cases through the edge, fog, and cloud as well as their evaluation on many software, hardware, and networking platforms; and a detailed description of the architecture and case study.

- The existing works on distributed AI either focus on distributed AI methodologies [24,25] or distributed applications development [26–28], or application migration to fog and edge [29,30]. In contrast, this paper broadly aims to provide theoretical and applied contributions on decoupling application development from AI by using the distributed AI as a Service (DAIaaS) concept to coordinate, standardize, and streamline existing research on distributed AI and application migration to the edge. The decoupling of application development from AI is needed because it allows application, sensor, and IoT developers to focus on the various domain-specific details, relieve them from worries related to the how-to of distributed training and inference, and help systemize and mass-produce technologies for smarter environments. The Imtidad reference architecture and case study, given in this paper, outlines the whole process and roadmap of developing a service catalog using distributed AI as a Service, and, essentially, this provides a blueprint and procedure for decoupling applications and AI, enabling smart application development as a foundation for smarter societies. The approach allows development of unified interfaces to facilitate both independent and collaborative software development across different application domains. This is a continuation of our earlier research, where a DAIaaS concept was proposed and investigated using simulations [13].

The rest of the paper is organized as follows. Section 2 reviews the related works. Section 3 describes the reference architecture, methodology, and service catalog. Section 4 details the system architecture and design for the skin disease diagnosis case study. Section 5 provides results and their analysis. Section 6 concludes the paper and provides future lines of research.

2. Related Works

This section reviews the literature related to topics of this paper, distributed AI for skin diseases diagnosis over the edge. Section 2.1 discusses the works related to distributed artificial intelligence over cloud, fog, and edge. Section 2.2 reviews the works related to skin disease diagnosis using AI and Section 2.3 discusses the research gap.

2.1. Distributed Artificial Intelligence (DAI) over Cloud, Fog, and Edge

Distributed Artificial Intelligence (DAI) allows AI to be distributed across multiple agents, processes, cores, physical, or virtual, computational nodes with the aim of sharing data, improving data processing capabilities, and providing faster, privacy-preserved, node-local, global, or system-wide solutions [13]. Distributed AI on clouds has been the focus of many proposals, [31,32], for intensive computation or global knowledge sharing. Edge Intelligence (EdgeAI) and fog intelligence are among the main DAI approaches where AI models are distributed across fog nodes (intermediate nodes between edge and cloud layers) or network edges [33]. Models can be pre-trained on powerful machines (cloud), then modified and optimized to run in the resource-constrained edges. Edges, fogs, and cloud can also collaborate where some of the pre-processing and less-intensive computations are performed in edges and global processing performed in the cloud [13].

Several research studies have discussed the convergence of edge, fog, and AI, as well as their various architectures [30,34–40]. Pattnaik et al. [41] have proposed and evaluated different approaches to distribute ML across cloud and edge layers, including a variety of distributed edge and cloud-based training and inference with either local or global knowledge. Muhammed et al. [33] proposed UbiPriSEQ, a framework to optimize privacy, security, energy efficiency, and quality of service (QoS). UbiPriSEQ uses Deep Reinforcement Learning to optimize local processing and offloading on edge, fog, and cloud. Sparse matrix-vector multiplication (SpMV) is used as an application to implement and evaluate the proposed framework UbiPriSEQ. In our earlier work, Janbi et al. [13], we proposed a DAIaaS framework to standardize distributed AI provisioning across all layers (edge, fog, and cloud) aiming to facilitate the process of generic software development across different application domains and allow for developers to focus on the domain-specific details rather than how-to develop and deploy distributed AI. To this end, multiple case studies and several scenarios, applications, distributed AI delivery models, sensing modules, and software modules were developed to explore various architectures and understand performance barriers.

Another recently emerging direction is Federated Learning (FL), where edge devices collaborate to train ML models. Model aggregation can be performed centrally in the cloud or be distributed between nodes. Gao et al. [42] have proposed a cloud-edge collaborative learning framework with an elastic local update method. In addition, the n-soft synchronization approach has been proposed that combines both synchronous and asynchronous approaches. Chen et al. [43] have proposed a federated transfer learning approach for healthcare wearables to train global models across different organizations securely. Fully decentralized FL approaches, where no central server and models are aggregated directly by edge devices, have also been proposed in the literature. Hegedűs et al. [44] have provided a comparison of central FL and decentralized FL as well as introduced two optimization techniques for decentralized FL, a token-based flow control and partitioned models subsampling. Kim et al. [45] have proposed an architecture of FL based on blockchain technology to enable secure local model exchange. Both verification and rewards systems are designed to support the exchange process between edges. The existing works on federated learning have focused on federated training over distributed devices, while our work differs from it and complements it, both in the broad aims of our research and the specific contributions of this paper (as highlighted in Section 2.3 (Research Gap) and elsewhere in the paper).

2.1.1. Tiny AI and Edge: Research and Frameworks

Table 1 gives a summary of research papers that utilized Tiny AI models, i.e., lighter versions of AI models on edge devices. Tiny AI models are customized AI models that are optimized or compressed to minimize the requirements for model memory and computation power. All the listed research has used TensorFlow Lite [46] to optimize and deploy the AI models locally. For each research in the table, the application domain, the specific application under that domain, and the adopted AI model are specified. Zebin et al. [47] have designed and implemented a tiny CNN model to optimally monitor human activity recognition using mobile devices. In the domain of the autonomous vehicles, a traffic sign recognition Tiny DL model based on Single Shot MultiBox Detector (SSD) has been developed by Benhamida et al. [48]. Alsing [49] has evaluated different tiny AI models for note detections in a smart home environment. For the security domain, Zeroual et al. [50] have developed a face recognition authentication model on mobile devices to authenticate users before accessing cloud services. Alternatively, Ahmadi et al. [51] have proposed an intelligent local malware detection approach for android devices based on random forests classifier. Soltani et al. [52] have developed a Tiny Deep CNN model for Signal Modulation Classification that identifies signals SNR region for wireless networks. A Tiny AI model on Unmanned Aerial Vehicles (UAV) has been proposed by Domozi et al. [53] to detect objects in search and rescue missions.

Regarding the deployment of AI at the edges, a few frameworks have been proposed and developed to run AI models on edge devices. These include Caffe2 [54], TensorFlow Lite [46], and PyTorch Mobile [55]. These frameworks support various edge platforms such as Android, iOS, and Linux and customize AI models to fit within the resource-constrained edge.

2.1.2. Distributed AI in Healthcare

EdgeAI is still in its infancy and attracting more researchers and companies to bring AI closer to users [34]. It aims to provide distributed, low-latency, reliable, scalable, and private AI services [35]. Many applications that require real-time responses can utilize edgeAI, such as autonomous vehicles, smart homes, smart cities, and security [47–53]. There are some works that have considered distributed AI for healthcare, which is the focus of this work too. Zebin et al. [47] have proposed a human activity recognition framework to run on mobile devices. They used batch normalization for CNN recognition tasks using data from wearable sensors. Isakov et al. [31] have developed a monitoring and detection system that aims to detect falls accurately through the use of mobile devices. The mobile devices are used for preprocessing and they perform a non-linear analysis on the cloud. Hassan et al. [32] proposed a remote pain monitoring system based on a fog-based architecture to process patient biopotential signals locally and detect pain in a real-time manner. They offloaded some of the processing to the cloud in case of local resource shortage and provided remote access through a web application. Muhammed et al. [56] have addressed the challenges of meeting network quality of service (QoS) requirements including network latency, bandwidth, and reliability challenges for delivering real-time mobile healthcare services.

Table 1. Related works: Tiny AI at the edge.

Reference	Application Domain	Application	AI Model
Zebin et al. [47]	Monitoring and Healthcare Systems	Human Activity Recognition	Custom CNN Model
Benhamida et al. [48]	Autonomous Vehicles	Traffic Sign Recognition	SSD MobileNetV2
Zeroual et al. [50]	Security (Authentication)	Face Recognition	VggNET

Table 1. *Cont.*

Reference	Application Domain	Application	AI Model
Alsing [49]	Smart Homes	Object (Notes) Detection	R-CNN, SSD, and Tiny YOLO
Soltani et al. [52]	Wireless Networking	Signal Modulation Classification	DeepSig CNN
Domozi et al. [53]	UAVs Search and Rescue	Object Detection	SSD
Ahmadi et al. [51]	Security	Malware Detection	IntelliAV (Random Forests)

2.2. Skin Lesion Diagnosis

Health information technology systems such as clinical decision support (CDS) systems are designed to support physicians and other health professionals in their decision-making tasks. AI based Computer-Aided Diagnosis (CAD) systems have been subject to rapidly growing interest for the diagnosis of skin disease [57]. They are used as a "second opinion" tool that assists radiologists and physicians in image interpretations and diseases diagnosis. There has been a continuous increase in skin cancer cases rates around the world, so, given that it is the most common cancer in the United States and worldwide [58], more research must be done in this area. Especially, since an accurate and early diagnosis of skin cancer would improve treatment and survival rates [59]. Computer vision algorithms are used to analyze images and identify abnormal structures. This helps professionals to detect the earliest signs of abnormality and support their evaluation. Clinical imaging and dermatoscopy are now considered to be an essential part of the dermatology clinics for diagnosis, treatment, follow-up, and documentation [60,61]. Skin diagnosis (and identifying benign and malignant skin lesions) is an important factor in the early detection and prevention of skin cancer. Automated skin diagnosis using dermoscopy and AI might also let patients avoid skin biopsy [62]. DL is one of the AI approaches that are becoming very popular for dermoscopic images classification problem. This has been boosted by the introduction of many dermoscopic datasets that are publicly available [57]. These datasets consist of labeled images belonging to various types of benign and cancerous skin lesions. Training DL model with such datasets would create an appropriate and accurate model for CAD systems.

Several research studies have been proposed in the literature aiming to improve the accuracy of skin diagnosis [63–72]. Convolutional neural networks (CNN) are adopted in most proposals [63–71], except in [72] where the authors proposed fuzzy classification for skin lesion segmentation. Some proposals have considered other information or data in the diagnosis process such as demographic and medical history [66] and sonification (audio) [73]. Pretrained CNN models have been retrained and evaluated in [63,65–69,71] and multiple CNN models have been ensembled in [64,66,69,70,73]. A review of DL segmentation, classification, and pre-processing techniques for skin lesion detection is provided in [74]. Table 2 summarizes the literature that has been reviewed in this subsection, related to skin disease diagnosis.

Table 2. Related works: skin disease diagnosis.

Reference	AI	Approach	Classes	Datasets
Jha et al. [63]	Double U-Net	Segmentation	7	MICCAI 2015, CVC-ClinicDB, ISIC-2018, and Science Bowl 2018
Bajwa et al. [64]	CNN Ensemble	Classification	7–23	ISIC 2018 and DermNet

Table 2. *Cont.*

Reference	AI	Approach	Classes	Datasets
Zhang et al. [65]	CNN	Classification	4	Clinical dataset Peking Union Medical College Hospital
Wei et al. [66]	CNN Ensemble	Classification	7	ISIC 2018
Liu et al. [67]	CNN on multi-image, demographic information and medical history	Classification	27	Collected from teledermatology practice in U.S.
Gavrilov et al. [68]	CNN	Classification	2	ISIC 10,000 expanded to 1,000,000 using distortions
Garcia et al. [72]	Fuzzy algorithm	Segmentation	3	ISIC 2016 and ISIC 2017

2.3. Research Gap

The literature review presented in this section has evidenced the current research gap with no earlier reference architectures on DAIaaS and no implementations of skin disease diagnosis on fog and edge. This is the first research where a reference architecture for DAIaaS is proposed and implemented, and a healthcare disease diagnosis service is developed and studied in great detail, with a catalog containing several AI and Tiny AI services supported on multiple software, hardware, and networking platforms, as well as several use cases evaluated using multiple benchmarks. The services are designed to enable different use cases such as a patient at home taking images of their skin lesion and performing the diagnosis by themself with the help of a service, or a travelling medical professional requesting a diagnosis from a fog device or cloud. The users of the service can be patients, medical professionals, the family members of the patient, or any other stakeholder. Similarly, the services can be used by someone who has the disease diagnosis model, or the image, or both, by requesting the required resource (image or the model) from other providers. The novelty and high impact of this research lies in the developed reference architecture, the service catalog offering many services, the potential for the implementation of innovative use cases through the edge, fog, and cloud, and their evaluation on many software, hardware, and networking platforms, as well as a detailed description of the architecture and case study.

Commenting on the specific application we have selected for this paper, i.e., skin disease diagnosis (this comment applies to similar applications), it is important to note that having an accurate disease diagnosis model is not enough; the deployment of the model for real-time usage is an essential part of the AI system development. This includes where and how the model is going to be installed. First, both model size and complexity will influence the processing or inference time, especially with resource constrained devices. In addition, the emerging trend of virtual and mobile services including healthcare services, which are required as a result of the current COVID-19 pandemic, will require innovative and flexible architectures to support them. Therefore, the development of quick and accurate diagnosis methods for physicians must intrinsically consider in their designs the distributed architectures that these diagnosis methods will be deployed on.

3. Imtidad Reference Architecture, Methodology, and Service Catalog

This section describes our proposed Imtidad reference architecture for creating distributed AI services over the cloud, fog, and edge layers and describes the service catalog, service use cases, and the service evaluation benchmarks. The section is organized as follows. The reference architecture overview is provided and elaborated in Section 3.1. A series of use cases (e.g., a user takes a photo of a lesion on their skin and instantaneously attempts to diagnose it using their preferred service from the service catalog) are outlined in Section 3.2. An implementation of the reference architecture using a service catalog, designed as part of this research, is described in Section 3.3. A description of execution platforms is provided in Section 3.4. The metrics that have been used to evaluate and compare the services are defined and explained (service energy consumption and service values) in Section 3.5.

3.1. Reference Architecture and Methodology Overview

The Imtidad reference architecture is proposed as a blueprint and procedure for decoupling applications and AI and streamlining the design and deployment of distributed AI services over the cloud, fog, and edge layers. Figure 2 depicts the Imtidad reference architecture for the skin disease diagnosis case study. The figure can be considered an insanitation or refinement of the Imtidad reference architecture for a given application; in this case skin disease diagnosis. The architecture lists all required services to create new DAIaaS services from the selection of the application to service production and operations. Each of the rectangular blocks (e.g., Service Design) in the figure can be considered a component or a service, and these services can independently and asynchronously talk to each other to create services and service catalogs.

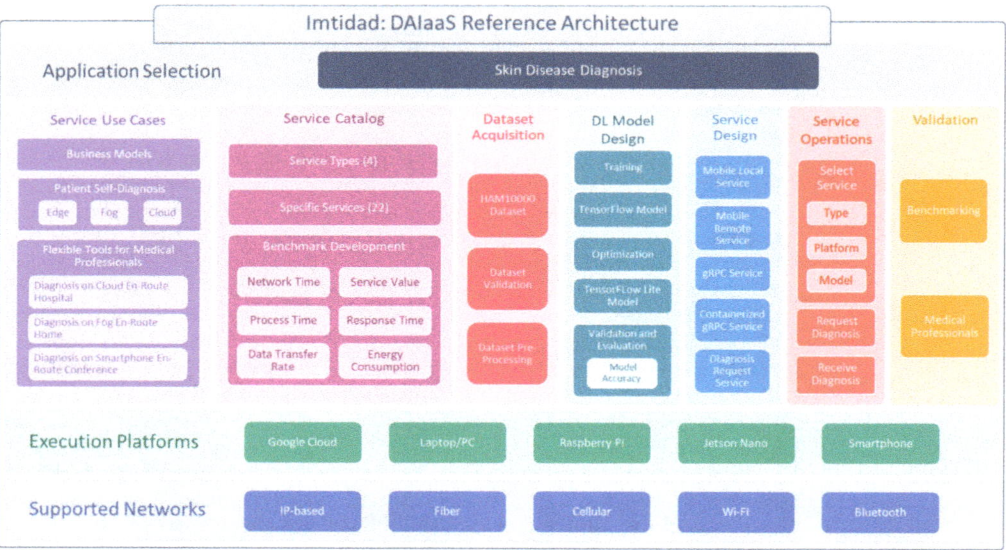

Figure 2. Imtidad reference architecture.

Figure 3 depicts a sequential workflow diagram for creating a skin disease diagnosis catalog. It is created by refining Imtidad Reference Architecture. The service development and deployment process begins with a selection of an application domain, in this case, skin disease diagnosis. A dataset is required for the selected application, so that the designed model may be trained and validated. The dataset acquisition process includes dataset validation and pre-processing in preparation for training. Then, Deep Learning models are designed, trained, optimized, and validated. First, the TensorFlow (TF) model is generated, then, an optimized version is created, which, in this case, was the TensorFlow Lite (TFLite) model. Use cases are determined considering possible scenarios and business models. After that, different types of services may be designed to provide support in a series of scenarios. A service catalog is created to communicate and present various service models to users (see Table 3 and Section 3.2 for details). In addition, service providers need to find a way to benchmark services by developing evaluation metrics such as service values, energy consumption, and response time. Several execution platforms and networks are selected, and the designed services are deployed. When the services are ready for operation, the users can choose one of the services from the catalog and send their diagnosis request. External opinion might be required for validation, in this case healthcare professional opinion can be used to validate the predicted diagnosis. Validation can be done by users, service designers and providers, or a third party such as auditors.

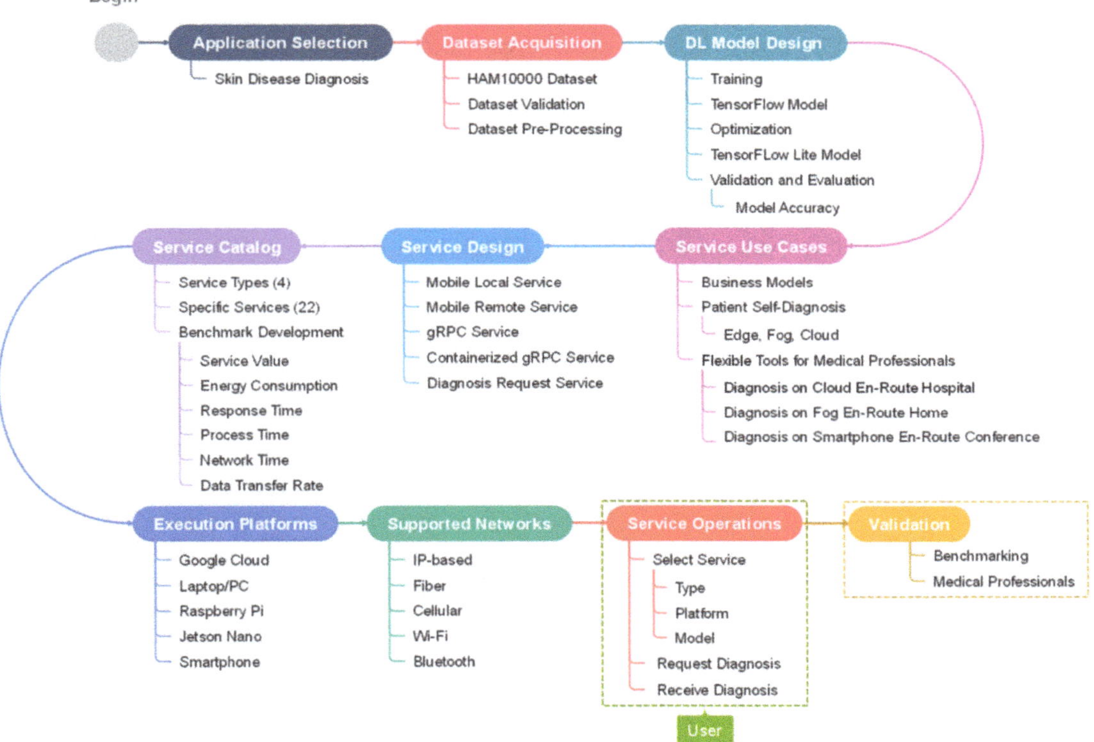

Figure 3. Workflow Diagram for creating a skin disease diagnosis catalog refined from Imtidad Reference Architecture.

3.2. Service Use Cases

Use cases are identified considering possible scenarios and business models for provisioning distributed AI services and skin disease diagnosis services, over the cloud, fog, and edge layers. These have been used to design a variety of services that suit different conditions and requirements. Services are listed in a services catalog for the user to select one of them and use it to diagnose a lesion image. The design of skin disease diagnosis services involves and concerns all parties including patients, patients' families, medical professionals, and, even, service providers. Patients and medical professionals are the direct users of the system and they are looking for instantaneous results and services available all the time and everywhere, while service providers aim for users' satisfaction by providing high QoS and at the same time protecting their product and copyrights.

Local services in smartphones, where model and image classification tasks are performed locally in the user device, guarantee a real-time response with no requirement for an Internet connection, and will preserve the user's privacy as the images stay on the user's device. This kind of service can be used by patients or doctors anywhere using their own smartphones. However, this will only work if the user's device has the required resources needed to store and run AI models, and model accuracy may be compromised when converted into the Tiny version. On the other hand, remote services in smartphones, would extend the service capability and enable collaboration between edge devices. Services from nearby devices can be used when the users' devices are either unable to process the image locally or they are looking for more accurate results. In this case, users can collaborate and provide services to each other without having to share their models. In addition, the DL model service providers may also want to keep their model's copyrights and not share them, and at the same time, they want to guarantee service availability. To accomplish that, the service provider can provide a secure device (smartphone) in the facility (e.g., clinic) or with the medical professional to carry anywhere. In this case, skin images will be sent to the local device in the local network but not through the internet, which will provide some level of privacy for the users.

Table 3. The Imtidad service catalog.

	Service Type	Layers	Platform	Platform Specifications	Platform Energy	Network Specification	Latency	Model
1 2 3 4	Containerized gRPC Service	Cloud or Fog	Google Cloud Compute Node	2 CPUs 8 GB Memory 80 Concurrency	100 W [75]	Fiber or Cellular	Depends on Network	A B ALite BLite
5 6 7 8		Fog or edge	HP Pavilion Laptop	CPU: Intel® Core™ i7-8550U @ 1.80 GHz (Turbo up to 4.00 Ghz) 8 GB Memory	Idle: 10.2 W Working: 66.3 W	Wireless LAN Frequency Band: 2.4 GHz/5 GHz Data rate < 450 Mbps	Partially Configurable	A B ALite BLite
9 10 11 12	gRPC Service	Fog or Edge	NVIDIA Jetson nano	GPU: 128-core Maxwell CPU: Quad-core ARM A57 @ 1.43 GHz 4 GB Memory	5–10 W	Wireless LAN Frequency Band: 2.4 GHz/5 GHz Data rate < 450 Mbps	Partially Configurable	A B ALite BLite
13 14 15 16		Fog or Edge	Raspberry Pi Model B (8 GB)	1.5 GHz Quad-core ARM Cortex-A72 8 GB Memory 2.4/5.0 GHz IEEE 802.11ac wireless	Idle: 2.7 W Working: 5.1 W	Wireless LAN Frequency Band: 2.4 GHz/5 GHz Data rate < 450 Mbps	Partially Configurable	A B ALite BLite
17 18		Fog or Edge	Raspberry Pi Model B (4 GB)	1.5 GHz Quad-core ARM Cortex-A72 4 GB Memory 2.4/5.0 GHz IEEE 802.11ac wireless	Idle: 2.7 W Working: 5.1 W	Wireless LAN Frequency Band: 2.4 GHz/5 GHz Data rate < 450 Mbps	Partially Configurable	ALite BLite
19 20	Mobile Local Service	Edge	Samsung Galaxy S9	GPU: ARM Mali-G72 MP18 Octa-Core, 2 CPUs: 2.7 Ghz Quad-Core Mongoose M3 1.8 Ghz Quad-Core ARM Cortex-A55 4 GB Memory	Idle: 1.09 W Working: 5.16 W	Wireless LAN Frequency Band: 2.4 GHz/5 GHz Data rate < 450 Mbps	Partially Configurable	ALite BLite
21 22	Mobile Remote Service	Edge	Samsung Galaxy Note 4	GPU: ARM Mali-T760 MP6 Octa-Core, 2 CPUs: 1.9 GHz Quad-core ARM Cortex-A57 1.3 GHz Quad-core ARM Cortex-A53 4 GB Memory	Idle: 1.4 W Working: 9.4 W	Wireless LAN Frequency Band: 2.4 GHz/5 GHz Data rate < 450 Mbps	Partially Configurable	ALite BLite

Mobile devices (smartphones) are limited in their capabilities, therefore, devices such as laptops, NVIDIA Jetson nano, and Raspberry Pi can be used in edge or fog layers to run more complicated models or serve a large number of users simultaneously. These devices can be provided by service providers can and placed in hospitals, clinics, or, even, homes, to serve medical professionals and other users. Devices at the edge or fog layers would increase service availability and the level of user privacy and security. Nevertheless, they are incomparable with the cloud where resources are almost unlimited. The cloud is the original service provisioning platform for AI applications though services provided from the cloud have a higher latency and more congested networks. Services at the cloud can be used in case other local services at edge or fog layers are busy or absent. Moreover, DL model service can be resides in the cloud, and data or local models can be uploaded to it for model retraining to improve the global model accuracy.

3.3. Service Catalog

The service catalog lists all diagnosis services with their characteristics for the users to choose from. Diagnosis services are responsible for image classification. A total of 22 services are produced from a combination of various types of services, devices, and models (see Table 3) that suit different purposes. For each service, the service type, layers, devices, network, and models are listed. There are four different skin disease diagnosis service types, namely, local mobile service, remote mobile service, gRPC service, and containerized gRPC service. These services can be run on different layers of the network architecture including cloud, fog, and edge. Seven different devices are used for evaluation that varies in their capabilities. Google cloud virtual machines (VMs), a laptop, an NVIDIA Jetson nano, two Raspberry pi (4G and 8G), and two mobile devices (Samsung Galaxy S9 and Samsung Galaxy Note 4). Wi-Fi local area network (LAN) and the Internet wide area network (WAN) are both considered, including fiber and cellular networks. An Internet connection is required for cloud communications, but all other levels are deployed in the local network which means that their traffics is going through a Wi-Fi modem. Nevertheless, they may be deployed farther than this on a base station on other LANs close to the user. The four developed models (A, ALite, B, and BLite) are considered for all devices, though only ALite and BLite are possible for some devices due to device capability limitations. This service catalog is designed for our specific case study to show a practical example of service catalogs. This means that all sorts of devices and networks could be used to design the user's services, and they are not limited to what is specified here. Table 4 lists the acronyms and their definitions that have been used use throughout the paper for the 22 services in the service catalog.

Table 4. The acronyms used for the services and service definitions.

	Acronym	**Definition**
1	CloudA	Model A (Executed on Cloud)
2	CloudB	Model B (Executed on Cloud)
3	CouldALite	Model ALite (Executed on Cloud)
4	CloudBLite	Model BLite (Executed on Cloud)
5	FogA	Model A (Executed on Fog—HP Laptop)
6	FogB	Model B (Executed on Fog—HP Laptop)
7	FogALite	Model ALite (Executed on Fog—HP Laptop)
8	FogBLite	Model BLite (Executed on Fog—HP Laptop)
9	JetsonA	Model A (Executed on Jetson)
10	JetsonB	Model B (Executed on Jetson)
11	JetsonALite	Model ALite (Executed on Jetson)
12	JetsonBLite	Model BLite (Executed on Jetson)
13	Rasp8A	Model A (Executed on Raspberry pi 8 GB)

Table 4. *Cont.*

	Acronym	**Definition**
14	Rasp8B	Model B (Executed on Raspberry pi 8 GB)
15	Rasp8ALite	Model ALite (Executed on Raspberry pi 8 GB)
16	Rasp8BLite	Model BLite (Executed on Raspberry pi 8 GB)
17	Rasp4ALite	Model ALite (Executed on Raspberry pi 4 GB)
18	Rasp4BLite	Model BLite (Executed on Raspberry pi 4 GB)
19	MobileLALite	Model ALite (Executed on Local Mobile—Galaxy S9)
20	MobileLBLite	Model BLite (Executed on Local Mobile—Galaxy S9)
21	MobileRALite	Model ALite (Executed on Remote Mobile—Galaxy Note 4)
22	MobileRBLite	Model BLite (Executed on Remote Mobile—Galaxy Note 4)

3.4. Devices and Hardware Platforms

Seven different execution platforms are adopted in the service catalog. Google Cloud Run is selected for the cloud services which is a serverless platform that facilitates running invocable Docker container images via requests or events. Services are the main resources of the Cloud Run and each has a unique and permanent URL. Services are created by deploying a container image on infrastructure that is fully managed and optimized by Google. Service configuration includes maximum allocated memory limit, number of assigned virtual CPUs (vCPUs), and maximum number of requests (concurrent requests). An HP Pavilion laptop has been used as the fog node in our experiments. It comprises an Intel® Core™ i7-8550U CPU and 8 GB Memory. The CPU has a total of 4 cores and 8 threads with a base frequency of 1.80 Ghz and a maximum single-core turbo frequency of 4.00 Ghz. Two types of single-board computers have been used NVIDIA Jetson nano and Raspberry Pi. NVIDIA Jetson nano is a platform designed by NVIDIA to run AI applications at the edge. The used Jetson Developer Kit is equipped with 128-core NVIDIA Maxwell™ architecture-based GPU, Quad-core ARM® A57, and 4 GB 64-bit Memory. Figure 4 gives a brief of Jetson nano specifications and a picture of the device. Raspberry Pi is a tiny and low-cost single-board computer. Several generations of Raspberry Pi have been released during the years. In this research, two Raspberry Pi 4 Model Bs have been used. Both cards have the same Quad-core ARM Cortex-A72 processor, but one has 4 GB memory and the other has 8 GB memory. Figure 5 gives a brief of Raspberry Pi specifications and a picture of the device. Two Samsung smartphones have been used, Galaxy S9 and Galaxy Note 4. Samsung Galaxy S9 comes with ARM Mali-G72 GPU and Octa-Core CPU (Quad-Core Mongoose M3 and Quad-Core ARM Cortex-A55), Samsung Galaxy Note 4 comes with ARM Mali-T760 GPU and Octa-Core CPU (Quad-core ARM Cortex-A57 and Quad-core ARM Cortex-A53), and both have 4 GB memory. Figure 6 gives a brief of the smartphone's specifications and provides pictures for both smartphones. A full depiction of the Imtidad testbed is given in Section 4.

Figure 4. NVIDIA Jetson Nano.

Raspberry Pi 4 Model B

Two Raspberry Pi 4 Model B

- Processor
    - A Broadcom BCM2711 system-on-chip, and it runs on a 1.5-GHz quad-core 64-bit ARM Cortex-A72 CPU @ 1.5 GHz

- Memory
    - 4GB LPDDR4-3200 SDRAM (Device Type 1)
    - 8GB LPDDR4-3200 SDRAM (Device Type 2)

Figure 5. Raspberry Pi 4 Model B.

Samsung Galaxy S9

- GPU: ARM Mali-G72 MP18

- Octa-Core CPU
    - 2.7Ghz Quad-Core Mongoose M3
    - 1.8Ghz Quad-Core ARM Cortex-A55

- Memory: 4 GB

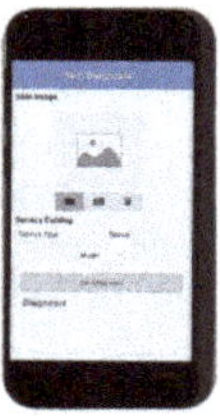
Samsung Galaxy S9

Samsung Galaxy Note4

- GPU: ARM Mali-T760 MP6

- Octa-Core CPU
    - 1.9GHz Quad-core ARM Cortex-A57
    - 1.3GHz Quad-core ARM Cortex-A53

- Memory: 4 GB

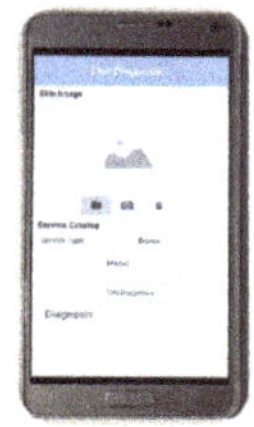
Samsung Galaxy Note4

Figure 6. Samsung Galaxy Smartphones.

These platforms can be located in different layers at cloud, fog, or edge. The main difference between these layers is the place where processing occurs. The cloud is located far away from the users on datacenter/s and accessed through an Internet connection, Wide Area Network (WAN). On the other hand, fog is located near users and the edge, on the same Local Area Network (LAN) or a near LAN, and it does not require an Internet connection. Fog devices might be located in streets, base stations, houses, cafes, hospitals, etc., to serve local users, while the cloud is designed to serve a large number of users. The cloud provides resources on-demand and can scale up easily. Though cloud and fog might have the same type of CPUs, cloud can increase the number of located CPUs on request or with high demands while fog resources are limited. In our case study, the cloud is the Google datacenter, specifically the Google Cloud Run platform. For the Containerized gRPC Service, two CPUs are allocated with an 8 GB memory limit and 80 concurrent requests at a time. The Fog is the HP Pavilion Laptop with an Intel® Core™ i7-8550U CPU and 8 GB Memory. Other devices on the LAN, such as NVIDIA Jetson nano and Raspberry Pi, can also be referred to as fog but for simplicity, we only refer to the laptop as Fog.

3.5. Service Evaluation

To provide a way to evaluate various services in the service catalog, service energy consumptions and service values have been used as evaluation metrics. The estimated service energy consumption (e_t) for each task is calculated as an aggregated value of the data transfer energy consumption and the device energy consumption (Equation (1)).

$$e_t = (\varepsilon_n * d * t) + (\eta * p) \tag{1}$$

The first part of Equation (1) calculates the data transfer energy consumption where ε_n is the estimated energy of a gigabyte transfer on a network of type n. Andrae and Edler [76] energy consumption estimations of wired fixed access network, wireless access network, and Wi-Fi for 2020 have been used in the calculation. The used energy consumption averages are 0.195 kWh/GB, 0.5435 kWh/GB, and 0.12 kWh/GB for network types Fiber, 4G, and Wi-Fi, respectively. The term d is the size of the transferred data for each task, including both request and response packets. The term t is the average network time which is calculated as the difference between the response and processing time. The second part of Equation (1) calculates processing energy consumption for the service device, where η is the estimated device processing energy, which varies depending on the type of device and its specification (see Table 3 for the devices' energy-related data). The term p is the average processing time for each request. The terms d, t, and p are all averages of data collected from the experiments.

Relative values are calculated to compare two absolute values to each other, which in return provides a better way to compare service-to-service values than the absolute values such as response time, process time, energy consumption, etc. Two relative values are computed service energy value (eValue) and service speed value (sValue), as a way to benchmark different services in terms of their accuracy, energy consumption, and speed (response time). Service eValue provides accuracy-to-energy relative value, considering model accuracy and service energy consumption. Equation (2) is used to calculate the services eValue, where e_t is the estimated service energy for each task using Equation (1) and a is the model accuracy, which represents the percentage of true disease prediction. The model accuracy is discussed in detail, for each model, in Section 4.3. Service sValue provides accuracy-to-speed relative value considering model accuracy and service response time. Service sValue is calculated using Equation (3), where r_t is the average response time for each task and a is the model accuracy.

$$\text{eValue} = \frac{a}{e_t} \tag{2}$$

$$\text{sValue} = \frac{a}{r_t} \tag{3}$$

Note that the purpose of computing service value is to define a method for benchmarking services and it can be considered independent of the parametric values in the equations, such as e_t, ε_t, ε_n, η, etc., as they can be replaced by more accurate and specific values.

4. System Architecture and Design (Skin Lesion Diagnosis Services)

This section describes the design of the proposed distributed skin disease diagnosis services. Figure 7 gives a depiction of Imtidad testbed including its devices and platforms both hardware and software. The testbed consists of one NVIDIA Jetson nano card, two Raspberry Pi cards, two Samsung smartphones, one HP Pavilion Laptop, and access to the Google Cloud Run platform. All these are connected through a wireless connection and equipped with the required software platforms. The white box on the bottom lists the software platforms used in the Imtidad testbed. The specifications of each device have been discussed in detail in Section 3.4, and the rest of this section will explain the whole system architecture and its components in detail.

This section is organized as follows. First, an overview of the system is provided and elaborated in Section 4.1, then each service is discussed in detail in the rest of the section. Section 4.2 discusses available skin datasets and the selected dataset for model training. The DL model service and model design and evaluation are described in Section 4.3. The following sections discuss each service as follows: Section 4.4 the mobile local service, Section 4.5 the mobile remote service, Section 4.6 the gRPC service, Section 4.7 the containerized gRPC service, and Section 4.8 the diagnosis request service.

4.1. System Overview

The case study presented in this paper focused on the classification of the diagnoses of common pigmented skin lesions through Deep Learning-based analysis of multi-source dermatoscopic images, to elaborate on our distributed Deep Learning DL-as-a-service reference architecture. A service catalog, containing 22 different services, has been designed and implemented to investigate the proposed Imtidad reference architecture. These services belong to four service classes (or service types) that are distinguished by their varying communication and software platforms (containerized gRPC, gRPC, Android, and Android

Nearby). Android service class is referred to as "Mobile Local" and the Android Nearby service class as "Mobile Remote". The services are executed on a range of platforms or devices (both terms are used, platforms, and devices, interchangeably according to the context) including Google Cloud (Compute Node), HP Pavilion Laptop, NVIDIA Jetson nano, Raspberry Pi Model B (8 GB), Raspberry Pi Model B (4 GB), Samsung Galaxy S9, and Samsung Galaxy Note 4. These devices could exist in one or multiple of the three distributed system layers, cloud, fog, and edge. Service performance has been evaluated on fiber, cellular, Wi-Fi, and Bluetooth networks, although the designed services are IP-based and can use any IP-based networks. The 22 distributed AI services are based on four different Deep Learning models for skin cancer diagnosis, two of these are standard Deep Learning models, called Deep Learning "Model A" and "Model B". The other two models are the lighter versions of the Deep Learning models A and B called "ALite" and "BLite". The lighter models are Tiny AI models created using the Google platform TensorFlow Lite. The performance of all four models has been evaluated for all the devices, except for Raspberry Pi Model B (4 GB) and the mobile devices that were unable to execute standard models (A and B) due to the device resource limitations.

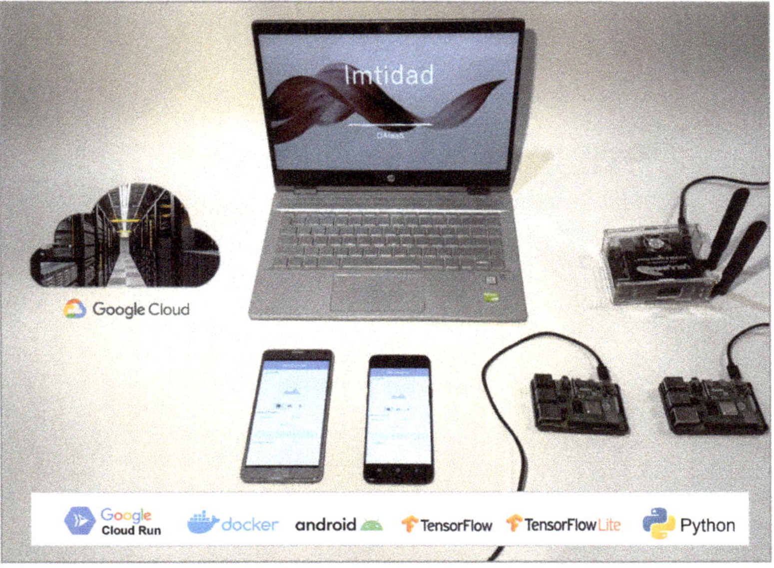

Figure 7. Imtidad testbed: devices and platforms.

The developed system follows a service-based design architecture rather than a component-based architecture. As services are self-contained, loosely coupled, reusable, and programming language-independent components, they provide flexibility and are easy to deploy on various platforms. Figure 8 shows the system architecture, consisting of six different services: DL model service, mobile local service, mobile remote service, gRPC service, containerized gRPC service, and diagnosis request service. The arrows linking various services show the communication among them. The DL model service is responsible for designing, implementing, training, retraining, and optimizing DL models using TensorFlow. It provides two types of models: the TF_model and the TFLight_model. Four different types of services have been designed that provide skin image diagnosis (classification) services, namely, mobile local service, mobile remote service, gRPC service, and containerized gRPC service, which are explained in detail in later sections. The diagnosis request service is used by users to request skin disease diagnosis from one of the diagnosis services. The user takes or selects a skin image from their drive. Then, one of the services is selected from the provided service catalog, and a request is sent to it. Depending on the service type, a connection is established with the provider and the image is sent to the provider for classification (diagnosis). When the results are sent back, they are presented to the user.

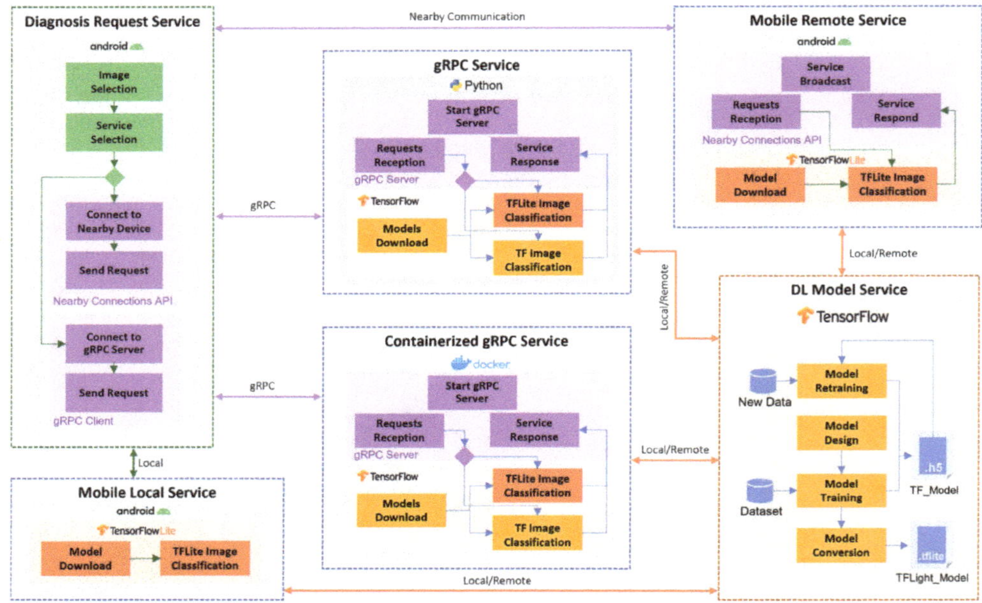

Figure 8. System architecture and design (skin lesion diagnosis services).

Algorithm 1 is the master algorithm for creating new DAI services following the proposed reference architecture (see Figure 2). The algorithm comprises a list of six services that are designed and instantiated. They are shown in Figure 8, in addition to dataset acquisition and service catalog creation. The parametrization of services is used to show the instantiation of services on different devices. For instance, mobile local services are only instantiated on mobile devices while gRPC services are instantiated on various devices including PCs, laptops, Jetson Nanos, and Raspberry Pis.

Algorithm 1: The Master Algorithm: Create_Services(skin_disease_diagnosis)

Input: ServiceClass skin_disease_diagnosis
Output: service_catalog
1   dataset_acquisition(skin_disease)
2   deep_learing_model_service ← design_deep_learing_model(tf_model, tf_lite_model)
3   instantiate(deep_learing_model_service)
4   service_catalog ← create_service_catalog (skin_disease_diagnosis)
5   mobile_local_service ← design_mobile_local (mobile)
6   instantiate (mobile_local_service)
7   mobile_remote_service ← design_mobile_remote (mobile)
8   instantiate (mobile_remote_service)
9   grpc_service ← design_grpc (pc, laptop, jetson, raspberry)
10   instantiate (grpc_service)
11   containerized_grpc_service ←design_container_grpc (cloud, pc, laptop)
12   instantiate (container_grpc_service)
13   diagnosis_request_service ←design_diagnosis_request (mobile, pc, laptop)
14   instantiate (diagnosis_request_service)

Algorithm 2 is a generalized algorithm for the four types of skin image diagnosis (classification) services: mobile local service, mobile remote service, gRPC service, and containerized gRPC service. It explains the service provisioning procedure followed by diagnosis services. The main function is get_diagnosis, which is called by the diagnosis request service. It takes a skin image as input and returns a list of probabilities of each class of skin disease.

Algorithm 2: Diagnosis_Service

Input: skin_image
Output: P[p_0, . . . ,p_C] $\in$ R //C is the number of skin disease classes
1   Function: get_diagnosis (skin_image)
2   Init: size $\leftarrow$ model input dimension, std $\leftarrow$ model normalization factor
      //Image pre-processing
3   skin_image $\leftarrow$ skin_image.rsize(size, size)//Resize the image to size x size
4   img_array[size, size] $\leftarrow$ convert_to_array(skin_image)//Convert image to an array
      //Normalization
5   For x in img_array
6   For y in img_array[x]
7   img_array[x, y] $\leftarrow$ img_array[x, y]/std
8   End For
9   End For
      //Classification
10   model $\leftarrow$ load_model() //load trained model
11   P $\leftarrow$ model.predict(img_array)
12   Return P

4.2. Dataset

There are several open skin datasets available. The International Skin Imaging Collaboration (ISIC) [77] has introduced many datasets from different sources as part of their annual challenge including ISBI, HAM10000, BCN_20000, and MSK Datasets. Interactive Atlas of Dermoscopy (IAD) [78] and PH2 [79] have also provided a dataset of dermoscopy images. He et al. [71] have collected two datasets, Skin-10 and Skin-100, as part of their research, but they have not been made publicly available. In this research, the HAM10000 (Human Against Machine with 10,000 training images) [80] dataset has been used to train the designed models. Table 5 lists the dataset characteristic including the number of images and classes of diagnoses. The dataset has been published in the Harvard Dataverse data repository and consists of 10,015 dermatoscopic images belonging to seven different diagnostic categories of common skin pigmented lesions. The last column in the table shows examples of dermatoscopic images that belong to different diagnosis classes.

4.3. DL Models Service

The DL model service is responsible for model design, training, retraining, and optimization (see Figure 8). This service may be located locally or remotely on cloud, fog, or edge devices. However, retrieving models from different layers of the network would affect the response time. New models can be retrieved on an interval basis or as the services agreement specifies and depending on the user preferences. TensorFlow, an ML open-source tool developed by Google, is used for model development. Algorithm 3 shows the procedure that this service follows to design a model. First, the TF model is designed and trained using the given dataset. Some pre-processing is performed on the dataset images including image resizing and normalization. After training, the TF model is saved in a Hierarchical data format version 5 (H5) file which stores model weights and configuration so they can be restored anytime. Then, the TF model is converted to a TensorFlow Lite (TFLite) model which is an optimized version of the TF model to run on mobile, embedded, and IoT devices. The TFLite model is saved in a file with the (.tflite) extension. The subsections that follow present a discussion on the design, training, evaluation, and conversion of the two models used in this paper.

Table 5. HAM10000 dataset characteristics.

Class	Diagnostic Categories	Code	Images	Sample
0	Actinic Keratoses and Intraepithelial Carcinoma/Bowen's Disease	akiec	327	
1	Basal Cell Carcinoma	bcc	514	
2	Benign Keratosis-Like Lesions (Solar Lentigines/Seborrheic Keratoses and Lichen-Planus-Like Keratoses)	bkl	1099	
3	Dermatofibroma	df	115	
4	Melanoma	mel	1113	
5	Melanocytic Nevi	nv	6705	
6	Vascular Lesions (Angiomas, Angiokeratomas, Pyogenic Granulomas, and Hemorrhage)	vasc	142	
Total Number of Images			10,015	

Algorithm 3: DL_Models_Service

Input: skin_image_dataset
Output: tf_model, tf_lite_model files

```
1    Function: design_models (skin_image_dataset)
2      Init: size ← input dimension, std ← normalization factor
3      tf_file_name ← "TF_model", tf_lite_file_name ← "TFLite_model"
       //Model designing
4      tf_model ← design_tf_model()
       //Model training
5      For skin_image in skin_image_dataset
          //Image pre-processing
6          skin_image ← skin_image.rsize(size, size)//Resize the image to size x size
7          img_array[size, size] ← convert_to_array(skin_image)//Convert image to an array
           //Normalization
8          For x in img_array
9              For y in img_array[x]
10                 img_array[x, y] ← img_array[x, y]/std
11             End For
12         End For
13         tf_model.train_model(img_array)
14     End For//end training
       //save TF model
15     tf_model.save_model(tf_file_name)
       //Convert to TFLite model
16     tf_lite_model ← convert_to_tflite(tf_model)
17     tf_lite_model.save_model(tf_lite_file_name)
```

4.3.1. TensorFlow Model Design

Two models have been designed, implemented, trained, evaluated, and converted to smaller models for edge devices. The first model (A) is based on the pre-trained model Inception v3, while the second model (B) is a pure CNN model. Figure 9 shows model (A) architecture, starting with the Inception v3 model and ending with a dense layer that has seven nodes representing each class of diagnosis. Inception v3 is a pre-trained CNN model consisting of 48 layers and trained using the ImageNet database. Multiple layers have been added to the Inception v3 model to improve its performance when it is trained with the dermatoscopic images, including 2D Convolution (Conv2D), 2D Maximum Pooling (MaxPooling2D), Dropout, Flatten, and Dense. Figure 10 shows model B architecture consisting of a series of 19 layers including 2D Convolution (Conv2D), 2D Maximum Pooling (MaxPooling2D), Dropout, Flatten, and Dense layers. The first layer, Conv2D, receives the input image of shape (299,299,3), and the last layer is a dense layer that has seven nodes representing each class of the diagnosis.

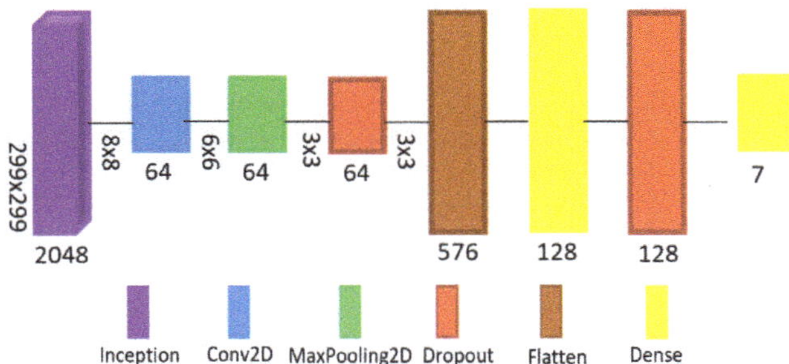

Figure 9. Model (A) architecture.

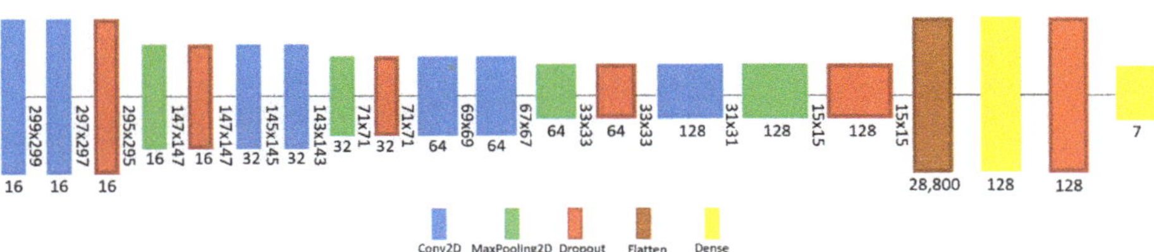

Figure 10. Model (B) architecture.

Both models were trained using the HAM10000 dataset. The dataset was split with 60:20:20 percentages for training, validation, and testing, respectively. Model accuracy (a) was calculated for each subset of data as the percentage of true disease prediction. Model (A) had 0.96, 0.83, and 0.82 accuracies, while model (B) had 0.79, 0.78, and 0.77 accuracies for training, validation, and testing, respectively. To evaluate the accuracy of models A and B in terms of various disease classes, the heatmaps have been used to plot the confusion matrix of the test dataset predictions. Figure 11 present the heatmaps that illustrate the accuracy of classification results for the seven classes. The darker diagonal line in Figure 11a shows that Model A classification results for various classes of disease are more accurate than Model B. The nv class had the highest level of accuracy on both models and model A outperformed model B in akiec, bcc, mel, and vasc classes.

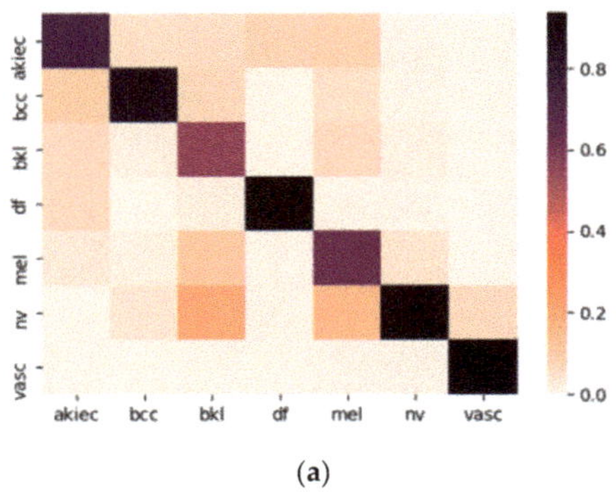 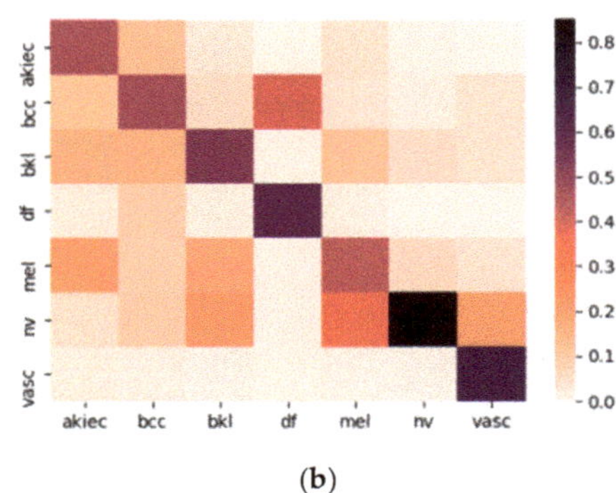

(a) **(b)**

Figure 11. The accuracy heatmap of different classes: (**a**) Model A; (**b**) Model B.

4.3.2. TensorFlow Lite (TFLite) Model

After training and validating both models, TFLite Converter has been used to convert the saved TF models into TFLite models. TFLite Converter generates optimized TFLite models in a FlatBuffer serializable format identified by the (.tflite) file extension. To evaluate both models, the four model versions (A, ALite, B, and BLite) were run for the training, validation, and testing datasets. Table 6 lists the characteristic of models A and B and compares the original (TensorFlow) model and TFLite model in terms of memory footprint and accuracy. After conversion, both A and B models were reduced in size by around three-fold, with no reduction in model accuracy.

Table 6. Comparison between original TF and TFLite Models.

Model	Layers	Total Parameters	TensorFlow				TensorFlow Lite		Size Ratio
			RAM (MB)	Training (%)	Validation (%)	Testing (%)	RAM (MB)	Testing (%)	
A	55	23,057,255	264	95.64	83.23	82.38	87.8	82.38	3.01
B	19	3,833,367	43.9	79.05	78.28	77.33	14.6	77.33	

4.4. Mobile Local Service

In the mobile local service, both diagnosis service and diagnosis request service reside in the user device. Therefore, the user's mobile device should have the required resources to save and run the model locally. As shown in Figure 8, the TensorFlow Lite model is provided by the DL model service in a (.tflite) format. In the diagnosis request service, the user selects a skin image and chooses the mobile local service from their catalog. The mobile local service uses the local TFLite Interpreter in the mobile device to load the model and perform image classification tasks. This type of service guarantees a real-time response and preserves user privacy as the images do not have to be sent across the network to a remote service.

4.5. Mobile Remote Service

The mobile remote service is located in mobile devices and is responsible for providing classification services to nearby devices. As shown in Figure 8, this service is equipped with a TFLite Interpreter, Android Nearby Connections API, and downloads the model from the DL model service. Android Nearby Connections API is used for service connection and management. It is a networking API provided by Android for peer-to-peer service and connection management with nearby devices using technologies such as Bluetooth, Wi-Fi, IP, and audio. This includes service advertising, discovery, connection, and data exchange in a real-time manner. Figure 12 shows messages exchanges between the mobile remote service and the diagnosis request service for the service provisioning process. The mobile remote service starts service advertisement by periodically broadcasting messages that include the service name and service ID. The diagnosis requests service listens to broadcast messages for service discovery and when the required service provider is found, the connection is requested. This invokes the connection establishment process, which includes connection acceptance from both sides and connection result acknowledgment. When the connection establishment is successful, the user can start requesting diagnosis services by sending a skin image to the provider, who uses the TFLite Interpreter to classify the image and return the result.

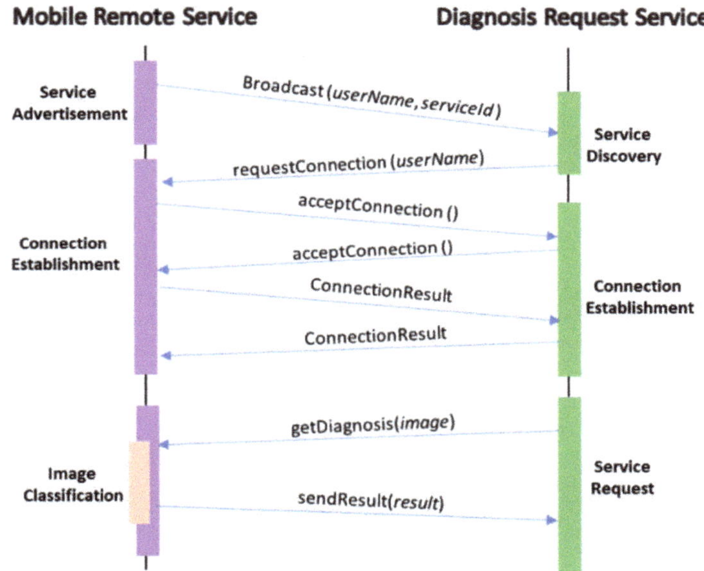

Figure 12. Mobile remote service nearby connection workflow.

4.6. gRPC Service

The gRPC service is implemented using remote procedure calls, specifically Google Remote Procedure Call (gRPC). gRPC is a framework for building platform-independent services and providing various utilities to facilitate service implementation and deployment. Proto syntax is used to define the request and response messages that are passed between gRPC servers and clients. As shown in Figure 8, gRPC services support both TF and TFLite models for skin diagnosis. These models are provided by the DL model service. Secure Sockets Layer (SSL) protocol is used to provide secured communications between the server and client. The diagnosis request service first establishes a secure channel with the gRPC service and then sends the diagnosis request, including the skin image. When the gRPC service receives the request, it passes the image to either the TensorFlow or TFlite Interpreter to classify the image and returns the result. The result is then sent back as a gRPC response including classification probabilities.

4.7. Containerized gRPC Service

The containerized gRPC service is a version of the gRPC service that is containerized as a Docker container (see Figure 8). Docker containers provide an executable, lightweight, and standalone container image that encapsulates everything the gRPC service needs in order to run. This service image is deployed in Google Cloud using the Cloud Run platform. Containerized gRPC service reduces efforts in deploying gRPC service into the cloud especially when they are already supported by the cloud platform, such as the Google Cloud Run platform that have been used here. Cloud Run provides a fully managed serverless platform to deploy highly scalable containerized applications. The containerized gRPC service could not replace the gRPC service as Docker containers do not have full support for many of the AI libraries for different processor architectures such as armv7 and aarch64 in Raspberry Pi and Jetson. Therefore, offering this variety of technologies and software platform allows services to be instantiated anywhere in cloud, fog, and edge layers.

4.8. Diagnosis Request Service

The diagnosis request service has been developed using Android studio, so that it could run on Android devices. This service is responsible for image selection and communication with various diagnosis services. Algorithm 4 shows the procedure that the diagnosis request service follows to get a skin diagnosis prediction from one of the skin image diagnosis services and present the final result.

The algorithm takes, as an input, the user selected skin image and the chosen service type from the provided service catalog. In the case of mobile local service, the local service installed in the device will be used for skin image classification directly. In other cases, the diagnosis request service first establishes a connection with the required service. If the mobile remote service is chosen, the application listens to the nearby service broadcasts and establishes a connection with a nearby mobile device. For gRPC-based services (gRPC and containerized gRPC), the application uses gRPC stubs to communicate with the services. When the connection is ready, the diagnosis request is sent along with the skin image to be classified (diagnosed) by the chosen diagnosis service and when the results are sent back, they are presented to the user. Figure 13 shows screenshots of the user interface for the skin diagnosis application, which enables the user to request a diagnosis service. The screenshots are numbered from 1 to 5 to show the steps involved in selecting a service and obtaining a diagnosis on the application.

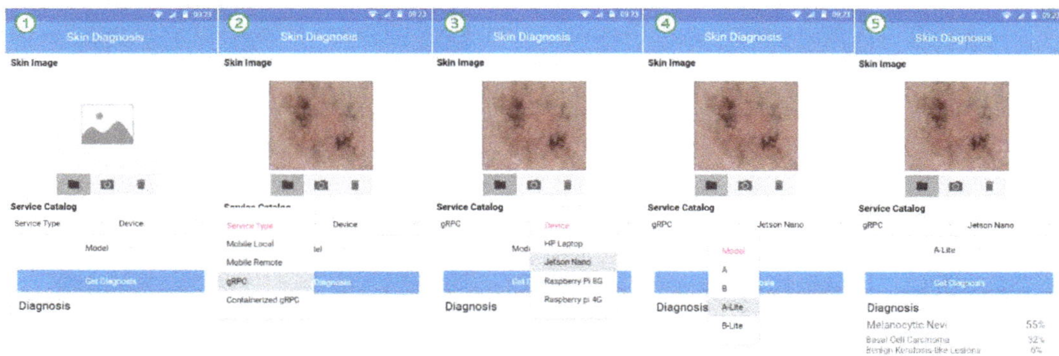

Figure 13. Mobile application user interface for access to skin diagnosis services.

Algorithm 4: Diagnosis_Request_Service
Input: skin_image, service_type
Output: class of skin disease

```
1    Function: skin_diagnosis (skin_image, service_type)
2      Init: ip ← grpc server ip address, crt ← server certificate, url ← cloud service URL,
3            service_id ← nearby service ID, user_name ← given device name
4      Try
5          If service_type = = mobile_local_service Then
                //request from the local service
6              P[p₀, … ,p_C] ← mobile_local_service.get_diagnosis (skin_image)
7          Else If service_type = = mobile_remote_service Then
                //Create connection with the nearby device
8              connection ← request_connection(user_name)
                //Send a request to the diagnosis service
9              P[p₀, … ,p_C] ← connection.get_diagnosis (skin_image)
10         Else If service_type = = grpc_service Then
                //Create a secure channel with the diagnosis service
11             channel ← create_secure_channel (ip, crt)
12             P[p₀, … ,p_C] ← channel.get_diagnosis (skin_image)
13         Else If service_type = = containerized_grpc_service Then
                //Create a secure channel with the cloud
14             channel ← create_secure_channel (url)
15             P[p₀, … ,p_C] ← channel.get_diagnosis (skin_image)
16         End If
                //Find the largest probability value
17         probability ← p₀
18         prediction_class ← 0
19         For c in C
20             If P[c] > probability Then
21                 probability ← P[c]
22                 prediction_class ← c
23             End If
24         End For
                //Find corresponding labels
25         prediction_label ← get_prediction_label(prediction_class)
26         Return prediction_label
27     Catch exception
28         Return error_message
29     End Try
```

5. Service Evaluation and Analysis

This section presents and discusses our experiments and results. First, the experiment settings are presented (Section 5.1). Then, every evaluation metric has been discussed and evaluated, including processing time (Section 5.2), response time (Section 5.3), network time (Section 5.4), service data transfer rate (Section 5.5), and the services' energy consumption (Section 5.6) and values (Section 5.7).

5.1. Experimental Settings

The experiments were conducted in a real-life environment in a typical family home setting to represent everyday city life. They took place over a period of several weeks. Every week, they were conducted for four consecutive days (from Saturday to Tuesday), at three different times of the day. Unfortunately, limited human, and other, resources, made it impossible to conduct the experiments more frequently (every three h) and for the seven weekdays. Table 7 lists the various evaluation variables for which data had been collected during the experiments and those were recorded as testing logs. The table lists the variable names, definitions, units, and an example of collected data.

Figure 14 shows the networking setup for the experiments. All edge devices are connected to a WiFi router that provides a local connection between them and a connection to the Cloud through the Fiber and 4G networks. Two WiFi routers have been used separately for the two different experiment settings. One is the Fiber WiFi router which is both a fiber optic modem and WiFi router that is connected to the fiber optic cable provided by Internet Service Provider (ISP). The second is a 4G WiFi router connected to the 4G cellular network via a SIM card provided by ISP. The smartphones use the Android Nearby Connections API to create a peer-to-peer (P2P) connection between them, which uses either WiFi or Bluetooth for communication. The figure shows the Fog node connects to the edge WiFi network through the 4G and Fiber networks. This is depicted to show how it should be connected in reality and to avoid confusion for the reader. However, the Fog device in our case is connected to the edge devices through the same two routers. This is done due to the human and infrastructure resource limitations since having the fog node in a separate network requires a separate physical space and human support for conducting experiments. In our case, this is an acceptable setup because in studying fog node performance we have focused on the computational performance of the fog node which depends on the device compute capability and is virtually independent of the network performance.

Table 7. Evaluation data variables and examples.

Variable	Definition	Unit	Example
Date	Date of the request	Date	16/02/2021
Day	Day of the week	Date	Tuesday
Hour	Time of the request	Time	06:15:49.333
ImageName	Image name from the dataset	String	ISIC_0033458
ModelVersion	Model used for classification (A = 1, B = 2, ALite = 3, BLite = 4)	Number	1
DeviceName	The name of the device: Cloud, Laptop, Jetson, Rasp8, Rasp4, S9, Note 4	String	Cloud
RequestSize	The packet size of the request message	Bytes	171,316
RequestSentTimestamp	The timestamp when the request is sent by the user	Milliseconds	1,613,445,211,566
RequestReceiveTimestamp	The timestamp when the request is received at the service	Milliseconds	1,613,445,342,877
ServiceProcessTime	The service processing time (inference or compute)	Milliseconds	6334
ResponseSentTimestamp	The timestamp when the response is sent from the service	Milliseconds	1,613,445,349,211
ResponseReceiveTimestamp	The timestamp when the response is received at the user device	Milliseconds	1,613,445,349,325
ResponseTime	The total time taken to obtain a response, from the moment the request had been made	Milliseconds	137,759
ResponseSize	The packet size of the response message	Bytes	112
Result	Diagnosis results as a probability of each class of the diseases (7 values for 7 classes separated by commas)	List of floats	$[4.025427 \times 10^{-7}, 1.1340192 \times 10^{-7}, 4.8968374 \times 10^{-8}, 6.5097774 \times 10^{-6}, 3.8256036 \times 10^{-7}, 0.00020890325, 0.9997837]$

5.2. Service Processing Time

The processing time is the time that the diagnosis service needs to process an image and predict the skin disease category. It depends on both model complexity and device resources. The processing time was recorded at different times of the day during the week. Figures 15 and 16 show processing times for all service types, devices, and models. Services and devices specifications can be referred to in Table 3.

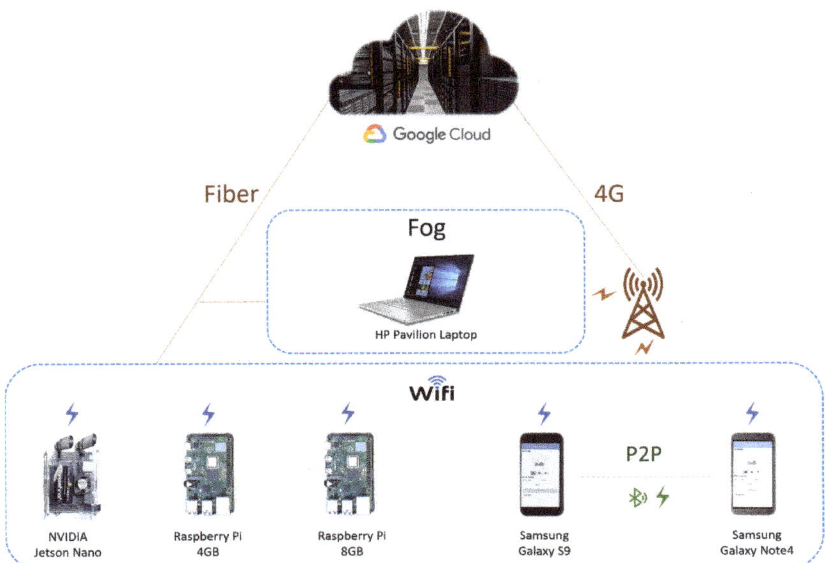

Figure 14. Networking setup.

Figure 15 compares models processing time behavior for each service type and device. The bar chart presents the average processing time where the horizontal axis represents devices, the vertical axis represents the average processing time in seconds, and bars represent model types. For all devices, model A average processing time is higher than that of model B, even for the TFLite versions, which was excepted considering the complexity and size of model A. Jetson device has the highest average processing time for all models compared to other devices and this is related to both Jetson memory limitation and device capability. On Jetson, the average processing times were 49 s, 10 s, 2 s, and 0.5 s for models A, B, ALite, and BLite, respectively. The lowest average processing times were for the Fog device with 7.7 s, 0.8 s, 0.5 s, and 0.1 s for models A, B, ALite, and BLite, respectively.

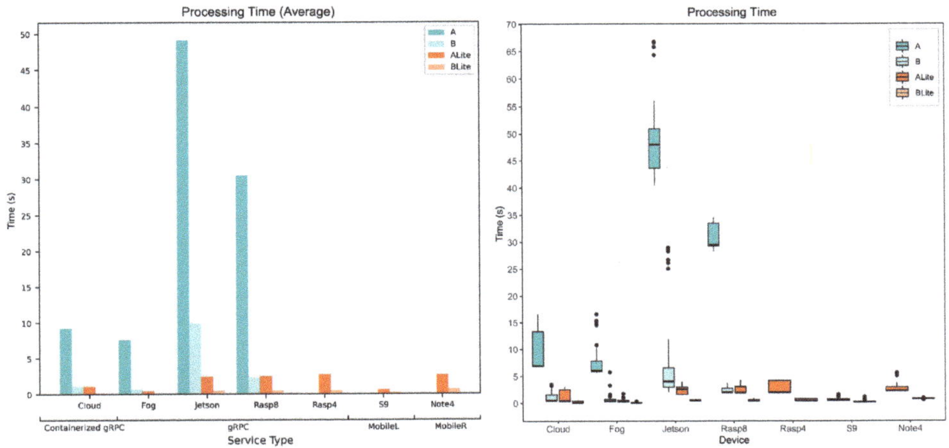

Figure 15. Service processing time (by device).

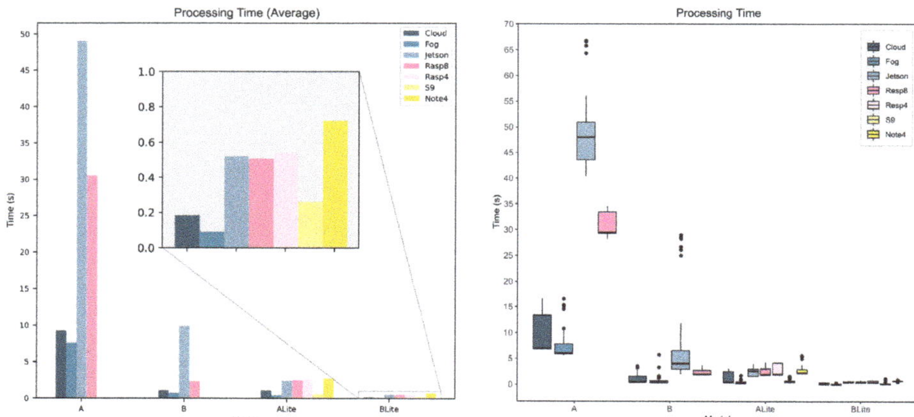

Figure 16. Service processing time (by model).

The boxplot in Figure 15 depicts the processing time data distribution for the whole data collected in our experiments. Boxplots show five statistical measurements the minimum, first quartile (Q1), median (Q2), third quartile (Q3), and maximum. The first quartile (Q1) is the 25th percentile, the median (Q2) is the 50th percentile, and the third quartile (Q3) is the 75th percentile of data. These are depicted as the bottom line in the colored box, the middle thick line inside, and the top line of the box, respectively. The distance between Q1 and Q3 (the height of the colored box) is called the interquartile range (IQR). The maximum and minimum values are the highest and lowest points of the vertical lines on the top and the bottom of the colored boxes. They are calculated using the quartiles and IQR, Q3 + 1.5 × IQR for the maximum and Q1 − 1.5 × IQR for the minimum. Any value more than maximum or less than minimum values is considered as an outlier and depicted as a circle outside the boxplot. Jetson minimum, Q1, Q2, Q3, and maximum processing times for model A were 41 s, 43 s, 48 s, 51 s, and 55 s, respectively, with some outliers over 65 s (see Figure 15 boxplot). This shows that all recorded values for Jetson model A are greater than any other devices or models. Model B processing times distribution on Jetson was much better with 13 s as the maximum value though there were some outliers over 25 s.

Figure 16 compares the device's processing time behavior for each model. The bar chart presents the average processing time where the horizontal axis represents models, the vertical axis represents the average processing time in seconds, and the bars represent the seven devices that were being evaluated. The Fog had the lowest average processing times among all devices for all models and this can be related to the resources of the fog device. An HP Pavilion Laptop had been used here as a fog device which has an Intel® Core™ i7-8550U processer and 8 GB memory. The Fog has, even, outperformed the Cloud average processing time, 9.3 s, 1.1 s, 1.15 s, and 0.18 s (Cloud) compared to 7.7 s, 0.8 s, 0.5 s, and 0.1 s (Fog) for models A, B, ALite, and BLite, respectively. It seems that the vCPU assigned by Google for the containerized gRPC service is less powerful than the Intel® Core™ i7-8550U processer in the Fog device (gRPC service). The exact physical CPU the containerized gRPC service run on is unspecified by Google on the Google Cloud Run platform.

For models A and B, the Cloud provided the second-best average processing time (9.3 s and 1.1 s), followed by Rasp8 (30.5 s and 2.4 s), and Jetson (49.1 s and 9.9 s), respectively. Rasp8 and Rasp4 average processing times were almost identical for TFLite models (around 3 s ALite and 0.5 s BLite), though Rasp4 could not process original TF models due to memory shortage. For the ALite model, after the Fog (0.46 s), the S9 was the fastest (0.61 s) followed by the Cloud (1.15 s), Jetson (2.48 s), Rasp8 (2.52 s), Note 4 (2.72 s), and Rasp4 (2.74 s). For the BLite model, after the Fog (0.09 s), the Cloud was the fastest (0.18 s), followed by S9

(0.26 s), Jetson (0.52 s), Rasp8 (0.51 s), Rasp4 (0.53 s), and Note 4 (0.72 s). Although both S9 and Note 4 are mobile devices, S9 showed better results due to its processor capabilities (see Table 3).

Looking at previous observations, the processing time is closely related to both device capabilities and the model size and complexity. TFLite optimization greatly improved the processing time and there was no accuracy loss (in our case). In more complex models, the accuracy may lower to a certain level, which may jeopardize the application, depending on its criticality. Devices at the fog or edge layers showed acceptable results compared to the cloud which make them great candidates for local processing.

5.3. Service Response Time

The response time is the total time since the request was made until the result is returned; this includes the processing time. Figures 17 and 18 show the response time of all service types, devices, and models. Service and device specifications are referred to in Table 3. Figure 17 compares the models' response time behavior for each service type and device. The bar chart presents the average response time where the horizontal axis represents devices, the vertical axis represents the average response time in seconds, and the bars represent model types. The average processing time (see Figure 15) and the average response time (Figure 17) have similar behavior. Model A average response time is always more than model B's average response time for both TF and TFLite versions. Unlike other devices, the Cloud performed better for model B (3.36 s) than model ALite (3.7 s), and Rasp8 had an identical average response time (2.9 s) for both model B and ALite. Looking at the boxplot in Figure 17 which shows the data distribution of the response time for all the data collected in the experiments. There is a difference in the Cloud performance distribution of model B and ALite. Model ALite IQR is larger than the IQR of model B though both medians (Q2) are at around 1 s. CloudALite had an outliner above 25 s, while the highest CloudB outliner is at around 17 s. For Rasp8, both model B and ALite had a similar distribution, although the maximum value of ALite is more than that of B.

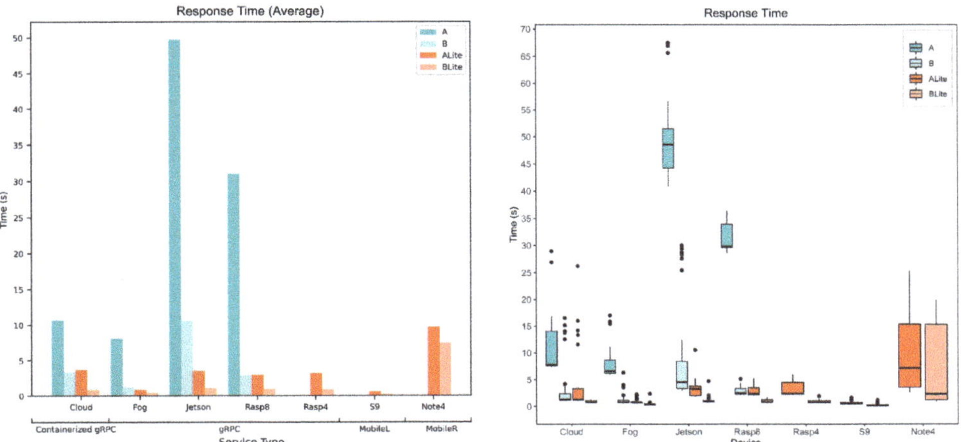

Figure 17. Service response time (by device).

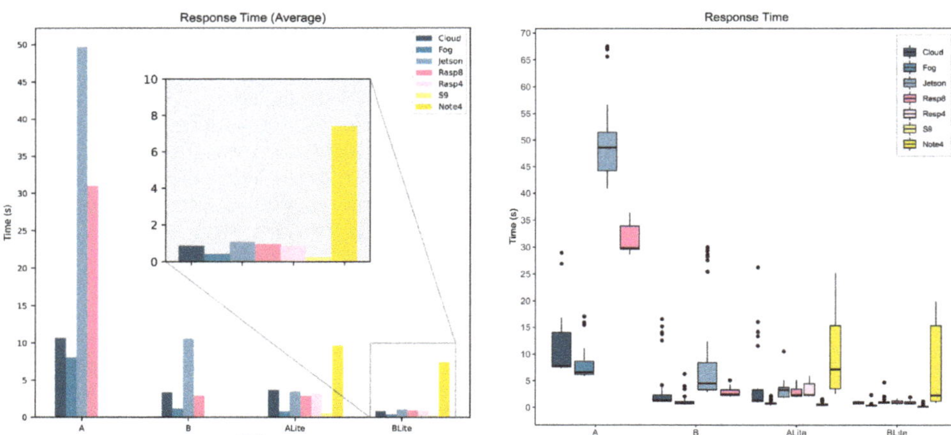

Figure 18. Service response time (by model).

Figure 18 compares the devices' response time behavior for each model. The bar chart presents the average response time, where the horizontal axis represents models, the vertical axis represents the average response time in seconds, and the bars represent the seven devices that were being evaluated. Although the Cloud and Fog had a close average processing time (see Figure 16), the difference between them is greater when it comes to response times. The difference between the Cloud and Fog average processing times is 1.6 s, 0.4 s, 0.7 s, and 0 s for models A, B, ALite, and BLite, respectively; the difference between their average response times is 2.6 s, 2.2 s, 2.8 s, and 0.4 s for models A, B, ALite, and BLite, respectively. The Cloud average response time is higher as it requires more time to transfer the image across the internet. The networking time metric that is covered in the next section clearly shows the load on the network of different services. The Jetson average processing time greatly affects its response time, especially for model A (48.7 s), with more than 18 s difference with the next highest response time, which is for Rasp8 (31 s). S9 had the lowest average response times for TFLite models 0.6 s (ALite) and 0.3 s (BLite), while Note 4 had the highest average response time among TFLite models (10 s and 7 s), which can be related to the Android Nearby Connections API that was used for mobile device services.

The boxplot in Figure 18 provides a deeper look at the service's response time values showing the distribution of all collected response times. Note 4 boxplots for ALite and BLite show that the maximum response times were 25 s and 20 s while the medians (Q2) were 7 s and 2 s. BLite median is close to the Q3 (1 s), which means that 50% of collected response times for Note 4 was below 1 s. This variation of data means that this type of service response time is highly unpredictable.

These results show that local processing at mobile devices with MobileL services had the best response time as they do not require any network communication, though only TFLite or small models can be accommodated. On the other hand, other fog and edge devices at the local network such as the Fog and Rasp8 can accommodate more complex models and provide fast responses to local service requests.

5.4. Service Network Time

The network time is calculated as the difference between the response time and the processing time, so it includes connection initialization time, devices network processing, and data transfer time (see Equation (4)). This metric shows both the load on the network for each service type and the performance of the different network connection technologies used to communicate with the diagnosis services. There are two basic types of communication protocols that have been used in this research, namely, the nearby connections and gRPC. The mobile remote services use the Android Nearby Connections API to create a peer-to-peer connection using various technologies such as Bluetooth, Wi-Fi, IP, and audio depending on the available connection. Other services use gRPC and SSL for communication and the mobile local service does not require any network communication. In addition, the network time gives an indication of the data transfer factor of the total response time.

$$NetworkTime = ResponseTime - ServiceProcessTime \tag{4}$$

5.4.1. Model and Device Behavior

In this section, the network time is evaluated for various devices and models. Figures 19 and 20 show the calculated network time values from the collected response and processing times data. Figure 19 focuses on models' behavior for each service type and device while Figure 20 focuses on the behavior of the devices for each model.

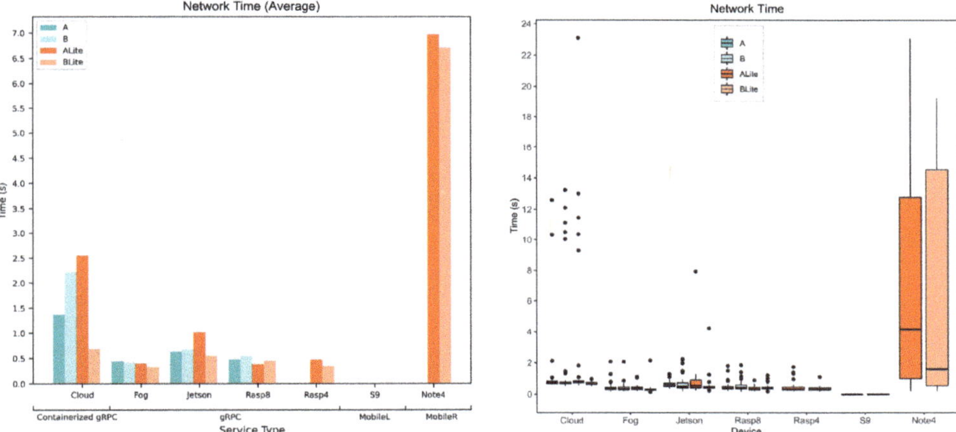

Figure 19. Service network time (by device).

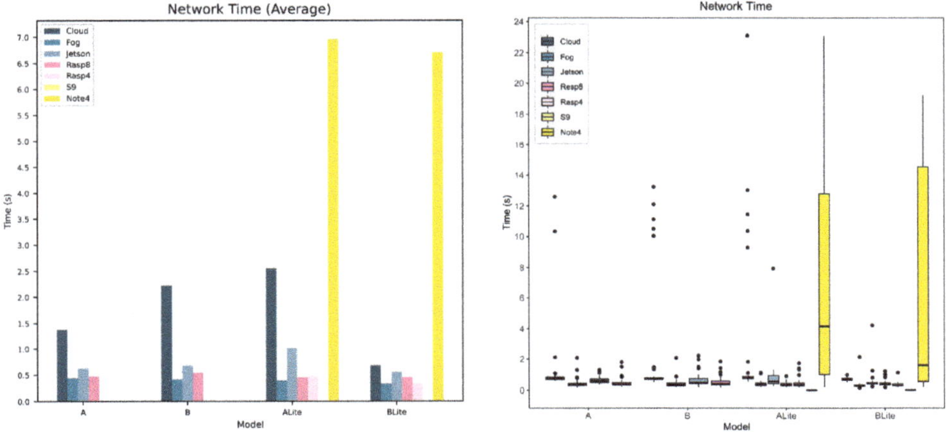

Figure 20. Service network time (by model).

The bar chart on the left hand side of Figure 19 presents the average network time for all service types and devices in terms of model types. MobileR services had the highest values of average network time (around 7 s), and this can be related to the nearby connections, especially since the network time includes the time required for connection initialization. The difference of average network time between different models on the same device is very small for all devices except the Cloud. The average network time of CloudBLite was 0.6 s, while for other models the network time was 1.4 s, 2.2 s, and 2.6 s (for A, B, and ALite, respectively). However, looking at the right-hand side of Figure 19, the network time boxplot shows the distribution of the data. The Q1, Q2, and Q3 of the Cloud network time for all models fall below 1 s, though there are a lot of anomalies above 9 s for models A, B, and BLite, which have affected the average value. All other devices have a similar distribution of network time but with fewer and much lower anomalies, except for the mobile remote. The maximum value of MobileRALite is 23 s and MobileRBLite is 19 s, both have a minimum of 0.5 s. MobileRALite median is 4 s and MobileRBLite is median 2 s. This high variation of network time on MobileR indicates that the nearby connections are more unpredictable.

Figure 20 compares calculated network time in terms of devices for each model. The bar chart on the left side shows the average network time. Despite the MobileR, whose behavior was explained earlier, the Cloud had the highest values among them. This is expected, as it is the only service that is located across WAN, and all other services are on LAN. Among devices on the LAN, the Fog had the best average network time for all models of around 0.39 s. Resp (Rasp4 0.41 s and Rasp8 0.46 s) was the second best followed by Jetson (0.72 s). The boxplot on the right side shows the distribution of these values. The Cloud had the highest anomalies followed by Jetson. All other device network times fall below 2 s, including all anomalies.

To summarize, the results confirm that both the type of connection and technique used for communication are affecting the networking time. Local services are always the best option if the available resources are sufficient for processing although the available network cards and other device specifications showed a variation of network times among devices on the same LAN.

5.4.2. Behavior over Weekdays

This section describes the network time, which has been evaluated over the whole period of the experiment to investigate the behavior of the devices. The data were collected for four days starting from Saturday to Tuesday, three times a day for all models and devices. Figure 21 shows the calculated network times plotted over a time series. In the scatter plot on the left side of Figure 21, the network times were plotted as colored dots where each color represents a different device. The vertical axis represents the network time in seconds, the horizontal axis represents the time series including days and hours, the dots represent the calculated network time for each device at a specific time, and the line shows the trend of the network time over time. The trend curve is plotted using the LOESS (Locally Estimated Scatterplot Smoothing) regression analysis method. The S9 mobile device has no network time as it runs a mobile local service that does not require any network communication.

The highest network time trend line (top line) is for Note 4, which runs a remote mobile service. The behavior of the Nearby Connections API has been observed in the previous section, which had a very high distribution of data (see the boxplot of Note 4 in Figure 19). Similarly, it can be seen that the Note 4 data points are spread all over the graph, with a maximum of 23 s on Sunday 00:00 and a minimum of around 0.1 s on Tuesday 13:00. The Cloud is the second-worst network time (second trend line from the top). However, there are eleven values over the Note 4 trend line, ten of them ranging from 9 s to 13 s and one 23 s on Tuesday 21:00. Other devices on LAN are showing a similar trend line, except for Jetson (the purple line), which went slightly higher on Saturday until Sunday afternoon. Saturday 15:00 was the highest with 8 s, and the second highest was on Saturday 22:00 with around 4 s network time. The Fog, Rasp8, and Rasp4 were more stable with one point over the Cloud trend line for the Fog at around 2 s on Sunday 18:00.

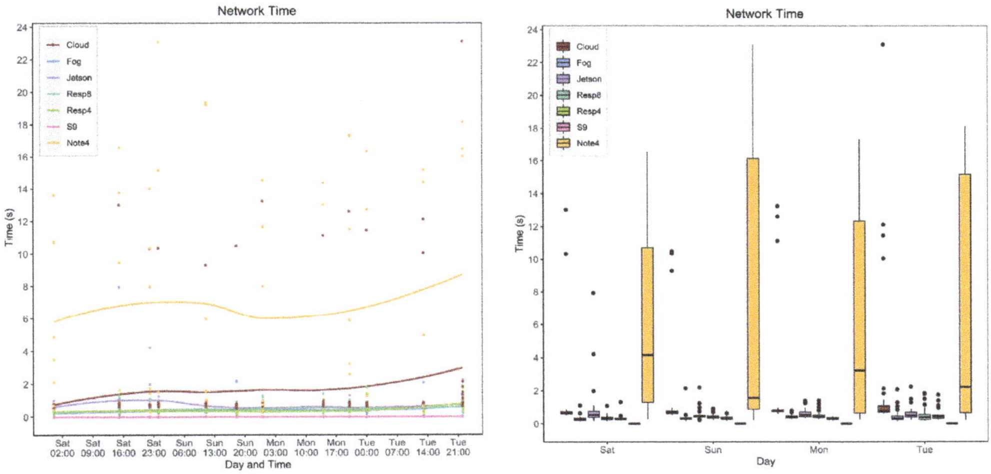

Figure 21. Network time at different times and days.

Looking at the shape of the trend lines over time, all the lines were lower on Sunday 00:00 and higher on Tuesday 22:00. Note 4 trend line fluctuated more than the other lines, the curve rose on Saturday afternoon until Sunday night. On Monday at daytime, the network time was lower, and then the curve started to rise again from Monday evening until the end of the period. The Cloud network time trend started with a low network time at 1 s on Saturday 00:00, then started to build up, and stabilized at around 2 s from Saturday evening to Monday evening, before it rose again from Monday night to Tuesday night reaching 3 s.

The boxplot on the right side of Figure 21 shows the distribution of calculated network times on different days for different devices. Due to the space limitation in the figure, only the distribution of days has been plotted, not specific times. The boxplot confirms the earlier observation made from the scatter plot. The large boxes of Note 4 confirm the high distribution of network times on the days shown in the scatter plot. Similarly, the Cloud had many outliers over the maximum values on all days and the high outliers for Jetson on Saturday confirm the curve in the Jetson trend line.

To summarize, the results showed there are changes in the device's network times on different times and days. These changes could be related to the user's network usage trend at different times of the day and during weekends and weekdays. Further investigation is needed to find trends in network usage. Such information could be used for network and service placement planning which could improve the QoS.

5.4.3. Cellular (4G) vs. Fiber Networks

In this section, a comparative study is made of fiber-optic and cellular 4G internet connections. An experiment has been conducted over three days, from Sunday 28 March 2021 to Tuesday 31 March 2021. The data were collected for both fiber and 4G at two different times of the day, and both Internet connections were from the same network provider. The Cloud services are the ones that require the internet connection to connect to them as they were installed in the Google datacenter. All other services do not require an internet connection as they were installed in the LAN.

Figure 22 shows the network time of all Cloud services (for all models) for both fiber and 4G Internet connections. The vertical axis represents the network time in seconds and the horizontal axis represents the time series including days and hours. In the scatter plot on the left side of Figure 22, the dots represent the calculated network time for each connection at a specific time, and the line shows the trend of the network time over time. The trend curve is plotted using the LOESS regression analysis method. As expected, the fiber connection had a better network time (around 2 s) than 4G (ranging from 3 s to 10 s). The fiber connection is more stable over time with a slight rise at the end of the period to around 2.5 s. However, there are a few (seven points total) higher values between 9 s and 13 s. The cellular (4G) connection is less stable over time as the trend line fluctuates over time with many high and low values. The lowest value was 1.5 s on Sunday 28 March 2021 at 11:00, and the highest value was 30 s on Tuesday 31 March 2021 at 00:00. It appears that there was higher demand on the cellular network from Monday night to Tuesday afternoon and lower demand on Sunday afternoon to Monday afternoon, which produced these variations.

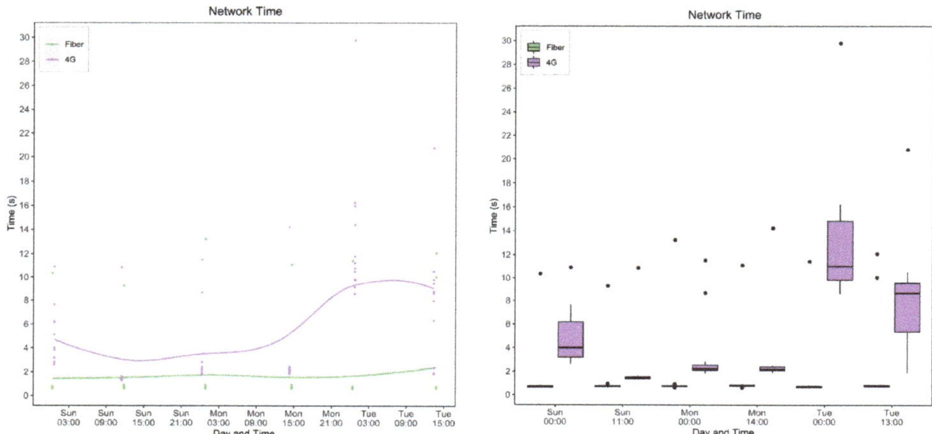

Figure 22. Network time for fiber and 4G of the cloud services.

The boxplot on the right side of Figure 22 shows the distribution of calculated network times for both fiber and 4G internet connections over time. The boxplot confirms our earlier observation from the scatter plot. The large boxes of the 4G network on Sunday 00:00, Tuesday 00:00, and Tuesday 13:00 are aligned with the curve in the 4G trend line in the scatter plot. The fiber network was much more stable, with smaller IQRs, consistent medians, and few outliners over the whole period.

5.5. Service Data Transfer Rate

The service data transfer rate metric is the rate at which the data are being transferred from the request service to the diagnosis service and back again. It includes the time needed for the operating system to initialize the connection, prepare the packets, and send them across the network. Mobile local services do not have a service data transfer rate, as they do not require network communications. The service data transfer rate is calculated as the total size of the transferred data divided by the network time (see Equation (5)). The RequestSize and the ResponseSize are sizes of the request and response packets in bits. Figures 23 and 24 show the calculated service data transfer rate from the collected packet sizes and calculated network times.

$$ServiceDataTransfrerRate = \frac{ResponseSize + RequestSize}{NetworkTime} \tag{5}$$

Figure 23 compares the service data transfer rate of different models for each service type and device. The bar chart presents the average service data transfer rate where the horizontal axis represents devices, the vertical axis represents the average service data transfer rate in Kbps, and bars represent model types. The gRPC service on the Fog device had the highest average service data transfer rate for all models 4 Kbps for A, B, and ALite as well as 5 Kbps for BLite, which is aligned with the average network time discussed earlier. The Cloud service had the lowest service data transfer rate among all models, 1.7 Kbps, 1.7 Kbps, 1.6 Kbps, and 2 Kbps for models A, B, ALite, and BLite, respectively. This was expected, as the Cloud services are the only services that require the data to be transferred across WAN. The boxplot on the right side of Figure 23 shows the distribution of the service data transfer rates. All devices show larger boxplots than the Cloud's boxplots, this means that the service data transfer rate for all local devices varies in its values more than the Cloud's values. Note 4 showed a very low service data transfer rate, with minimum and Q1 values of around 0.1 Kbps. In addition, the medians of both MobileRALite (0.6 Kbps) and MobileRBLite (1.5 Kbps) are lower than those of CloudALite (1.9 Kbps) and CloudBLite (2.1 Kbps).

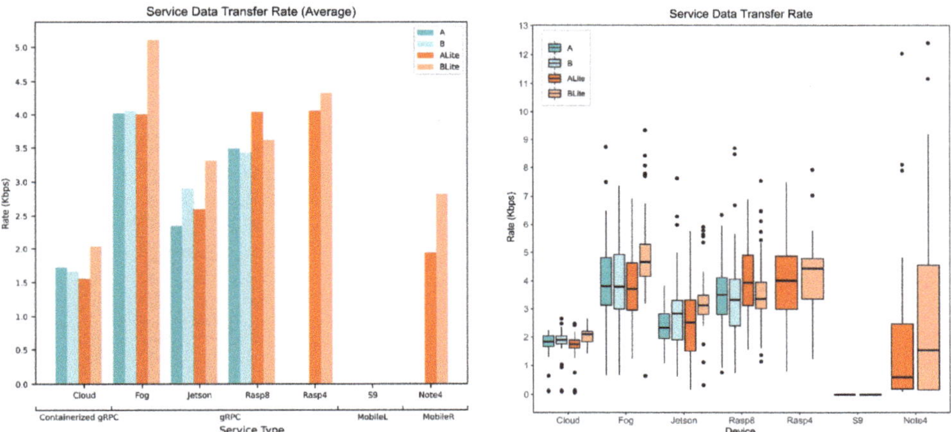

Figure 23. Service data transfer rate (by device).

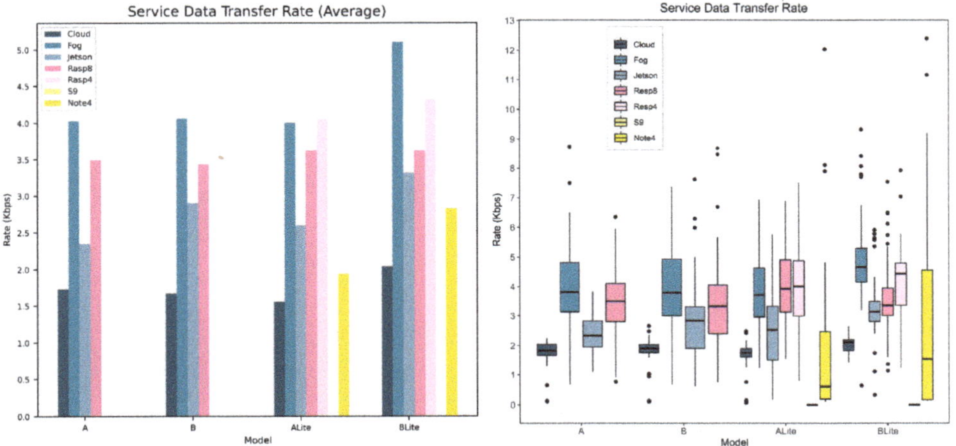

Figure 24. Service data transfer rate (by model).

Figure 24 compares the service data transfer rate of different devices for each model type. For all original TF models, the Fog had the best data service transfer rate followed by Resp8, Jetson, and Cloud. For the ALite model, Rasp4 was better than the Fog by 0.06 Kbps, and they were followed by Resp8, Jetson, Note 4, and Cloud. For the BLite model, the Fog was the best followed by Rasp4, Resp8, Jetson, Note 4, and Cloud. The boxplot on the right side of Figure 24 shows the distribution of the service data transfer rates. The medians of the original TF models show the same pattern as the average values; however, the TFLite models showed a slightly different pattern. Unlike the averages, Note 4 medians were lower than the Clouds, and Rasp4 and Rasp8 both had a similar median of 4 Kbps for ALite model.

5.6. Service Energy Consumption

Energy is a key factor for system efficiency in terms of cost and environmental sustainability. Therefore, services that consume less energy are favorable. Figure 25 shows the estimated average energy consumption per task for all service types presented in service catalog (see Table 3). The bar chart on the left side shows energy consumption grouped in terms of devices, while the one on the right side shows energy consumption grouped in terms of models. MobileL had the lowest energy consumption for both ALite (0.0009 Wh) and BLite (0.0004 Wh), as no energy is used on data transfer in those models. The Cloud had the highest energy consumption for all models, 0.26 Wh, 0.03 Wh, 0.03 Wh, and 0.01 Wh for models A, B, ALite, and BLite, respectively. The BLite model consumed the least energy for all service types, compared to other models which was expected, considering the characteristics of this model. On the other hand, model A had the highest energy consumption due to its computation and memory requirements. The CloudA had the highest energy consumption of 0.26 Wh followed by FogA (0.14 Wh), JetsonA (0.14 Wh), and Rasp8 A (0.04 Wh).

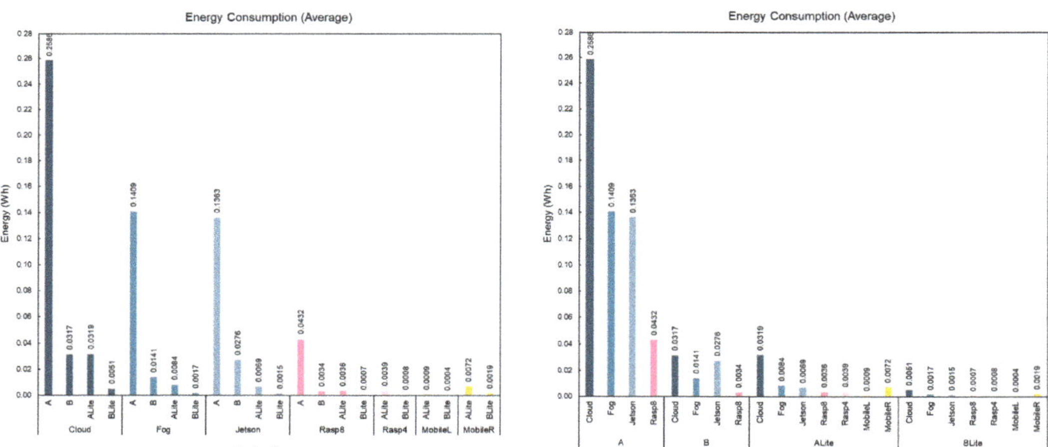

Figure 25. Service energy consumption.

5.7. Service Value (eValue and sValue)

Two relative values are calculated, one for energy (eValue) and the other for speed (sValue) (see Section 3.5). These service values are used to compare the 22 different service types in terms of their accuracy, energy, and speed (response time). We only used the Fiber network in these calculations (the same applies to the energy consumption values presented in the previous section). The service values are computed using appropriate energy consumption parameters (see Section 3.5). For example, the Cloud eValue uses both Fiber and Wi-Fi energy consumption values. For Bluetooth, in the figures, we used the same energy consumption as for the Wi-Fi but this could easily be replaced by precise Bluetooth energy values. Note that there are also no problems in computing and plotting service values for the 4G network, but this will lengthen the paper and unnecessarily add to its complexity. The comparison provided for 4G versus Fiber in Section 5.4.3 only presents a comparison between network times; all other values, such as the service values, can be drawn from it. This is to bring another design dimension to the reader's attention, while keeping the article complexity to a minimum.

Figure 26 shows normalized service eValues as an integer between 0 and 100 for all service types. The bar chart on the left side shows the service eValues grouped in terms of devices, while the one on the right side shows the service eValues grouped in terms of models. MobileLBLite had the highest service eValue, and CloudA had the lowest eValue, which is aligned with their energy consumption. In general, the BLite model had the highest values among other models, and model A had the lowest values. When it comes to devices, MobileL services had the best service eValues, though they can only run TFLite models. MobileL services do not require network communication, which eliminates the network data transfer energy from the energy equation (see Equation (1)), reduces their energy consumption, and increases their eValues. The Rasp8 services had the best service eValue among services that run original TF models, and they are the second best for TFLite models after MobileL. This can be related to the energy consumption of the Raspberry Pi devices, which is the lowest among all devices used in the experiments (see Table 3). The Cloud services had the worst eValues due to both devices and data transfer energy consumptions.

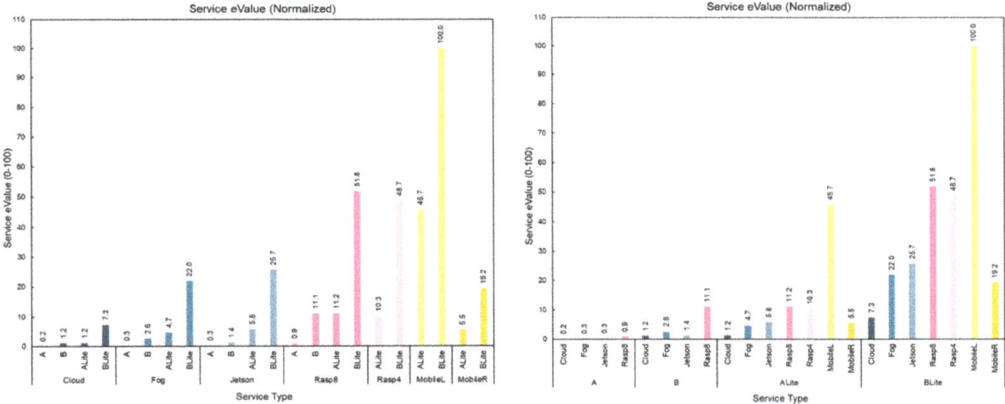

Figure 26. Service eValue.

Figure 27 shows normalized service sValues as an integer between 0 and 100 for all service types. The bar chart on the left side shows the service sValues grouped in terms of devices, while the one on the right side shows the service sValues grouped in terms of models. MobileLBLite had the highest service eValue, and JetsonA had the lowest sValue. For devices running TFLite models, MobileR had the lowest sValues, and for devices running TF models, Jetson had the lowest sValues. In general, MobileL had the best sValues, and the Fog services came in second place. Rasp8 and Rasp4 had similar sValues, and the Cloud services' were better than those for A and BLite models. The sValue is strongly related to the services' response times, which have been discussed extensively in Section 5.3.

To summarize, MobileL services had the highest eValue and sValue, as they are using less energy and provide faster responses. The only concern with MobileL services is that they are limited in their resources and cannot accommodate large and complex models or large volumes of data. The Cloud services were much better in terms of sValues but not eValues due to their high energy consumption. The Fog also performed very well in terms of sValues (they are the second-best), but Rasp8 outperformed them when it came to eValues. Jetson services had closer eValue and sValue, as their high processing time affected both energy and response time.

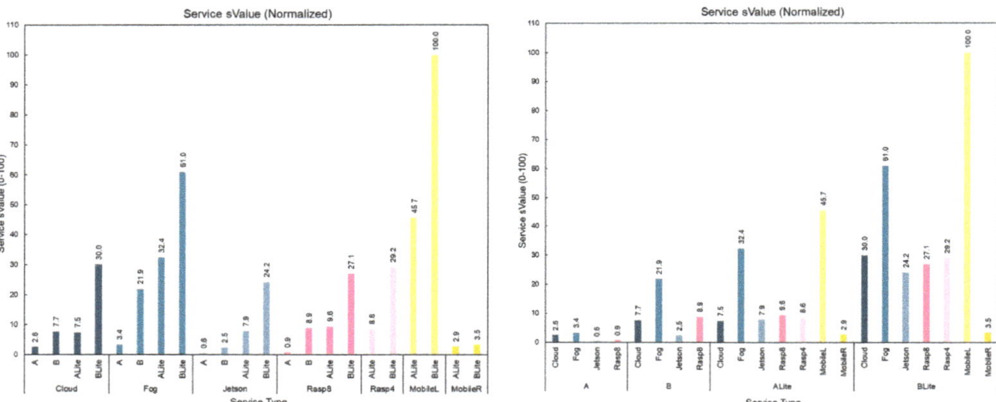

Figure 27. Service sValue.

6. Conclusions and Future Work

Digital services are the fundamental building blocks of technology-driven smart cities and societies. There has been an increasing need for distributed services that provide intelligence near the fog and edge for reasons such as privacy, security, performance, and costs. The healthcare sector is not an exception; not only does it require such distributed services, but also it is also driven by many other factors including declining public health, increase in chronic diseases, ageing population, rising healthcare costs, and COVID-19.

In this paper, the Imtidad reference architecture is proposed, implemented, and evaluated. It provides DAIaaS over the cloud, fog, and edge using a service catalog case study containing 22 AI skin disease diagnosis services. These services belong to four service classes that are distinguished by software platforms (containerized gRPC, etc.) and are executed on a range of hardware platforms (NVIDIA Jetson nano, etc.) and four network types (Fiber, etc.). The AI models for diagnosis included two standard and two Tiny AI Deep Neural Networks to enable their execution at the edge. They were trained and tested using 10,015 real-life dermatoscopic images.

A detailed evaluation of the DAIaaS skin lesion diagnosis services was provided using several benchmarks. A DL service on a local smartphone provides the best service in terms of energy followed by a Raspberry Pi edge device. A DL service on a local smartphone provides the best service also in terms of speed followed by a laptop device in the fog layer. DL services in the edge layer on local smartphones are the best in terms of energy and response time (speed) as they do not require any network communication, though they can only accommodate TFLite or small models. TFLite optimization provided a great improvement in terms of processing time and compatibility with edge devices. However, it could reduce model accuracy to some levels that could be tolerated depending on the criticality of the application and user preferences. Therefore, we considered the accuracy of the model in both eValue and sValue, to provide a way for the user to choose and trade-off between these factors, energy, and speed. Other devices in the fog and edge layers, such as a laptop and Raspberry Pi (8 GB), can accommodate more complex models and at the same time provide fast responses to local service requests. DL service on a remote smartphone provided unpredictable behavior in terms of network time compared to other edge and fog services due to the Android Nearby Connections API, which is used for nearby smartphone communication. The Cloud services' processing time is close to the Fog services, though the response time is higher as it requires more time to transfer the image across the internet. This would depend on particular scenarios, such as those requiring heavy computations, which would render the cloud to have much faster responses because in those cases the processing time would be a bottleneck for low-resource fog devices. DL services in the cloud layer also depend on the type of internet connection used. Our evaluation of both Fiber and Cellular (4G) internet connections on the Cloud services confirmed that the fiber network connection is more stable and has lower network time than the cellular connection (4G in this case, but this may change for 5G and 6G). Obviously, while fiber connection was shown to be more stable, it has limitations in terms of user mobility. The Cloud services eValue and sValue are both affected by the required network communication over WAN.

The novelty and the high impact of this research lies in the developed reference architecture, the service catalog offering a large number of services, the potential for the implementation of innovative use cases through the edge, fog, and cloud, and their evaluation on many software, hardware, and networking platforms, as well as a detailed description of the architecture and case study. To the best of the authors' knowledge, this is the first research paper in which a reference architecture for DAIaaS is proposed and implemented, as well as in which a healthcare application (skin lesion diagnosis) is developed and studied in detail. This work is expected to have an extensive impact on developing smart distributed service infrastructures for healthcare and other sectors.

Future research on distributed services will focus on improving the accuracy and other performance aspects of the skin disease AI model and services. While the design, implementation, and evaluation of the proposed reference architecture and DAIaaS services is detailed and diverse, human, computer, and network resource limitations impeded a higher diversity of hardware, networks, and more frequent measurements. Future lines of research will be oriented towards improving the granularity of the measurements as well as adding to the diversity of the software, hardware, and communication platforms.

Future work will also consider improving and refining the reference architecture, extending it through the development of services in other application domains and sectors including many smart city applications that we have developed over the years including smart cities [2,3,81], big data [8,20], improving computing algorithms [82,83], education [1], spam detection [84], accident and disaster management [85,86], autonomous vehicles and transportation [87–91], and healthcare [6,56,92,93].

AI will be an important parameter in the evolution of the 5th Generation (5G) networks and the conceptualization and design of the 6th Generation (6G) networks. Technologies such as network function virtualization (NFV), software-defined networking (SDN), 3D network architectures, and energy harvesting strategies will play important roles in delivering the promises of 5G and 6G networks. However, it is AI that is expected to be the main player in network design and operations, not only in terms of the use of AI for the optimization of network functions, but also due to the expectations that AI, being a fundamental ingredient of smart applications, will be a major workload to be supported by next-generation networks. While 5G promises us high-speed mobile internet, 6G pledges to support ubiquitous AI services through next-generation softwarization, heterogeneity, and configurability of networks [13]. The work on 6G is in its infancy and requires the community to conceptualize and develop its design, implementation, deployment, and use cases [13]. This paper is part of our broader work on distributed AI as a Service and is a timely contribution to this area of developing next-generation infrastructure, including the network infrastructure, needed to support smart societies of the future. Our earlier work [13] proposed a framework for provisioning Distributed AI as a service in IoE (Internet of Everything) and 6G environments and evaluated it using three case studies on distributed AI as service delivery in smart environments, including a smart airport and a smart district. This paper adds to the earlier work by extending another case study on developing a service catalog of distributed services.

Author Contributions: Conceptualization, N.J. and R.M.; methodology, N.J. and R.M.; software, N.J.; validation, N.J. and R.M.; formal analysis, N.J., R.M., J.M.C. and T.Y.; investigation, N.J., R.M., I.K., A.A., T.Y. and J.M.C.; resources, R.M., I.K. and A.A.; data curation, N.J.; writing–original draft preparation, N.J. and R.M.; writing–review and editing, R.M., I.K., A.A., T.Y. and J.M.C.; visualization, N.J.; supervision, R.M. and I.K.; project administration, R.M., I.K. and A.A.; funding acquisition, R.M., A.A. and I.K. All authors have read and agreed to the published version of the manuscript.

Funding: The authors acknowledge with thanks the technical and financial support from the Deanship of Scientific Research (DSR) at King Abdulaziz University (KAU), Jeddah, Saudi Arabia, under Grant No. RG-10-611-38. The experiments reported in this paper were performed on the Aziz supercomputer at KAU.

Institutional Review Board Statement: Not applicable.

Informed Consent Statement: Not applicable.

Data Availability Statement: The HAM10000 dataset is a public dataset available from the link provided in the article.

Acknowledgments: The work carried out in this paper is supported by the HPC Center at King Abdulaziz University. The training and software development work reported in this paper was carried out on the Aziz supercomputer.

Conflicts of Interest: The authors declare no conflict of interest.

References

1. Mehmood, R.; Alam, F.; Albogami, N.N.; Katib, I.; Albeshri, A.; Altowaijri, S.M. UTiLearn: A Personalised Ubiquitous Teaching and Learning System for Smart Societies. *IEEE Access* **2017**, *5*, 2615–2635. [CrossRef]
2. Yigitcanlar, T.; Butler, L.; Windle, E.; Desouza, K.C.; Mehmood, R.; Corchado, J.M. Can Building 'Artificially Intelligent Cities' Safeguard Humanity from Natural Disasters, Pandemics, and Other Catastrophes? An Urban Scholar's Perspective. *Sensors* **2020**, *20*, 2988. [CrossRef] [PubMed]
3. Yigitcanlar, T.; Kankanamge, N.; Regona, M.; Maldonado, A.; Rowan, B.; Ryu, A.; DeSouza, K.C.; Corchado, J.M.; Mehmood, R.; Li, R.Y.M. Artificial Intelligence Technologies and Related Urban Planning and Development Concepts: How Are They Perceived and Utilized in Australia? *J. Open Innov. Technol. Mark. Complex.* **2020**, *6*, 187. [CrossRef]
4. AlOmari, E.; Katib, I.; Albeshri, A.; Mehmood, R. COVID-19: Detecting Government Pandemic Measures and Public Concerns from Twitter Arabic Data Using Distributed Machine Learning. *Int. J. Environ. Res. Public Health* **2021**, *18*, 282. [CrossRef] [PubMed]
5. Yigitcanlar, T.; Corchado, J.; Mehmood, R.; Li, R.; Mossberger, K.; Desouza, K. Responsible Urban Innovation with Local Government Artificial Intelligence (AI): A Conceptual Framework and Research Agenda. *J. Open Innov. Technol. Mark. Complex.* **2021**, *7*, 71. [CrossRef]
6. Alotaibi, S.; Mehmood, R.; Katib, I.; Rana, O.; Albeshri, A. Sehaa: A Big Data Analytics Tool for Healthcare Symptoms and Diseases Detection Using Twitter, Apache Spark, and Machine Learning. *Appl. Sci.* **2020**, *10*, 1398. [CrossRef]
7. Yigitcanlar, T.; Mehmood, R.; Corchado, J.M. Green Artificial Intelligence: Towards an Efficient, Sustainable and Equitable Technology for Smart Cities and Futures. *Sustainability* **2021**, *13*, 8952. [CrossRef]
8. Alam, F.; Mehmood, R.; Katib, I.; Albogami, N.N.; Albeshri, A. Data Fusion and IoT for Smart Ubiquitous Environments: A Survey. *IEEE Access* **2017**, *5*, 9533–9554. [CrossRef]
9. Mehmood, R.; Faisal, M.A.; Altowaijri, S. Future networked healthcare systems: A review and case study. In *Handbook of Research on Redesigning the Future of Internet Architectures*; IGI Global: Hershey, PA, USA, 2015; pp. 531–555.
10. Greco, L.; Percannella, G.; Ritrovato, P.; Tortorella, F.; Vento, M. Trends in IoT based solutions for health care: Moving AI to the edge. *Pattern Recognit. Lett.* **2020**, *135*, 346–353. [CrossRef]
11. Mukherjee, A.; Ghosh, S.; Behere, A.; Ghosh, S.K.; Buyya, R. Internet of Health Things (IoHT) for personalized health care using integrated edge-fog-cloud network. *J. Ambient Intell. Humaniz. Comput.* **2020**, *12*, 943–959. [CrossRef]
12. Farahani, B.; Barzegari, M.; Aliee, F.S.; Shaik, K.A. Towards collaborative intelligent IoT eHealth: From device to fog, and cloud. *Microprocess. Microsyst.* **2019**, *72*, 102938. [CrossRef]
13. Janbi, N.; Katib, I.; Albeshri, A.; Mehmood, R. Distributed Artificial Intelligence-as-a-Service (DAIaaS) for Smarter IoE and 6G Environments. *Sensors* **2020**, *20*, 5796. [CrossRef] [PubMed]
14. Usman, S.; Mehmood, R.; Katib, I. Big Data and HPC Convergence for Smart Infrastructures: A Review and Proposed Architecture. *Smart Infrastruct. Appl.* **2020**, 561–586.
15. Al-Dhubhani, R.; Mehmood, R.; Katib, I.; Algarni, A. Location privacy in smart cities era. In *Lecture Notes of the Institute for Computer Sciences, Social-Informatics and Telecommunications Engineering*; Springer Publishing Company: New York, NY, USA, 2018; Volume 224, pp. 123–138.

16. Assiri, F.Y.; Mehmood, R. *Software Quality in the Era of Big Data, IoT and Smart Cities*; Springer: Cham, Switzerland, 2020; pp. 519–536.
17. Singh, J.; Kad, S.; Singh, P.D. *Implementing Fog Computing for Detecting Primary Tumors Using Hybrid Approach of Data Mining*; Springer: Singapore, 2021; pp. 1067–1080.
18. Amin, S.U.; Hossain, M.S. Edge Intelligence and Internet of Things in Healthcare: A Survey. *IEEE Access* **2020**, *9*, 45–59. [CrossRef]
19. Arfat, Y.; Aqib, M.; Mehmood, R.; Albeshri, A.; Katib, I.; Albogami, N.; Alzahrani, A. Enabling Smarter Societies through Mobile Big Data Fogs and Clouds. *Procedia Comput. Sci.* **2017**, *109*, 1128–1133. [CrossRef]
20. Arfat, Y.; Usman, S.; Mehmood, R.; Katib, I. Big Data for Smart Infrastructure Design: Opportunities and Challenges. *Smart Infrastruct. Appl.* **2020**, 491–518.
21. Ahmad, A.; Fahmideh, M.; Altamimi, A.B.; Katib, I.; Albeshri, A.; Alreshidi, A.; Mehmood, R. Software Engineering for IoT-Driven Data Analytics Applications. *IEEE Access* **2021**, *9*, 48197–48217. [CrossRef]
22. Tang, P.; Dong, Y.; Chen, Y.; Mao, S.; Halgamuge, S. QoE-Aware Traffic Aggregation Using Preference Logic for Edge Intelligence. *IEEE Trans. Wirel. Commun.* **2021**, *20*, 6093–6106. [CrossRef]
23. Tsaur, W.-J.; Yeh, L.-Y. DANS: A Secure and Efficient Driver-Abnormal Notification Scheme With IoT Devices Over IoV. *IEEE Syst. J.* **2018**, *13*, 1628–1639. [CrossRef]
24. Cui, X.; Zhang, W.; Finkler, U.; Saon, G.; Picheny, M.; Kung, D. Distributed Training of Deep Neural Network Acoustic Models for Automatic Speech Recognition: A comparison of current training strategies. *IEEE Signal Process. Mag.* **2020**, *37*, 39–49. [CrossRef]
25. Langer, M.; He, Z.; Rahayu, W.; Xue, Y. Distributed Training of Deep Learning Models: A Taxonomic Perspective. *IEEE Trans. Parallel Distrib. Syst.* **2020**, *31*, 2802–2818. [CrossRef]
26. Aspri, M.; Tsagkatakis, G.; Tsakalides, P. Distributed Training and Inference of Deep Learning Models for Multi-Modal Land Cover Classification. *Remote Sens.* **2020**, *12*, 2670. [CrossRef]
27. Sugi, S.S.S.; Ratna, S.R. A novel distributed training on fog node in IoT backbone networks for security. *Soft Comput.* **2020**, *24*, 18399–18410. [CrossRef]
28. Li, C.; Wang, S.; Li, X.; Zhao, F.; Yu, R. Distributed perception and model inference with intelligent connected vehicles in smart cities. *Ad Hoc Netw.* **2020**, *103*, 102152. [CrossRef]
29. Hosseinalipour, S.; Brinton, C.G.; Aggarwal, V.; Dai, H.; Chiang, M. From Federated to Fog Learning: Distributed Machine Learning over Heterogeneous Wireless Networks. *IEEE Commun. Mag.* **2020**, *58*, 41–47. [CrossRef]
30. Zhou, Z.; Chen, X.; Li, E.; Zeng, L.; Luo, K.; Zhang, J. Edge Intelligence: Paving the Last Mile of Artificial Intelligence With Edge Computing. *Proc. IEEE* **2019**, *8*, 1738–1762. [CrossRef]
31. Cao, Y.; Chen, S.; Hou, P.; Brown, D. FAST: A fog computing assisted distributed analytics system to monitor fall for stroke mitigation. In Proceedings of the 2015 IEEE International Conference on Networking, Architecture and Storage, NAS 2015, Boston, MA, USA, 6–7 August 2015; pp. 2–11.
32. Hassan, S.R.; Ahmad, I.; Ahmad, S.; AlFaify, A.; Shafiq, M. Remote Pain Monitoring Using Fog Computing for e-Healthcare: An Efficient Architecture. *Sensors* **2020**, *20*, 6574. [CrossRef]
33. Mohammed, T.; Albeshri, A.; Katib, I.; Mehmood, R. UbiPriSEQ—Deep reinforcement learning to manage privacy, security, energy, and QoS in 5G IoT hetnets. *Appl. Sci.* **2020**, *10*, 7120. [CrossRef]
34. Deng, S.; Zhao, H.; Fang, W.; Yin, J.; Dustdar, S.; Zomaya, A.Y. Edge Intelligence: The Confluence of Edge Computing and Artificial Intelligence. *IEEE Internet Things J.* **2020**, *7*, 7457–7469. [CrossRef]
35. Wang, X.; Han, Y.; Leung, V.C.M.; Niyato, D.; Yan, X.; Chen, X. Convergence of Edge Computing and Deep Learning: A Comprehensive Survey. *IEEE Commun. Surv. Tutorials* **2020**, *22*, 869–904. [CrossRef]
36. Park, J.; Samarakoon, S.; Bennis, M.; Debbah, M. Wireless Network Intelligence at the Edge. *Proc. IEEE* **2019**, *107*, 2204–2239. [CrossRef]
37. Chen, J.; Ran, X. Deep Learning With Edge Computing: A Review. *Proc. IEEE* **2019**, *107*, 1655–1674. [CrossRef]
38. Isakov, M.; Gadepally, V.; Gettings, K.M.; Kinsy, M.A. Survey of Attacks and Defenses on Edge-Deployed Neural Networks. In Proceedings of the 2019 IEEE High Performance Extreme Computing Conference (HPEC), Waltham, MA, USA, 24–26 September 2019; pp. 1–8.
39. Rausch, T.; Dustdar, S. Edge intelligence: The convergence of humans, things, and AI. In Proceedings of the 2019 IEEE International Conference on Cloud Engineering, IC2E 2019, Prague, Czech Republic, 24–27 June 2019; pp. 86–96.
40. Shi, Z. *Advanced Artificial Intelligence*, 2nd ed.; World Scientific: Singapore, 2019.
41. Pattnaik, B.S.; Pattanayak, A.S.; Udgata, S.K.; Panda, A.K. Advanced centralized and distributed SVM models over different IoT levels for edge layer intelligence and control. *Evol. Intell.* **2020**, 1–15. [CrossRef]
42. Gao, Y.; Liu, L.; Zheng, X.; Zhang, C.; Ma, H. Federated Sensing: Edge-Cloud Elastic Collaborative Learning for Intelligent Sensing. *IEEE Internet Things J.* **2021**, *8*, 11100–11111. [CrossRef]
43. Chen, Y.; Qin, X.; Wang, J.; Yu, C.; Gao, W. FedHealth: A Federated Transfer Learning Framework for Wearable Healthcare. *IEEE Intell. Syst.* **2020**, *35*, 83–93. [CrossRef]
44. Hegedűs, I.; Danner, G.; Jelasity, M. Decentralized learning works: An empirical comparison of gossip learning and federated learning. *J. Parallel Distrib. Comput.* **2020**, *148*, 109–124. [CrossRef]
45. Kim, H.; Park, J.; Bennis, M.; Kim, S.-L. Blockchained On-Device Federated Learning. *IEEE Commun. Lett.* **2019**, *24*, 1279–1283. [CrossRef]

46. TensorFlow Lite | ML for Mobile and Edge Devices. Available online: https://www.tensorflow.org/lite (accessed on 12 October 2020).
47. Zebin, T.; Scully, P.J.; Peek, N.; Casson, A.J.; Ozanyan, K.B. Design and Implementation of a Convolutional Neural Network on an Edge Computing Smartphone for Human Activity Recognition. *IEEE Access* **2019**, *7*, 133509–133520. [CrossRef]
48. Benhamida, A.; Varkonyi-Koczy, A.R.; Kozlovszky, M. Traffic Signs Recognition in a mobile-based application using TensorFlow and Transfer Learning technics. In Proceedings of the 2020 IEEE 15th International Conference of System of Systems Engineering (SoSE), Budapest, Hungary, 2–4 June 2020; pp. 000537–000542.
49. Alsing, O. *Mobile Object Detection Using TensorFlow Lite and Transfer Learning*; KTH Royal Institute of Technology: Stockholm, Sweden, 2018.
50. Zeroual, A.; Derdour, M.; Amroune, M.; Bentahar, A. Using a Fine-Tuning Method for a Deep Authentication in Mobile Cloud Computing Based on Tensorflow Lite Framework. In Proceedings of the ICNAS 2019: 4th International Conference on Networking and Advanced Systems, Annaba, Algeria, 26–27 June 2019.
51. Ahmadi, M.; Sotgiu, A.; Giacinto, G. IntelliAV: Toward the feasibility of building intelligent anti-malware on android devices. In Proceedings of the Lecture Notes in Computer Science (Including Subseries Lecture Notes in Artificial Intelligence and Lecture Notes in Bioinformatics), Reggio, Italy, 29 August–1 September 2017; Volume 10410, pp. 137–154.
52. Soltani, N.; Sankhe, K.; Ioannidis, S.; Jaisinghani, D.; Chowdhury, K. Spectrum Awareness at the Edge: Modulation Classification using Smartphones. In Proceedings of the 2019 IEEE International Symposium on Dynamic Spectrum Access Networks, DySPAN 2019, Newark, NJ, USA, 11–14 November 2019.
53. Domozi, Z.; Stojcsics, D.; Benhamida, A.; Kozlovszky, M.; Molnar, A. Real time object detection for aerial search and rescue missions for missing persons. In Proceedings of the 2020 IEEE 15th International Conference of System of Systems Engineering (SoSE), Budapest, Hungary, 2–4 June 2020; pp. 000519–000524.
54. Caffe2 Deep Learning Framework | NVIDIA Developer. Available online: https://developer.nvidia.com/caffe2 (accessed on 12 October 2020).
55. Home | PyTorch. Available online: https://pytorch.org/mobile/home/ (accessed on 12 October 2020).
56. Muhammed, T.; Mehmood, R.; Albeshri, A.; Katib, I. UbeHealth: A Personalized Ubiquitous Cloud and Edge-Enabled Networked Healthcare System for Smart Cities. *IEEE Access* **2018**, *6*, 32258–32285. [CrossRef]
57. Goyal, M.; Knackstedt, T.; Yan, S.; Hassanpour, S. Artificial intelligence-based image classification methods for diagnosis of skin cancer: Challenges and opportunities. *Comput. Biol. Med.* **2020**, *127*, 104065. [CrossRef]
58. Skin Cancer Facts & Statistics—The Skin Cancer Foundation. Available online: https://www.skincancer.org/skin-cancer-information/skin-cancer-facts/ (accessed on 24 March 2020).
59. Bray, F.; Ferlay, J.; Soerjomataram, I.; Siegel, R.L.; Torre, L.A.; Jemal, A. Global cancer statistics 2018: GLOBOCAN estimates of incidence and mortality worldwide for 36 cancers in 185 countries. *CA Cancer J. Clin.* **2018**, *68*, 394–424. [CrossRef] [PubMed]
60. Claudiu, P.; Ion, B.; Tudorel, P.; Iordache, I.; Nida, B.; Abdulazis, T.; Panculescu, F. The Value of Digital Dermatoscopy in the Diagnosis and Treatment of Precancerous Skin Lesions. *ARS Med. Tomitana* **2018**, *24*, 40–45. [CrossRef]
61. Ashique, K.T.; Aurangabadkar, S.J.; Kaliyadan, F. Clinical photography in dermatology using smartphones: An overview. *Indian Dermatol. Online J.* **2015**, *6*, 158–163. [CrossRef] [PubMed]
62. Sonthalia, S.; Kaliyadan, F. *Dermoscopy Overview and Extradiagnostic Applications*; StatPearls Publishing: Treasure Island, FL, USA, 2020.
63. Jha, D.; Riegler, M.A.; Johansen, D.; Halvorsen, P.; Johansen, H.D. DoubleU-Net: A Deep Convolutional Neural Network for Medical Image Segmentation. In Proceedings of the IEEE Symposium on Computer-Based Medical Systems, Rochester, MA, USA, 28–30 July 2020; pp. 558–564.
64. Bajwa, M.N.; Muta, K.; Malik, M.I.; Siddiqui, S.A.; Braun, S.A.; Homey, B.; Dengel, A.; Ahmed, S. Computer-Aided Diagnosis of Skin Diseases Using Deep Neural Networks. *Appl. Sci.* **2020**, *10*, 2488. [CrossRef]
65. Zhang, X.; Wang, S.; Liu, J.; Tao, C. Towards improving diagnosis of skin diseases by combining deep neural network and human knowledge. *BMC Med Informatics Decis. Mak.* **2018**, *18*, 69–76. [CrossRef]
66. Wei, L.-S.; Gan, Q.; Ji, T. Skin Disease Recognition Method Based on Image Color and Texture Features. *Comput. Math. Methods Med.* **2018**, *2018*, 1–10. [CrossRef]
67. Liu, Y.; Jain, A.; Eng, C.; Way, D.H.; Lee, K.; Bui, P.; Kanada, K.; Marinho, G.D.O.; Gallegos, J.; Gabriele, S.; et al. A deep learning system for differential diagnosis of skin diseases. *Nat. Med.* **2020**, *26*, 900–908. [CrossRef]
68. Gavrilov, D.A.; Melerzanov, A.V.; Shchelkunov, N.N.; Zakirov, E.I. Use of Neural Network-Based Deep Learning Techniques for the Diagnostics of Skin Diseases. *Biomed. Eng.* **2019**, *52*, 348–352. [CrossRef]
69. Goyal, M.; Oakley, A.; Bansal, P.; Dancey, D.; Yap, M.H. Skin Lesion Segmentation in Dermoscopic Images With Ensemble Deep Learning Methods. *IEEE Access* **2019**, *8*, 4171–4181. [CrossRef]
70. Ünver, H.M.; Ayan, E. Skin Lesion Segmentation in Dermoscopic Images with Combination of YOLO and GrabCut Algorithm. *Diagnostics* **2019**, *9*, 72. [CrossRef]
71. He, X.; Wang, S.; Shi, S.; Tang, Z.; Wang, Y.; Zhao, Z.; Chu, X. Computer-Aided Clinical Skin Disease Diagnosis Using CNN and Object Detection Models. In Proceedings of the 2019 IEEE International Conference on Big Data, Big Data 2019, Los Angeles, CA, USA, 9–12 December 2019; pp. 4839–4844.

72. Garcia-Arroyo, J.L.; Garcia-Zapirain, B. Segmentation of skin lesions in dermoscopy images using fuzzy classification of pixels and histogram thresholding. *Comput. Methods Programs Biomed.* **2018**, *168*, 11–19. [CrossRef] [PubMed]
73. Walker, B.; Rehg, J.; Kalra, A.; Winters, R.M.; Drews, P.; Dascalu, J.; David, E. Dermoscopy diagnosis of cancerous lesions utilizing dual deep learning algorithms via visual and audio (sonification) outputs: Laboratory and prospective observational studies. *eBioMedicine* **2019**, *40*, 176–183. [CrossRef] [PubMed]
74. Baig, R.; Bibi, M.; Hamid, A.; Kausar, S.; Khalid, S. Deep Learning Approaches Towards Skin Lesion Segmentation and Classification from Dermoscopic Images—A Review. *Curr. Med Imaging Former. Curr. Med Imaging Rev.* **2020**, *16*, 513–533. [CrossRef] [PubMed]
75. Kurpicz, M.; Orgerie, A.C.; Sobe, A. How Much Does a VM Cost? Energy-Proportional Accounting in VM-Based Environments. In Proceedings of the 24th Euromicro International Conference on Parallel, Distributed, and Network-Based Processing, PDP 2016, Heraklion, Greece, 17–19 February 2016; pp. 651–658.
76. Andrae, A.S.G.; Edler, T. On Global Electricity Usage of Communication Technology: Trends to 2030. *Challenges* **2015**, *6*, 117–157. [CrossRef]
77. ISIC Challenge. Available online: https://challenge.isic-archive.com/data (accessed on 11 October 2020).
78. Lio, P.A.; Nghiem, P. Interactive Atlas of Dermoscopy. *J. Am. Acad. Dermatol.* **2004**, *50*, 807–808. [CrossRef]
79. ADDI—Automatic Computer-Based Diagnosis System for Dermoscopy Images. Available online: https://www.fc.up.pt/addi/ph2database.html (accessed on 11 October 2020).
80. Tschandl, P.; Rosendahl, C.; Kittler, H. The HAM10000 dataset, a large collection of multi-source dermatoscopic images of common pigmented skin lesions. *Sci. Data* **2018**, *5*, 180161. [CrossRef]
81. Mehmood, R.; See, S.; Katib, I.; Chlamtac, I. *Smart Infrastructure and Applications: Foundations for Smarter Cities and Societies*; Springer International Publishing; Springer Nature Switzerland AG: Cham, Switzerland, 2020.
82. Mohammed, T.; Albeshri, A.; Katib, I.; Mehmood, R. DIESEL: A novel deep learning-based tool for SpMV computations and solving sparse linear equation systems. *J. Supercomput.* **2020**, *77*, 6313–6355. [CrossRef]
83. Muhammed, T.; Mehmood, R.; Albeshri, A.; Katib, I. SURAA: A Novel Method and Tool for Loadbalanced and Coalesced SpMV Computations on GPUs. *Appl. Sci.* **2019**, *9*, 947. [CrossRef]
84. Bosaeed, S.; Katib, I.; Mehmood, R. A Fog-Augmented Machine Learning based SMS Spam Detection and Classification System. In Proceedings of the 2020 5th International Conference on Fog and Mobile Edge Computing, FMEC 2020, Paris, France, 20–23 April 2020; pp. 325–330.
85. Yigitcanlar, T.; Regona, M.; Kankanamge, N.; Mehmood, R.; D'Costa, J.; Lindsay, S.; Nelson, S.; Brhane, A. Detecting Natural Hazard-Related Disaster Impacts with Social Media Analytics: The Case of Australian States and Territories. *Sustainability* **2022**, *14*, 810. [CrossRef]
86. Aqib, M.; Mehmood, R.; Alzahrani, A.; Katib, I. *In-Memory Deep Learning Computations on GPUs for Prediction of Road Traffic Incidents Using Big Data Fusion*; Springer: Cham, Switzerland, 2020; pp. 79–114.
87. Alomari, E.; Katib, I.; Albeshri, A.; Yigitcanlar, T.; Mehmood, R. Iktishaf+: A Big Data Tool with Automatic Labeling for Road Traffic Social Sensing and Event Detection Using Distributed Machine Learning. *Sensors* **2021**, *21*, 2993. [CrossRef]
88. Alkhamisi, A.O.; Mehmood, R. An Ensemble Machine and Deep Learning Model for Risk Prediction in Aviation Systems. In Proceedings of the 2020 6th Conference on Data Science and Machine Learning Applications (CDMA), Riyadh, Saudi Arabia, 4–5 March 2020; pp. 54–59.
89. Aqib, M.; Mehmood, R.; Alzahrani, A.; Katib, I.; Albeshri, A.; Altowaijri, S.M. Rapid Transit Systems: Smarter Urban Planning Using Big Data, In-Memory Computing, Deep Learning, and GPUs. *Sustainability* **2019**, *11*, 2736. [CrossRef]
90. Alam, F.; Mehmood, R.; Katib, I.; Altowaijri, S.M.; Albeshri, A. TAAWUN: A Decision Fusion and Feature Specific Road Detection Approach for Connected Autonomous Vehicles. *Mob. Networks Appl.* **2019**, 1–17. [CrossRef]
91. AlOmari, E.; Katib, I.; Mehmood, R. Iktishaf: A Big Data Road-Traffic Event Detection Tool Using Twitter and Spark Machine Learning. *Mob. Netw. Appl.* **2020**, 1–16. [CrossRef]
92. Alam, F.; Almaghthawi, A.; Katib, I.; Albeshri, A.; Mehmood, R. iResponse: An AI and IoT-Enabled Framework for Autonomous COVID-19 Pandemic Management. *Sustainability* **2021**, *13*, 3797. [CrossRef]
93. Alotaibi, H.; Alsolami, F.; Mehmood, R. DNA Profiling: An Investigation of Six Machine Learning Algorithms for Estimating the Number of Contributors in DNA Mixtures. *Int. J. Adv. Comput. Sci. Appl.* **2021**, *12*, 130–136. [CrossRef]

Viewpoint

Can Building "Artificially Intelligent Cities" Safeguard Humanity from Natural Disasters, Pandemics, and Other Catastrophes? An Urban Scholar's Perspective

Tan Yigitcanlar [1,*], **Luke Butler** [1], **Emily Windle** [1], **Kevin C. Desouza** [2], **Rashid Mehmood** [3] **and Juan M. Corchado** [4,5,6,7]

[1] School of Built Environment, Queensland University of Technology, 2 George Street, Brisbane, QLD 4000, Australia; luke.butler@hdr.qut.edu.au (L.B.); emily.windle@connect.qut.edu.au (E.W.)

[2] QUT Business School, Queensland University of Technology, 2 George Street, Brisbane, QLD 4000, Australia; kevin.desouza@qut.edu.au

[3] High Performance Computing Center, King Abdulaziz University, Al Ehtifalat St, Jeddah 21589, Saudi Arabia; rmehmood@kau.edu.sa

[4] Bisite Research Group, University of Salamanca, 37007 Salamanca, Spain; corchado@usal.es

[5] Air Institute, IoT Digital Innovation Hub, 37188 Salamanca, Spain

[6] Department of Electronics, Information and Communication, Faculty of Engineering, Osaka Institute of Technology, Osaka 535-8585, Japan

[7] Pusat Komputeran dan Informatik, Universiti Malaysia Kelantan, Kelantan 16100, Malaysia

* Correspondence: tan.yigitcanlar@qut.edu.au; Tel.: +61-7-3138-2418

Received: 28 April 2020; Accepted: 22 May 2020; Published: 25 May 2020

Abstract: In recent years, artificial intelligence (AI) has started to manifest itself at an unprecedented pace. With highly sophisticated capabilities, AI has the potential to dramatically change our cities and societies. Despite its growing importance, the urban and social implications of AI are still an understudied area. In order to contribute to the ongoing efforts to address this research gap, this paper introduces the notion of an artificially intelligent city as the potential successor of the popular smart city brand—where the smartness of a city has come to be strongly associated with the use of viable technological solutions, including AI. The study explores whether building artificially intelligent cities can safeguard humanity from natural disasters, pandemics, and other catastrophes. All of the statements in this viewpoint are based on a thorough review of the current status of AI literature, research, developments, trends, and applications. This paper generates insights and identifies prospective research questions by charting the evolution of AI and the potential impacts of the systematic adoption of AI in cities and societies. The generated insights inform urban policymakers, managers, and planners on how to ensure the correct uptake of AI in our cities, and the identified critical questions offer scholars directions for prospective research and development.

Keywords: artificial intelligence (AI); artificially intelligent city; artificially intelligence commons; smart city; smart urban technology; urban informatics; sustainable urban development; climate change; pandemics; natural disasters

---

1. Introduction

What Is an Artificially Intelligent City?

During the current Anthropocene era—the geological epoch which has had significant human impact on Earth's geology and ecosystems—we have developed technological capabilities that have enabled us to greedily use limited natural resources for economic profit [1,2]. This ruthless capitalist practice not only brought about anthropogenic climate change, but also caused socioeconomic inequalities to soar globally [3].

In recent years, technology, as part of knowledge-based development efforts [4], has been viewed as the solution to severe environmental, economic, and social crises [5,6]. Consequently, the smart city concept has come to the forefront of discourses on urban planning and development [7]. Accordingly, many see emerging technologies, particularly artificial intelligence (AI), as a way to safeguard our civilization from the catastrophic consequences of climate change [8], biodiversity loss [9], natural disasters [10], unsustainable development [11], pandemics [12], and so on.

Simply, AI is defined as "machines or computers that mimic cognitive functions that humans associate with the human mind, such as learning and problem solving" [13]. AI-driven computational techniques are diverse and range from rule-based systems to deep learning systems. A popular AI knowledge map was created by Corea [14]. His conceptualization brings together the AI paradigms and problem domains (Figure 1).

The AI paradigm and its technology-enabled solutions—whether it is autonomous driving, home automation (so-called domotics), robotics, chatbots, or advanced data analytic tools—have opened up new opportunities for cities, where most of the world population resides, where most of the production and consumption activities take place, and also where most of the negative environmental externalities are generated [15,16]. While some scholars see AI as an opportunity to advance smart cities (or smartness of cities) [17–21], others see AI generating a whole new city brand, especially when the AI applications become mainstream in our cities [22]. In other words, in the near future, we will see a trend to build 'artificially intelligent cities' from scratch, or to retrofit traditional cities, converting them into artificially intelligent ones.

We define an artificially intelligent city as an urban locality functioning as a robust system of systems, and whose economic, societal, environmental, and governmental activities are based on sustainable practices driven by AI technologies, helping us achieve social good and other desired outcomes and futures for all humans and non-humans.

In the age of smart cities—where urban locations are starting to be wired with smart technologies including sensor networks—and given the highly sophisticated capabilities of AI, we foresee a potential dramatic change in our cities and societies [23]. There is, hence, an increasing need to investigate the urban and social implications of AI. This is an understudied area of research, and a gap in the literature on AI and city/society.

We also note that there are different levels of AI, including: (a) reactive machines (e.g., IBM's Deep Blue); (b) limited memory AI (e.g., chatbots, virtual assistants, self-driving vehicles); (c) theory of mind AI (a concept that is in progress at the moment); and (d) self-aware AI (only hypothetical at this stage) [24]. There is also another categorization of levels of AI, such as: (a) artificial narrow intelligence (represents all of the existing AI today); (b) artificial general intelligence (its main idea is that AI agents can learn, perceive, understand, and function completely like a human being); and (c) artificial superintelligence (an idea that AI replicates the multifaceted intelligence of human beings and becomes exceedingly better at everything it does) [25]. The disruption of each level of AI will be different in our cities and societies. Throughout this paper, we focus on the current level of AI: artificial narrow intelligence.

Against this backdrop, we prepared this viewpoint in order to help in bridging this gap along with promoting further research on the topic. In this paper, we introduce a provocative artificially intelligent city notion as the potential successor of the currently popular smart city concept, where city smartness today is increasingly depending on the use of viable technology solutions, including AI. The paper, by placing the AI literature, developments, trends, and applications under the microscope, provides a commentary on whether building artificially intelligent cities can safeguard humanity from natural disasters, pandemics, and other catastrophes.

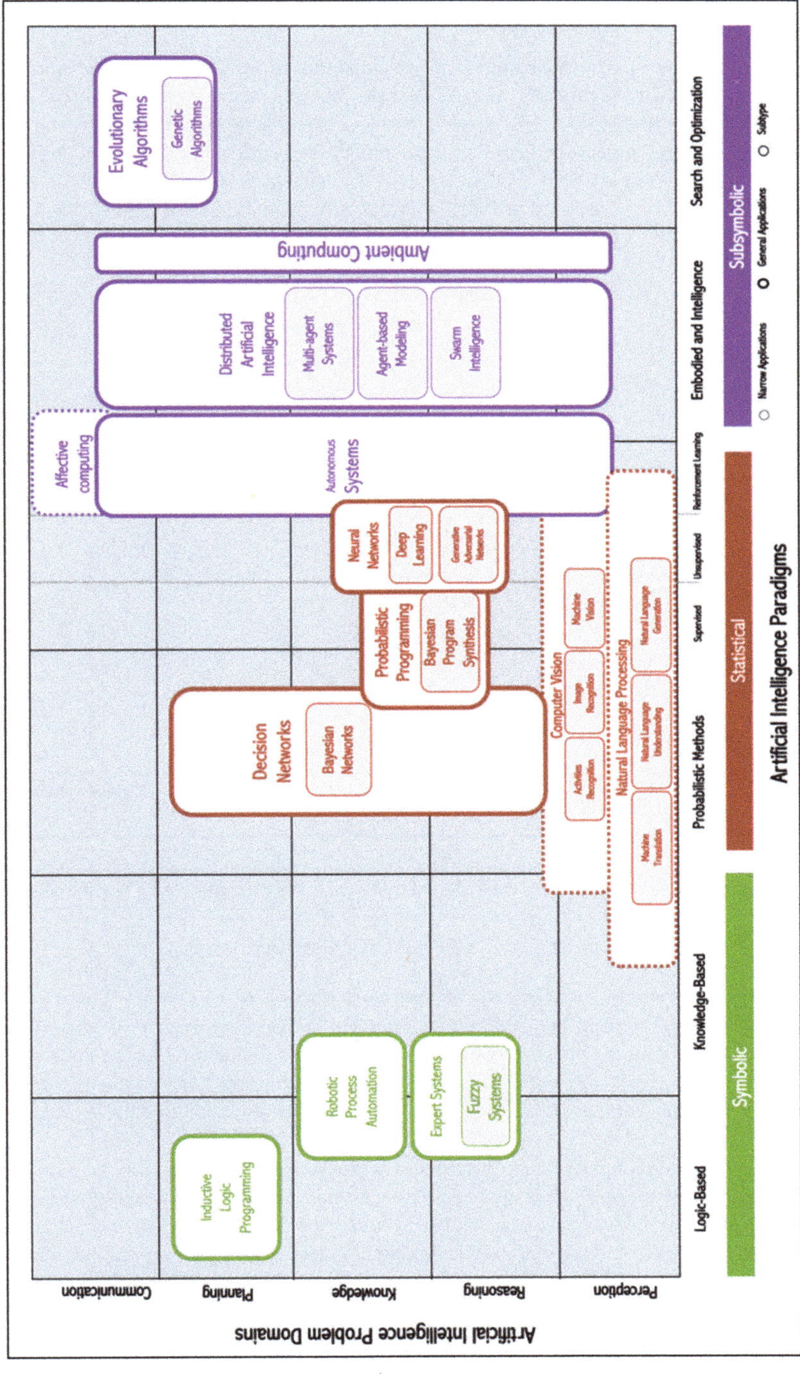

Figure 1. Classification of AI-driven computational techniques, derived from Corea [14].

2. Conceptual and Practical Background

2.1. Has the Artificial Intelligence Era Already Begun?

AI is one of the most disruptive technologies of our time and its capabilities have progressed rapidly [26]. The uptake of AI in organizations is on the rise. For instance, between 2018 and 2019, the number of organizations that deployed AI grew from 4% to 14%, and among the AI applications, conversational AI is at the top of corporate agendas spurred by the worldwide success of Amazon Alexa, Google Assistant, and Apple's Siri [27]. Gartner [28] provides insights into the hype cycle for AI applications, which reflects the growing popularity of machine learning, intelligent applications, and AI-as-a-Service (AIaaS) or AI-Platform-as-a-Service (AI-PaaS) (Figure 2).

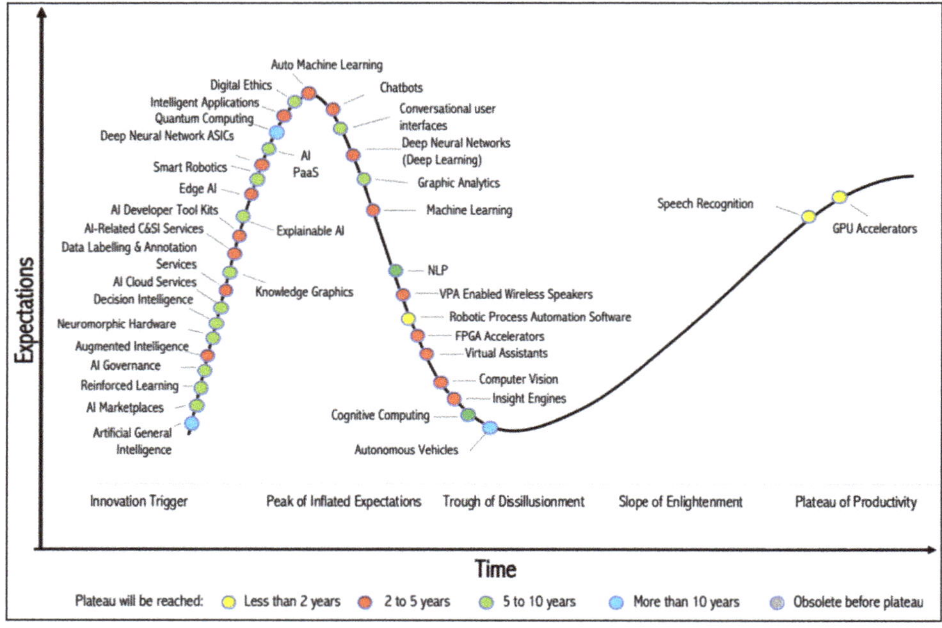

Figure 2. Hype cycle of AI applications, derived from Gartner [28].

In recent years, governments around the world have started to see AI as a nation-defining and global economic competitiveness-increasing capability [29]. In recognition of the increasing importance of AI, as of February 2020, 50 countries have already developed specific national AI strategies—where these countries represent 90% of global gross domestic product (GDP). Figure 3 illustrates the location of these countries; a brief further info on each country's national strategy is provided in Appendix A (Figure A1).

AI-driven computational techniques are diverse and wide-ranging. For example, AI has been in use for quite some time in the tasks that are risky or cause harm to humans. This includes the use of automated robots for bomb detection or combat of unmanned aerial vehicles, and the use of autonomous trucks in the mining industry or mobile reconnaissance units for space exploration [30].

AI-enabled applications include robotic processes [31] for automating public sector tasks, and autonomous delivery bots [32] and chatbots [33] for enhancing business intelligence, stakeholder engagement experience, and customer service quality. Today, AI is rapidly changing the nature of jobs. Many of the services that have been offered by human workers are now being revolutionized by technology. For example, chatbots automate the work of information technology (IT) professionals [34] and human resource (HR) departments [35], so that they can focus on higher value tasks.

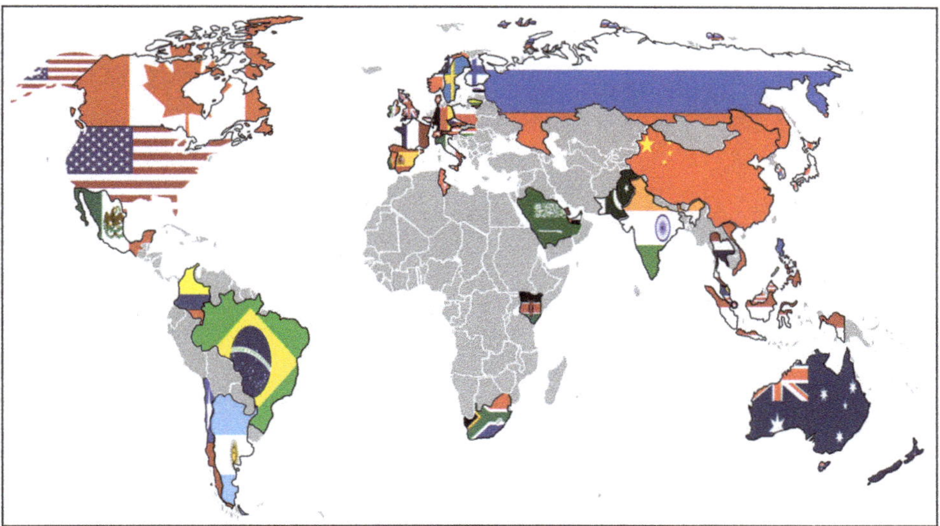

Figure 3. Countries with a national AI strategy, derived from Holon IQ [29].

Autonomous vehicles and driverless shuttle buses are being trialed worldwide. Driverless shuttle bus services are expected to start carrying fare-paying customers in Scotland later in 2020 [36]. Likewise, robot police services are planned to be launched in Houston, Texas to curb petty crime and free up law enforcement resources in 2020 [37].

AI-based systems are providing various solutions. These solutions facilitate the creation of new products and services in many different fields. Particularly, sensor networks are undergoing great expansion and development and the combination of both AI and sensor networks has now become a reality to change our lives and our cities. The integration of these two prominent technologies—including AIoT (AI-of-Things)—also benefits other areas such as Industry 4.0, Internet-of-Things (IoT), demotic systems, and so on [38,39].

AI is being employed to model the spread of COVID-19 to assist decision makers in understanding the future implications of the virus and the measures that should be taken to limit its spread [40]. For instance, in China, AI is being used to minimize the spread of COVID-19 by mobilizing robots that do cleaning and food preparation tasks [41]. Moreover, the European Union [42] launched the EU vs. Virus challenge via a Pan-European hackathon to find ways to tackle COVID-19 via AI and other applications.

AI also has the potential to help in addressing some of the planetary challenges (Table 1). The World Economic Forum [43] underlines the following eight AI applications as "game changers": (a) autonomous and connected electric vehicles; (b) distributed energy grids; (c) smart agriculture and food systems; (d) next-generation weather and climate prediction; (e) smart disaster response; (f) AI-designed intelligent, connected, and livable cities; (g) a transparent digital earth; and (h) reinforcement learning for earth sciences breakthroughs.

Table 1. AI application areas for addressing planetary challenges, derived from World Economic Forum [43].

Planetary Challenges	AI Application Areas
Climate change	Clean power
	Smart transport options
	Sustainable production and consumption
	Sustainable land use
	Smart cities and homes
Healthy oceans	Fishing sustainability
	Preventing pollution
	Protecting habitats
	Protecting species
	Impacts from climate change (including acidification)
Clean air	Filtering and capture
	Monitoring and prevention
	Early warning
	Clean fuels
	Real-time, integrated, adaptive urban management
Biodiversity and conversation	Habitat protection and restoration
	Sustainable trade
	Pollution control
	Invasive species and disease control
	Realizing natural capital
Water security	Water supply
	Catchment control
	Water efficiency
	Adequate sanitation
	Drought planning
Weather and disaster resilience	Prediction and forecasting
	Early warning systems
	Resilient infrastructure
	Financial instruments
	Resilience planning

2.2. How Is Artificial Intelligence Being Utilized in Cities?

In the previous section, we have provided some examples of the use of AI in cities. Here in this section, we share a few more examples to cover some of the other aspects of AI for cities. In particular, the AI solutions implemented in Australia have been taken as an example. Like many other advanced knowledge and innovation economies, AI is a rapidly growing field in Australia. Furthermore, the country has been an early adopter of smart technologies [44], particularly for targeting industrial and urban sustainability outcomes [45–47]. Some of the existing AI applications and experienced challenges in the country are discussed as follows:

- Autonomous vehicles and driverless shuttle buses are being trialed throughout Australia, in all capital cities and some regional centers [48]. Nevertheless, the regulation efforts of autonomous vehicles are yet to follow the autonomous driving trials and developments.
- State of NSW police have been using AI systems to identify drivers illegally using mobile phones [49]. These systems review images, detect offences, and then exclude non-offenders. Nonetheless, images are then authorized following a review by human operators.
- The importance of review of AI outputs by human operators was highlighted by the Australian federal government's incorrect use of AI for automatic detection of Centrelink debt and issuing of infringement notices without human input [50]. The process resulted in some individuals receiving notices incorrectly and placed the onus of proof onto the accused.

- Other issues have resulted from facial recognition software used in surveillance and crime prevention, which may have unfairly discriminated against Aboriginal and Torres Strait Islanders [51].

Despite these issues, development of AI continues in a variety of fields in Australia, and has been investigated for its use in product/goods delivery [52], environmental and transport monitoring [53], disaster prediction [54], healthcare [55], infrastructure [56], data privacy [57], and agriculture [58]. Just to provide some examples, AI's contributions to healthcare practice are listed in Table 2. Additionally, AI applications have been used in big data analytics, such as its use in social media analytics to aid natural disaster management. Figure 4 is an example of the disaster severity map generated for the 2010–2011 Queensland Floods with the help of machine learning technology [59].

Table 2. AI applications and motivation for adoption in healthcare practice, derived from Park [60].

Application	Motivation for Adoption
Robot-assisted surgery	Technological advances in robotic solutions for more types of surgery
Virtual nursing assistants	Increasing pressure caused by medical labor shortage
Administrative workflow	Easier integration with existing technology infrastructure
Fraud detection	Need to address complex service and payment fraud attempts
Dosage error reduction	Prevalence of medical errors, which leads to tangible penalties
Connected machines	Proliferation of connected machines and devices
Clinical trial participation	Client cliff, plethora of data, outcomes-driven approach
Preliminary diagnosis	Interoperability and data architecture to enhance accuracy
Automated image diagnosis	Storage capacity, greater trust in AI technology
Cybersecurity	Increase in breaches, pressure to protect health data

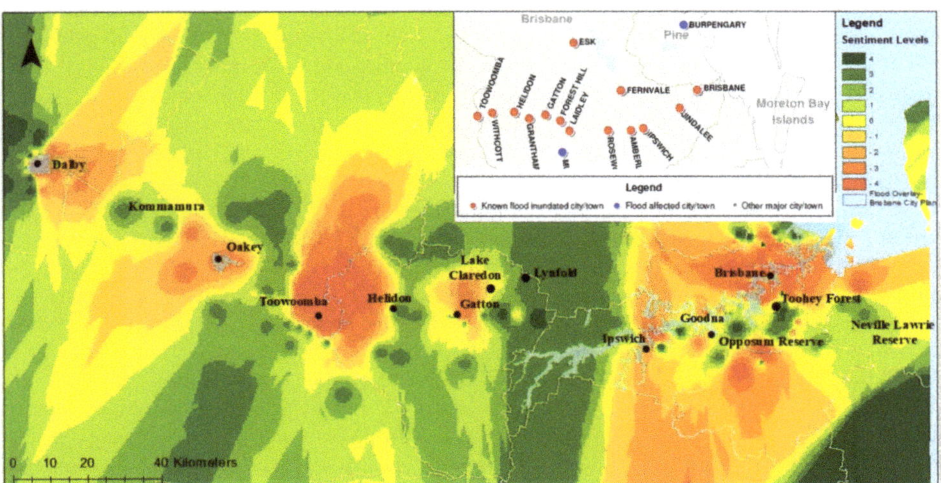

Figure 4. AI and big data analytics in natural disaster management, derived from Kankanamge et al. [59].

In terms of strategizing AI, there have been some promising developments in Australia. The most notable one is the AI roadmap, codeveloped by CSIRO's Data61 and the Australian Government Department of Industry, Innovation and Science. The roadmap identifies strategies to help develop a national AI capability to boost the productivity of Australian industry, create jobs and economic growth, and improve the quality of life for current and future generations. The roadmap emphasizes the need to concentrate on the three key domains: (a) natural resources and the environment; (b) health, aging, and disability; and (c) cities, towns, and infrastructure [61]. Table 3 below elaborates the objectives of these AI domains. Additionally, OECD's [62] AI policy observatory provides a useful repository of AI in Australia.

Table 3. Priority AI specialization domains and their objectives, derived from Data61 [61].

Domain	Objective
Natural resources and the environment	Developing AI solutions for enhanced natural resource management to reduce the costs and improve the productivity of agriculture, mining, fisheries, forestry, and environmental management
Health, aging, and disability	Developing AI solutions for health, aging, and disability support to reduce costs, improve wellbeing, and make quality care accessible for all Australians
Cities, towns, and infrastructure	Developing AI solutions for better towns, cities, and infrastructure to improve the safety, efficiency, cost-effectiveness, and quality of the built environment

3. Discussion

3.1. Can Artificial Intelligence Help Cities Become Smarter?

Cities are complex organisms and their complexity increases exponentially as they continue to grow [63]. With computational abilities vastly superior to humans, when it comes to ingesting large swaths of data, AI systems are among the core elements of most smart city projects [64].

Other smart technologies such as internet-of-things (IoT) [65], autonomous vehicles (AV) [66,67], big data [68], 5G wireless communication [69], robotics [70], blockchain [71], cloud computing [72], 3D printing [73], virtual reality (VR) [74], augmented reality (AR) [75], digital twins [76], and so on are also transforming our cities [77].

For instance, it is increasingly common to combine machine learning with other emerging technologies to generate advanced urban solutions. Examples include: the use of deep learning and high-performance computing (HPC) for traffic predictions using sensor data [78], incident prediction [79], disaster management [80], and rapid transit systems designed to optimize urban mobility systems [81]. Machine learning has also been used with big data technologies and social media for logistics and urban planning [82,83], event detection for urban governance [84], disease detection [85], and identifying the sources of noise pollution at the city scale [86].

Additionally, machine learning has been applied along with distributed computing to improve basic scientific computing operations that are fundamental to urban design modeling methodologies [87]. Moreover, machine learning is paired with IoT for human activity recognition [88], smart farming [89], and developing next-generation distance learning systems [90]. Furthermore, machine learning benefits from data fusion in ubiquitous IoT environments [91], where this creates a potential to significantly enhance AV decision capabilities [92]. Figure 5 lists AI capabilities and their use by domains.

Nevertheless, it is when AI is combined with these technologies that we can really see its big potential to address complex challenges and harness opportunities within our urban environments—given that some ethical issues are adequately addressed. Despite the AI and ethics issue being discussed in academic and government circles, so far only limited guiding principles have been produced and legislated [94]. In that regard, the European Parliament's [95] initiative on guidelines for the European Union (EU) on ethics in AI is a commendable but limited attempt, as ethical rules on AI are so far essentially of a self-regulatory nature, and there is growing demand for more government oversight.

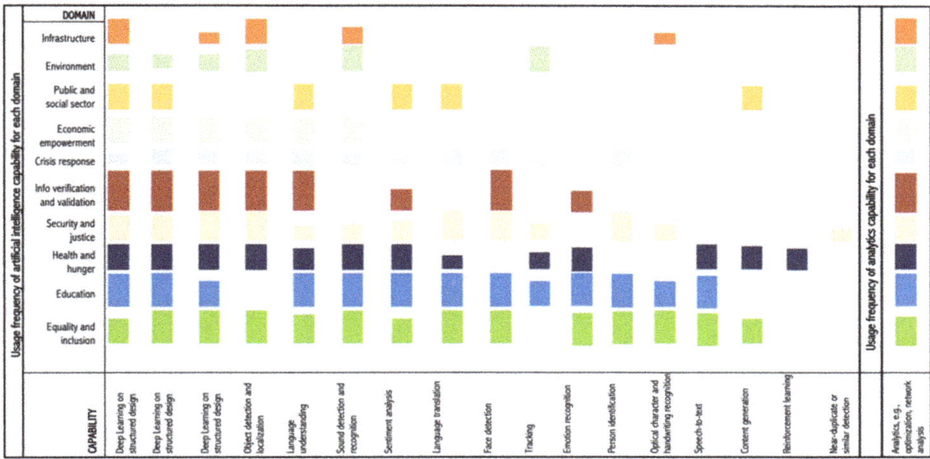

Figure 5. AI capabilities and their use by domains, derived from McKinsey Global Research Institute [93].

3.2. What Are the Promises and Pitfalls of Artificial Intelligence for Cities?

A recent study [96] that evaluated the levels of smartness of Australian local government areas advocated for the importance of integrating urban technologies, including AI, into local service delivery and governance, for instance, the use of AI in tasks that enhance environmental sustainability, such as sorting waste for recycling [97]. Additionally, the practice review conducted by McKinsey Global Research Institute [93] discloses projects from across the globe where AI is utilized for achieving UN's sustainable development goals (SDG) (Figure 6). Nevertheless, before AI is implemented on a wider scale, it is important to understand how this technology can contribute to making our cities (and the planet) smarter. Conversely, understanding the pitfalls of AI will enable us to ensure AI delivers the desired outcomes in urban areas and beyond.

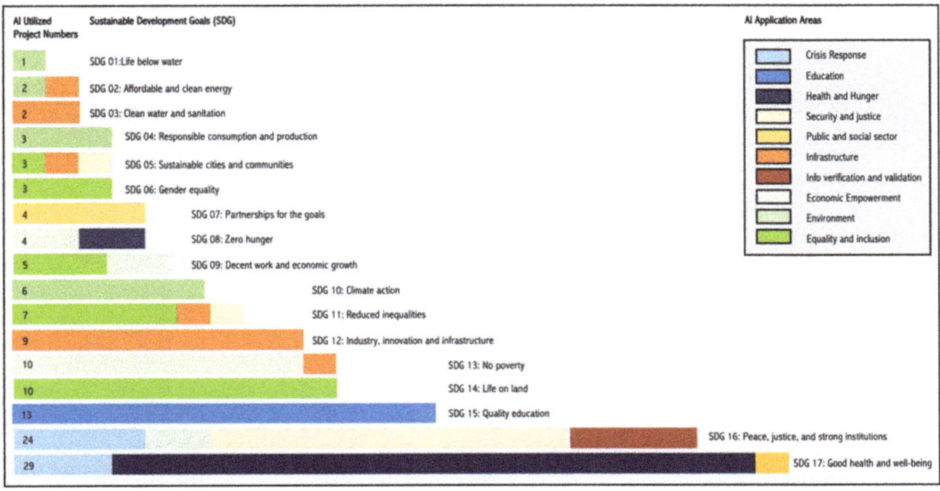

Figure 6. AI utilization for achieving sustainable development goals, derived from McKinsey Global Research Institute [93].

With the above-mentioned issue in mind, our team in another study [22] evaluated the promises and pitfalls of AI for cities according to the main smart city dimensions of economy, society, environment, and governance [98]. Table 4 below summarizes the key findings of the study.

Table 4. Promises and pitfalls of AI for cities, derived from Yigitcanlar et al. [22].

Domains	Promises	Pitfalls
Economy	▪ Enhance productivity and innovation ▪ Reduce costs and increase resources ▪ Support the decision-making process ▪ Automate decision-making	▪ Biased decision-making ▪ Unstable job market ▪ Loss of revenue streams ▪ Loss of employment ▪ Economic inequality
Society	▪ Improve healthcare monitoring ▪ Enhance health diagnosis outcomes ▪ More adaptable education system ▪ More personalized teaching ▪ Task optimization	▪ Biased decision-making ▪ Misdiagnosis ▪ Unstable job market ▪ Loss of employment ▪ Data privacy and security
Environment	▪ Assist environmental monitoring ▪ Optimize energy consumption ▪ Optimize energy production ▪ Optimize transport systems ▪ Assist in developing more environmentally efficient transport systems	▪ Biased decision-making ▪ Increased urban sprawl ▪ More kilometers traveled by motor vehicles ▪ Changed property values ▪ Energy intensive technology ▪ Increased carbon footprint
Governance	▪ Enhance surveillance systems ▪ Improve cyber safety ▪ Aid in disaster management planning and operations ▪ Assist citizens with new technologies	▪ Biased decision-making ▪ Racial bias and discrimination ▪ Suppression of public voice/protest ▪ Violation of civil liberties ▪ Privacy concern ▪ Unethical use of technology ▪ Risk of misinformation ▪ Cybersecurity concern

3.3. What Are the Ways to Maximize Artificial Intelligence Promises and Minimize Pitfalls?

The biggest pitfalls of AI-enabled solutions are that they may aggravate the existing socioeconomic disparity [99] and have privacy [100] (for example, increased government surveillance during COVID-19) and cybersecurity [101] issues. Most of our cities are already fragile and inattention to how local governments maintain social compacts will only increase their fragility [102]. It is imperative that technological progress does not accelerate the widening of existing fractures, or incubate new sources of fractures, in our cities [103].

Given the fast-paced implementation of AI, it is important that we act now and find ways to minimize the pitfalls of AI while maximizing its promises [104]. Some of the useful actions are presented below.

- The first step should be to engage multiple stakeholders [105]. Active collaboration among people from a wide range of industries and backgrounds can help highlight the promises of AI technology, identify pitfalls, and improve trust. This will also contribute to humanizing AI.
- Secondly, paramount to developing trust is demonstrating the ability of AI technology to ensure data security and reduce vulnerabilities [106], including hacking and misinformation.
- Thirdly, AI technology should be agile, so that it can cope with uncertainty [107]. It must also be frugal so it can be implemented in a way that does not lead to wasting public resources on failed attempts and does not become obsolete.

- Additionally, regulation is crucial for controlled implementation [108]—standards and ethical frameworks help ensure AI is deployed responsibly and in keeping with public values.
- Furthermore, more research and development (R&D) is required to ensure the cascading effects of AI, across the various levels of a city (local, neighborhood, city, and the larger regional ecosystem) and society. Deploying AI systems calls for an assessment of their impact on a system of systems [109].
- Next, it is critical to develop AI solutions with a public research consortium to ensure that technology is not solely used as a means of gaining profit.
- Finally, it is also important to consider the intended, as well as the unintended, consequences of AI [110] that will arise not only within one system (e.g., economic) but across the collection of interrelated systems (e.g., interaction between economic, social, and physical infrastructures).

4. Conclusions

4.1. Are Artificially Intelligent Cities on the Horizon?

According to Andrew Ng, cofounder of Google Brain, "AI is the new electricity. Just as 100 years ago electricity transformed industry after industry, AI will now do the same." The impact of AI will go beyond the industry; it is set to change the world [111]. An internationally conducted survey [112] highlighted that "the prospect of an AI future both excites and concerns people around the globe. Nonetheless, fears around the drawbacks of AI are offset by the benefits, and the net result is positive. AI will likely to change society for the better." AI applications have also significant potential to transform our cities. This may lead to the next-generation smart cities [113] being coined as "artificially intelligent cities". Building artificially intelligent cities may save our civilization from the earlier mentioned catastrophes, but it all depends on how we design and use AI, and on who will profit from it [114]. The risk here is for AI to become a vehicle for increasing the wealth of the top 1% of income earners (i.e., top 10 wealthiest people in the world and monopolistic multinational corporations) and the power of biased and unethical politicians [115].

Time will tell if AI systems make our cities "smart enough" to provide better living conditions for all (i.e., people, flora, and fauna coexisting in urban ecosystems). To date, while there have been significant technological advances, these have not been matched with innovations in governance mechanisms. In addition, the policy apparatuses of most local governments need significant modernization to take full advantage of technology affordances in an agile manner [116].

If there is one thing that cities and local governments have learned from the ongoing COVID-19 pandemic, caused by the SARS-CoV-2 virus, is that when they are willing, they can respond in a proactive and agile manner to the changing environmental conditions. We hope that cities will keep on this track after the current crises pass, modernizing their governance mechanisms and policy frameworks to take full advantage of emerging technologies—particularly AI.

The COVID-19 pandemic has also demonstrated that local governments need to seriously consider their digital infrastructure capabilities and capacities. For example, when the Queensland state government (in Australia) decided to make education available online, its infrastructure failed to deliver the public service (i.e., provision of online education) due to significant web traffic. A few schools had backups in place with paper resources but the problem highlighted significant issues with existing networks [117]. Online education has also highlighted social equity issues associated with the digital divide with some lower income students struggling to meet the required technology capabilities [118], and special needs students, including those who speak a language other than English, struggling to receive the required one-on-one assistance [119].

In this paper, we mainly focused on the artificial narrow intelligence level of AI. Nevertheless, if somehow one day we manage to build artificial general intelligence or artificial superintelligence (these two AI levels also correspond to singularity, that is, in simple terms, the intelligence explosion), we need to do all it takes for it to be, as Tegmark [120] calls it, a "Friendly-AI" (a superintelligence whose goals are aligned with ours). Speculation on how to build artificial general intelligence or artificial superintelligence or singularity that would reshape our cities, societies, and civilization is beyond the scope of this viewpoint.

4.2. What Are the Key Lines of Research Concerning Artificial Intelligence and Cities?

There are some important issues, in the context of AI and cities, that prospective research must address in order to provide our cities and societies with the best technological outcomes. We strongly believe that further investigating some of the critical issues in prospective research projects by scholars of this highly interdisciplinary field will shed light on the better conceptualization and practice of AI (artificial narrow intelligence level) in the context of cities and societies. These issues are listed below:

- How can AI systems be developed for cities that are robust, less hackable, and are not used to manipulate and control populations (e.g., voting for a politician/political party)?
- How can we best tackle the AI pitfalls to assure positive outcomes for cities and societies (e.g., security, privacy, regulations, and inequality)?
- How can we avoid heavy reliance on automated decision-making systems, making the society passive or inactive in determining its goals?
- How can AI-induced decisions or solutions in cities be more participatory, democratic, and transparent?
- How can AI be utilized best in cities to achieve desired urban outcomes for all (i.e., human and non-human)?
- How can we determine the best possible scenarios and factors of success and failure in implementing AI in cities?
- How can we determine the best approach to start building artificially intelligent cities (e.g., from scratch, retrofitting, or a combination of both)?
- How can the uniqueness, image, or character of each city and society be maintained given AI is in the play and there might be one best solution?
- How can we design AI systems for cities that preserve, and even promote, societal values and cultural heritage and historic artifacts (e.g., embrace legacy), while simultaneously exploiting emerging technologies and contemporary platforms?
- How can we form the AI commons and ensure that AI can achieve its potential for social good?
- How can local governments meet the need for rich, real-time, location- and context-specific data and preserve privacy and security, while designing AI systems?
- How can the negative environmental externalities of large AI technology and systems be minimized?
- How can the blueprints be developed for the next global transformation of cities to create carbon-free and adaptive futures for humanity?

Considering AI's current ability to ingest big data for exploratory studies and real-time decision-making, it would be worthwhile to address the following research question (in addition to the above list): How can AI be used to find what we may have missed in terms of developing better (e.g., fairer and more productive) social structures, social geography, social good, political structures, economic structures, energy sources, modes of transportation, design of living structures and spaces (i.e., in normal and disaster times, such as those that COVID-19 and similar pandemics could bring on us), and so on?

The concept of AI advising us on human sociology and similar matters may sound very offensive to some, but when properly done, AI is merely a tool that can be used by humans for their advantage (in the sense of artificial narrow intelligence). Humans tend to learn and incrementally apply the acquired knowledge into practice. AI can analyze ideas for us faster and more in depth, and together with other developments in technologies (e.g., AR, high-performance computing, IoT, and big data), it could allow us to study and predict the potential harms and benefits of alternative ideologies, and develop better futures. Moreover, AI, in the context of artificially intelligent cities, can also help us transform our cities into smarter and more prosperous and creative ones [121–123].

Lastly, we conclude the paper by elaborating on the question we raised in the title of this paper—Can building artificially intelligent cities safeguard humanity from natural disasters, pandemics, and other catastrophes? The existing AI literature reviewed in this paper, unfortunately, does not allow us to answer it with a confident "yes". The answer to this question depends on the findings of the studies focusing on the above-listed critical questions. While we continue to have hope that AI technology will help fix or at least ease the problems created by us, perhaps another important issue is whether we will be able to use AI for the common good of all—rather than the so-called 1% [124] that is already in control of the world economy.

On that very point, at his Turing Lecture on deep learning for AI, Yoshua Bengio [125] highlighted the critical importance of using AI for social good and introduced two actionable items: (a) favoring machine learning applications to help the poorest countries fight climate change, improve healthcare and education, and so on; and (b) forming the concept of AI commons and coordinate, prioritize, and channel funding for the use of AI for social good. As stated by Hager et al. [126], "AI can be a major force for social good; but it depends on how we shape this new technology, and the questions we use to inspire young researchers."

Author Contributions: T.Y. designed and supervised the study, and finalized the manuscript. K.C.D., R.M. and J.M.C. contributed to the write-up of the manuscript, and improved the rigor, relevance, and reach of the study. L.B. and E.W. prepared the first draft of the manuscript and assisted in data collection. All authors have read and agreed to the published version of the manuscript.

Funding: This research received no external funding.

Acknowledgments: This research did not receive any specific grant from funding agencies in the public, commercial, or not-for-profit sectors. The authors thank the managing editor and three anonymous referees for their invaluable comments on an earlier version of the manuscript.

Conflicts of Interest: The authors declare no conflict of interest.

Appendix A

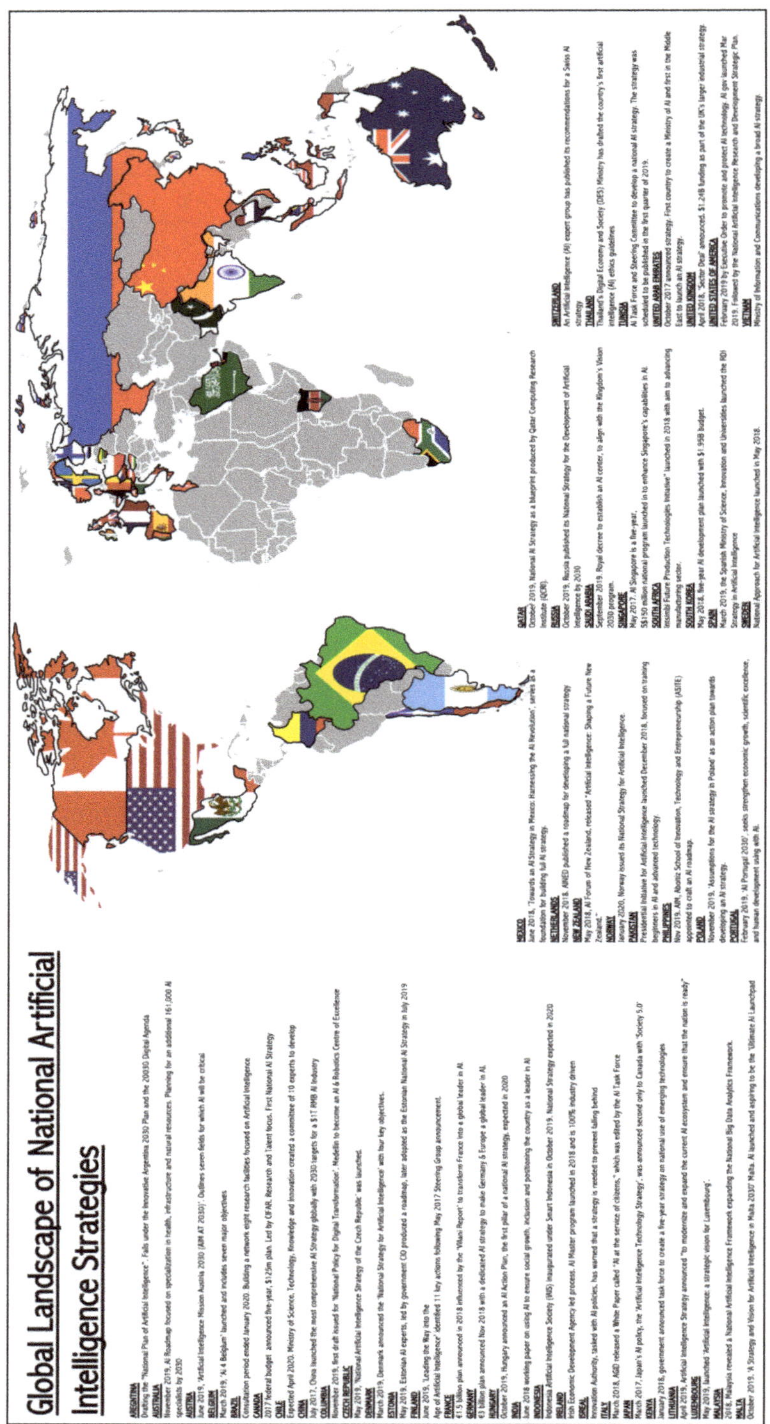

Figure A1. Global landscape of national artificial intelligence strategies, derived from Holon IQ [29].

References

1. Dizdaroglu, D.; Yigitcanlar, T.; Dawes, L. A micro-level indexing model for assessing urban ecosystem sustainability. *Smart. Sustain. Built Environ.* **2012**, *1*, 291–315. [CrossRef]
2. Yigitcanlar, T.; Foth, M.; Kamruzzaman, M. Towards post-anthropocentric cities: Reconceptualizing smart cities to evade urban ecocide. *J. Urban Technol.* **2019**, *26*, 147–152. [CrossRef]
3. Speth, J.G. *The Bridge at the Edge of the World: Capitalism, the Environment, and Crossing from Crisis to Sustainability*; Yale University Press: New Haven, CT, USA, 2009.
4. Metaxiotis, K.; Carrillo, F.; Yigitcanlar, T. *Knowledge-Based Development for Cities and Societies: Integrated Multi-Level Approaches: Integrated Multi-Level Approaches*; IGI Global: Hersey, PA, USA, 2010.
5. Poumadere, M.; Bertoldo, R.; Samadi, J. Public perceptions and governance of controversial technologies to tackle climate change: Nuclear power, carbon capture and storage, wind, and geoengineering. *Wiley Interdiscip. Rev. Clim. Chang.* **2011**, *2*, 712–727. [CrossRef]
6. Adenle, A.A.; Azadi, H.; Arbiol, J. Global assessment of technological innovation for climate change adaptation and mitigation in developing world. *J. Environ. Manag.* **2015**, *161*, 261–275. [CrossRef] [PubMed]
7. Yigitcanlar, T. *Technology and the City: Systems, Applications and Implications*; Routledge: New York, NY, USA, 2016.
8. Huntingford, C.; Jeffers, E.S.; Bonsall, M.B.; Christensen, H.M.; Lees, T.; Yang, H. Machine learning and artificial intelligence to aid climate change research and preparedness. *Environ. Res. Lett.* **2019**, *14*, 124007. [CrossRef]
9. Zilli, D.; Parson, O.; Merrett, G.V.; Rogers, A. A hidden Markov model-based acoustic cicada detector for crowdsourced smartphone biodiversity monitoring. *J. Artif. Intell. Res.* **2014**, *51*, 805–827. [CrossRef]
10. Shah, H.; Ghazali, R.; Hassim, Y.M. Honey bees inspired learning algorithm: Nature intelligence can predict natural disaster. In *Recent Advances on Soft Computing and Data Mining*; Springer: Cham, Switzerland, 2014; pp. 215–225.
11. Vinuesa, R.; Azizpour, H.; Leite, I.; Balaam, M.; Dignum, V.; Domisch, S.; Nerini, F.F. The role of artificial intelligence in achieving the sustainable development goals. *Nat. Commun.* **2020**, *11*, 1–10. [CrossRef]
12. Vaishya, R.; Javaid, M.; Khan, I.H.; Haleem, A. Artificial intelligence (AI) applications for COVID-19 pandemic. *Diabetes Metab. Syndr. Clin. Res. Rev.* **2020**, *14*, 337–339. [CrossRef]
13. Schalkoff, R.J. *Artificial Intelligence: An Engineering Approach*; McGraw-Hill: New York, NY, USA, 1990.
14. Corea, F. AI Knowledge Map: How to Classify AI Technologies. 2018. Available online: https://www.forbes.com/sites/cognitiveworld/2018/08/22/ai-knowledge-map-how-to-classify-ai-technologies/#5e99db627773 (accessed on 11 May 2020).
15. Yun, J.; Lee, D.; Ahn, H.; Park, K.; Lee, S.; Yigitcanlar, T. Not deep learning but autonomous learning of open innovation for sustainable artificial intelligence. *Sustainability* **2016**, *8*, 797. [CrossRef]
16. Allam, Z.; Dhunny, Z.A. On big data, artificial intelligence and smart cities. *Cities* **2019**, *89*, 80–91. [CrossRef]
17. Desouza, K.C. Governing in the Age of the Artificially Intelligent City. 2017. Available online: https://www.governing.com/commentary/col-governing-age-artificially-intelligent-city.html (accessed on 23 April 2020).
18. Batty, M. Artificial intelligence and smart cities. *Environ. Plan. B* **2018**, *45*, 3–6. [CrossRef]
19. Swindell, D.; Desouza, K.C.; Hudgens, R. Dubai Offers Lessons for Using Artificial Intelligence in Local Government. 2018. Available online: https://www.brookings.edu/blog/techtank/2018/09/28/dubai-offers-lessons-for-using-artificial-intelligence-in-local-government (accessed on 23 April 2020).
20. Voda, A.I.; Radu, L.D. Artificial intelligence and the future of smart cities. *Broad Res. Artif. Intell. Neurosci.* **2018**, *9*, 110–127.
21. Nikitas, A.; Michalakopoulou, K.; Njoya, E.T.; Karampatzakis, D. Artificial intelligence, transport and the smart city: Definitions and dimensions of a new mobility era. *Sustainability* **2020**, *12*, 2789. [CrossRef]
22. Yigitcanlar, T.; Desouza, K.C.; Butler, L.; Roozkhosh, F. Contributions and risks of artificial intelligence (AI) in building smarter cities: Insights from a systematic review of the literature. *Energies* **2020**, *13*, 1473. [CrossRef]
23. Boyd, R.; Holton, R.J. Technology, innovation, employment and power: Does robotics and artificial intelligence really mean social transformation? *J. Sociol.* **2018**, *54*, 331–345. [CrossRef]
24. Hintze, A. Understanding the Four Types of AI, from Reactive Robots to Self-Aware Beings. 2016. Available online: https://theconversation.com/understanding-the-four-types-of-ai-from-reactive-robots-to-self-aware-beings-6761 (accessed on 22 April 2020).

25. Joshi, N. 7 Types of Artificial Intelligence. 2019. Available online: https://www.forbes.com/sites/cognitiveworld/2019/06/19/7-types-of-artificial-intelligence/#5ad56f62233e (accessed on 22 April 2020).

26. Echeverría, J.; Tabarés, R. Artificial intelligence, cybercities and technosocieties. *Minds. Mach.* **2017**, *27*, 473–493. [CrossRef]

27. Columbus, R. What's New in Gartner's Hype Cycle for AI. 2019. Available online: https://www.enterpriseirregulars.com/144131/whats-new-in-gartners-hype-cycle-for-ai-2019 (accessed on 12 May 2020).

28. Gartner. Top Trends on the Gartner Hype Cycle for Artificial Intelligence. 2019. Available online: https://www.gartner.com/smarterwithgartner/top-trends-on-the-gartner-hype-cycle-for-artificial-intelligence-2019?utm_campaign=RM_NA_2019_SWG_NL_NL38_IT&utm_medium=email&utm_source=Eloqua&cm_mmc=Eloqua-_-Email-_-LM_RM_NA_2019_SWG_NL_NL38_IT-_-0000 (accessed on 12 May 2020).

29. Holon, I.Q. 50 National Artificial Intelligence Strategies Shaping the Future of Humanity. 2020. Available online: https://www.holoniq.com/notes/50-national-ai-strategies-the-2020-ai-strategy-landscape (accessed on 12 May 2020).

30. Kaplan, J. *Artificial Intelligence: What Everyone Needs to Know*; Oxford University Press: London, UK, 2016.

31. Deloitte. The New Machinery of Government: Robotic Process Automation in the Public Sector. 2017. Available online: https://www2.deloitte.com/content/dam/Deloitte/uk/Documents/Innovation/deloitte-uk-innovation-the-new-machinery-of-govt.pdf (accessed on 22 April 2020).

32. Murrer, S. Starship's Obliging Robots Extend Their Delivery Area to Bring Lunch or Dinner to More People in Milton Keynes. 2020. Available online: https://www.miltonkeynes.co.uk/news/people/starships-obliging-robots-extend-their-delivery-area-bring-lunch-or-dinner-more-people-milton-keynes-2448849 (accessed on 22 April 2020).

33. Polani, D. Emotionless Chatbots are Taking over Customer Service and It is Bad News for Consumers. 2017. Available online: https://theconversation.com/emotionless-chatbots-are-taking-over-customer-service-and-its-bad-news-for-consumers-82962 (accessed on 22 April 2020).

34. Brown, J. Chatbots Debut in North Carolina, Allow IT Personnel to Focus on Strategic Tasks. 2016. Available online: https://www.govtech.com/Chatbots-Debut-in-North-Carolina-Allow-IT-Personnel-to-Focus-on-Strategic-Tasks.html (accessed on 22 April 2020).

35. Sheth, B. Chatbots are the New HR Managers: Want to Use Chatbots to Automate the Majority of HR Services? 2019. Available online: https://www.thebalancecareers.com/use-chatbots-to-automate-hr-many-services-4171964 (accessed on 22 April 2020).

36. Nassir, N. Driverless Buses Can Help End the Suburbs' Public Transport Woes. 2019. Available online: https://theconversation.com/driverless-buses-can-help-end-the-suburbs-public-transport-woes-117258 (accessed on 22 April 2020).

37. Begley, D. Robot Police Coming to Houston Transit Center, Rail Platform, Park and Ride Lot. 2020. Available online: https://www.houstonchronicle.com/news/transportation/article/Robot-police-coming-to-Houston-transit-center-14999004.php (accessed on 22 April 2020).

38. Alrajeh, N.A.; Lloret, J. Intrusion detection systems based on artificial intelligence techniques in wireless sensor networks. *Int. J. Distrib. Sens. Netw.* **2013**, *9*, 351047. [CrossRef]

39. Lu, J.; Feng, L.; Yang, J.; Hassan, M.M.; Alelaiwi, A.; Humar, I. Artificial agent: The fusion of artificial intelligence and a mobile agent for energy-efficient traffic control in wireless sensor networks. *Future Gener. Comput. Syst.* **2019**, *95*, 45–51. [CrossRef]

40. Alsinglawi, B.; Elkhodr, M.; Mubin, O. COVID-19 Death Toll Estimated to Reach 3,900 by Next Friday, According to AI Modelling. 2020. Available online: https://theconversation.com/covid-19-death-toll-estimated-to-reach-3-900-by-next-friday-according-to-ai-modelling-133052 (accessed on 22 April 2020).

41. Meisenzahl, M. These Robots are Fighting the Coronavirus in China by Disinfecting Hospitals, Taking Temperatures, and Preparing Meals. 2020. Available online: https://www.businessinsider.com.au/see-chinese-robots-fighting-the-coronavirus-in-photos-2020-3?r=US&IR=T (accessed on 22 April 2020).

42. European Union. EuvsVirus Challenge: Pan-European Hackathon. 2020. Available online: https://euvsvirus.orghtml (accessed on 11 May 2020).

43. World Economic Forum. 8 Ways AI Can Help Save the Planet. 2018. Available online: http://www3.weforum.org/docs/Harnessing_Artificial_Intelligence_for_the_Earth_report_2018.pdf (accessed on 11 May 2020).

44. Yigitcanlar, T. Australian Local Governments' Practice and Prospects with Online Planning. *URISA J.* **2006**, *18*, 7–17.
45. Arbolino, R.; De Simone, L.; Carlucci, F.; Yigitcanlar, T.; Ioppolo, G. Towards a sustainable industrial ecology: Implementation of a novel approach in the performance evaluation of Italian regions. *J. Clean. Prod.* **2018**, *178*, 220–236. [CrossRef]
46. Ingrao, C.; Messineo, A.; Beltramo, R.; Yigitcanlar, T.; Ioppolo, G. How can life cycle thinking support sustainability of buildings? Investigating life cycle assessment applications for energy efficiency and environmental performance. *J. Clean. Prod.* **2018**, *201*, 556–569. [CrossRef]
47. Yigitcanlar, T.; Sabatini-Marques, J.; da-Costa, E.M.; Kamruzzaman, M.; Ioppolo, G. Stimulating technological innovation through incentives: Perceptions of Australian and Brazilian firms. *Technol. Forecast. Soc. Chang.* **2019**, *146*, 403–412. [CrossRef]
48. Austroads. Autonomous Vehicle Trials in Australia. 2020. Available online: https://austroads.com.au/drivers-and-vehicles/future-vehicles-and-technology/trials (accessed on 22 April 2020).
49. NSW Government. Mobile Phone Detection Cameras: Cameras Targeting Illegal Phone Use across NSW. 2020. Available online: https://roadsafety.transport.nsw.gov.au/stayingsafe/mobilephones/technology.html (accessed on 22 April 2020).
50. Carney, T. Robodebt Failed Its Day in Court, What Now? 2019. Available online: https://theconversation.com/robodebt-failed-its-day-in-court-what-now-127984 (accessed on 22 April 2020).
51. Bogle, A. Technology's Potential to Help or Harm 'Almost Limitless', Human Rights Commission Warns. 2018. Available online: https://www.abc.net.au/news/science/2018-07-24/technology-surveillance-potential-harm-human-rights-commission/10024808 (accessed on 22 April 2020).
52. Letheren, K. Experiments in Robotics Could Help Amazon Beat Australia's Slow Delivery Problem. 2017. Available online: https://theconversation.com/experiments-in-robotics-could-help-amazon-beat-australias-slow-delivery-problem-87598 (accessed on 22 April 2020).
53. Ossola, A.; Staas, L.; Leishman, M. A Solution to Cut Extreme Heat by up to 6 Degrees is in Our Own Backyards. 2020. Available online: https://theconversation.com/a-solution-to-cut-extreme-heat-by-up-to-6-degrees-is-in-our-own-backyards-133082 (accessed on 22 April 2020).
54. Data61. Spark: Predicting Bushfire Spread. 2020. Available online: https://data61.csiro.au/en/Our-Research/Our-Work/Safety-and-Security/Disaster-Management/Spark (accessed on 22 April 2020).
55. Data61. Assessing Breast Density Automatically. 2020. Available online: https://data61.csiro.au/en/Our-Research/Our-Work/Health-and-Communities/Precision-health/AutoDensity (accessed on 22 April 2020).
56. Data61. Water Pipe Failure Prediction. 2020. Available online: https://data61.csiro.au/en/Our-Research/Our-Work/Future-Cities/Planning-sustainable-infrastructure/Water-pipes (accessed on 22 April 2020).
57. Data61. Artificial Intelligence and Machine Learning. 2020. Available online: https://data61.csiro.au/en/Our-Research/Focus-Areas/AI-and-Machine-Learning (accessed on 22 April 2020).
58. CSIRO. Machine-Learning for Crop Breeding. 2020. Available online: https://www.csiro.au/en/Research/AF/Areas/Crops/Grains/Machine-learning (accessed on 22 April 2020).
59. Kankanamge, N.; Yigitcanlar, T.; Goonetilleke, A. How engaging are disaster management related social media channels? The case of Australian state emergency organisations. *Int. J. Disaster Risk Reduct.* **2020**, *48*, 101571. [CrossRef]
60. Park, A. Top 10 AI Applications for Healthcare in 2020: Accenture Report. 2020. Available online: https://www.beckershospitalreview.com/artificial-intelligence/top-10-ai-applications-for-healthcare-in-2020-accenture-report.html (accessed on 12 May 2020).
61. Data61. Artificial Intelligence Roadmap. 2020. Available online: https://data61.csiro.au/en/Our-Research/Our-Work/AI-Roadmap (accessed on 15 May 2020).
62. OECD. AI in Australia. 2020. Available online: https://oecd.ai/dashboards/countries/Australia (accessed on 15 May 2020).
63. Li, X.; Yeh, A.G. Calibration of cellular automata by using neural networks for the simulation of complex urban systems. *Environ. Plan.* **2001**, *33*, 1445–1462. [CrossRef]
64. Yigitcanlar, T.; Kamruzzaman, M.; Foth, M.; Sabatini-Marques, J.; da Costa, E.; Loppolo, G. Can cities become smart without being sustainable? A systematic review of the literature. *Sustain. Cities Soc.* **2019**, *45*, 348–365. [CrossRef]

65. Jin, J.; Gubbi, J.; Marusic, S.; Palaniswami, M. An information framework for creating a smart city through internet of things. *IEEE IoT J.* **2014**, *1*, 112–121. [CrossRef]

66. Faisal, A.; Yigitcanlar, T.; Kamruzzaman, M.; Currie, G. Understanding autonomous vehicles: A systematic literature review on capability, impact, planning and policy. *J. Transp. Land Use* **2019**, *12*, 45–72. [CrossRef]

67. Yigitcanlar, T.; Wilson, M.; Kamruzzaman, M. Disruptive impacts of automated driving systems on the built environment and land use: An urban planner's perspective. *J. Open Innov. Technol. Mark. Complex.* **2019**, *5*, 24. [CrossRef]

68. Al Nuaimi, E.; Al Neyadi, H.; Mohamed, N.; Al-Jaroodi, J. Applications of big data to smart cities. *J. Internet Serv. Appl.* **2015**, *6*, 25. [CrossRef]

69. Gozalvez, J. 5G worldwide developments [mobile radio]. *IEEE Veh. Technol Mag.* **2017**, *12*, 4–11. [CrossRef]

70. Grieco, L.A.; Rizzo, A.; Colucci, S.; Sicari, S.; Piro, G.; Di Paola, D.; Boggia, G. IoT-aided robotics applications: Technological implications, target domains and open issues. *Comput. Commun.* **2014**, *54*, 32–47. [CrossRef]

71. Marsal-Llacuna, M.L. Future living framework: Is blockchain the next enabling network? *Technol. Forecast. Soc. Chang.* **2018**, *128*, 226–234. [CrossRef]

72. Mazza, D.; Tarchi, D.; Corazza, G.E. A unified urban mobile cloud computing offloading mechanism for smart cities. *IEEE Commun. Mag.* **2017**, *55*, 30–37. [CrossRef]

73. Van Wijk, A.J.; Van Wijk, I. *3D Printing with Biomaterials: Towards a Sustainable and Circular Economy*; IOS Press: Amsterdam, The Netherlands, 2015.

74. Portman, M.E.; Natapov, A.; Fisher-Gewirtzman, D. To go where no man has gone before: Virtual reality in architecture, landscape architecture and environmental planning. *Comput. Environ. Urban Syst.* **2015**, *54*, 376–384. [CrossRef]

75. Carozza, L.; Tingdahl, D.; Bosché, F.; Van Gool, L. Markerless vision-based augmented reality for urban planning. *Comput. Aided Civ. Infrastruct. Eng.* **2014**, *29*, 2–17. [CrossRef]

76. Dembski, F.; Wössner, U.; Letzgus, M.; Ruddat, M.; Yamu, C. Urban digital twins for smart cities and citizens: The case study of Herrenberg, Germany. *Sustainability* **2020**, *12*, 2307. [CrossRef]

77. Yigitcanlar, T.; Kankanamge, N.; Vella, K. How are the smart city concepts and technologies perceived and utilized? A systematic geo-twitter analysis of smart cities in Australia. *J. Urban Technol.* **2020**. [CrossRef]

78. Aqib, M.; Mehmood, R.; Alzahrani, A.; Katib, I.; Albeshri, A.; Altowaijri, S.M. Smarter traffic prediction using big data, in-memory computing, deep learning and GPUs. *Sensors* **2019**, *19*, 2206. [CrossRef]

79. Aqib, M.; Mehmood, R.; Alzahrani, A.; Katib, I. In-memory deep learning computations on GPUs for prediction of road traffic incidents using big data fusion. In *Smart Infrastructure and Applications*; Springer: Cham, Switzerland, 2020; pp. 79–114.

80. Aqib, M.; Mehmood, R.; Alzahrani, A.; Katib, I. A smart disaster management system for future cities using deep learning, GPUs, and in-memory computing. In *Smart Infrastructure and Applications*; Springer: Cham, Switzerland, 2020; pp. 159–184.

81. Aqib, M.; Mehmood, R.; Alzahrani, A.; Katib, I.; Albeshri, A.; Altowaijri, S.M. Rapid transit systems: Smarter urban planning using big data, in-memory computing, deep learning, and GPUs. *Sustainability* **2019**, *11*, 2736. [CrossRef]

82. Suma, S.; Mehmood, R.; Albugami, N.; Katib, I.; Albeshri, A. Enabling next generation logistics and planning for smarter societies. *Procedia Comput. Sci.* **2017**, *109*, 1122–1127. [CrossRef]

83. Amaxilatis, D.; Mylonas, G.; Theodoridis, E.; Diez, L.; Deligiannidou, K. LearningCity: Knowledge Generation for Smart Cities. In *Smart Cities Performability, Cognition, & Security*; Springer: Cham, Switzerland, 2020; pp. 17–41.

84. Alomari, E.; Mehmood, R.; Katib, I. Road traffic event detection using twitter data, machine learning, and apache spark. In Proceedings of the 2019 IEEE SmartWorld, Ubiquitous Intelligence & Computing, Advanced & Trusted Computing, Scalable Computing & Communications, Cloud & Big Data Computing, Internet of People and Smart City Innovation, Leicester, UK, 19–23 August 2019; pp. 1888–1895.

85. Alotaibi, S.; Mehmood, R.; Katib, I.; Rana, O.; Albeshri, A. Sehaa: A big data analytics tool for healthcare symptoms and diseases detection using Twitter, Apache Spark, and Machine Learning. *Appl. Sci.* **2020**, *10*, 1398. [CrossRef]

86. Bello, J.P.; Silva, C.; Nov, O.; Dubois, R.L.; Arora, A.; Salamon, J.; Mydlarz, C.; Doraiswamy, H. Sonyc: A system for monitoring, analyzing, and mitigating urban noise pollution. *Commun. ACM* **2019**, *62*, 68–77. [CrossRef]

87. Usman, S.; Mehmood, R.; Katib, I.; Albeshri, A. ZAKI+: A machine learning based process mapping tool for SpMV computations on distributed memory architectures. *IEEE Access* **2019**, *7*, 81279–81296. [CrossRef]

88. Alam, F.; Mehmood, R.; Katib, I.; Albeshri, A. Analysis of eight data mining algorithms for smarter Internet of Things (IoT). *Procedia Comput. Sci.* **2016**, *98*, 437–442. [CrossRef]

89. Khanum, A.; Alvi, A.; Mehmood, R. Towards a semantically enriched computational intelligence (SECI) framework for smart farming. In Proceedings of the International Conference on Smart Cities, Infrastructure, Technologies and Applications, Jeddah, Saudi Arabia, 27–29 November 2017; Springer: Cham, Switzerland, 2017; pp. 247–257.

90. Mehmood, R.; Alam, F.; Albogami, N.N.; Katib, I.; Albeshri, A.; Altowaijri, S.M. UtiLearn: A sersonalized ubiquitous teaching and learning system for smart societies. *IEEE Access* **2017**, *5*, 2615–2635. [CrossRef]

91. Alam, F.; Mehmood, R.; Katib, I.; Albogami, N.N.; Albeshri, A. Data fusion and IoT for smart ubiquitous environments: A survey. *IEEE Access* **2017**, *5*, 9533–9554. [CrossRef]

92. Alam, F.; Mehmood, R.; Katib, I.; Altowaijri, S.M.; Albeshri, A. TAAWUN: A decision fusion and feature specific road detection approach for connected autonomous vehicles. *Mob. Netw. Appl.* **2019**. [CrossRef]

93. McKinsey Global Research Institute. Applying Artificial Intelligence for Social Good. 2018. Available online: https://www.mckinsey.com/featured-insights/artificial-intelligence/applying-artificial-intelligence-for-social-good (accessed on 12 May 2020).

94. Data61. Artificial Intelligence: Australia's Ethics Framework—A Discussion Paper. 2020. Available online: https://consult.industry.gov.au/strategic-policy/artificial-intelligence-ethics-framework/supporting_documents/ArtificialIntelligenceethicsframeworkdiscussionpaper.pdf (accessed on 15 May 2020).

95. European Parliament. EU guidelines on Ethics in Artificial Intelligence: Context and Implementation. 2019. Available online: https://www.europarl.europa.eu/RegData/etudes/BRIE/2019/640163/EPRS_BRI(2019)640163_EN.pdf (accessed on 11 May 2020).

96. Yigitcanlar, T.; Kankanamge, N.; Butler, L.; Vella, K.; Desouza, K. Smart Cities down under: Performance of the Australian Local Government Areas. 2020. Available online: https://eprints.qut.edu.au/136873 (accessed on 23 April 2020).

97. Harwood, S. 'Oscar' the Garbage bot Sorts Waste at YVR Using Artificial Intelligence. 2019. Available online: https://bc.ctvnews.ca/oscar-the-garbage-bot-sorts-waste-at-yvr-using-artificial-intelligence-1.4400398 (accessed on 22 April 2020).

98. Yigitcanlar, T.; Kamruzzaman, M.; Buys, L.; Ioppolo, G.; Sabatini-Marques, J.; da Costa, E.M.; Yun, J.J. Understanding 'smart cities': Intertwining development drivers with desired outcomes in a multidimensional framework. *Cities* **2018**, *81*, 145–160. [CrossRef]

99. Makridakis, S. The forthcoming Artificial Intelligence (AI) revolution: Its impact on society and firms. *Futures* **2017**, *90*, 46–60. [CrossRef]

100. Agusti, S.; Antoni, M.B. *Advances in Artificial Intelligence for Privacy Protection and Security*; World Scientific: Singapore, 2009.

101. Wilner, A.S. Cybersecurity and its discontents: Artificial intelligence, the internet of things, and digital misinformation. *Int. J.* **2018**, *73*, 308–316. [CrossRef]

102. Selby, J.D.; Desouza, K.C. Fragile cities in the developed world: A conceptual framework. *Cities* **2019**, *91*, 180–192. [CrossRef]

103. Desouza, K.C.; Selby, J.D. How Technological Progress Can Cause Urban Fragility. 2019. Available online: https://www.brookings.edu/blog/techtank/2019/02/07/how-technological-progress-can-cause-urban-fragility (accessed on 22 April 2020).

104. Salmon, P.; Hancock, P.; Carden, T. To Protect Us from the Risks of Advanced Artificial Intelligence, We Need to Act Now. 2019. Available online: https://theconversation.com/to-protect-us-from-the-risks-of-advanced-artificial-intelligence-we-need-to-act-now-107615 (accessed on 22 April 2020).

105. David, N.; Subic, A. Hope and Fear Surround Emerging Technologies, But All of Us Must Contribute to Stronger Governance. 2018. Available online: https://theconversation.com/hope-and-fear-surround-emerging-technologies-but-all-of-us-must-contribute-to-stronger-governance-96122 (accessed on 22 April 2020).

106. Bao, H.; He, H.; Liu, Z.; Liu, Z. Research on information security situation awareness system based on big data and artificial intelligence technology. In Proceedings of the 2019 International Conference on Robots & Intelligent System, Macau, China, 3–8 November 2019; pp. 318–322.

107. Yager, R.R. Using fuzzy measures for modeling human perception of uncertainty in artificial intelligence. *Eng. Appl. Artif. Intell.* **2020**, *87*, 103228. [CrossRef]

108. Hoffmann-Riem, W. Artificial intelligence as a challenge for law and regulation. In *Regulating Artificial Intelligence*; Springer: Cham, Switzerland, 2020; pp. 1–29.

109. Berry, B.J. Cities as systems within systems of cities. *Pap. Reg. Sci.* **1964**, *13*, 147–163. [CrossRef]

110. Cheatham, B.; Javanmardian, K.; Samandari, H. Confronting the Risks of Artificial Intelligence. 2019. Available online: https://www.healthindustryhub.com.au/wp-content/uploads/2019/05/Confronting-the-risks-of-AI-2019.pdf (accessed on 22 April 2020).

111. ACS (Australian Computer Society). Artificial Intelligence: A Starter Guide to the Future of Business. 2018. Available online: https://www.acs.org.au/insightsandpublications/reports-publications/artificial-intelligence.html (accessed on 15 May 2020).

112. ARM. AI Today, AI Tomorrow: The ARM 2020 Global AI Survey. 2020. Available online: https://pages.arm.com/artificial-intelligence-survey.html (accessed on 21 May 2020).

113. Yigitcanlar, T.; Han, H.; Kamruzzaman, M.; Ioppolo, G.; Sabatini-Marques, J. The making of smart cities: Are Songdo, Masdar, Amsterdam, San Francisco and Brisbane the best we could build? *Land Use Policy* **2019**, *88*, 104187. [CrossRef]

114. King, B.A.; Hammond, T.; Harrington, J. Disruptive technology: Economic consequences of artificial intelligence and the robotics revolution. *J. Strateg. Innov. Sustain.* **2017**, *12*, 53–67.

115. Klien, E. Who are the 99 Percent? 2011. Available online: https://www.washingtonpost.com/blogs/ezra-klein/post/who-are-the-99-percent/2011/08/25/gIQAt87jKL_blog.html (accessed on 22 April 2020).

116. Desouza, K.C.; Hunter, M.; Yigitcanlar, T. Under the hood: A look at techno-centric smart city development. *Pub Manag.* **2019**, *101*, 30–35.

117. ABC. Queensland Home Schooling Website Glitch Blocks Students on First Day of Online Learning. 2020. Available online: https://www.abc.net.au/news/2020-04-20/queensland-home-schooling-technical-issues/12163934 (accessed on 23 April 2020).

118. SBS. Three Kids and no Computer: The Families Hit Hardest by Australia's School Closures. 2020. Available online: https://www.sbs.com.au/news/three-kids-and-no-computer-the-families-hit-hardest-by-australia-s-school-closures (accessed on 23 April 2020).

119. ABC. Migrant Parents in Australia Face Challenges Posed by Home Learning Model Amid Coronavirus Pandemic. 2020. Available online: https://www.abc.net.au/news/2020-04-17/migrant-parents-face-challenges-during-coronavirus-home-learning/12154036 (accessed on 23 April 2020).

120. Tegmark, M. *Life 3.0: Being Human in the Age of Artificial Intelligence*; Knopf: New York, NY, USA, 2017.

121. Chang, D.L.; Sabatini-Marques, J.; Da Costa, E.M.; Selig, P.M.; Yigitcanlar, T. Knowledge-based, smart and sustainable cities: A provocation for a conceptual framework. *J. Open Innov. Technol. Mark. Complex.* **2018**, *4*, 5. [CrossRef]

122. Yigitcanlar, T.; Metaxiotis, K.; Carrillo, F.J. *Building Prosperous Knowledge Cities: Policies, Plans and Metrics*; Edward Elgar Publishing: Cheltenham, UK, 2012.

123. Yigitcanlar, T.; Velibeyoglu, K.; Baum, S. *Creative Urban Regions: Harnessing Urban Technologies to Support Knowledge City Initiatives*; IGI Global: Hersey, PA, USA, 2008.

124. Matthews, D. Are 26 Billionaires Worth More Than Half the Planet? The Debate, Explained. 2019. Available online: https://www.vox.com/future-perfect/2019/1/22/18192774/oxfam-inequality-report-2019-davos-wealth (accessed on 22 April 2020).

125. Bengio, Y. Turing Lecture: Deep Learning for AI. 2019. Available online: https://www.heidelberg-laureate-forum.org/video/turing-lecture.html (accessed on 11 May 2020).

126. Hager, G.D.; Drobnis, A.; Fang, F.; Ghani, R.; Greenwald, A.; Lyons, T.; Parkes, D.C.; Schultz, J.; Saria, S.; Smith, S.F.; et al. Artificial Intelligence for Social Good. 2017. Available online: https://arxiv.org/pdf/1901.05406.pdf (accessed on 12 May 2020).

 *sensors*

Article

Understanding Sensor Cities: Insights from Technology Giant Company Driven Smart Urbanism Practices

Gaspare D'Amico [1,*], Pasqua L'Abbate [2], Wenjie Liao [3], Tan Yigitcanlar [4] and Giuseppe Ioppolo [1,*]

[1] Department of Economics, University of Messina, Via dei Verdi, 75, 98122 Messina, Italy
[2] Department of Civil, Environmental, Building Engineering and Chemistry, Polytechnic of Bari, 70125 Bari, BA, Italy; p.labbate@tiscali.it
[3] Institute of New Energy and Low-Carbon Technology, Sichuan University, Chengdu 610065, China; wenjieliao@outlook.com
[4] School of Built Environment, Queensland University of Technology, 2 George Street, Brisbane, QLD 4000, Australia; tan.yigitcanlar@qut.edu.au
* Correspondence: gasdamico@unime.it (G.D.); ioppolog@unime.it (G.I.)

Received: 3 July 2020; Accepted: 3 August 2020; Published: 6 August 2020

Abstract: The data-driven approach to sustainable urban development is becoming increasingly popular among the cities across the world. This is due to cities' attention in supporting smart and sustainable urbanism practices. In an era of digitalization of urban services and processes, which is upon us, platform urbanism is becoming a fundamental tool to support smart urban governance, and helping in the formation of a new version of cities—i.e., City 4.0. This new version utilizes urban dashboards and platforms in its operations and management tasks of its complex urban metabolism. These intelligent systems help in maintaining the robustness of our cities, integrating various sensors (e.g., internet-of-things) and big data analysis technologies (e.g., artificial intelligence) with the aim of optimizing urban infrastructures and services (e.g., water, waste, energy), and turning the urban system into a smart one. The study generates insights from the sensor city best practices by placing some of renowned projects, implemented by Huawei, Cisco, Google, Ericsson, Microsoft, and Alibaba, under the microscope. The investigation findings reveal that the sensor city approach: (a) Has the potential to increase the smartness and sustainability level of cities; (b) Manages to engage citizens and companies in the process of planning, monitoring and analyzing urban processes; (c) Raises awareness on the local environmental, social and economic issues, and; (d) Provides a novel city blueprint for urban administrators, managers and planners. Nonetheless, the use of advanced technologies—e.g., real-time monitoring stations, cloud computing, surveillance cameras—poses a multitude of challenges related to: (a) Quality of the data used; (b) Level of protection of traditional and cybernetic urban security; (c) Necessary integration between the various urban infrastructure, and; (d) Ability to transform feedback from stakeholders into innovative urban policies.

Keywords: sensor city; City 4.0; sustainable urban development; smart city; smart urbanism; smart governance; disruptive urban transition; Internet-of-Things (IoT); technology giants; sensors

---

1. Introduction

Today, cities are at the forefront of increasing urbanization and digitalization pressures [1–3], where they play a crucial role in supporting the transition towards a sustainable and smart urbanism practice [4–6]. The current urban context is associated with numerous economic, social and environmental issues, such as waste management [7–10], energy efficiency [11–13], renewable energy sources [14,15], water management [16], social, cultural, and health aspects [17–22], material flows [23], biodiversity [24], transport [25,26], land use optimization [27], air and noise pollution prevention [28–30], infrastructure mishaps [31], economic growth [32]. Policymakers, thus, need a paradigm shift, developing innovative, sustainable, and intelligent solutions to optimize urban processes and improve citizens' quality of life and sustainability of the city [33–36].

In recent years, the notion of 'sensor city' has emerged as a response to the future challenges of growing urbanization and datafication [37,38]. This new version of the city—i.e., City 4.0,—(thanks to the urban dashboards and platforms integrates Internet-of-Things (IoT) infrastructure [39], sensors [40], real-time monitoring stations [41,42], digital cameras [43], actuators [44], real-time tracking systems [45,46], big data analytical techniques [47–49], information and communication technologies (ICTs) [50,51], cloud computing [52], smart grid [53,54], artificial intelligence (AI) [55–57], autonomous shuttles [58], and other digital appliances with physical objects that characterize urban context) improves the efficiency of resources usage.

Specifically, city dashboards accommodate visual/graphical and dynamic analysis suite capable of holistically combining urban infrastructures to view, integrate and communicate real-time information on performance, trends, and future urban scenarios [59,60]. These dashboards are characterized by a high degree of interactivity with users, capable of combining, filtering, querying and overlapping large amounts of urban data [61]. Indeed, urban dashboards are implemented to facilitate an understanding on major urban issues and provide stakeholders with a sense of accountability and engagement on smart urban governance activities [62–64].

According to the forecasts of the "World Smart Cities Spending Guide" provided by the International Data Corporation [65], the total expenditure for smart urban solutions this year alone amounted to almost USD 124 billion. This is an increase of 18.9% compared to 2019 [65]. Specifically, global cities such as Singapore, Tokyo, New York City and London occupy the top of the ranks in terms of investments in smart urban initiatives [65]. Furthermore, cities such as Toronto, Adelaide, Hamburg, Kansas City, Dallas and Stockholm (see Table A1 in Appendix A) have implemented participatory and intelligent platforms that use ICTs to connect companies, local authorities, universities, start-ups, citizens, associations, and so on in order to support the decision-making process, allowing the collection, processing, monitoring and analysis of large amounts of urban data [66–68]. In this sense, Bibri [34] described cities as complex networks of holistic relationships that integrate smart and sustainable solutions in order to provide a suitable context for long-term urban strategic development.

Indeed, the rapid and pervasive development of ICTs taking place all over the world is transforming cities into centers of economic, social, environment and technological development with the aim of providing increasingly efficient, sustainable and smart urban services [37,50,69,70]. In this regard, the United Nations (UN) Agenda 2030 defines ICTs as necessary tools to facilitate the transition towards sustainable development [71]. Hence, policymakers use data and information sharing systems through IoT technologies for planning, monitoring, and evaluating the performance of urban policies, and for improving transparency, active participation of citizens and awareness of urban issues [72–74]. For example, cities such as Singapore, Zurich, Oslo, Geneva, Copenhagen, Auckland, Melbourne, Taipei, Helsinki, Bilbao and Düsseldorf represent forward-looking cases regarding the use of ICTs in the urban area, occupying the first places in the ranking developed by the IMD World Competitiveness Center Smart City Observatory [75].

The current urban theoretical and managerial debates increasingly focus on the role of ICTs and their integration with various aspects related to sustainable urban development [76,77]. Indeed, the literature provides several synonymous for sensor city such as digital city [78–80], smart city [81–83], ubiquitous city [84–86], knowledge city [87–89], intelligent city [90,91], techno-centric city [92], creative city [93,94], sustainable city [95,96], informational city [97,98], smart sustainable city [99–102], and artificially intelligent city [57], which express the importance of ICTs in the management of the cities of future.

Nevertheless, technology alone cannot be a panacea for all urban issues related to growing urbanization [103–105]. In particular, cities excessively connected to IoT and big data analytical solutions have often been criticized for being too techno-centric and for underestimating social and environmental aspects [106–108]. At the same time, sustainable cities struggle to integrate the technological approach with the social, environmental and economic dimensions of sustainable development [102,109]. Specifically, the use of advanced techniques (e.g., real-time monitoring stations for energy consumption, location systems to guide urban traffic, cloud computing systems for sharing sensitive data between government departments, urban infrastructures such as smart bins, smart street lamps, and surveillance cameras) highlights a multitude of challenges related to the quality of the data used, level of protection of traditional and cybernetic urban security, necessary data integration between the various urban infrastructures, and ability to transform feedback from citizens and other stakeholders into innovative urban policies. Consequently, sensors and related ICT infrastructures are rapidly gaining strategic importance for sustainable and disruptive urban development [110]. They are not only enhancing in terms of technological aspects, but also social, environmental, and economic ones.

This paper aims to explore the main challenges related to sensor cities, emphasizing the opportunities and critical issues of this growing datafication of urban contexts. In this sense, an integrated and holistic framework is proposed which includes a theoretical and managerial review of the main disruptive technological applications. The study, thus, identifies and compares IoT solutions based on sensors, big data analysis and other technologies related to ICTs adopted by different cities, to manage urban development in an innovative and computerized manner. This paper generates insights from the current sensor city best practices by placing some renowned projects, implemented by Huawei, Cisco, Google, Ericsson, Microsoft, and Alibaba, under the microscope. With the objective to offer a detailed overview, the paper adopts a mixed research approach able to integrate a literature review and an in-depth analysis of several case studies. In this regard, the proposed framework provides users with greater knowledge and awareness of sensor cities' development.

The paper is structured as follows. Section 2 introduces the mixed methodological approach used. Section 3 analytically describes the proposed framework, highlighting the technological factors such as IoT, big data analysis, AI, ICTs, real-time monitoring stations, sensors, cloud computing, digital platforms, and urban challenges that characterize sensor cities. Finally, Section 4 provides conclusions and some considerations on the contribution of sensor cities to future urban challenges.

2. Materials and Methods

The development of this study is structured according to a mixed approach, characterize by a literature review and a detailed analysis on several case studies, in order to create a framework for policymakers that collect the forward-looking ICT urban initiatives, emphasizing a data-focused method in the assessment of urban development. The implementation of this study is carried out in different phases and in this sense the methodological approach followed is illustrated in Figure 1.

In Phase 1: Identification, the research question, keywords, and research databases are defined. Regarding the objective and the research question, the study aims to identify and analyze the main characteristics of sensor cities, highlighting the impacts of their technological applications on urban development.

The attention of the study has focused that have provided an interdisciplinary and/or transdisciplinary perspective to the development of sensor city. Specifically, these scientific disciplines include technology and innovation management, urban planning, policy, sustainable development, environmental management, data science, urban informatics, geography, urban development, strategic management, and urban statistics. Moreover, the paper—through a qualitative approach—analyzes several sensor cities considered successful examples of disruptive urban actors, capable of integrating sensor strategies with the economic, social, and environmental aspects of urban development in detail. As a result, the study includes peer-reviewed journal articles, book chapters, and conference proceedings, grey literature such as government documents, and industry technical reports. In terms of databases, ScienceDirect, Google Scholar and cities' websites were utilized to achieve the analysis. Furthermore, the keywords searched include ((("sensor city" OR "sensor cities") AND ("ubiquitous city" OR "ubiquitous cities") AND ("digital city" OR "digital cities") AND ("real-time city" OR "real-time cities") AND ("sentient city" OR "sentient cities") AND ("intelligent city" OR "intelligent cities") AND ("data-driven city" OR "data-driven cities") AND ("smart city" OR "smart cities") AND ("sustainable city" OR "sustainable cities") AND ("sustainable development") AND ("smart urban applications") AND ("urban IoT") AND ("urban sustainability") AND ("urban development") AND ("big data applications") AND ("urban sensors") AND ("knowledge city" OR "knowledge cities") AND ("disruptive urban development").

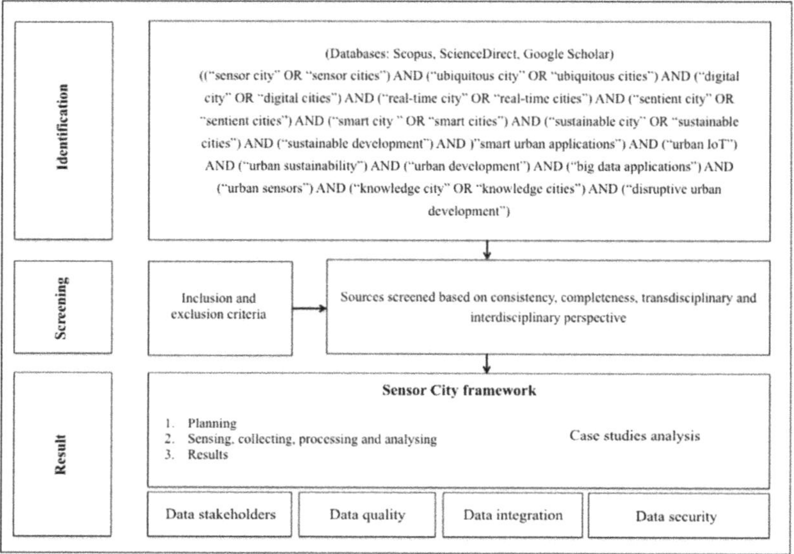

Figure 1. Methodological approach adopted (source: authors).

In Phase 2: Screening, the review aims to provide a clear and comprehensive definition of the concept of sensor city, introducing urban dimensions (e.g., governance, economy, environment, mobility, people, and living) and technological solutions (e.g., IoT, sensors, AI, ICTs, big data analytics) necessary for its implementation. In order to further refine the research, all the selected sources were screened following a set of inclusion and exclusion criteria, in line with the objective and the research question of the study. As exclusion criteria, sources with partial information and inconsistent with the topic of the study were not included in the search.

In Phase 3: Result, the integration between literature review and case study analysis provides a detailed framework useful for developing a theoretical approach to sensor cities, underlining the holistic relationships between urban sustainability and computerization aspects. Nonetheless, it is important to consider that sensor cities taken into consideration in the case studies analysis use non-uniform terminology and lack detailed quantitative data on sensor infrastructures. In this sense, a greater understanding and awareness of urban sensing is needed to develop forward-looking projects in line with future urban challenges. This research focused on 20 sensor cities, with a high quality in social, economic, environmental and technological infrastructures. Consequently, the sample analyzed does not represent the different types of cities globally. Most cities around the world are not equipped to collect, monitor, analyze and evaluate urban performance through innovative platforms and dashboards. In sum, this review provides: (a) A clear definition of sensor city; (b) A sensor city framework; (c) A detailed analysis of several sensor cities, and; (d) Various different sensing policy and actions that currently policymakers have in place.

3. Results

Given the growing importance of sensor city solutions in the world, technology giant companies (e.g., Huawei, Ericsson, Microsoft, Oracle, Alibaba, Deutsche Telekom, Samsung) are involved in numerous urban technological projects in order to provide platforms, products and services in line with the necessary paradigm shift [111]. Thus, sensor cities aim to involve citizens and companies in the process of planning, monitoring, and analyzing of urban processes and raising awareness and comprehension on environmental, social, and economic issues [112,113]. Table A1 in the Appendix A presents some examples of cities equipped with real-time monitoring stations and sensors that analyze and collect urban performance.

Few [114] in his book 'Information Dashboard Design: The Effective Visual Communication of Data' defines the dashboard as a visual, consolidated and organized display on a single screen of the most important information needed to monitor, analyze and achieve one or more urban objectives. Similarly, Kitchin [115] explains urban dashboards as digital, physical or mixed interfaces that allow users to actively or passively interact in urban data monitoring and management in order to improve understanding of urban systems. Kitchin and McArdle [116] and Petit and Leoa [117] describe urban dashboards as platforms that use dynamic and interactive graphic interfaces, maps, 3D models, augmented reality, bar charts, and so on, to support urban decision-making with the aim of monitoring, analyzing and interpreting performance and trends of cities. Thus, the sensor city is generally accepted as a digital platform where users are completely immersed.

Nevertheless, theoretically, the ideals that characterize sensor cities can be defined as universal and homogeneous, while contexts tend to be specific and heterogeneous [118]. For example, cities have different social, economic, environmental and technological infrastructures and are governed by different political and bureaucratic systems with interests that are difficult to combine in a single urban project. Furthermore, while urban data is increasingly easy to access, technical and digital skills are difficult to find in some urban contexts around the world, in order to manage, analyze and interpret urban data.

Urban contexts using IoT and/or ICT platforms require more initial efforts due to the assembly and maintenance of the infrastructure, in order to adapt it to increasingly complex urban scenarios [119,120]. In detail, it is a matter of integrating devices that are: (a) Heterogeneous, capable of generating different types of data; (b) Powered by different energy sources, such as battery or renewable energy; (c) Dynamic and flexible, capable of analyzing constantly changing urban scenarios, and; (d) Unpredictable, in the sense that technological applications can provide conflicting results by analyzing the same data, as they use different protocols and standards.

The significant costs of designing, installing, and maintaining monitor stations and sensor technologies represent a barrier to entry for smaller cities or those located in less development regions [121,122]. The success of an urban sensor strategy depends essentially on the economic, social and environmental characteristics of the urban context taken into consideration together with organizational, ethical and transdisciplinary factors of the actors involved [118].

The sensor city framework is illustrated in Figure 2. In detail, the operational phases, integrated in a holistic perspective of the urban context, are divided into planning, sensing, collecting, processing and analysis of urban data and results.

In Phase 1, or planning, the role of ICTs is highlighted, which allows citizens to participate in the decision-making process and to enhance systemic collaboration between stakeholders involved, contributing significantly to greater comprehension, transparency and accountability [123–125]. Hence, planning activities through smart solutions permits a holistic and integrated approach of the various urban dimensions (e.g., governance, economy, environment, mobility, living, and people), reducing costs and time of the bureaucratic collaborations between departments and improving quality and efficiency of urban services [126]. Nonetheless, most cities do not work like companies (e.g., IBM, Cisco, Google). They tend to be disorganized, e.g., departments do not collaborate on solutions [127]. On this point,

Cugurullo [128] elaborates that urban contexts promoted as examples of integrated and holistic urban planning are often fragmented and disconnected, characterized by several incompatible components.

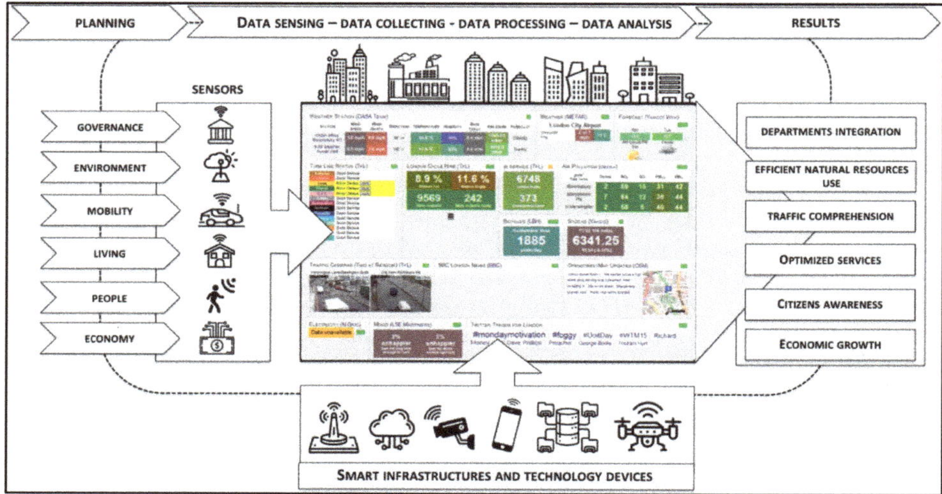

Figure 2. Sensor city framework (source: authors).

The Digital Single Market strategy adopted by the European Commission (EC) represents one of the fundamental pillars of the policies for creating a digital single environment [129]. Indeed, the strategy aims to ensure a better, secure and uniform access for citizens to digital networks in order to improve technological knowledge and encourage greater social inclusion.

The governance dimension highlights the spread of social media (e.g., Facebook, Twitter, Instagram) and other data sharing platforms that allow increasingly integrated and structured communications between stakeholders [130]. Thus, by adopting a bottom-up approach, these feedback-generation, sharing and management tools transform citizens, companies, local authorities, and so on, into active participants in the governance of sensor cities [131–133].

The environmental dimension involves various aspects of urban context such as:

- Waste management, capable of providing real-time type and quantity of waste via cloud solutions, to optimize time and resources [10,134]. With the development of new technology devices, electronic waste (e-waste) represents an environmental and health challenge due to its difficult disposal and potential impacts on the environment.

- Energy efficiency, which refers to all technologies able to reduce energy consumption. For example, street lighting networks equipped with sensors and public and private buildings equipped with real-time consumption monitoring systems represent the forward-looking urban infrastructures needed to manage urban activities efficiently [12,13,135].

- Renewable energy sources, such as solar, wind, thermal and biogas, which represent a sustainable and efficient alternative to fossil fuels, capable of guaranteeing a stabilization of energy prices and a sustainable source of energy supply for urban processes (e.g., public transport, heating of public and private buildings) [14,15].

- Water management, which includes solutions capable of providing real-time data on the consumption and quality of water distribution system, fundamental for the sustainability and efficiency of the urban system [16].

- Material flows, those underline the exchange and transformation of resources (e.g., raw materials, by-products, waste) between various interested actors [23].

- Biodiversity conservation, which includes revitalization actions for abandoned urban and industrial areas and the safeguarding of green and urban spaces [24].
- Land use optimization, as support for urban agriculture and infrastructure [27].
- Air and noise pollution prevention, to reduce pollutant emissions through ICTs, sensors, and real-time monitoring stations, capable of providing high-definition videos and images to check environmental quality [30].

The mobility dimension includes, amongst others, smart and sustainable public transport system, availability of urban infrastructure suitable for autonomous vehicles, car sharing stations for electric vehicles and ICTs relating to traffic and road monitoring [25,136–138].

The aspects related to the living dimension include actions to improve the use of cultural and entertainment facilities (e.g., libraries, museums, schools, public parks, sport facilities) [20].

The people dimension includes policies aimed at promoting a 'sustainable' community, improving digital literacy, providing various assistance programs for citizens with special needs and a quality healthcare system, ensuring gender equality in terms of pay and office positions, safe and healthy work environments, and so on [139].

The economic dimension refers to the ability of the urban context to favor a path of growth through technological innovation, entrepreneurship and sustainability, attracting the most innovative companies, start-ups and talents, and capable of promoting a digitalized and collaborative development [32].

In Phase 2, with the objective of sensing, collecting, processing and analyzing urban data, the Chinese telecom company Huawei, for example, has developed 'Smart City Solution', a platform used in over 160 cities in 40 countries in Asia and Europe, capable of analyzing large volumes of real-time urban data and providing policymakers a method of predictive analysis of future urban scenarios (Figure 3). Specifically, Lanzhou New Area represents the first new state-level development area in northwest China which, through the support of Huawei's network, was able to build the nation's first governmental IoT and wireless sensors network that integrates both broadband and narrowband communications, integrating 31 departments (e.g., public safety, finance, energy, transport, healthcare, education) with 45 eLTE stations. The program aims to improve the quality of life by optimizing urban resources in a smart and sustainable manner [140].

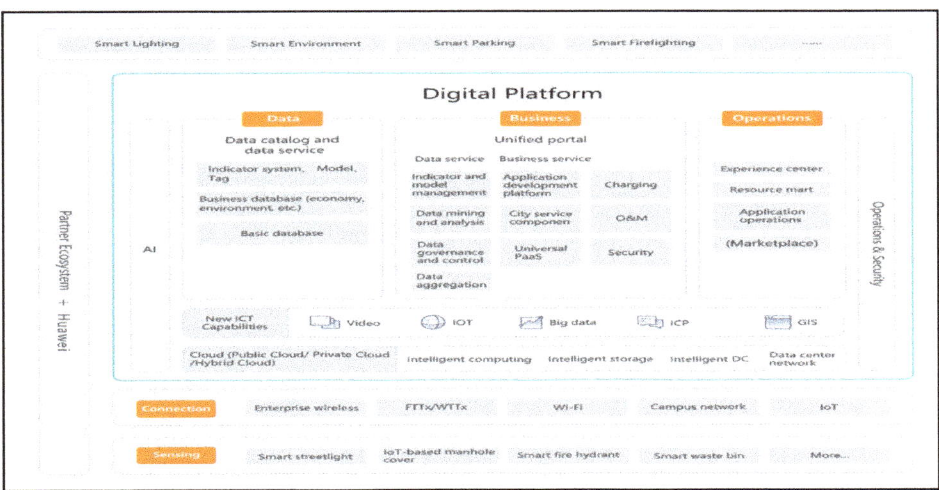

Figure 3. Huawei Digital Platform, derived from [141].

Similarly, Cisco has developed the 'Cisco Kinetic for Cities' framework, an open and easy-to-use urban data sharing platform that combines data provided by sensors, applications and other third-party devices to create a dynamic sensor city infrastructure, encouraging the exchange of innovative urban initiatives between policymakers, companies and citizens (Figure 4) [142].

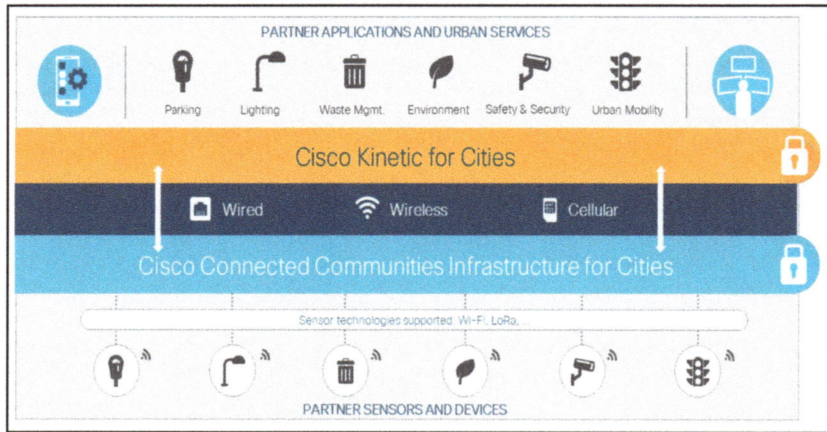

Figure 4. The Cisco Smart + Connected Digital Platform, derived from [143].

At the same time, Alibaba has developed the 'City Brain' program that has also become a useful tool for city managers, providing a holistic dashboard that can improve the perception of urban data and real-time processing capacity [144]. Figure 5 shows the real-time detection and analysis platform for city events. By integrating the data from the surveillance rooms, it is possible to coordinate traffic lights to give priority passage to response vehicles (e.g., police, fire-fighters, rescue and other vehicles) in case of emergency. Likewise, the program has been launched in several Chinese cities such as Hangzhou, Shanghai, Chongqing, Suzhou, Haikou, Beijing, Chengdu, Quzhou, and Jiaxing. Projects as CityBrain highlight the continuous interaction between local authorities and technology companies in managing urban governance. In fact, the Chinese platforms are largely owned by national technology companies [60].

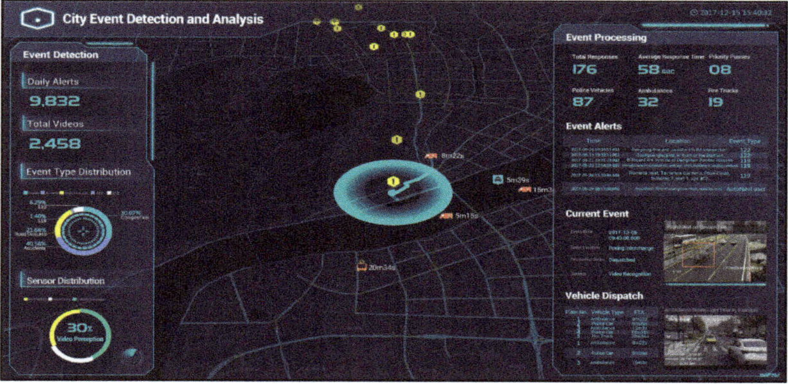

Figure 5. Alibaba City Event detection and smart processing, derived from [145].

Barcelona has changed its sensor city approach from top-down to bottom-up, involving its citizens in participating in innovative urban projects. For example, the Smart Citizen Kit is a dashboard that collects data on the environmental such as air composition, temperature, light intensity, sound levels and humidity through sensors and ICTs [146]. The data collected in real-time are sent via Wi-Fi to an open data platform and are used to create maps that display environmental conditions, equipping public and private stakeholders with urban data in order to develop and/or improve services for citizens [147]. Telefonica—through the Valencia Smart City Project—aims to transform the city of Valencia into an intelligent ecosystem and fully connected via 350 sensors, allowing the management of public resources through a single ICT platform and improving several urban areas such as transport, energy, efficiency and environmental services. The city of Santander has installed around 20,000 parking sensors in the streets, introducing intelligent waste containers capable of monitoring and measuring air pollution, rainfall, and traffic density. Through the integration and interpretation of the corresponding data, the Santander municipal administration optimizes waste truck routes to save staff and fuel costs, or controls the irrigation of city parks to save water [148].

Deutsche Telekom have implemented similar dashboard, equipping urban physical object (e.g., street lamps, bins, parking lots, traffic lights) with software, sensors and connectivity systems integrated into a shared network. In this sense, the collection, monitoring, analysis and interpretation of urban data allows the implementation of innovative service and business models (Figure 6).

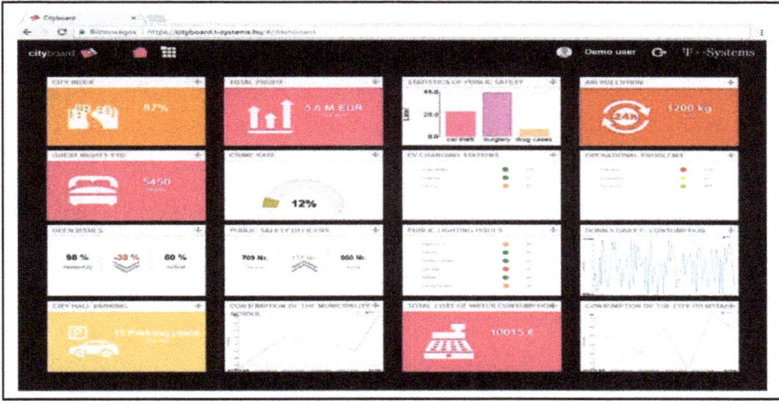

Figure 6. Smart City Dashboard developed by Deutsche Telekom for Hungary subsidiary, derived from [149].

In 2018, Deutsche Telekom and the city of Hamburg implemented around 11,000 parking spaces equipped with sensors that can provide the current availability status to users via the app [150]. The local authorities of Gelsenkirchen (Germany) have decided to collaborate with Huawei and GELSEN-NET in order to implement a new open and shared urban governance system, developing an ICT infrastructure capable of integrating key data to improve the efficiency of the public services and the accuracy of the decision-making process. In particular, the 'Safe City' program uses Huawei's extensive wired and wireless broadband network and the IoT that connects industrial parks, hospitals, schools, pedestrian areas and urban centers, creating a sustainable ICT ecosystem. As part of the T-City project in Friedrichshafen (Germany), Deutsche Telekom has partnered with Alcatel-Lucent to provide hardware solutions and network devices for T-City. Specifically, Deutsche Telekom has developed over 40 pilot projects in several categories such as mobility, research, tourism, culture, health, and employment.

The LuxTurrim5G ecosystem in Espoo (Finland) aims to transform the Kera area into a sustainable and digitized urban neighborhood, involving Nokia and several partners in order to create a multitude of urban digital services. The LuxTurrim5G platform is able to collect, store, manage and share large amounts of urban data and solutions in a safe and efficient way between local authorities, companies and citizens. Specifically, the network developed by Nokia includes, among other things, over 50 Wi-Fi devices, 75 cameras, 49 different sensors that monitor air quality, climate, temperature, road surface conditions as well as CO_2 levels, 9 radar devices, 7 information screens, a charging station for electric vehicles and a charging and landing station for drones.

Oracle through a partnership with the Indian state of Maharashtra has implemented a platform to design, develop and analyze government-to-citizen and smart-business government services across the state. Maharashtra also aims to connect all its 113 million residents via fiber, including those in the more than 300 cities of Maharashtra and its 29,000 villages. Dallas has chosen to collaborate with Ericsson to implement an advanced traffic management system through a dashboard capable of monitoring, managing, analyzing and aggregating different data in real-time from traffic sensors and cameras to dynamically control the traffic lights. The goal of the city is to have an intuitive and shared analysis tool capable of integrating data from the various departments and agencies.

According to the report 'The future of street lighting', street lighting represents the main part (about 40%) of a city's overall electricity costs. In this sense, the replacement of conventional lamps with LED bulbs can reduce energy consumption by up to 80% when using a centralized management system. For example, intelligent lighting platform of Cisco and Sensity in Kansas transforms each lamppost into a sensor connected to a broadband wireless network, creating a true interconnected lighting network capable of collecting real-time data such as intelligent parking systems, electricity consumption, or air quality level. In this sense, the smart pole developed by Deutsche Telekom and Nokia represents a paradigm shift in the use of lighting poles as it goes beyond simple urban lighting and constitutes an essential component of the infrastructure of sensor cities (Figure 7).

The Hamburg Port Authority (HPA) through Cisco systems has developed sensors that monitor the use of resources (e.g., trucks, cranes, means of transport, ships) and infrastructures (e.g., roads, parking lots, storage warehouses) in one of the busiest and most important ports in Europe. In this sense, modernization through IoT technological innovation allows the Port Road Management Center to plan future investments in the traffic infrastructure, optimizing the flow of traffic and minimizing the externalities of the port on the inhabitants of the city.

Ericsson aims to make Stockholm the smartest city in the world by 2040. With the water Monitoring Network project, Ericsson has the ability to design and implement a real-time water quality-monitoring network using an IoT sensor system located in the Stockholm water system. In addition, big data analytics are used that can analyze the data produced by the sensors and provide more information on changes in water quality such as pH and temperature.

In many cases, small towns do not need sophisticated public transportation solutions. In this sense, the challenge is to combine the departure time of buses, subways, railways, and so on, with data from traffic jams and road works on highways, showing the best alternative connections. As a partner of the Kooperation Östliches Ruhrgebiet, Deutsche Telekom helps to connect local public transport in the German state of North Rhine-Westphalia. In this regard, Deutsche Telekom's strategy allows not only meeting the needs of passengers in terms of both transparency in information (e.g., timetables, delays) and efficiency, sharing the same platform for multiple tasks. In Croatia, Deutsche Telekom's subsidiary, Hrvatski Telekom, has developed an electric vehicle charging network of 145 charging points in 101 charging stations in 70 cities. The project also integrates an ICT infrastructure that helps users to book and pay for vehicles top-ups and receive real-time availability information.

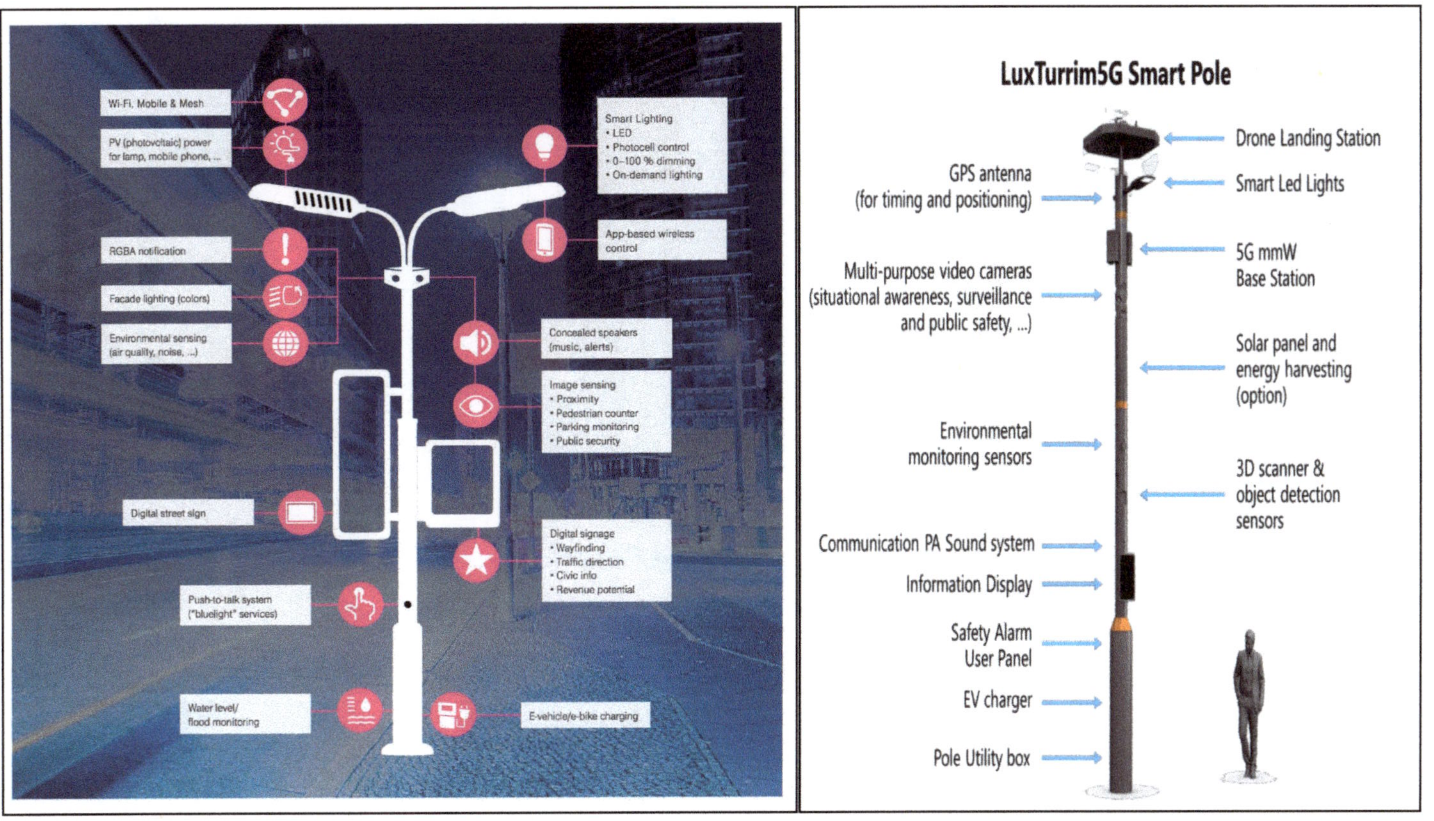

Figure 7. Smart Street Lighting models developed by Deutsche Telekom and Nokia, derived from [151,152].

The final phase concerning the results explaining the decisive role of sensor solutions in terms of implementation of innovative urban processes (e.g., efficient use of natural resources, cross departmental integration, real-monitoring of traffic congestion and energy consumption, smart management of infrastructures), in order to strategically improve the contribution of sensor cities to disruptive urban development. Consequently, the optimization of Quality of Experience (QoE) and Quality of Services (QoS) has become a crucial aspect in the implementation of urban services and processes [153]. Particularly, QoE represents an evaluation of human experience when interacting with technology solutions and stakeholders in a given context. Therefore, a detailed analysis of the QoE should consider several actors interacting with each other at different levels (environmental, social, economic, technological) and with different goals. To do this, it is necessary to define the main interactions between citizens, companies, local authorities, social organizations, and so on. In sum, some QoE aspects related to sensor city are listed in Table 1.

Table 1. Impact of quality of experience on sensor city, derived from [153].

	Economic, Social and Privacy Implications	E-government	Health and Assisted Living	Intelligent Transportation Systems	Smart Grids, Energy Efficiency, and Environment
	Usability	Usability	Usability	Usability	Usefulness
	Personalization	Personalization	Availability	Usefulness	Accessibility
Quality of Experience	Transparency	Transparency	Personalization	Effectiveness	Personalization
			Effectiveness	Accessibility	
			Accessibility	Efficiency	
			Efficiency		

In this regard, the challenge for policymakers will be the widespread use of several tools, such as devices, platforms, algorithms, and networks, to access, share and integrate the largest number and type of urban data. The big data evolution process requires significant data gathering, highly specialized personnel, and infrastructures owned by different municipalities, agencies, corporations, and private companies [5]. This attention of ICT leads urbanist mark Swilling to describe the sensor city as a form of algorithmic urbanism [154]. In this sense, concepts, such as big data analysis, ICT, sensors, real-time monitoring systems, smart infrastructures, IoT, represent necessary but not sufficient solutions [105,115,155]. The reconfiguration of the urban system expresses the need to integrate the environmental, social, economic dimensions of smart technologies into an integrated, efficient, and computerized urban context [6,49,156,157].

4. Discussion

The perspective of implementing sensor cities based on IoT and on different urban technologies that permit the collection, monitoring, analysis and integration of large amounts of urban data has developed new opportunities for policymakers regarding the ability to plan, combine and evaluate social, environmental and economic aspects in a holistic manner [34,158,159].

The expansion of urban technologies has stimulated local authorities, technology companies, start-ups, citizens, municipal companies, and other actors involved, to develop increasingly innovative and sophisticated projects, devices, platforms, algorithms, systems, initiatives, and so on, especially in technologically and ecologically developed cities. Consequently, the use of advanced techniques such as real-time monitoring stations for energy consumption, location systems to guide urban traffic, cloud computing systems for sharing sensitive data between government departments, urban infrastructures such as smart bins, smart street lamps, and surveillance cameras, have great potential in transforming the way we understand and evaluate the urban context and, at the same time, highlight a multitude of challenges related to the quality of the data used, the level of protection of traditional and cybernetic urban security, the necessary data integration between the various urban infrastructures and the ability to transform feedback from citizens and other stakeholders into innovative urban policies (Figure 8). In that perspective, sensor cities represent a sophisticated paradigm shift in the concept of the technical-urban context, necessary for a transition towards disruptive urban development [160–162].

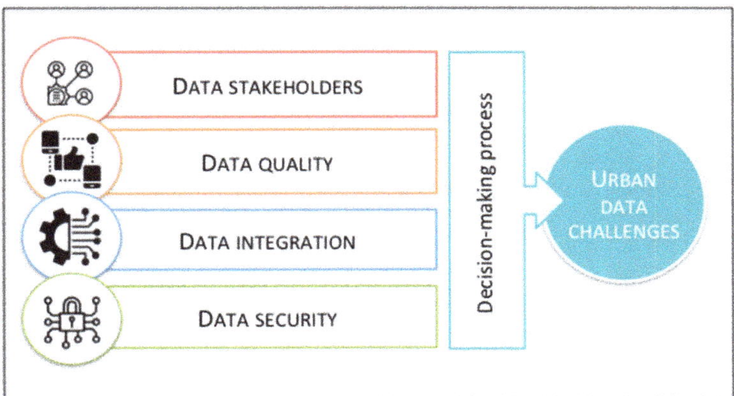

Figure 8. Future urban data challenges (source: authors).

The first challenge refers to the development of innovative urban policies in line with the needs of citizens, local authorities, companies, organizations, and so on. Hence, to use a significant amount of urban data according to the needs of the stakeholders involved, it is necessary to identify the relevant information and understand its consistency and validity. For example, in a survey, citizens of the well-known German tourist destination, Heidelberg, identified traffic as the most urgent issue to be solved. In this sense, the efforts of local policymakers have been focused on how to improve urban mobility. Thus, the identification, collection and processing of information provided by the actors involved are crucial in the development of urban sensor projects [163–165].

The second challenge concerns the need to guarantee a certain level of data quality in order to develop efficient and innovative urban policies. For example, data collection sensors can produce incorrect, partial or missing values as they use different standards or protocols, generating inconsistencies and differences between the data. For example, CityPulse described by Puiu et al. [166] is a real-time monitoring framework supported by the European Union (EU) that processes, integrates and adapts uncertain and incomplete data through quality of information techniques in order to develop reliable information capable of satisfying user requests. In this regard, the urban project represents a practical model of how to move from vertically to horizontally interconnected services [167].

Cities with concentrated IoT and sensor networks may provide unreliable communications due to incorrect data transmission. Specifically, the incorrect transmission of urban data does not only require time, but also retransmits the information flow, which negatively affects the quality of the data [168]. As a result, the quality of data and the technologies to overcome problems of inconsistency, partiality, and unreliability should be considered in the development and implementation of sensor city projects [169,170].

The third challenge concerns the integration of different types of data, collected, processed and analyzed by different institutions, companies and/or independent authorities. One of the main tasks of public decision-makers, and urban managers, planners, and policymakers is to implement infrastructures, models, and networks capable of connecting data deriving from different urban sectors, promoting better communication between the various stakeholders involved [171]. For example, the dashboard implemented by Alibaba, illustrated in Figure 5, requires an information flow and a systemic communication between government, security and emergency service institutions. Hence, the holistic combination of data is necessary to plan and better understand the potential of sensor cities.

The fourth challenge involves the security of sensitive data. In this regard, ensuring maximum privacy protection is indispensable for developing urban policies based on sensors, video surveillance cameras, monitoring stations, cloud computing, and so on. Data security issues not only have isolated effects, but also often affect all the urban dimensions (analyzed in the previous section). Thus, data security is treated as a fundamental issue in the management of sensor cities [172–174]. Through the collection and use of large amounts of data and technological solutions that analyze human behavior, it is possible to influence not only the fight against crime through sensors that analyze movements or facial expressions, but also urban governance in terms of planning and the attractiveness of the city [175–177]. Nonetheless, the corresponding interventions are particularly controversial in terms of privacy, emphasizing ethical requirements in the urban safety planning process [178,179]. Therefore, integrating the problems related to traditional and cybernetic urban security into the planning and implementation projects of sensor cities is necessary to guarantee safe and digitalized urban development [180,181].

The challenges explained in this section are related because the collection, processing and analysis of data through IoT solutions, sensors and big data analysis refer to interdependent urban dimensions (e.g., governance, environment, mobility, economy, life and people). Nevertheless, this perspective requires a change in the structural paradigm of technical-scientific skills, greater organizational flexibility of government institutions and a more aware and involved citizenship in the urban administration.

5. Conclusions

Sensor and IoT technologies are acquiring ever-increasing importance in the techno-spatial context by collecting, processing, analyzing, and integrating large amount of data to improve the healthy functioning of our cities. In this regard, the importance of the corresponding urban assessment tools is evidenced by the vast investment made by technology giants, such as Cisco, Google, Microsoft, Ericsson, Alibaba, Oracle, in prototype sensor cities. These initiatives are helping cities, such as Toronto, Valencia, Dallas, Singapore, Kansas City, Hamburg, Shenzhen, Adelaide, Dublin, integrate IoT and related applications for big data, and actively engaging in city sensor development. The sensing city model allows policymakers to benefit from big data and urban analytics to improve urban policies, services, and operations.

The purpose of this paper is to explore and integrate different sensor cities taken as case studies consistent with the research demand and to discuss technological solutions (e.g., sensors, devices, IoT, AI, platforms, digital infrastructures, computer models, ICTs), emphasizing the economic, social and environmental benefits of their practical application. Hence, the technological applications enabled by the IoT and ICTs in general have the potential to improve daily urban activities, providing a series of opportunities concerning urban dimensions such as governance, economy, environment, mobility, people and life. In this perspective, the social, environmental and economic challenges arising from the use of these tools have been explained. The study finds that disruptive urban technologies not only promote highly efficient and computerized urban processes and efficiency, but also improve the understanding of planning, monitoring and analysis of the performance of sensor cities by increasing awareness of citizens, businesses, local authorities, and so on.

The framework of sensor cities (see Figure 2) illustrates and defines the process of detection, collection, processing, analysis of urban data up to their transformation into information necessary for urban policymakers. In this regard, the use of IoT platforms equipped with systems, devices, sensors, algorithms, platforms, models, and so on provides a complete and exhaustive view of the future trends of urban performance measurement methodologies. Furthermore, these methodologies and approaches give rise to some important issues such as data quality and integrity, cyber security, digital data and information ethics, and regulations. These are among the critical issues that prospective research must tackle.

To recap the findings and conclude the paper, we list the following highlights of the study: (a) The proposed framework improves the understanding of the cities' progress towards becoming sensing localities to address rising issues effectively and efficiently; (b) Achieving a truly smart urbanism requires continuous monitoring trough urban platforms and dashboards; (c) The data-driven approach highlights a multitude of challenges related to urban governance; (d) Urban dashboards improve citizen accountability and awareness on urban issues, and; (e) Data security, quality, integration and stakeholders must be involved in the urban sensor strategy.

Author Contributions: G.D. designed and supervised the study, and finalized the manuscript. G.I., P.L., W.L. and T.Y. contributed to the write-up of the manuscript, and improved the rigor, relevance, and reach of the study. G.D. and G.I. prepared the first draft of the manuscript and assisted in data collection. All authors have read and agreed to the published version of the manuscript.

Funding: This research received no external funding.

Acknowledgments: This research did not receive any specific grant from funding agencies in the public, commercial, or not-for-profit sectors. The authors thank the managing editor and anonymous referees for their invaluable comments on an earlier version of the manuscript.

Conflicts of Interest: The authors declare no conflict of interest.

Appendix A

Table A1. Sensor city projects of high-tech companies investigated in this study (source: authors).

Location		High-Tech Company Partner	Sensor City Project	Focus	Source
City	Nation				
Toronto	Canada	Google	Sidewalk Labs	Building, energy, waste, environment, water	[182]
Hamburg Port	Germany	Cisco	smartROAD	Building, energy, water, mobility	[183]
Gelsenkirchen	Germany	Huawei	Gelsenkirchen: A Small, Smart City with Big Plans	Mobility, environment, energy	[184]
Adelaide	Australia	Cisco	Lighthouse City	Mobility, environment	[185]
Kansas City	U.S.A.	Cisco	Smart+Connected Communities	Mobility, water, energy	[186]
Singapore	Singapore	Google	Smarter Digital City 3.0	Mobility, living, people, finance, retail	[187]
Friedrichshafen	Germany	Deutsche Telekom	T-City Friedrichshafen	Energy, mobility, healthcare, building	[188]
Charlotte	U.S.A.	Microsoft	Charlotte sustainability city	Energy, mobility, education, safety	[189]
Dublin	Ireland	IBM	SMART Dublin	Mobility, environment, building, water, energy	[190,191]
Aspern Seestadt	Austria	Siemens	Aspern Smart City Research	Building, energy, environment, water	[192]
Stockholm Port	Sweden	Ericsson	Stockholm Royal Seaport's Smart Energy City project	Water, environment, energy	[193]
Dallas	U.S.A.	Ericsson	Factory of the Future	Mobility	[194]
Espoo	Finland	Nokia	Luxturrim5G Project	Environment, mobility, safety, energy	[195]
Hangzhou	China	Alibaba	City Brain	Mobility, water, environment, healthcare	[144]
Maharashtra	India	Oracle	Centre of Excellence	Mobility, water, education	[196]
Bari Matera	Italy	TIM	Bari Matera 5G	Tourism, culture, mobility, safety, environment, healthcare, agriculture	[197]
Saemangeum	South Korea	Samsung	Green Energy Industrial Complex	Energy, education, agriculture, tourism, environment	[198]
Shenzhen	China	Tencent	Tencent Campus	Energy, environment, mobility, education	[199]
Guiyang	China	Alibaba, Tencent, Apple	Guiyang Sunac City	Environment, mobility, recreation	[200]
Valencia	Spain	Telefonica	Valencia Smart City Platform	Mobility, waste, environment, energy	[201]

References

1. Yigitcanlar, T. Australian local governments' practice and prospects with online planning. *URISA J.* **2006**, *18*, 7–17.
2. Arbolino, R.; De Simone, L.; Carlucci, F.; Yigitcanlar, T.; Ioppolo, G. Towards a sustainable industrial ecology: Implementation of a novel approach in the performance evaluation of Italian regions. *J. Clean. Prod.* **2018**, *178*, 220–236. [CrossRef]
3. Ingrao, C.; Messineo, A.; Beltramo, R.; Yigitcanlar, T.; Ioppolo, G. How can life cycle thinking support sustainability of buildings? Investigating life cycle assessment applications for energy efficiency and environmental performance. *J. Clean. Prod.* **2018**, *201*, 556–569. [CrossRef]
4. Zheng, Y.; Liu, F.; Hsieh, H. U-Air: When urban air quality inference meets big data. In Proceedings of the 19th ACM SIGKDD International Conference on Knowledge Discovery and Data Mining, Chicago, IL, USA, 11–14 August 2013; Association for Computing Machinery: New York, NY, USA, 2013.
5. Bettencourt, L. The Uses of Big Data in Cities. *Big Data* **2014**, *2*, 12–22. [CrossRef] [PubMed]
6. Bibri, S.E.; Krogstie, J. Smart sustainable cities of the future: An extensive interdisciplinary literature review. *Sustain. Cities Soc.* **2017**, *31*, 183–212. [CrossRef]
7. Huang, S.-L.; Hsu, W.-L. Materials flow analysis and emergy evaluation of Taipei's urban construction. *Landsc. Urban Plan.* **2013**, *63*, 61–74. [CrossRef]
8. Costi, P.; Minciardi, R.; Robba, M.; Rovatti, M.; Sacile, R. An environmentally sustainable decision model for urban solid waste management. *Waste Manag.* **2004**, *24*, 277–295. [CrossRef]
9. Zaman, A.U.; Lehmann, S. The zero-waste index: A performance measurement tool for waste management systems in a 'zero waste city'. *J. Clean. Prod.* **2013**, *50*, 123–132. [CrossRef]
10. Aazam, M.; St-Hilaire, M.; Lung, C.-H.; Lambadaris, I. Cloud-Based Smart Waste Management for Smart Cities. In Proceedings of the IEEE International Workshop on Computer Aided Modeling and Design of Communication Links and Networks, CAMAD, Toronto, ON, Canada, 23–25 October 2016; pp. 188–193.
11. Glazebrook, G.; Newman, P. The city of the future. *Urban Plan.* **2018**, *3*, 20. [CrossRef]
12. Chui, K.; Lytras, M.; Visvizi, A. Energy Sustainability in Smart Cities: Artificial Intelligence, Smart Monitoring, and Optimization of Energy Consumption. *Energies* **2018**, *11*, 2869. [CrossRef]
13. Ghiani, E.; Serpi, A.; Pilloni, V.; Sias, G.; Simone, M.; Marcialis, G.; Armano, G.; Pegoraro, P.A. A Multidisciplinary Approach for the development of Smart Distribution Networks. *Energies* **2018**, *11*, 2530. [CrossRef]
14. Lund, H.; Mathiesen, B.; Connolly, D.; Østergaard, P. Renewable energy systems—A smart energy systems approach to the choice and modelling of 100% renewable solutions. *Chem. Eng. Trans.* **2014**, *39*, 1–6.
15. Strielkowski, W.; Streimikiene, D.; Fomina, A.; Semenova, E. Internet of Energy (IoE) and High-Renewables Electricity System Market Design. *Energies* **2019**, *12*, 4790. [CrossRef]
16. Gourbesville, P.; Ler, L. Framework Implementation for Smart Water Management. *EPIC Ser. Eng.* **2018**, *3*, 1139–1146.
17. Nitschke, U.; Ouan, N.; Peters, G. Megacities—A challenge for (German) Development Cooperation. *ASIEN* **2007**, *103*, 79–87.
18. Lin, J. Urbanization and Inequality in China's Megacities: A perspective from Chinese Industrial Workers. In *Dialogues of Sustainable Urbanisation: Social Science Research and Transitions to Urban Contexts*; Condie, J., Cooper, A.M., Eds.; University of Western Sydney: Penrith, Australia, 2015; pp. 164–168.
19. Jowell, A.; Zhou, B.; Barry, M. The impact of megacities on health: Preparing for a resilient future. *Lancet Planet. Health* **2017**, *1*, e176–e178. [CrossRef]
20. Macke, J.; Sarate, J.; De Atayde Moschen, S. Smart Sustainable Cities Evaluation and Sense of Community. *J. Clean. Prod.* **2019**, *239*, 118103. [CrossRef]
21. Muzammal, M.; Talat, R.; Sodhro, A.H.; Pirbhulal, S. A multi-sensor data fusion enabled ensemble approach for medical data from body sensor networks. *Inf. Fusion* **2020**, *53*, 155–164. [CrossRef]
22. Sodhro, A.H.; Sangaiah, A.K.; Sodhro, G.H.; Lohano, S.; Pirbhulal, S. An Energy-Efficient Algorithm for Wearable Electrocardiogram Signal Processing in Ubiquitous Healthcare Applications. *Sensors* **2018**, *18*, 923. [CrossRef] [PubMed]
23. Kennedy, C.; Cuddihy, J.; Engel-Yan, J. The changing metabolism of cities. *J. Ind. Ecol.* **2007**, *11*, 43–59. [CrossRef]

24. Morimoto, Y. Biodiversity and ecosystem services in urban areas for smart adaptation to climate change: "Do you Kyoto"? *Landsc. Ecol. Eng.* **2010**, *7*, 9–16. [CrossRef]
25. Arndt, W.-H.; Schäfer, T.; Emberger, G.; Tomaschek, J.; Lah, O. Transport in Megacities—Development of Sustainable Transportation Systems. In Proceedings of the 13th World Conference on Transport Research (WCTR), Rio de Janeiro, Brazil, 15–18 July 2013.
26. Sadiku, M.; Shadare, A.; Musa, S. Smart Transportation: A Primer. *Int. J. Adv. Res. Comput. Sci. Softw. Eng.* **2017**, *7*, 6–7. [CrossRef]
27. Kim, G.; Miller, P.A.; Nowak, D.J. Urban vacant land typology: A tool for managing urban vacant land. *Sustain. Cities Soc.* **2018**, *36*, 144–156. [CrossRef]
28. Kamal-Chaoui, L. The Implementation of the Korean Green Growth Strategy in Urban Areas. In *OECD Regional Development Working Papers 2011/02*; OECD Publishing: Paris, France, 2011.
29. OECD. Compact City Policies: Korea Towards Sustainable and Inclusive Growth. In *OECD Green Growth Studies*; OECD Publishing: Paris, France, 2014.
30. Peng, H.; Bohong, Z.; Qinpei, K. Smart City Environmental Pollution Prevention and Control Design based on Internet of Things. IOP Conference Series. *Earth Environ. Sci.* **2017**, *94*, 012174.
31. Thomson, G.; Newman, P. Urban fabrics and urban metabolism—From sustainable to regenerative cities. *Resour. Conserv. Recycl.* **2018**, *132*, 218–229. [CrossRef]
32. Kumar, V.T.M.; Dahiya, B. Smart Economy in Smart Cities. In *Smart Economy in Smart Cities*, 1st ed.; Springer Singapore: Singapore, 2017; Chapter 1.
33. Batty, M. Big data, smart cities and city planning. *Dialogues Hum. Geogr.* **2013**, *2*, 274–279. [CrossRef]
34. Bibri, S.E. A foundational framework for smart sustainable city development: Theoretical, disciplinary, and discursive dimensions and their synergies. *Sustain. Cities Soc.* **2018**, *38*, 758–794. [CrossRef]
35. Bibri, S.E.; Krogstie, J. Smart Eco-City Strategies and Solutions for Sustainability: The Cases of Royal Seaport, Stockholm, and Western Harbor, Malmö, Sweden. *Urban Sci.* **2020**, *4*, 11. [CrossRef]
36. Young, G.W.; Kitchin, R.; Naji, J. Building City Dashboards for Different Types of Users. *J. Urban Technol.* **2020**, 1–21. [CrossRef]
37. Batty, M. Smart Cities, Big Data. *Environ. Plan. B Plan. Des.* **2012**, *39*, 191–193. [CrossRef]
38. Postránecký, M.; Svítek, M. *Smart City Near to 4.0—An Adoption of Industry 4.0 Conceptual Model*; Smart City Symposium Prague (SCSP): Prague, Czech Republic, 2017; pp. 1–5.
39. Wirtz, B.W.; Weyerer, J.C.; Schichtel, F.T. An integrative public IoT framework for smart government. *Gov. Inf. Q.* **2018**, *36*, 333–345. [CrossRef]
40. Arkian, H.; Diyanat, A.; Pourkhalili, A. MIST: Fog-based Data Analytics Scheme with Cost-Efficient Resource Provisioning for IoT Crowdsensing Applications. *J. Netw. Comput. Appl.* **2017**, *82*, 152–165. [CrossRef]
41. Goncalves, R.J.; Sgurev, V.; Jotsov, V.; Kacprzyk, J. *Intelligent Systems: Theory, Research and Innovation in Applications*; Springer International Publishing: New York, NY, USA, 2020; p. 864.
42. Tekouabou, S.C.K.; Alaoui, E.A.A.; Cherif, W.; Silkan, H. Improving parking availability prediction in smart cities with IoT and ensemble-based model. *J. King Saud Univ. Comput. Inf. Sci.* **2020**. [CrossRef]
43. Liu, W.; Xu, Z. Some practical constraints and solutions for optical camera communication. *Philos. Trans. A Math. Phys. Eng. Sci.* **2020**, *378*, 20190191. [CrossRef] [PubMed]
44. Bonomi, F.; Milito, R.; Zhu, J.; Addepalli, S. Fog Computing and its Role in the Internet of Things. In Proceedings of the First Edition of the MCC Workshop on Mobile Cloud Computing (MCC'12), Helsinki, Finland, 17 August 2012; Association for Computing Machinery: New York, NY, USA, 2012; pp. 13–16.
45. Uhlemann, T.H.-J.; Lehmann, C.; Steinhilper, R. The Digital Twin: Realizing the Cyber-Physical production System for Industry 4.0. *Procedia CIRP* **2017**, *61*, 335–340. [CrossRef]
46. Rathore, M.M.U.; Paul, A.; Hong, W.H. Exploiting IoT and Big Data Analytics: Defining Smart Digital City using Real-Time Urban Data. *Sustain. Cities Soc.* **2017**, *40*, 600–610. [CrossRef]
47. Al Nuaimi, E.; Al Neyadi, H.; Mohamed, N.; Al-Jaroodi, J. Applications of big data to smart cities. *J. Internet Serv. Appl.* **2015**, *6*, 25. [CrossRef]
48. Hashem, I.A.T.; Chang, V.; Anuar, N.B. The Role of Big Data in Smart City. *Int. J. Inf. Manag.* **2016**, *36*, 748–758. [CrossRef]
49. Bibri, S.E. Big Data Science and Analytics for Smart Sustainable Urbanism. In *Unprecedented Paradigmatic Shifts and Practical Advancements*; Springer: Berlin, Germany, 2019.

50. Kramers, A.; Höjer, M.; Lovehagen, N.; Wangel, J. Smart sustainable cities—Exploring ICT solutions for reduced energy use in cities. *Environ. Model. Softw.* **2014**, *56*, 52–62. [CrossRef]
51. Palvia, P.; Baqir, N.; Nemati, H. ICT for socio-economic development: A citizens' perspective. *Inf. Manag.* **2017**, *55*, 160–176. [CrossRef]
52. Baucas, M.J.; Spachos, P. Using cloud and fog computing for large scale IoT-based urban sound classification. *Simul. Model. Pract. Theory* **2020**, *101*, 102013. [CrossRef]
53. Witkowski, K. Internet of Things, Big Data, Industry 4.0—Innovative Solutions in Logistics and Supply Chains Management. *Procedia Eng.* **2017**, *182*, 763–769. [CrossRef]
54. Daissaoui, A.; Boulmakoul, A.; Karim, L.; Lbath, A. IoT and Big Data Analytics for Smart Buildings: A Survey. *Procedia Comput. Sci.* **2020**, *170*, 161–168. [CrossRef]
55. Ullah, Z.; Al-Turjman, F.; Mostarda, L. Cognition in UAV-aided 5G and beyond communications: A survey. *IEEE Trans. Cogn. Commun. Netw.* **2020**. [CrossRef]
56. Yigitcanlar, T.; Desouza, K.C.; Butler, L.; Roozkhosh, F. Contributions and risks of artificial intelligence (AI) in building smarter cities: Insights from a systematic review of the literature. *Energies* **2020**, *13*, 1473. [CrossRef]
57. Yigitcanlar, T.; Butler, L.; Windle, E.; Desouza, K.C.; Mehmood, R.; Corchado, J.M. Can Building "Artificially Intelligent Cities" Safeguard Humanity from Natural Disasters, Pandemics, and Other Catastrophes? An Urban Scholar's Perspective. *Sensors* **2020**, *20*, 2988. [CrossRef] [PubMed]
58. Faisal, A.; Yigitcanlar, T.; Kamruzzaman, M.; Paz, A. Mapping two decades of autonomous vehicle research: A systematic scientometric analysis. *J. Urban Technol.* **2020**, in press. [CrossRef]
59. Eicker, U.; Weiler, V.; Schumacher, J.; Braun, R. On the design of an urban data and modelling platform and its application to urban district analyses. *Energy Build.* **2020**, *217*, 109954. [CrossRef]
60. Caprotti, F.; Liu, D. Emerging platform urbanism in China: Reconfigurations of data, citizenship and materialities. *Technol. Forecast. Soc. Chang.* **2020**, *151*. [CrossRef]
61. Bissell, D. Affective platform urbanism: Changing habits of digital on-demand consumption. *Geoforum* **2020**. [CrossRef]
62. Young, G.W.; Kitchin, R. Creating design guidelines for building city dashboards from a user's perspectives. *Int. J. Hum. Comput. Stud.* **2020**, *140*, 102429. [CrossRef]
63. Batty, M. A perspective on city dashboards. *Reg. Stud. Reg. Sci.* **2015**, *2*, 29–32. [CrossRef]
64. Young, G.W.; Naji, J.; Charlton, M.; Brunsdon, C.; Kitchin, R. Future cities and multimodalities: How multimodal technologies can improve smart-citizen engagement with city dashboards. In Proceedings of the Institute of Sustainable Urbanism Talks #05: Future Cities, Braunschweig, Germany, 14 November 2017.
65. International Data Corporation (IDC). *Worldwide Smart Cities Spending Guide*; International Data Corporation: Framingham, MA, USA, 2020.
66. Lane, N.D.; Eisenman, S.B.; Musolesi, M.; Miluzzo, E.; Campbell, A.T. Urban sensing: Opportunistic or participatory? In Proceedings of the 9th Workshop on Mobile Computing Systems and Applications, Napa Valley, CA, USA, 25–26 February 2008; HotMobile: Hevel Modi'in Regional Council, Israel, 2008; pp. 11–16.
67. Perng, S.Y.; Kitchin, R.; Donncha, D.M. Hackathons, entrepreneurial life and the making of smart cities. *Geoforum* **2018**, *97*, 189–197. [CrossRef]
68. Kankanamge, N.; Yigitcanlar, T.; Goonetilleke, A.; Kamruzzaman, M. Determining disaster severity through social media analysis: Testing the methodology with South East Queensland Flood Tweets. *Int. J. Disaster Risk Reduct.* **2020**, *42*, 101360. [CrossRef]
69. Piro, G.; Cianci, I.; Grieco, L.A.; Boggia, G.; Camarda, P. Information centric services in Smart Cities. *J. Syst. Softw.* **2013**, *88*, 169–188. [CrossRef]
70. Sharifi, A. A typology of smart city assessment tools and indicator sets. *Sustain. Cities Soc.* **2020**, *53*, 101936. [CrossRef]
71. United Nations (UN). *Transforming Our World: The 2030 Agenda for Sustainable Development*; UN General Assembly: New York, NY, USA, 2015.
72. Gabrys, J. Programming Environments: Environmentality and Citizen Sensing in the Smart City. *Environ. Plan. D Soc. Space* **2014**, *32*, 30–48. [CrossRef]
73. Marsal-Llacuna, L.; Segal, M.E. The Intelligenter method (I) for making "smarter" city projects and plans. *Cities* **2016**, *55*, 127–138. [CrossRef]

74. Yigitcanlar, T. *Sustainable Urban and Regional Infrastructure Development: Technologies, Applications and Management*; IGI Global: Hersey, PA, USA, 2010.

75. International Institute for Management Development (IMD). *IMD Smart City Index 2019*; World Competitiveness Center: Lausanne, Switzerland, 2019.

76. Hilty, L.M.; Aebischer, B.; Rizzoli, A.E. Modeling and evaluating the sustainability of smart solutions. *Environ. Model. Softw.* **2014**, *56*, 1–5. [CrossRef]

77. Kamble, S.; Gunasekaran, A.; Gawankar, S. Sustainable Industry 4.0 framework: A systematic literature review identifying the current trends and future perspectives. *Process. Saf. Environ. Prot.* **2018**, *117*, 408–425. [CrossRef]

78. Ishida, T.; Isbister, K. *Digital Cities—Technologies, Experiences, and Future Perspectives*; Springer: Berlin/Heidelberg, Germany, 2000; p. 1765.

79. Stollmann, J. Digital Cities. In *The Wiley Blackwell Encyclopedia of Urban and Regional Studies*; Wiley Blackwell: Hoboken, NJ, USA, 2019.

80. Colding, J.; Colding, M.; Barthel, S. Applying seven resilience principles on the Vision of the Digital City. *Cities* **2020**, *103*. [CrossRef]

81. Talari, S.; Shafie-khah, M.; Siano, P.; Loia, V.; Tommasetti, A.; Catalao, J.P.S. A Review of Smart Cities Based on the Internet of Things Concept. *Energies* **2017**, *10*, 421. [CrossRef]

82. Yigitcanlar, T.; Kamruzzaman, M.D.; Buys, L.; Ioppolo, G.; Marques, J.; Da Costa, M.E.; Yun, J.J. Understanding 'smart cities': Intertwining development drivers with desired outcomes in a multidimensional framework. *Cities* **2018**, *81*, 145–160. [CrossRef]

83. Saborido, R.; Alba, E. Software systems from smart city vendors. *Cities* **2020**, *101*, 102690. [CrossRef]

84. Lee, S.H.; Yigitcanlar, T.; Han, J.H.; Leem, Y.T. Ubiquitous urban infrastructure: Infrastructure planning and development in Korea. *Innovation* **2018**, *10*, 282–292. [CrossRef]

85. Shin, D.H. Ubiquitous city: Urban technologies, urban infrastructure and urban informatics. *J. Inf. Sci.* **2009**, *35*, 515–526. [CrossRef]

86. Wang, J.; Hui, L.C.K.; Yiu, S.M.; Wang, E.K.; Fang, J. A survey on cyber-attacks against nonlinear state estimation in power systems of ubiquitous cities. *Pervasive Mob. Comput.* **2017**, *39*, 52–64. [CrossRef]

87. Pancholi, S.; Yigitcanlar, T.; Guaralda, M. Public space design of knowledge and innovation spaces: Learnings from Kelvin Grove Urban Village, Brisbane. *J. Open Innov. Technol. Mark. Complex.* **2015**, *1*, 13. [CrossRef]

88. Lopez-Ruiz, V.R.; Alfaro-Navarro, J.L.; Nevado-Pena, D. Knowledge-city index construction: An intellectual capital perspective. *Expert Syst. Appl.* **2014**, *41*, 5560–5572. [CrossRef]

89. Penco, L.; Ivaldi, E.; Bruzzi, C.; Musso, E. Knowledge-based urban environments and entrepreneurship: Inside EU cities. *Cities* **2020**, *96*, 102443. [CrossRef]

90. Komninos, N. Intelligent cities: Towards interactive and global innovation environments. *Int. J. Innov. Reg. Dev.* **2009**, *1*, 337–355. [CrossRef]

91. Liugailaitė-Radzvickienė, L.; Jucevicius, R. Going to be an Intelligent City. *Procedia Soc. Behav. Sci.* **2014**, *156*, 116–120. [CrossRef]

92. Willis, K.S.; Aurigi, A. *Digital and Smart Cities*; Routledge: New York, NY, USA, 2017.

93. Landry, C. *The Creative City—A Toolkit for Urban Innovators*; Earthscan: Sterling, VA, USA, 2008.

94. Baum, S.; O'Connor, K.; Yigitcanlar, T. The implications of creative industries for regional outcomes. *Int. J. Foresight Innov. Policy* **2009**, *5*, 44–64. [CrossRef]

95. Li, X.; Fong, P.S.W.; Dai, S.; Li, Y. Towards sustainable smart cities: An empirical comparative assessment and development pattern optimization in China. *J. Clean. Prod.* **2019**, *215*, 730–743. [CrossRef]

96. Sodiq, A.; Baloch, A.A.B.; Khan, S.A.; Sezer, N.; Mahmoud, S.; Jama, M.; Abdelaal, A. Towards Modern Sustainable Cities: Review of Sustainability principles and Trends. *J. Clean. Prod.* **2019**, *227*, 972–1001. [CrossRef]

97. Mainka, A.; Khveshchanka, S.; Stock, W.G. Dimensions of Informational City Research. In *Conference: Digital Cities 7—Real World Experiences*; State Library of Queensland: Brisbane, Australia, 2011.

98. Rutherford, J. *Informational City. International Encyclopedia of Human Geography*, 2nd ed.; Elsevier: Amsterdam, The Netherlands, 2020; pp. 315–320.

99. Bibri, S.E. The IoT for smart sustainable cities of the future: An analytical framework for sensor-based big data applications for environmental sustainability. *Sustain. Cities Soc.* **2018**, *38*, 230–253. [CrossRef]

100. Chang, D.L.; Marques, J.; da Costa, E.M.; Selig, P.M.; Yigitcanlar, T. Knowledge-based, smart and sustainable cities: A provocation for a conceptual framework. *J. Open Innov. Technol. Mark. Complex.* **2018**, *4*, 5. [CrossRef]

101. Akande, A.; Cabral, P.; Gomes, P.; Casteleyn, S. The Lisbon Ranking for Smart Sustainable Cities in Europe. *Sustain. Cities Soc.* **2018**, *44*, 475–487. [CrossRef]

102. Huovila, A.; Bosch, P.; Airaksinen, M. Comparative analysis of standardized indicators for Smart sustainable cities: What indicators and standards to use and when? *Cities* **2019**, *89*, 141–153. [CrossRef]

103. Ahvenniemi, H.; Huovila, A.; Pinto-Seppä, I.; Airaksinen, M. What are the differences between sustainable and smart cities? *Cities* **2017**, *60*, 234–245. [CrossRef]

104. Allam, Z.; Dhunny, Z.A. On big data, artificial intelligence and smart cities. *Cities* **2019**, *89*, 80–91. [CrossRef]

105. Yigitcanlar, T.; Han, H.; Kamruzzaman, M.; Ioppolo, G.; Sabatini-Marques, J. The making of smart cities: Are Songdo, Masdar, Amsterdam, San Francisco and Brisbane the best we could build? *Land Use Policy* **2019**, *88*, 104187. [CrossRef]

106. Höjer, M.; Wangel, J. Smart Sustainable Cities: Definition and Challenges. In *ICT Innovations for Sustainability, Advances in Intelligent Systems and Computing, 310*; Hilty, L.M., Aebischer, B., Eds.; Springer: Cham, Switzerland, 2015; pp. 333–349.

107. Yigitcanlar, T.; Kamruzzaman, M.D.; Foth, M.; Marques, J.; da Costa, E.; Ioppolo, G. Can cities become smart without being sustainable? A systematic review of the literature. *Sustain. Cities Soc.* **2018**, *45*, 348–365. [CrossRef]

108. Yigitcanlar, T.; Foth, M.; Kamruzzaman, M. Towards post-anthropocentric cities: Reconceptualizing smart cities to evade urban ecocide. *J. Urban Technol.* **2019**, *26*, 147–152. [CrossRef]

109. Bibri, S.E.; Krogstie, J. On the Social Shaping Dimensions of Smart Sustainable Cities: A Study in Science, Technology, and Society. *Sustain. Cities Soc.* **2017**, *29*, 219–246. [CrossRef]

110. Hancke, G.P.; Silva, B.; Hancke, G.P., Jr. The Role of Advanced Sensing in Smart Cities. *Sensors* **2012**, *13*, 393–425. [CrossRef] [PubMed]

111. Yigitcanlar, T.; Kankanamge, N.; Vella, K. How are the smart city concepts and technologies perceived and utilized? A systematic geo-twitter analysis of smart cities in Australia. *J. Urban Technol.* **2020**, 1–20. [CrossRef]

112. Hasegawa, Y.; Sekimoto, Y.; Seto, T.; Fukushima, Y.; Maeda, M. My City Forecast: Urban planning communication tool for citizen with national open data. *Comput. Environ. Urban Syst.* **2019**, *77*. [CrossRef]

113. Yamagata, Y.; Yang, P.P.J.; Chang, S.; Tobey, M.B.; Binder, R.B.; Fourie, P.J.; Jittrapirom, P.; Kobashi, T.; Yoshida, T.; Aleksejeva, J. *Urban systems and the role of big data*; Urban Systems Design: Bermondsey, UK; London, UK, 2020.

114. Few, S. *Information Dashboard Design: The Effective Visual Communication of Data*; O'Reilly Media: Newton, MA, USA, 2006.

115. Kitchin, R. The real time city? Big data and smart urbanis. *GeoJournal* **2014**, *79*, 1–14. [CrossRef]

116. Kitchin, R.; McArdle, G. Urban Data and City Dashboards: Six Key Issues. In *Data and the City*; Routledge: London, UK, 2016; pp. 1–21.

117. Pettit, C.; Leao, S.Z. Dashboard. In *Encyclopedia of Big Data*; Schintler, L.A., McNeely, C.L., Eds.; Springer: Cham, Switzerland, 2017.

118. Karvonen, A.; Cugurullo, F.; Caprotti, F. Introduction: Situating Smart Cities. In *Inside Smart Cities: Place, Politics and Urban Innovation*; Karvonen, A., Cugurullo, F., Caprotti, F., Eds.; Routledge: London, UK, 2019; pp. 1–12.

119. Sharma, A.; Singh, P.K.; Kumar, Y. An Integrated Fire Detection System using IoT and Image processing Technique for Smart Cities. *Sustain. Cities Soc.* **2020**, *61*, 102332. [CrossRef]

120. Mocnej, J.; Pekar, A.; Seah, W.K.G.; Papcun, P.; Kajati, E.; Cupkova, D.; Koziorek, J.; Zolotova, I. Quality-enabled decentralized IoT architecture with efficient resources utilization. *Robot. Comput. Integr. Manuf.* **2021**, *67*, 102001. [CrossRef]

121. Castell, N.; Dauge, F.R.; Schneider, P.; Vogt, M.; Lerner, U.; Fishbain, B.; Broday, D.; Bartonova, A. Can commercial low-cost sensor platforms contribute to air quality monitoring and exposure estimates? *Environ. Int.* **2017**, *99*, 293–302. [CrossRef]

122. Yigitcanlar, T.; Velibeyoglu, K.; Baum, S. *Creative Urban Regions: Harnessing Urban Technologies to Support Knowledge City Initiatives*; IGI Global: Hersey, PA, USA, 2008.

123. Ioppolo, G.; Cucurachi, S.; Salomone, R.; Saija, G.; Shi, L. Sustainable Local Development and Environmental Governance: A Strategic Planning Experience. *Sustainability* **2016**, *8*, 180. [CrossRef]
124. Wijs, L.; Witte, P.; Geertman, S. How smart is smart? Theoretical and empirical considerations on implementing smart city objectives—A case study of Dutch railway station areas. *Eur. J. Soc. Sci. Res.* **2016**, *29*, 424–441. [CrossRef]
125. Gil, O.; Cortés-Cediel, M.E.; Cantador, I. Citizen participation and the rise of digital media platforms in smart governance and smart cities. *Int. J. E Plan. Res.* **2019**, *8*, 19–34. [CrossRef]
126. Ruhlandt, R.W.S. The governance of smart cities: A systematic literature review. *Cities* **2018**, *81*, 1–23. [CrossRef]
127. Dowling, R.; McGuirk, P.; Gillon, C. Strategic or Piecemeal? Smart City Initiatives in Sydney and Melbourne. *Urban Policy Res.* **2019**, *37*, 429–441. [CrossRef]
128. Cugurullo, F. Exposing smart cities and eco-cities: Frankenstein urbanism and the sustainability challenges of the experimental city. *Environ. Plan. A Econ. Space* **2018**, *50*, 73–92. [CrossRef]
129. European Commission (EC). Available online: www.ec.europa.eu (accessed on 6 June 2020).
130. Johnson, P.A.; Robinson, P.J.; Philpot, S. Type, tweet, tap, and pass: How smart city technology is creating a transactional citizen. *Gov. Inf. Q.* **2020**, *37*, 101414. [CrossRef]
131. Yeh, H. The effects of successful ICT-based smart city services: From citizens' perspectives. *Gov. Inf. Q.* **2017**, *34*, 556–565. [CrossRef]
132. Axelsson, K.; Granath, M. Stakeholders' stake and relation to smartness in smart city development: Insights from a Swedish city planning project. *Gov. Inf. Q.* **2018**, *35*, 693–702. [CrossRef]
133. Cardullo, P.; Kitchin, R. Being a 'citizen' in the smart city: Up and down the scaffold of smart citizen participation in Dublin, Ireland. *GeoJournal* **2019**, *84*, 1–13. [CrossRef]
134. Ahad, M.A.; Paiva, S.; Tripathi, G.; Feroz, N. Enabling technologies and sustainable smart cities. *Sustain. Cities Soc.* **2020**, *61*, 102301. [CrossRef]
135. Shahidehpour, M.; Li, Z.; Ganji, M. Smart cities for a sustainable urbanization: Illuminating the need for establishing smart urban infrastructures. *IEEE Electrif. Mag.* **2018**, *6*, 16–33. [CrossRef]
136. De, M.; Sikarwar, S.; Kumar, V. Strategies for Inducing Intelligent Technologies to Enhance Last Mile Connectivity for Smart Mobility in Indian cities. In *Progress in Advanced Computing and Intelligent Engineering*; Springer: Singapore, 2019; pp. 373–384. [CrossRef]
137. Cugurullo, F.; Acheampong, R.A.; Gueriau, M.; Dusparic, I. The transition to autonomous cars, the redesign of cities and the future of urban sustainability. *Urban Geogr.* **2020**, 1–27. [CrossRef]
138. Acheampong, R.A.; Cugurullo, F. Capturing the behavioural determinants behind the adoption of autonomous vehicles: Conceptual frameworks and measurement models to predict public transport, sharing and ownership trends of self-driving cars. *Transp. Res. Part F Traffic Psychol. Behav.* **2019**, *62*, 349–375. [CrossRef]
139. Syed, L.; Jabeen, S.; Manimala, S.; Alsaeedi, A. Smart healthcare framework for ambient assisted living using IoMT and big data analytics techniques. *Future Gener. Comput. Syst.* **2019**, *101*, 136–151. [CrossRef]
140. Huawei. Available online: www.e.huawei.com (accessed on 6 June 2020).
141. Huawei Digital Platform. Available online: https://e.huawei.com/en/digital-platform/smart-city (accessed on 6 June 2020).
142. Cisco Kinetic for Cities. Available online: www.blogs.cisco.com (accessed on 6 June 2020).
143. The Cisco Smart + Connected Digital Platform. Available online: www.cisco.com (accessed on 6 June 2020).
144. Zhang, J.; Hua, X.S.; Huang, J.; Shen, X.; Chen, X.; Zhou, Q.; Fu, Z.; Zhao, Y. City Brain: Practice of Large-Scale Artificial Intelligence in the Real World. In *IET Smart Cities*; The Institution of Engineering and Technology: Beijing, China, 2019.
145. Alibaba City Event Detection and Smart Processing. Available online: www.alibabacloud.com (accessed on 6 June 2020).
146. Smart Citizen Kit. Available online: www.smartcitizen.me (accessed on 6 June 2020).
147. Camprodon, G.; Gonzalez, O.; Barberan, V.; Pérez, M.; Smari, V.; de Heras, M.A.; Bizzotto, A. Smart Citizen Kit and Station: An open environmental monitoring system for citizen participation and scientific experimentation. *HardwareX* **2019**, *6*, e00070. [CrossRef]
148. Gutiérrez Bayo, J. *International Case Studies of Smart Cities: Santander, Spain*; Inter-American Development Bank: Washington, DC, USA, 2016.

149. Smart City Dashboard Developed by Deutsche Telekom for Hungary Subsidiary. Available online: www.t-system.hu (accessed on 7 June 2020).
150. Deutsche Telekom City of Hamburg. Available online: www.telekom.com (accessed on 7 June 2020).
151. Smart Street Lighting Models Developed by Deutsche Telekom. Available online: www.deutschetelekom.com (accessed on 7 June 2020).
152. LuxTurrim5G Smart Pole. Available online: www.luxturrim5g.com (accessed on 8 June 2020).
153. Alvarez, O.; Markendahl, J.; Martinez, L. Quality of Experience (QoE)—based service differentiation in the smart cities context: Business Analysis. In Proceedings of the ICCS 2015—International Conference on City Sciences. New Architectures, Infrastructures and Services for Future Cities, Tongji University, Shanghai, China, 4–5 June 2015.
154. Allen, A.; Lampis, A.; Swilling, M. *Untamed Urbanisms, Routledge Advances in Regional Economics, Science and Policy*; Taylor & Francis: Abingdon, UK, 2016.
155. Hollands, R.G. Critical interventions into the corporate smart city. *Camb. J. Reg. Econ. Soc.* **2015**, *8*, 61–77. [CrossRef]
156. Shahrokni, H.; Lazarevic, D.; Brandt, N. Smart urban metabolism: Towards a real-time understanding of the energy and material flows of a city and its citizens. *J. Urban Technol.* **2015**, *22*, 65–86. [CrossRef]
157. Bifulco, F.; Tregua, M.; Amitrano, C.C.; D'Auria, A. ICT and sustainability in smart cities management. *Int. J. Public Sect. Manag.* **2016**, *29*, 132–147. [CrossRef]
158. Trilles, S.; Calia, A.; Belmonte, O.; Torres-Sospedra, J.; Montoliu, R.; Huerta, J. Deployment of an open sensorized platform in a smart city context. *Future Gener. Comput. Syst.* **2017**, *76*, 221–233. [CrossRef]
159. Jain, B.; Brar, G.; Malhotra, J.; Rani, S. A novel approach for smart cities in convergence to wireless sensor networks. *Sustain. Cities Soc.* **2017**, *35*, 440–448. [CrossRef]
160. Li, A.; Shahidehpour, M. Deployment of cybersecurity for managing traffic efficiency and safety in smart cities. *Electr. J.* **2017**, *30*, 52–61. [CrossRef]
161. Chatfield, A.T.; Reddick, C.G. A framework for Internet of Things-enabled smart government: A case of IoT cybersecurity policies and use cases in U.S. *Gov. Inf. Q.* **2019**, *36*, 346–357. [CrossRef]
162. Habibzadeh, H.; Nussbaum, B.H.; Anjomshoa, F.; Kantarci, B.; Soyata, T. A survey on cybersecurity, data privacy, and policy issues in cyber-physical system deployments in smart cities. *Sustain. Cities Soc.* **2019**, *50*, 101660. [CrossRef]
163. Certomà, C.; Corsini, F.; Frey, M. Hyperconnected, receptive and do-it-yourself city. An investigation into the European "imaginary" of crowdsourcing for urban governance. *Technol. Soc.* **2020**, *61*, 101229. [CrossRef]
164. De Guimarães, J.C.F.; Severo, E.A.; Junior, L.A.F.; Da Costa, W.P.L.B.; Salmonia, F.T. Governance and quality of life in smart cities: Towards sustainable development goals. *J. Clean. Prod.* **2020**, *253*, 119926. [CrossRef]
165. Wu, W.N. Determinants of citizen-generated data in a smart city: Analysis of 311 system user behavior. *Sustain. Cities Soc.* **2020**, *59*, 102167. [CrossRef]
166. Puiu, D.; Barnaghi, P.; Tonjes, R.; Kumper, D.; Ali, M.I.; Mileo, A.; Parreira, J.X.; Fischer, M.; Kolozali, S.; Farajidavar, N. Citypulse: Large scale data analytics framework for smart cities. *IEEE Access* **2016**, *4*, 1086–1108. [CrossRef]
167. CityPulse. Available online: www.ict-citypulse.eu/page/ (accessed on 8 June 2020).
168. Jiang, D. The construction of smart city information system based on the Internet of Things and cloud computing. *Comput. Commun.* **2020**, *150*, 158–166. [CrossRef]
169. Babar, M.; Arif, F.; Jan, M.A.; Tan, Z.; Khan, F. Urban data management system: Towards Big Data analytics for Internet of Things based smart urban environment using customized Hadoop. *Future Gener. Comput. Syst.* **2019**, *96*, 398–409. [CrossRef]
170. Engin, Z.; van Dijk, J.; Lan, T.; Longley, P.A.; Treleaven, P.; Batty, M.; Penn, A. Data-driven urban management: Mapping the landscape. *J. Urban Manag.* **2020**, *9*, 140–150. [CrossRef]
171. Yigitcanlar, T. *Rethinking Sustainable Development: Urban Management, Engineering, and Design*; IGI Global: Hersey, PA, USA, 2010.
172. Sadgali, I.; Sael, N.; Benabbou, F. Detection of credit card fraud: State of art. *Int. J. Comput. Sci. Netw. Secur.* **2018**, *18*, 76–83.
173. Borrion, H.; Ekblom, P.; Alrajeh, D.; Borrion, A.L.; Keane, A.; Koch, D.; Toubaline, S. The Problem with Crime Problem-Solving: Towards a Second Generation Pop? *Br. J. Criminol.* **2019**, *60*, 219–240. [CrossRef]

174. Laufs, J.; Borrion, H.; Bradford, B. Security and the smart city: A systematic review. *Sustain. Cities Soc.* **2020**, *55*, 102023. [CrossRef]

175. Cagliero, L.; Cerquitelli, T.; Chiusano, S.; Garino, P.; Nardone, M.; Pralio, B.; Venturini, L. Monitoring the citizens' perception on urban security in smart City environments. In Proceedings of the DAta Mining And Smart Cities Applications Workshop 2015 Co-Located with the 31st IEEE International Conference on Data Engineering (ICDE 2015), Seoul, Korea, 13–17 April 2015; pp. 112–116.

176. Rothkrantz, L. *Person Identification by Smart Cameras*; 2017 Smart City Symposium Prague (SCSP): Prague, Czech Republic, 2017; pp. 1–6.

177. Sajjad, M.; Nasir, M.; Muhammad, K.; Khan, S.; Jan, Z.; Sangaiah, A.K.; Baik, S.W. Raspberry Pi assisted face recognition framework for enhanced law-enforcement services in smart cities. *Future Gener. Comput. Syst.* **2020**, *108*, 995–1007. [CrossRef]

178. Parra, J.; Lopez, R. Application of Predictive Analytics for Crime Prevention: The Case of the City of San Francisco. In *Police: Global Perceptions, Performance and Ethical Challenges*; Nova Science Publishers, Inc.: New York, NY, USA, 2017; pp. 85–109.

179. Calzada, I. (Smart) citizens from data providers to decision-makers? The case study of Barcelona. *Sustainability* **2018**, *10*, 3252. [CrossRef]

180. Carter, D.M. *Cyberspace and Cyberculture. International Encyclopedia of Human Geography*, 2nd ed.; Elsevier: Amsterdam, The Netherlands, 2020; pp. 143–147.

181. Kumar, H.; Singh, M.K.; Gupta, M.P.; Madaan, J. Moving towards smart cities: Solutions that lead to the Smart City Transformation Framework. *Technol. Forecast. Soc. Chang.* **2020**, *153*, 119281. [CrossRef]

182. Sidewalk Labs. Available online: www.sidewalklabs.com (accessed on 11 June 2020).

183. SmartROAD. Available online: www.hamburg-port-authority.de/en/hpa-360/smartport/ (accessed on 11 June 2020).

184. Gelsenkirchen: A Small, Smart City with Big Plans. Available online: e.huawei.com/en/case-studies/global/2017/201709071445 (accessed on 11 June 2020).

185. Lighthouse City. Available online: www.iotjournal.com (accessed on 11 June 2020).

186. Smart+Connected Communities. Available online: www.newsroom.cisco.com (accessed on 11 June 2020).

187. Ipsos. (2019). Smarter Digital City 3.0. Report commissioned by Google. Available online: https://services.google.com/fh/files/misc/google_smarter_digital_city_3_whitepaper.pdf (accessed on 2 July 2020).

188. T-City Friedrichshafen. Available online: www.t-city.de/en (accessed on 11 June 2020).

189. Charlotte Sustainability City. Available online: www.americaninno.com (accessed on 12 June 2020).

190. *Smarter Cities Challenge Report*; IBM: Dublin, Ireland, 2015.

191. Coletta, C.; Heaphy, L.; Kitchin, R. From the accidental to articulated smart city: The creation and work of 'Smart Dublin'. *Eur. Urban Reg. Stud.* **2019**, *26*, 349–364. [CrossRef]

192. Aspern Smart City Research. Available online: www.ascr.at (accessed on 12 June 2020).

193. Stockholm Royal Seaport's Smart Energy City Project. Available online: www.ericsson.com (accessed on 12 June 2020).

194. Factory of the Future. Available online: www.ericsson.com (accessed on 12 June 2020).

195. Espoo Innovation Garden. Available online: www.espooinnovationgarden.fi/en/espoo (accessed on 24 July 2020).

196. Centre of Excellence. Available online: www.blogs.oracle.com (accessed on 15 June 2020).

197. Bari Matera 5G. Available online: www.barimatera5g.it (accessed on 15 June 2020).

198. Green Energy Industrial Complex. Available online: www.tecnologiaericerca.com (accessed on 15 June 2020).

199. Hu, R. The State of Smart Cities in China: The Case of Shenzhen. *Energies* **2019**, *12*, 4375. [CrossRef]

200. Guiyang Sunac City. Available online: www.ekistics.com (accessed on 16 June 2020).

201. Valencia Smart City Platform. Available online: www.smartcity.valencia.es (accessed on 16 June 2020).

sensors

Article

A Three-Dimensional Microstructure Reconstruction Framework for Permeable Pavement Analysis Based on 3D-IWGAN with Enhanced Gradient Penalty

Ludia Eka Feri [1], Jaehun Ahn [2], Shahrullohon Lutfillohonov [3] and Joonho Kwon [3,*

[1] Department of Big Data, Pusan National University, Busan 46241, Korea; ludia.ef@gmail.com
[2] Department of Civil and Environmental Engineering, Pusan National University, Busan 46241, Korea; jahn@pusan.ac.kr
[3] School of Computer Science and Engineering, Pusan National University, Busan 46241, Korea; shahrullo@pusan.ac.kr
* Correspondence: jhkwon@pusan.ac.kr; Tel.: +82-51-510-3149

Abstract: Owing to the increasing use of permeable pavement, there is a growing need for studies that can improve its design and durability. One of the most important factors that can reduce the functionality of permeable pavement is the clogging issue. Field experiments for investigating the clogging potential are relatively expensive owing to the high-cost testing equipment and materials. Moreover, a lot of time is required for conducting real physical experiments to obtain physical properties for permeable pavement. In this paper, to overcome these limitations, we propose a three-dimensional microstructure reconstruction framework based on 3D-IDWGAN with an enhanced gradient penalty, which is an image-based computational system for clogging analysis in permeable pavement. Our proposed system first takes a two-dimensional image as an input and extracts latent features from the 2D image. Then, it generates a 3D microstructure image through the generative adversarial network part of our model with the enhanced gradient penalty. For checking the effectiveness of our system, we utilize the reconstructed 3D image combined with the numerical method for pavement microstructure analysis. Our results show improvements in the three-dimensional image generation of the microstructure, compared with other generative adversarial network methods, and the values of physical properties extracted from our model are similar to those obtained via real pavement samples.

Keywords: 3D microstructure reconstruction; permeable pavement; deep learning; generative adversarial networks

Citation: Feri, L.E.; Ahn J.; Lutfillohonov, S.; Kwon, J. A Three-Dimensional Microstructure Reconstruction Framework for Permeable Pavement Analysis Based on 3D-IWGAN with Enhanced Gradient Penalty. *Sensors* **2021**, *21*, 3603. https://doi.org/10.3390/s21113603

Academic Editors: Rashid Mehmood, Juan M. Corchado and Tan Yigitcanlar

Received: 28 April 2021
Accepted: 20 May 2021
Published: 21 May 2021

Publisher's Note: MDPI stays neutral with regard to jurisdictional claims in published maps and institutional affiliations.

1. Introduction

In recent years, permeable pavement has been widely used in developed countries as part of low-impact development practices [1]. The main benefit of permeable pavement is that it can allow rainwater to pass through its pores into the ground below by filtering rainwater over the distinct layers, thereby reducing the excessive volume of waterlogging. The runoff volumes and discharge rates can be reduced from paved surfaces and the high risk of downstream flooding can be decreased. Additionally, permeable pavement contributes water quality enhancement. It can trap stormwater pollutants and prevent them from reaching downstream receiving waters [2]. Other benefits of using this special pavement are skid resistance, noise control, and surface temperature reduction [3].

However, there are some drawbacks to using permeable pavement. First, it is expensive owing to its special design. Permeable pavement consists of aggregate components with different materials such as concrete mixtures with cement, polymer, and plastic [4]. Second, it has a high maintenance cost, especially for cleaning the pavement. An industrial vacuum is needed to remove the particles that can block the spaces in the pavement. If it is not maintained properly, the water and pollutants can run off the surface, which can cause flooding. Finally, it is prone to the clogging problem that is closely related to the maintenance issue. Clogging is the reduction in porosity and permeability when fine particles, sand, or clay block the spaces or pores between the pavers. The quality of permeable pavement can be measured using the clogging potential value. Clogging potential is the ratio of porosity or permeability reduction because of clogging to the initial porosity or permeability in the unclogged state [5].

Several experimental studies related to clogging in permeable pavement have been reported by researchers over the years to develop an ideal pavement design [6–9]. A good permeable pavement should have a small clogging potential value. According to the researchers, we need to understand the effects of different pore structures to yield a permeable pavement that is less susceptible to clogging [5].

To analyze the clogging potential in permeable pavements, some important properties of pavement microstructure need to be measured. Microstructure is the small-scale structure of a material that can be viewed through a microscope. The properties of microstructure are very important for understanding the material structure and studying the mechanical behavior of a material. In permeable pavement, three crucial properties for analysis are (1) porosity, (2) permeability, and (3) hydraulic conductivity [5,10]. Porosity is a fraction of the volume of pores over the total volume of a material. It is considered as a most important property since it affects the flow rates in a permeable pavement. Permeability is defined as the ability of a porous material to transmit fluids, and hydraulic conductivity is the ease with which a fluid (water) can move through porous surfaces [11]. Along with these properties, some morphological aspects also need to be carefully measured to study the clogging phenomenon in a porous medium like the permeable pavement [12].

Traditionally, to measure the clogging potential properties, we needed to conduct field experiments [6,13] using real physical materials. These are high in cost because the experiments require several samples and instruments that must be installed in the real field. The setting of a test area is another aspect that can be expensive, especially for conducting a large-scale field experiment. It is also time consuming, especially in large-scale experiments, owing to the equipment setup, material selection, and test-area preparation, which involves an intricate process. Thus, a more efficient method to measure the clogging potential properties is needed. Alternatively, they can be measured through visual experiments with image analysis and computer simulation.

A virtual experiment is another alternative to estimate the clogging potential properties [14,15]. It is more efficient in comparison to field experiments because it uses fewer samples and equipment for the experiment, and the required time is also not high as required in a field experiment. For this type of experiment, the most important aspect is the preparation of a good three-dimensional model for computer simulation. Before the upsurge of deep learning in image processing, researchers used statistical methods and some basic machine learning techniques such as support vector machines and the genetic algorithm [16,17] to reconstruct a 3D microstructure. Another method uses stochastic models such as a Gaussian random field [18] combined with hybrid optimization that generates a 3D porous material structure using a two-point correlation function and a cluster correlation function. The most common method for 3D reconstruction is the two-point correlation method [19–21]. It characterizes the microstructure based on certain statistical features and then performs an optimization process to build a 3D structure that matches those statistical features.

However, there are some limitations to the basic machine learning techniques and statistical approach. They require predefined knowledge of the materials [19], many different samples, and the generated model sometimes does not resemble the real sample. In the statistical case, there is a possibility that two materials with different properties may share first- and second-order statistical information and chord-distribution functions. Therefore, it has failed to yield good estimates of the macroscopic properties [20]. In brief, previous approaches did not yield realistic 3D microstructure reconstruction due to lack of (1) many different samples and (2) a good generative model.

To overcome these limitations, we propose a deep-learning-based 3D microstructure reconstruction framework for permeable pavement analysis as a virtual experiment method. Our system first takes a two-dimensional image to extract the latent features of the image. Then, it generates the corresponding 3D microstructure image that can be obtained through the concept of the generative adversarial network [22,23]. Since it is hard to achieve stability in the original generative adversarial network during the training process, we decided to implement a 3D improved Wasserstein generative adversarial network (3D-IWGAN) [24,25] with an enhanced gradient penalty. One of the main advantages of utilizing this generative adversarial network is that it can produce sharper, even degenerate distributions, while other 3D GAN-based methods show instability in generated results. Thus, compared to other methods [22], our framework achieves high profit from the enhanced gradient penalty and keeps the stability during the training process.

The main contributions of this paper are summarized as follows:

- We propose a deep-learning-based 3D microstructure reconstruction framework for permeable pavement analysis. Our proposed framework can effectively reconstruct a 3D microstructure image after taking only one 2D image. For this purpose, our framework applies several image preprocessing techniques to the input 2D images and extracts the latent features from them via the variational autoencoder (VAE) method.
- In comparison to other generative adversarial network methods, we obtained more realistic 3D microstructure images of porous pavements and stable outputs. This stability is the main result of using 3D-IWGAN with an enhanced gradient penalty.
- The accuracy of our results has been verified by evaluating some physical properties extracted from the generated 3D samples. On observing nine different output samples, the average error difference in hydraulic computation was found to be less than 5%.

The remainder of this paper is organized as follows. Section 2 provides a brief overview of related research work. Section 3 presents some preliminaries of GANs. In Section 4, the 3D model generation and analysis of the generated model are explained in detail. The experiment and evaluation of the experimental results are discussed in Section 5. Finally, the conclusion of our work and future work are presented in Section 6.

2. Related Work

In this section, we provide a brief review of the related work. There are three categories to be discussed: (1) microstructure reconstruction approaches, (2) general 2D/3D object reconstruction approaches, and (3) deep-learning-based microstructure reconstruction approaches.

2.1. Microstructure Reconstruction

Microstructure is the small-scale structure of a material that can be viewed through a microscope. In the fields of materials science and civil engineering, microstructure properties are very important to understand the structure of a material and study its mechanical behavior.

The most common and earlier work for 3D microstructure reconstruction uses the two-point correlation method [19–21]. It characterizes the microstructure based on certain statistical features and then performs an optimization process to build a 3D structure that matches those statistical features. After this, statistical methods and basic machine learning algorithms were applied for reconstruction of 3D microstructure. Some of the well-known machine learning methods include support vector machines (SVMs) [16] and the genetic algorithm [17]. Among others, stochastic methods have shown better results for the given tasks. For example, Jiang et al. [18] proposed a method that exploits a Gaussian random field with the combination of hybrid optimization. It uses a two-point correlation function and a cluster correlation function to generate a 3D porous material structure.

2.2. General 2D/3D Object Reconstruction

After the successful results shown by convolutional neural networks (CNNs) in the field of image classification and computer vision, researchers started considering using CNNs and other deep learning methods in their research on permeable pavement.

Since the core functionality of GAN enables us to generate a set of realistic (2D) images from the given sample, thus it is extended to generate 3D images (objects) [26]. Although GAN can accurately visualize the results, it suffers from a major drawback: the instability of GAN training. To guarantee the stability of GAN training, researchers proposed to leverage the Wasserstein distance (WGAN) [27] and to further improve training of the WGAN (IWGAN) [24].

Table 1 lists the different methods for object reconstruction (microstructure and non-microstructure). It shows that the deep learning method is more common in 2D object reconstruction than 3D object reconstruction. More specifically, there is still no application of GAN in pavement microstructure. Therefore, we attempt to fill this gap by designing a GAN for this case.

Table 1. 2D/3D object reconstruction methods with statistical and deep learning approaches.

Approach	Method	Application	Output
Statistics	Two-point Correlation	geometry reconstruction, analysis	2D/3D microstructure
Deep Learning	CDBN	2D microstructure reconstruction	2D microstructure
Deep Learning	3DGAN	3D object reconstruction	3D objects (non-microstructure)
Deep Learning	DCGAN	2D object reconstruction	2D microstructure
Deep Learning	IWGAN	2D/3D object reconstruction	2D/3D objects (non-microstructure)
Deep Learning	3D-IWGAN (enhanced GP)	3D microstructure reconstruction	3D microstructure

2.3. Deep-Learning-Based Microstructure Reconstruction

There were a few research works on deep learning applications for microstructure reconstruction before the GAN technique. One example of such applications is a research work by Cang, Ruijin, et al. [28]. It proposed a convolutional deep belief network (CDBN) to reconstruct heterogeneous materials. Recently, Tran and Tran [29] introduced a 2D microstructure reconstruction framework based on the image inpainting method to solve the microstructure reconstruction problem in three different contexts.

Owing to the proliferation of GAN, several research approaches exploit the generative power of GAN. Since GAN is used for microstructure reconstruction [30], it suffers from the intrinsic instability issue. We published the preliminary results of a deep-learning-based microstructure reconstruction using IWGAN [31].

Recently, more complex methods have been combined with GAN-based methods [32–34]. An end-to-end three-dimensional reconstruction framework of porous media from a single two-dimensional dataset has been proposed [32]. Shams et al. [33] introduces coupled generative adversarial and autoencoder neural networks to overcome instability for the reconstruction of realizations of three-dimensional data. The model gains efficient results by applying a gradient-descent-based optimization method for training and stabilizing the neural networks. A transfer learning technique [34] is exploited to integrate statistical descriptors with feature maps from a pre-trained deep neural network into an overall loss function for an optimization-based reconstruction procedure.

Our proposed system is different from the aforementioned recent approaches in the following aspects. First, we focus on reconstructing the 3D microstructure of permeable pavements and obtaining 3D images with a given 2D image. In this regard, an end-to-end framework [32] has the same goal as our system. However, our system exploits the characteristics of 3D-IWGAN, whereas an end-to-end framework uses the modified version of BicycleGAN [35], which is mainly for 2D-to-2D translation. Second, we provide the more detailed steps for pre-processing and the pavement analysis methods for checking the effectiveness of generated 3D images.

3. Preliminaries

3.1. GAN

The generative adversarial network (GAN) is a deep learning method that learns the representation of high-dimensional probability from a given dataset, and was introduced by Ian Goodfellow et al. [22]. For image microstructure reconstruction, the dataset is a set of training sample images of the probability distribution underlying the image space.

GAN is a generative system that comprises two networks, a generator network and a discriminator network. The generator network converts latent vectors with a normal distribution into samples that can be classified as either real or fake by the discriminator network. The training process is stopped and the discriminator is discarded after a proper sample image is obtained from the generator. GAN follows a minmax game with a loss function that is formally defined as

$$\min_G \max_D \{\mathbb{E}_{x \sim p_{data}}[logD(x)] + \mathbb{E}_{z \sim p_{noise}}[log(1 - D(G(z)))]\} \tag{1}$$

where x is a real object input in discriminator D and z denotes the normally distributed latent vectors that represent the random input for generator G' as illustrated in Figure 1.

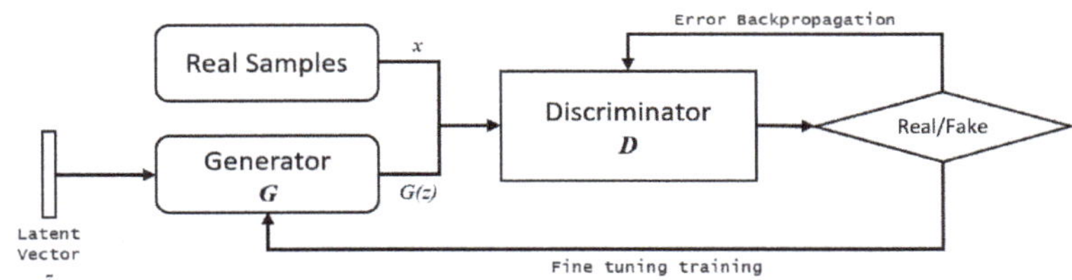

Figure 1. Generative adversarial networks. The discriminator network tries to distinguish between real and fake images, while the generator network tries to fool the discriminator by generating real-looking images.

3.2. 3D-IWGAN

In the original GAN method, the Kullback–Leibler divergence is minimized when training the networks to generate 3D objects. The drawback of using this method is the slow and unstable training process. A recent study introduced a method called 3D-improved Wasserstein GAN (3D-IWGAN) [25] that attempted to fix this issue and aimed to generate more realistic 3D objects by minimizing the Wasserstein distance between the data and generated distributions. The distance formula is

$$W(p_r, p_g) = \inf_{\gamma \sim \Pi(p_r, p_g)} \mathbb{E}_{(x,y) \sim \gamma}[\|x - y\|] \tag{2}$$

where $\Pi(p_r, p_g)$ is the set of all possible joint probability distributions between p_r and p_g and $\gamma \in \Pi(p_r, p_g)$ is the set of joint distributions in the continuous probability space. In the definition of Wasserstein distance, the infimum (greatest lower bound) indicates that we are only interested in the smallest cost.

The deviation of the discriminator's gradients is penalized from unity in 3D-IWGAN. It provides a more essential method for enforcing the Lipschitz constraint. The gradients of a differentiable function are at most one if and only if it is a 1-Lipschitz function. In the formal definition, the loss function for the discriminator in 3D-IWGAN is written as

$$\mathbb{E}_{\hat{x} \sim p_g}[D(\hat{x})] - \mathbb{E}_{x \sim p_r}[D(x)] + \lambda \mathbb{E}_{\hat{x} \sim p_x}[(\|\nabla_{\hat{x}} D(\hat{x})\|_2 - 1)^2] \tag{3}$$

where λ is the gradient penalty, p_g is the generator distribution, p_r is the target distribution, and p_x is the uniform distribution sampling.

The issue of the instability of GAN training can be tackled by employing the IW-GAN method. Our system utilized the IWGAN with an enhanced gradient penalty for 3D microstructure reconstruction. In comparison to previous approaches, our system demonstrated more stable results in training of the GAN. Table 2 displays the comparison between our enhanced gradient penalty and other GAN gradient penalties. A detailed explanation of this gradient penalty will be covered in Section 4.

Table 2. Gradient penalty comparison of different GAN methods. This measures the squared difference between the norm of the gradient of the predictions with respect to the input images and 1 in all cases.

Method	Gradient Penalty	Objective Function
3DGAN	none	$\mathbb{E}_{x \sim p_{data}}[logD(x)] + \mathbb{E}_{z \sim p_{noise}}[log(1 - D(G(z)))]$
3D-IWGAN (1-GP)	$\lambda\mathbb{E}_{\hat{x} \sim p_x}[(\|\nabla_{\hat{x}}D(\hat{x})\|_2 - 1)^2]$	$\mathbb{E}_{\hat{x} \sim p_g}[D(\hat{x})] - \mathbb{E}_{x \sim p_r}[D(x)] + \lambda\mathbb{E}_{\hat{x} \sim p_x}[(\|\nabla_{\hat{x}}D(\hat{x})\|_2 - 1)^2]$
3D-IWGAN (enhanced-GP)	$\lambda\mathbb{E}_{\hat{x} \sim p_x}[(\|\nabla_{\hat{x}}D(\hat{x})\|_2)^2]$	$\mathbb{E}_{\hat{x} \sim p_g}[D(\hat{x})] - \mathbb{E}_{x \sim p_r}[D(x)] + \lambda\mathbb{E}_{\hat{x} \sim p_x}[(\|\nabla_{\hat{x}}D(\hat{x})\|_2)^2]$

A latent vector is needed as an input to the generator network in the 3D-IWGAN system. In a general way, this latent vector is generated randomly using normal or uniform distribution. Meanwhile, our system generates this latent vector using a variational autoencoder (VAE). A VAE [36] consists of two parts: an encoder and a decoder. Its main task in our system is to encode material microstructures into a lower-dimensional latent space and to decode samples from the latent space back into microstructures. A full schematic view of the VAE architecture is given in Figure 2 Our VAE's decoder network is simultaneously used by the generator network to reproduce the original sample.

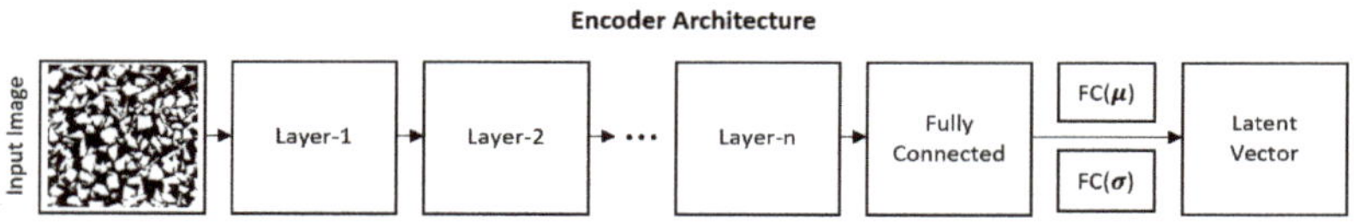

Figure 2. VAE encoder architecture. It encodes data into a lower-dimensional latent space.

4. Proposed System

4.1. System Architecture

The proposed system for the virtual experiment consists of three components: preprocessing, 3D model generator, and pavement analysis. The architecture of the system is illustrated in Figure 3.

In this section, we provide a brief overview of each component of the system. The preprocessing part has several tasks such as cropping, converting, downsizing, and resampling the images. In the model generation part, we use an adversarial network to generate a 3D image from the 2D images of the porous pavement microstructure. This 3D image is used as the input for pavement analysis, which further provides the numerical values of physical properties such as porosity, permeability, and hydraulic conductivity.

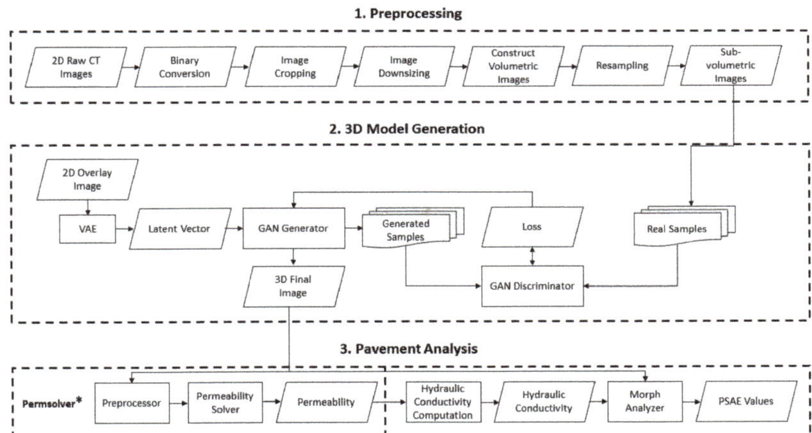

Figure 3. System architecture. This consists of three steps: preprocessing, 3D model generation, and pavement analysis.

4.2. Preprocessing

Before implementing the 3D-IWGAN, we need to prepare the image data to be fixed with the requirement of GAN input. The first step of preprocessing is binary image conversion. In this step, the raw CT images of the microstructure are converted into binary images with the dimension of 1130×1130 pixels. This binary conversion is based on Otsu's thresholding method.

In the second step, the image is cropped into a square with the dimension of 800×800 pixels, which is the maximum square size inside the circle of the converted binary image. The squared image is then resized into a smaller size of 300×300 pixels to fit the system requirement. This is the third step, known as image downsizing. The illustration of binary conversion, image cropping, and downsizing is shown in Figure 4.

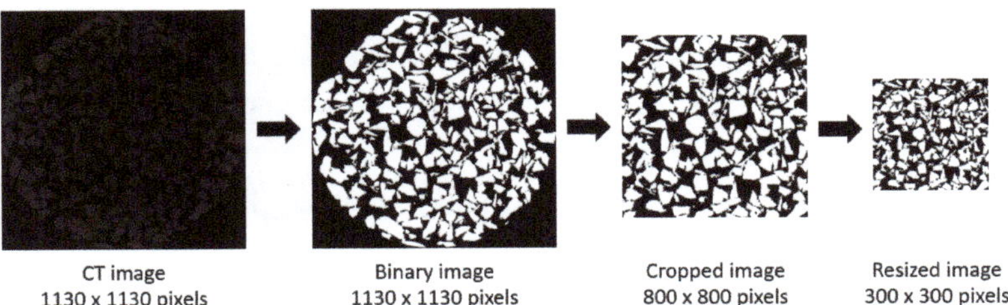

Figure 4. Binary conversion and image cropping. Input data are converted to binary versions, then converted and resized to fit the requirement of system.

The fourth step of preprocessing consists of volumetric image construction. In this step, a 3D microstructure image is created by stacking the binary images. The stack consists of 800 images of the permeable pavement microstructure and the size of each image is 800×800 pixels. The volumetric image of this generated microstructure is shown in Figure 5.

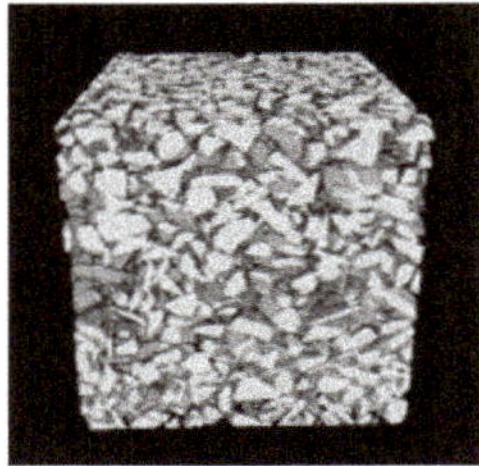

Figure 5. Volumetric image of permeable pavement sample. This consists of 800 images each of 800×800 pixel size.

Image resampling is the fifth step of preprocessing. Since we only have one training image per sample for the generator network, we need to create appropriate images for the training process by extracting sub-volumes from the voxelized binary images. For this experiment, we made approximately 24,389 images of 20^2 voxels from a single sample image of 300^3 voxels, illustrated in Figure 6. These generated images were used as the training images in the generator network.

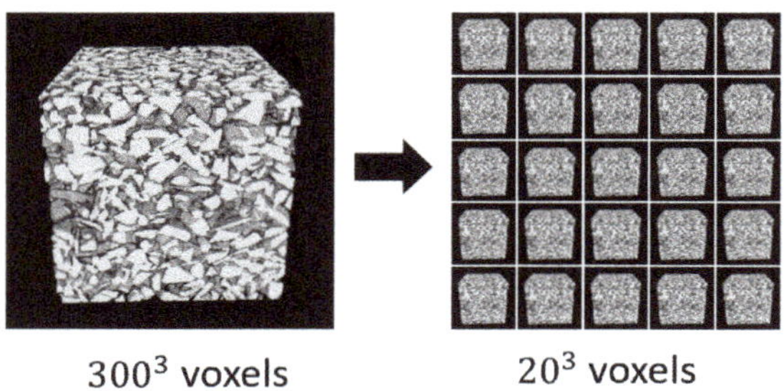

Figure 6. Volumetric image resampling. Overall, 24,389 images of 20^2 voxels are taken from a single 300^3 voxel image.

4.3. 3D Model Generation

In this subsection, we shall present details of the 3D model generation including the 3D-IWGAN process and enhanced gradient penalty.

4.3.1. 3D-IWGAN

The 3D model of the pervious pavement microstructure is generated using GAN in the first part of our system (3D model generator). The network takes a 2D image as the input in the generator network. This 2D image is transformed into a 3D image by the deconvolutional neural network. From the generator, the image is delivered to the discriminator network and it is tested against the real sample image of the microstructure to check whether the generated 3D model is real or fake using the discriminator.

The 3D-IWGAN network architecture used for 3D microstructure reconstruction in module one consists of two independent networks, the generator G and discriminator D. A 400-dimensional latent vector with normal distribution serves as the input for the generator network. Figure 7 shows the architecture of the generator network.

The latent vector was generated using the VAE, whose architecture is identical to that of discriminator network in the 3D image generation of GAN. Figure 8 illustrates the architecture of the VAE. This input passes through a fully connected layer with 2048 nodes. After getting through the fully connected layer, it passes through four deconvolutional layers whose length of stride and kernel size are 1 and 5, respectively. The first three deconvolutional layers in the generator network have a batch normalization layer and use the rectified linear unit (ReLU) as their activation function to address the vanishing gradient problem and promote sparse activations, while the fourth layer uses a hyperbolic tangent (tanh) as its activation function to make sure that our generating samples are in the range $[-1, 1]$. The output of the generator network is a 3D image of size $20 \times 20 \times 20$ voxels.

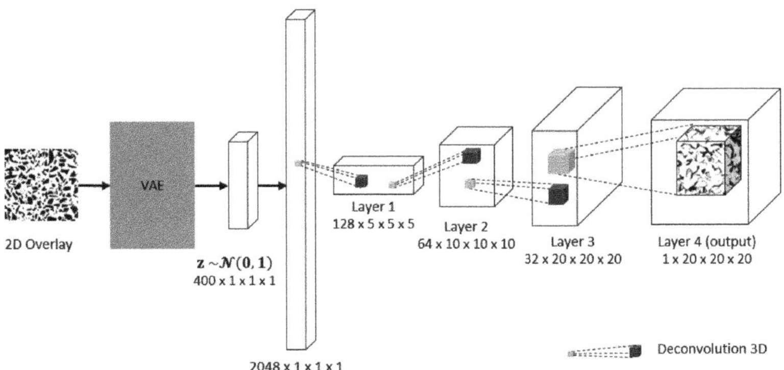

Figure 7. Generator network. It takes a fixed-length 400-dimensional latent vector as the input and generates a sample in the domain.

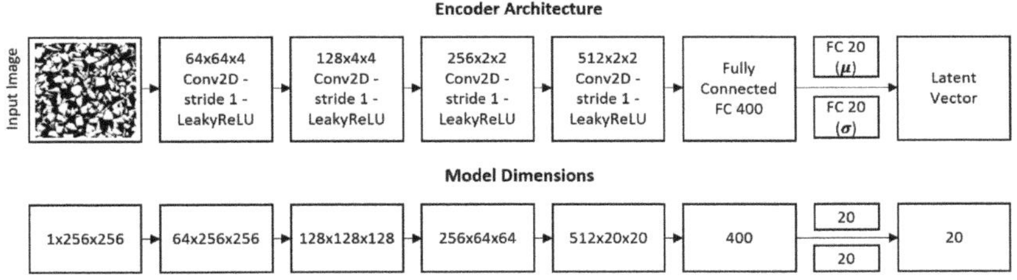

Figure 8. VAE architecture.

The discriminator network D takes an input image of size $20 \times 20 \times 20$ voxels. This input passes through four 3D convolutional layers, followed by a final fully connected layer. It is then condensed to a single value known as the discriminator output. The features of the architecture of the generator and discriminator networks are listed in Tables 3 and 4, respectively.

Table 3. Generator network architecture. The first three layers use ReLU activation functions and the last layer uses the Tanh activation function.

Layer	Type	Filters	Stride	Padding	Batchnorm	Activation Function
1	Deconvolution 3D	128	1	0	Yes	ReLU
2	Deconvolution 3D	64	2	1	Yes	ReLU
3	Deconvolution 3D	20	2	1	Yes	ReLU
4	Deconvolution 3D	1	1	1	No	Tanh

Table 4. Discriminator network architecture. All layers use the LeakyReLU activation function and no batch normalization is used.

Layer	Type	Filters	Stride	Padding	Batchnorm	Activation Function
1	Convolution 3D	1	1	1	No	LeakyReLU
2	Convolution 3D	32	2	1	No	LeakyReLU
3	Convolution 3D	64	2	1	No	LeakyReLU
4	Convolution 3D	128	1	0	No	LeakyReLU

4.3.2. Enhanced Gradient Penalty

Various types of gradient penalty are used to improve the convergence and stability of GAN training. One of the most commonly used gradient penalties is the one-centered gradient penalty [24], which is described as

$$\mathbb{E}_{\hat{x} \sim p_g}[D(\hat{x})] - \mathbb{E}_{x \sim p_r}[D(x)] + \lambda \mathbb{E}_{\hat{x} \sim p_x}[(\|\nabla_{\hat{x}} D(\hat{x})\|_2)^2] \tag{4}$$

In this paper, we used an enhanced gradient penalty for our 3D-IWGAN system. The original 3D-IWGAN gradient penalty is being modified to enhance the training stability of our GAN. Its formal definition is as follows:

$$\lambda \mathbb{E}_{\hat{x} \sim p_x}[(\|\nabla_{\hat{x}} D(\hat{x})\|_2 - 1)^2] \tag{5}$$

The gradient penalty used in this study is a form of zero-centered gradient penalty. In the zero-centered gradient penalty, we want to set the gradient to zero as the generator distribution (p_g) approaches the target distribution (p_r). When $p_g = p_r$, the gradient in connection to all datapoints on the line segment between a pair of real and fake samples should be zero.

4.4. Pavement Analysis

In the final part of our system, we analyzed the physical properties of the microstructure, such as porosity and permeability, using the 3D model output from the model generator part. The pavement analysis comprised two sub-parts: (1) *PermSolver* and (2) *MorphAnalyzer*.

PermSolver is an open-source framework developed by NIST [37]. It takes a 3D image of the microstructure as the input, converts it into x-y-z velocity and pressure components, then solves the Stokes equation using a finite difference scheme. By solving this equation, we can get the computed permeability of the porous microstructure.

Other than permeability values, we can also calculate the hydraulic conductivity (K) with this equation:

$$K = k\frac{\rho g}{\mu} \tag{6}$$

where k is the permeability value (from *Permsolver*), ρ is the density of fluid, g is the acceleration due to gravity, and μ is the dynamic viscosity of the fluid. In the following Section 5 the results of computing the hydraulic conductivity are described.

MorphAnalyzer is a tool that is used to compute the important characteristics of microstructure morphology. In our model we use three characteristics: porosity, specific area, and Euler characteristic. Porosity is calculated using the total number of pores based on the output of the 3D model reconstruction module. The specific area and Euler characteristic are obtained by using *ImageJ* [38]. The values extracted in this part are important in the investigation of the adsorption process and pore network connectivity of the permeable pavement microstructure.

5. Implementation and Evaluation

In this section, we present the performance evaluation of our system.

5.1. Experimental Setup

5.1.1. Hardware and Software

To evaluate the proposed system, we used one commodity machine. This machine is equipped with an Intel Xeon(R) CPU E5-2609 v4 @ 1.70 GHz x 16, with 64 GB memory, NVIDIA GeForce GTX 1060 6 GB, and runs on a 64-bit Ubuntu 16.04 operating system.

We implemented our approach in Python 2.7 using Tensorflow-GPU 1.2.1 performed with CUDA Toolkit 8.0 and cuDNN 6.0.

5.1.2. Dataset

The dataset used for this work was extracted from nine different samples of permeable pavement. The sample image was scanned using a CT machine with a voxel size of 88.194 µm. The size of one sample image after binary conversion to 8-bit resolution was 300^3 voxels. We resampled this image to 24,389 images with a size of 20^3 voxels. The size of this dataset was approximately 14.4 GB.

5.2. Experimental Results

5.2.1. 3D Pavement Reconstruction

The reconstruction process started by defining some hyperparameters for the generator and discriminator networks. We set 10^{-4} as the generator learning rate. Thus, it learns every 5 batches with batch normalization in all layers except the final layer. For the discriminator network, the learning rate is set to the same value as the generator network but it learns on each batch. The stochastic gradient descent using the Adam optimizer was used to perform the learning process, with momentum $\beta_1 = 0.5$ and $\beta_2 = 0.9$ for both networks.

We trained the network and calculated the loss for the discriminator network. The calculated loss is used to track the convergence and quality of generated objects, depicted in Figure 9. The first plot on the left is the GAN discriminator loss, the second plot in the center is the calculated loss for the original 3D-IWGAN, while the third plot is that of the 3D-IWGAN (with enhanced gradient penalty) discriminator loss. From the plots, we can see that the discriminator loss of 3D-IWGAN (enhanced GP) has a bouncing value in the beginning but it has relatively stable values after 600 iterations. Meanwhile, the discriminator loss of GAN and original GAN increase until 1000 iterations. Although our result shows that this value does not converge to zero, it is more stable in comparison to other methods. The stability in GAN's loss is also important to prevent the collapse and failure modes. To check the accuracy of our model, we added the computation of physical properties to display the similarities between our generated 3D model and the real sample. The result of this computation is presented in Permeable Pavement Analysis, Section 5.2.2.

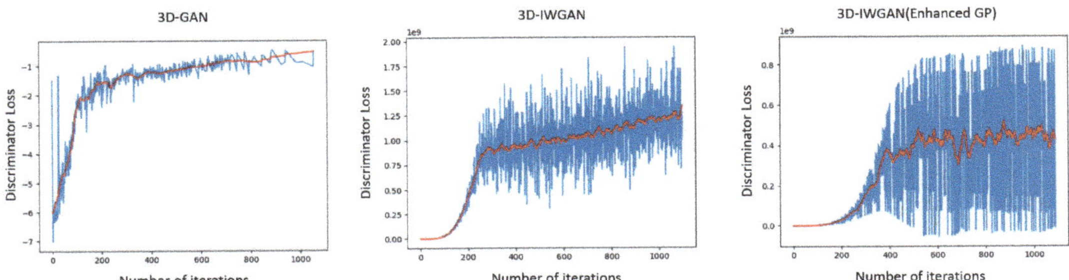

Figure 9. Discriminator loss comparison. The convergence and quality of generated objects are tracked by the calculated loss.

The result of the 3D reconstructed image using GAN is displayed in Figure 10a. Figure 10b illustrates the image generated using the original 3D-IWGAN, and the image generated using 3D-IWGAN (enhanced GP) is presented in Figure 10c. The images generated using the GAN method have more pores in comparison to those generated using the original 3D-IWGAN and our 3D-IWGAN, while the images generated by the original IW-GAN have more solid phases in comparison to other methods. Among the three methods, the 3D-IWGAN (enhanced GP) shows more realistic images of the pavement microstructure.

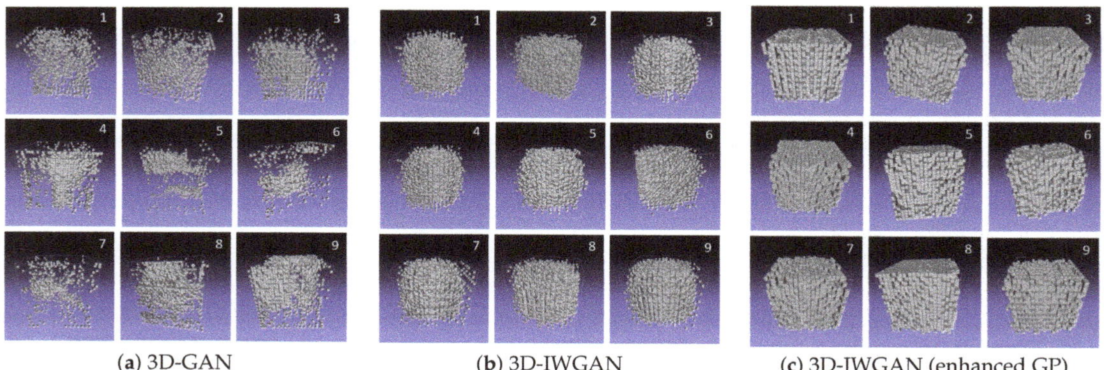

(**a**) 3D-GAN (**b**) 3D-IWGAN (**c**) 3D-IWGAN (enhanced GP)

Figure 10. Comparison of voxelized 3D Images. Images are reconstructed by GAN, 3D-IWGAN, and 3D-IWGAN (enhanced GP) methods.

We reconstructed the microstructure for nine different samples, whose results are presented in Figure 11.

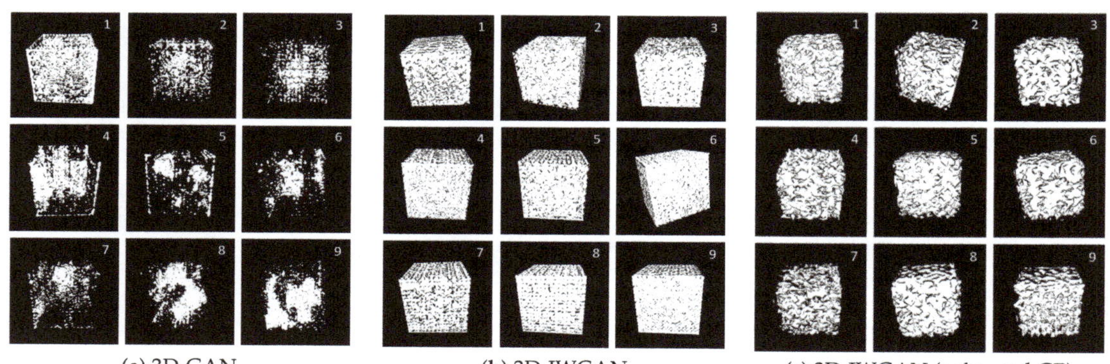

(**a**) 3D-GAN (**b**) 3D-IWGAN (**c**) 3D-IWGAN (enhanced GP)

Figure 11. Comparison of generated 3D images.

5.2.2. Permeable Pavement Analysis

The final part of our system is pavement analysis. After obtaining the 3D image of the pavement microstructure, we attempted to analyze the physical properties of the generated images. We attempted to evaluate five properties: porosity, permeability, hydraulic conductivity, specific surface area, and Euler characteristic.

In porous media such as permeable pavement, flow properties can be related to porosity, which is required to determine other properties such as permeability and hydraulic conductivity. Therefore, it needs to be estimated first. According to the ASTM standard, the porosity for permeable pavement is approximately 15–35% . Figure 12 shows the calculated porosity for the nine images generated using the IWGAN method. In comparison to the result of the previous GAN model, the proposed model shows better results for the values of porosity. The calculation results of the porosity from our generated 3D images shows that 66.67% of our samples are in the correct range of the standard value. Our method is better than the other methods, including none of the samples that fall within the appropriate range of standard values. The porosity values of the generated images from 3D-IWGAN (enhanced GP) illustrate a 66.67% improvement over the previous methods (3D-GAN and 3D-IWGAN).

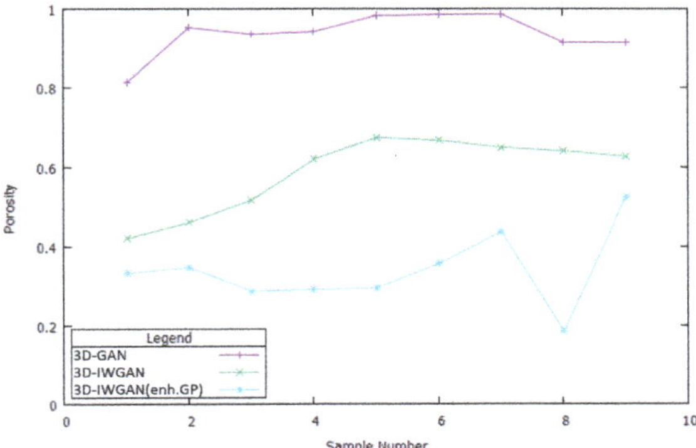

Figure 12. Computed porosity of 9 different samples. Our proposed method shows the highest result for the value of porosity among other methods.

We also calculated other properties such as the specific surface area and Euler number. The specific surface area is used to define the adsorption and dissolution processes in porous media, while the Euler number is used to characterize the connectivity of porous media. This is an important factor that determines the ability of fluids to flow. A less negative Euler number indicates a reduction in pore network connectivity. The values of surface area and Euler number are presented in Table 5. According to the ASTM standard, permeable pavement has a value of surface area approximately in the range 5–15%. Our results demonstrate that all samples are in the correct range (5–15)% of the surface area.

Table 5. Computed surface area and Euler number of 9 samples.

Sample	Specific Surface Area	Euler Number
1	7.37%	-1.27×10^3
2	8.93%	-4.53×10^2
3	9.45%	-3.21×10^2
4	8.77%	-5.25×10^2
5	10.93%	-2.19×10^2
6	8.65%	-4.16×10^2
7	7.91%	-5.98×10^2
8	14.26%	-1.09×10^2
9	6.17%	-6.26×10^2

After measuring the morphological values, we tackled the final part in our system: computing the permeability and hydraulic conductivity for each sample. The results of this computation can be used to determine the clogging potential of permeable pavement. The result for hydraulic conductivity is illustrated in Figure 13. It shows that the error difference in the nine different samples ranges between 1.9% and 5.6% or an average of approximately 3.54%.

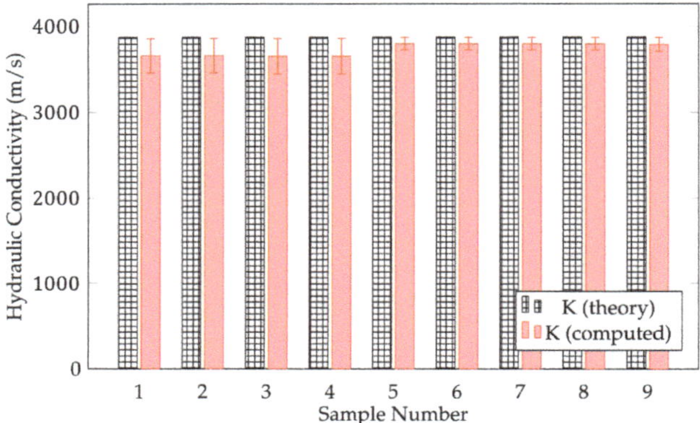

Figure 13. Computed hydraulic conductivity. The errors range between 1.9% and 5.6% among nine different samples.

The results of pavement analysis demonstrate that the proposed 3D model yielded better values than the model generated using 3D-GAN and original 3D-IWGAN. From the nine different samples, the average error difference in hydraulic conductivity computation was found to be below 5%. Moreover, the values of physical properties extracted through the proposed 3D model using our GAN method are closer to the baseline (theory) values than other GAN methods.

5.3. Discussion

We summarize our two major finding based on our experimental results. First, our framework can successfully generate 3D images of porous pavement microstructure from a single 2D image using 3D-IWGAN with enhanced GP. Figures 10 and 11 illustrate the generated 3D microstructure images as a visualization. Second, the generated 3D images are realistic in terms of the physical properties of permeable pavement extracted from our generated images. From Figure 12 and Table 5, we observe that the computed values of porosity and surface area fall in the correct ranges of the standard values. Figure 13 also displays a low error value of about 3.54% on average in nine different samples.

There are some limitations that restrain us from obtaining 100% accuracy despite getting decent results in 3D image generation and the computations of physical properties. Some limitations arise from the limited 2D image dataset as the input whereas others arise from the computational resources for our proposed framework. To provided a 2D input image for our framework, scanned images of pavement microstructure are needed at the pre-processing step. Since we only had a limited set of scanned images, we applied image resampling to obtain enough 2D images for our experiments. Powerful computational resources are important for generating 3D images with high resolution. For future experiments, several machines can be utilized as a distributed way to obtain large 3D images with higher resolutions.

6. Conclusions

Although permeable pavement has been extensively used to allow rainwater to pass through the pores into the ground, to control water quality, and to reduce the surface temperature, it has some obstacles during utilization: obtaining samples, maintenance cost, and clogging issues.

To overcome above-mentioned problems, in this paper, we proposed a three-dimensional microstructure reconstruction framework for permeable pavement analysis as one of the virtual experiment solutions. Since we utilized the generative model of 3D-IWGAN with an enhanced gradient penalty, our framework can effectively generate 3D microstructure reconstruction from a single 2D image of permeable pavement. To the best of our knowledge,

3D-IWGAN has not been used for the 3D model reconstruction of permeable pavement materials. From the visualization of the generated 3D microstructure images from nine samples, we have observed that our framework generates realistic 3D images.

When analyzing the flow in porous media, it is crucial to improve the values of porosity. Incorrect values of porosity will affect the other physical properties of permeable pavements such as permeability and hydraulic conductivity. These two properties are also important to assess the quality of generated images that are later used for pore network analysis. We have demonstrated that our framework generates more realistic 3D microstructure images, maintaining the values of these physical properties within the range of the standard values. In particular, the porosity values of the generated 3D images from 3D-IWGAN with enhanced GP were improved by 66.67% compared to the previous methods (3D-GAN and 3D-IWGAN).

Our proposed 3D microstructure reconstruction framework can be extended in two directions. First, we can add an additional module that generates and analyzes the pore network to conduct comprehensive virtual experiments. Second, we can combine some empirical methods with our framework to increase the quality of the generative model and avoid random errors.

Author Contributions: L.E.F. wrote the original version of the paper and the study, collected datasets, and conducted the experiments. J.A. provided the datasets, analyzed the data, and wrote the paper. S.L. contributed to revising the paper. J.K. supervised the work, and wrote and revised the paper. All authors have read and agreed to the published version of the manuscript.

Funding: This work was partly supported by an Institute of Information & communications Technology Planning & Evaluation (IITP) grant funded by the Korean government (MSIT) (No. 2020-0-00121, Development of data improvement and dataset correction technology based on data quality assessment), by the Institute of Information & Communications Technology Planning & Evaluation (IITP) grant funded by the Korean government (MSIT) (No.2019-0-01343, Regional strategic industry convergence security core talent training business) and by the Korea Agency for Infrastructure Technology Advancement (KAIA) grant funded by the Ministry of Land, Infrastructure, and Transport (Grant 21CTAP-C152124-03).

Institutional Review Board Statement: Not applicable.

Informed Consent Statement: Not applicable.

Acknowledgments: This work was partly supported by the Capacity Enhancement Program for Scientific and Cultural Exhibition Services through the National Research Foundation of Korea (NRF) funded by the Ministry of Science and ICT (NRF-2018X1A3A1069642) and by the National Research Foundation of Korea (NRF) grant funded by the Korean government (MSIT) (No. 2018R1A5A7059549).

Conflicts of Interest: The authors declare no conflict of interest.

References

1. Dietz, M.E. Low impact development practices: A review of current research and recommendations for future directions. *Water Air Soil Pollut.* **2007**, *186*, 351–363. [CrossRef]
2. Boogaard, F.; Lucke, T. Long-Term Infiltration Performance Evaluation of Dutch Permeable Pavements Using the Full-Scale Infiltration Method. *Water* **2019**, *11*, 320. [CrossRef]
3. Ahn, J.; Jung, J.; Kim, S.; Han, S.I. X-ray image analysis of porosity of pervious concretes. *Int. J. GEOMATE: Geotech. Constr. Mater. Environ.* **2014**, *6*, 796–799. [CrossRef]
4. Scholz, M.; Grabowiecki, P. Review of permeable pavement systems. *Build. Environ.* **2007**, *42*, 3830–3836. [CrossRef]
5. Deo, O.; Sumanasooriya, M.; Neithalath, N. Permeability reduction in pervious concretes due to clogging: Experiments and modeling. *J. Mater. Civ. Eng.* **2010**, *22*, 741–751. [CrossRef]
6. Pezzaniti, D.; Beecham, S.; Kandasamy, J. Influence of clogging on the effective life of permeable pavements. *Proc. Inst. Civ. Eng. Water Manag.* **2009**, *162*, 211–220. [CrossRef]
7. Lucke, T.; Beecham, S. Field investigation of clogging in a permeable pavement system. *Build. Res. Inf.* **2011**, *39*, 603–615. [CrossRef]
8. Yong, C.; McCarthy, D.; Deletic, A. Predicting physical clogging of porous and permeable pavements. *J. Hydrol.* **2013**, *481*, 48–55. [CrossRef]

9. Yong, C.F.; Deletic, A.; Fletcher, T.; Grace, M. The clogging behaviour and treatment efficiency of a range of porous pavements. In Proceedings of the 11th International Conference on Urban Drainage, Edinburgh, Scotland, UK, 31 August–5 September 2008.

10. Kia, A.; Wong, H.S.; Cheeseman, C.R. Clogging in permeable concrete: A review. *J. Environ. Manag.* **2017**, *193*, 221 –233. [CrossRef]

11. Şen, Z. Basic Porous Medium Concepts. In *Practical and Applied Hydrogeology*; Şen, Z., Ed.; Elsevier: Oxford, UK, 2015; Chapter 2, pp. 43–97.

12. Mays, D.C.; Hunt, J.R. Hydrodynamic aspects of particle clogging in porous media. *Environ. Sci. Technol.* **2005**, *39*, 577–584. [CrossRef]

13. Bean, E.Z.; Hunt, W.F.; Bidelspach, D.A. Field Survey of Permeable Pavement Surface Infiltration Rates. *J. Irrig. Drain. Eng.* **2007**, *133*, 249–255. [CrossRef]

14. Bentz, D.P. Virtual pervious concrete: Microstructure, percolation, and permeability. *ACI Mater. J.* **2008**, *105*, 297.

15. Manahiloh, K.N.; Muhunthan, B.; Kayhanian, M.; Gebremariam, S.Y. X-ray Computed Tomography and Nondestructive Evaluation of Clogging in Porous Concrete Field Samples. *J. Mater. Civ. Eng.* **2012**, *24*, 1103–1109. [CrossRef]

16. Sundararaghavan, V.; Zabaras, N. Classification and reconstruction of three-dimensional microstructures using support vector machines. *Comput. Mater. Sci.* **2005**, *32*, 223–239. [CrossRef]

17. Basanta, D.; Miodownik, M.A.; Holm, E.A.; Bentley, P.J. Using genetic algorithms to evolve three-dimensional microstructures from two-dimensional micrographs. *Metall. Mater. Trans. A* **2005**, *36*, 1643–1652. [CrossRef]

18. Jiang, Z.; Chen, W.; Burkhart, C. Efficient 3D porous microstructure reconstruction via Gaussian random field and hybrid optimization. *J. Microsc.* **2013**, *252*, 135–148. [CrossRef]

19. Quiblier, J.A. A new three-dimensional modeling technique for studying porous media. *J. Colloid Interface Sci.* **1984**, *98*, 84–102. [CrossRef]

20. Roberts, A.P. Statistical reconstruction of three-dimensional porous media from two-dimensional images. *Phys. Rev. E* **1997**, *56*, 3203. [CrossRef]

21. Yeong, C.; Torquato, S. Reconstructing random media. *Phys. Rev. E* **1998**, *57*, 495. [CrossRef]

22. Goodfellow, I.J.; Pouget-Abadie, J.; Mirza, M.; Xu, B.; Warde-Farley, D.; Ozair, S.; Courville, A.C.; Bengio, Y. Generative adversarial nets. In Proceedings of the 27th International Conference on Neural Information Processing Systems, Montreal, QC, Canada, 8–13 December 2014; pp. 2672–2680.

23. Goodfellow, I.J.; Bengio, Y.; Courville, A.C. *Deep Learning*; MIT Press: Cambridge, UK, 2016; Volume 1.

24. Gulrajani, I.; Ahmed, F.; Arjovsky, M.; Dumoulin, V.; Courville, A.C. Improved Training of Wasserstein GANs. In Proceedings of the Advances in Neural Information Processing Systems 30: Annual Conference on Neural Information Processing Systems 2017, Long Beach, CA, USA, 4–9 December 2017; pp. 5767–5777.

25. Smith, E.J.; Meger, D. Improved Adversarial Systems for 3D Object Generation and Reconstruction. In Proceedings of the 1st Annual Conference on Robot Learning, CoRL 2017, Mountain View, CA, USA, 13–15 November 2017; Volume 78, pp. 87–96.

26. Wu, J.; Zhang, C.; Xue, T.; Freeman, B.; Tenenbaum, J. Learning a probabilistic latent space of object shapes via 3d generative-adversarial modeling. In Proceedings of the Advances in Neural Information Processing Systems 29: Annual Conference on Neural Information Processing Systems 2016, Barcelona, Spain, 5–10 December 2016; pp. 82–90.

27. Arjovsky, M.; Chintala, S.; Bottou, L. Wasserstein Generative Adversarial Networks. In Proceedings of the 34th International Conference on Machine Learning, ICML 2017, Sydney, NSW, Australia, 6–11 August 2017; Volume 70, pp. 214–223.

28. Cang, R.; Xu, Y.; Chen, S.; Liu, Y.; Jiao, Y.; Ren, M.Y. Microstructure representation and reconstruction of heterogeneous materials via deep belief network for computational material design. *J. Mech. Des.* **2017**, *139*, 071404. [CrossRef]

29. Tran, A.; Tran, H. Data-driven high-fidelity 2D microstructure reconstruction via non-local patch-based image inpainting. *Acta Mater.* **2019**, *178*, 207–218. [CrossRef]

30. Mosser, L.; Dubrule, O.; Blunt, M.J. Reconstruction of three-dimensional porous media using generative adversarial neural networks. *Phys. Rev. E* **2017**, *96*, 043309. [CrossRef] [PubMed]

31. Feri, L.E.; Joonho Kwon, J.A. 3D microstructure reconstruction of Permeable Pavement using 3D-IWGAN. *Database Res.* **2014**, *34*, 22–33.

32. Feng, J.; Teng, Q.; Li, B.; He, X.; Chen, H.; Li, Y. An end-to-end three-dimensional reconstruction framework of porous media from a single two-dimensional image based on deep learning. *Comput. Methods Appl. Mech. Eng.* **2020**, *368*, 113043. [CrossRef]

33. Shams, R.; Masihi, M.; Boozarjomehry, R.B.; Blunt, M.J. Coupled generative adversarial and auto-encoder neural networks to reconstruct three-dimensional multi-scale porous media. *J. Pet. Sci. Eng.* **2020**, *186*, 106794. [CrossRef]

34. Bhaduri, A.; Gupta, A.; Olivier, A.; Graham-Brady, L. An efficient optimization based microstructure reconstruction approach with multiple loss functions. *arXiv* **2021**, arXiv:2102.02407.

35. Zhu, J.; Zhang, R.; Pathak, D.; Darrell, T.; Efros, A.A.; Wang, O.; Shechtman, E. Toward Multimodal Image-to-Image Translation. In Proceedings of the Advances in Neural Information Processing Systems 30: Annual Conference on Neural Information Processing Systems 2017, Long Beach, CA, USA, 4–9 December 2017; pp. 465–476.

36. Kingma, D.P.; Welling, M. Auto-Encoding Variational Bayes. *arXiv* **2014**, arXiv:1312.6114.

37. Bentz, D.P.; Martys, N.S. *A Stokes Permeability Solver for Three-Dimensional Porous Media*; U.S. Department of Commerce, Technology Administration, National Institute of Standards and Technology: Gaithersburg, MD, USA 2007.

38. Rasband, W. *ImageJ*; US National Institutes of Health: Bethesda, MD, USA, 1997.

sensors

Article

Performance Analysis of IoT and Long-Range Radio-Based Sensor Node and Gateway Architecture for Solid Waste Management

Shaik Vaseem Akram [1], Rajesh Singh [1], Mohammed A. AlZain [2,*], Anita Gehlot [1], Mamoon Rashid [3,*], Osama S. Faragallah [2], Walid El-Shafai [4] and Deepak Prashar [3]

[1] School of Electronics and Electrical Engineering, Lovely Professional University, Phagwara 144411, India; vaseem.11814442@lpu.in (S.V.A.); rajesh.23402@lpu.co.in (R.S.); anita.23401@lpu.co.in (A.G.)

[2] Department of Information Technology, College of Computers and Information Technology, Taif University, P.O. Box 11099, Taif 21944, Saudi Arabia; o.salah@tu.edu.sa

[3] School of Computer Science and Engineering, Lovely Professional University, Jalandhar 144411, India; deepak.prashar@lpu.co.in

[4] Department Electronics and Electrical Communications, Faculty of Electronic Engineering, Menoufia University, Menouf 32952, Egypt; walid.elshafai@el-eng.menofia.edu.eg

* Correspondence: m.alzain@tu.edu.sa (M.A.A.); mamoon.20574@lpu.co.in (M.R.)

Abstract: Long-range radio (LoRa) communication is a widespread communication protocol that offers long range transmission and low data rates with minimum power consumption. In the context of solid waste management, only a low amount of data needs to be sent to the remote server. With this advantage, we proposed architecture for designing and developing a customized sensor node and gateway based on LoRa technology for realizing the filling level of the bins with minimal energy consumption. We evaluated the energy consumption of the proposed architecture by simulating it on the Framework for LoRa (FLoRa) simulation by varying distinct fundamental parameters of LoRa communication. This paper also provides the distinct evaluation metrics of the the long-range data rate, time on-air (ToA), LoRa sensitivity, link budget, and battery life of sensor node. Finally, the paper concludes with a real-time experimental setup, where we can receive the sensor data on the cloud server with a customized sensor node and gateway.

Keywords: cloud server; customized sensor node; customized gateway; FLoRa simulation; LoRa range radio; solid waste management

Citation: Akram, S.V.; Singh, R.; AlZain, M.A.; Gehlot, A.; Rashid, M.; Faragallah, O.S.; El-Shafai, W.; Prashar, D. Performance Analysis of IoT and Long-Range Radio-Based Sensor Node and Gateway Architecture for Solid Waste Management. *Sensors* **2021**, *21*, 2774. https://doi.org/10.3390/s21082774

Academic Editor: Raffaele Bruno

Received: 15 March 2021
Accepted: 12 April 2021
Published: 14 April 2021

Publisher's Note: MDPI stays neutral with regard to jurisdictional claims in published maps and institutional affiliations.

1. Introduction

Waste management is the primary public utility for every community to ensure the protection of public health and the environment [1]. According to United Nations by 2025, 66% of the global population will be residing in urban cities, in comparison to 54% currently residing in the urban cities. It also estimates that the material consumption in cities will accelerate to nearly 90 billion tonnes by 2050, contrasted with 40 billion tonnes in 2010 [2]. The consumption of material significantly spikes up the generation of waste. The improper management, inadequate collection of waste, and dumping in the open-air adversely affect the environment [3]. However, the advancement in technology has progressively encouraged researchers to integrate the technology that can lower the effect of solid waste on the atmosphere [4]. At present, the Internet plays an imperative role in technological development and recent advancement in digital technologies had a tremendous impact on everyday lives, such as the Internet of Things (IoT) [5]. IoT is enhancing the quality of life as it authorizes the idea of smart cities by enhancing the interconnection between humans and objects [6]. The implementation of IoT in multiple sectors, such as as smart cities, intelligent wearable, smart homes, smart grids, and smart industries, have realized the real-time monitoring and controlling system [7]. The existing studies have primarily concentrated on the implementation of IoT in waste management for real-time monitoring and collection of garbage data [8]. In the waste collection system, one of the most promising technology is IoT [9,10], as IoT can track and handle the waste information from everywhere in the world over Internet Protocol (IP) [8]. In IoT, the role of wireless communication is crucial as the connectivity should be reliable and stable for continuous monitoring of the system [11]. At present, IoT is demanding minimum power consumption and long-range transmission features for wireless communication as the sensor nodes are energy-constrained devices [12]. With the evolution of Long-range (LoRa), the requirement of IoT is achieved as LoRa delivers long-range coverage with minimum power consumption [13]. The motivation of our study comes from Reference [14], in which a LoRa (Long-range) communication protocol is employed for the implementation of a smart waste management system to transmits real-time filling data with minimum energy consumption. With this advantage, LoRa Wide Area Network (LoRaWAN)-based IoT architecture is proposed with customized nodes for real-time monitoring of the bins. A customized sensor node and gateway are designed based on LoRa modulation technology for providing energy-efficient and stable connection establishment between the nodes and cloud server.

The main contribution of this study is divided as follows:

- We proposed architecture for designing and developing a customized sensor node and gateway based on LoRa technology for realizing the filling level of the bins with minimal energy consumption.
- We evaluated the energy consumption of the proposed architecture by simulating it on the Framework for LoRa (FLoRa) simulation by varying distinct fundamental parameters of LoRa communication.
- We provided the distinct evaluation metrics of the long-range data rate, time on-air (ToA), LoRa sensitivity, link budget, and battery life of sensor node.
- We concluded with a real-time experimental setup, where we received the sensor data on the cloud server with a customized sensor node and gateway.

The rest of the paper is categorized into various sections as defined. Section 2 addresses the work that has already been done in the field of solid waste management. Section 3 covers the overview of long range (LoRa) including evaluation metrics of LoRa. Section 4 covers the hardware implementation, where the design and development of customized sensor node and gateway are discussed. Section 5 covers proposed LoRa-based architecture for solid waste management. Performance analysis of LoRa is covered in Section 6. Result analysis are discussed in the Section 7, and the conclusion of this paper is finally expressed in Section 8.

2. Related Work

Smart waste management, in general, adopts advanced technology for enhancing the collection of waste management. The wireless technologies including wireless-fidelity (Wi-Fi), global Packet for Radio Service (GPRS), radio frequency identification (RFID), and global positioning system (GPS) are implemented for resolving the distinct issues in solid waste management. In regarding these many researchers have implemented the wireless sensor network (WSN) for monitoring the bin status wirelessly [15–17]. WSN implemented for scrutinizing the status of the bin and applying the closest vehicle first algorithm for optimizing the routes of municipal trucks [18]. Few researchers have integrated the web server and WSN for visualizing the filling of bins [19]. For transmitting the data to the web server, multiple communication technologies are embedded, one of these is GSM/GPRS module. AA smart collection system for solid waste is established with the integration of ultrasonic sensors, GSM protocol, and RFID technology [20]. The Vehicular Ad Hoc Networks (VANETs) framework was designed to provide a system for monitoring waste collections in real-time and almost every garbage bin has a sensor node, which detects the state of bins continuously and sends the signal to nearby vehicles when the bin is full [16]. Zigbee and Message Queuing Telemetry Transport (MQTT)-based waste management systems are also established for monitoring the bins, and MQTT protocol communicates the monitoring status from the ZigBee to the server [21]. iBags is a smart waste management system that is an integration of ZigBee and RFID for identifying the bags and communicating the status of the bins to the server [22]. The drawbacks of the ZigBee and GSM/GPRS are: GSM/GPRS communication consumes power consumption, and a prepaid amount needs to be recharged for transmitting the data [23,24]. The data transmission range of ZigBee is limited to short-range; however, the transmission range can be increased by installing Zigbee repeaters. Installation of Zigbee repeaters is having a higher possibility of connectivity failure. The evolution of LPWAN technologies has widened the opportunities for implementing IoT architecture to waste management, as this technology is capable of transmitting wireless data in the absence of remote cell technology with low power consumption [25,26]. The recent studies that have integrated into the smart waste management are: LoRa prototype is embedding for the sensor nodes for maximizing waste disposal routes in the Salamanca Area by using a low electrical energy consumption [27]. A low-powered sensor node is proposed for sensing the filling status of the bin with the assistance of LoRaWAN [28]. A recent study integrated the Raspberry Pi 3 Model B+ controller unit and LoRa communication protocol for monitoring the filling status of the waste in the bin [25]. In IoT applications, most designers have opted for microcontroller boards (Arduino board), or SOC (BeagleBone, Raspberry Pi, Lopy) for the platforms to connect sensor nodes to the internet [28]. From the above literature work, it is clear that there is a need of some architecture for designing and developing a customized sensor node and gateway that will be efficient enough for realizing the filling level of the bins with minimal energy consumption.

3. Overview of LoRa

Long-range (LoRa) is initiated in 2012, it can perform the communication with a high link budget with low power consumption [29]. LoRa is developed for supporting the LPWAN (Low Power Wide Area Network) that utilizes the ISM (Industrial, Scientific, Medical) bands of 433 MHz (Asia), 915 MHz (North America), and 868 MHz (Europe) [13]. LoRa modulation is formed from the CSS (Chirp Spread Spectrum) that provides robustness in case of multipath propagation interferences [30]. LoRa modulation is customized on distinct parameters, like spreading factor (SF), code rate (CR), and bandwidth (BW).

- SF is the number of chirps generated by each symbol, and its range is between 7 and 12. The higher the SF value, the better the receiver will eliminate the noise from the signal. The more time it takes to deliver a packet, the greater the amount it takes to transmit the packet.
- CF is the frequency of a carrier wave that is modulated to transmit signals; SX 1278 transceiver works on a carrier frequency of 433 MHz.
- BW depicts the frequency in the spectrum band, and it is chosen from these three bands: 500 kHz, 250 kHz, or 125 kHz. Large bandwidth represents speed transmission and the small bandwidth presents long-range transmission. The main parameter of the LoRa modulation is BW. The 2SF chirps that covers the entire frequency band is represented as LoRa symbol. Initially, it begins with a sequence of upward chirps, if the highest frequency band is achieved then frequency is wrapped and there will be a rise in the frequency again from the lowest frequency.

As LoRa is established on chirp spread spectrum modulation and the transmitted data is represented as chirp signal from a frequency range from fmin to fmax. Here, there is an opportunity of configuring the symbol by varying the SF and BW parameters. According to Reference [31], one symbol will take TS of second to transmit, which is a function of SF and BW are presented in the equation below:

$$\mathrm{TS} = \frac{2^{\mathrm{SF}}}{\mathrm{BW}}.$$

(1)

CR is implemented in communication for assuring short interference during transmission. In any data transmission, LoRa modulation includes a forward error correction (FEC). This is achieved by encoding 4-bit data with 5-bit, 6-bit, 7-bit, and even 8-bit redundancies. The LoRa signal can sustain small interferences with this redundancy. CR is expressed as

$$CR = \frac{4}{4+n},$$ (2)

where n has values in the range between 1 and 4. CR is proportional to transmission speed and inversely proportional to the time on-air (ToA) during the transmission of data. HigherCR represents high transmission speed and low ToA.

LoRa packet structure comprises a preamble, optional header, payload, and Cyclic Redundancy Check (CRC), as shown in Figure 1. A preamble is employed for communication between the receiver and the transmitter. The header contains the payload size and information of LoRa configuration, and it is encoded with CR = 4/8. The payload is determined with CR and CRC is transmitted at the end of the frame.

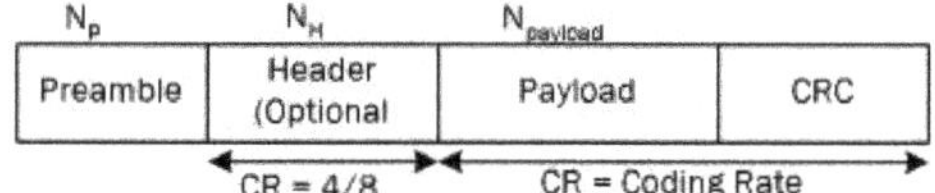

Figure 1. Long-range (LoRa) packet structure.

ToA is defined as the duration of time that is consumed before a receiver receives this signal. ToA of a packet is calculated with the combination of CR, SF, and BW. LoRa packet duration is defined as the combination of transmitted packet and preamble. The preamble length is expressed as follows.

$$T_{preamble} = (n_{preamble} + 4.25)T_{sym},$$ (3)

where npreamble indicates programmed preamble length. The payload is based upon the enabled header mode, and the number of payload symbol is calculated with this Equation (4).

$$n_{payload} = 8 + max\left(ciel\left\lceil \frac{(8PL - 4SF + 28 + 16RC - 20IH)}{4(SF - 2DE)} \right\rceil (CR + 4), 0 \right),$$ (4)

where PL indicates number of payload bytes, SF indicates spreading factor from 7 to 12, if IH = 0, it indicates header is enabled, if IH = 1, it indicates, if DE = 1, then Low Data Rate Optimize = 1, DE = 0; otherwise, CR is the coding rate (1 equivalent to 4/5, 4 to 4/8). The payload duration is the number of payload symbols multiplied by the symbol period, and it is expressed as:

$$T_{payload} = n_{payload} * Ts.$$ (5)

The time of the air or packet is the summation of the preamble and the payload duration, and it is expressed as:

$$T_{packet} = T_{preamble} + T_{payload}.$$ (6)

Bit Rate/data rate is defined as the rate at which bits are transferred from one location to another location. The bit rate (Rbit) of LoRa is expressed as:

$$R_{bit} = \frac{SF * BW}{2^{SF} * CR}.$$ (7)

Sensitivity is the inherent property of a system's ability to extract information from signals. It can also be quantified as the lowest possible signal strength that can trigger the system to its packet resolution. The first term is due to thermal noise in 1 Hz of bandwidth and can only be influenced by changing the temperature of the receiver. The second term, BW, is the receiver bandwidth. NF Is the receiver noise figure and is fixed for a given hardware implementation. Finally, SNR represents the signal to noise ratio required by the underling modulation scheme. It is the signal to noise ratio and bandwidth that are available as design variables to the LoRa designer. LoRa receiver sensitivity (S) is calculated with this expression as:

$$S = -174 + 10\log_{10}BW + NF + SNR, \tag{8}$$

where S = receiver sensitivity in dB, BW = Bandwidth in KHz, NF = Noise fig. of a receiver in dB, and SNR = Signal to Noise Ratio.

Signal-to-Noise Ratio (SNR) is the measure of received power signal and the noise floor power level. Generally, the range of SNR is between -20 dB and $+10$ dB. If the range is around the $+10$ dB, then the received signal is less corrupted. The SNR range of LoRa is between -7.5 dB to -20 dB.

A *link budget or received power* is the summation of all the losses and gains from the transmitter, over free space to the receiver [32]. Using a simple model, the link budget can be calculated by adding transmitter power (P_{Tx}), receiver sensitivity (Rx), antenna gain, and free space path loss (FSPL). The link budget is expressed as:

$$P_{Rx} = P_{Tx} - FSPL + G_{Tx} + G_{Rx}, \tag{9}$$

where P_{Rx} indicates received power or link budget (dBm), P_{Tx} denotes transmitter power (dBm), G_{Tx} denotes transmitting antenna gain (dB), G_{Rx} denotes es receiver antenna gain (dB), and FSPL denotes free space path loss (dB).

Free space path loss (FSPL) is defined as the amount of energy that is lost in free space during a communication between the transmitter (Tx) and receiver end (Rx). FSPL expression derives from friis transmission equation and friis transmission equation [33] is expressed as follows:

$$\frac{p_r}{p_t} = \left(\frac{\lambda}{4\pi R}\right)^2 G_{0t}G_{0r}, \tag{10}$$

where p_r and p_t denotes power receiver and power transmitted, λ indicates wave length, and $G_{0t}G_{0r}$ directivity of the transmitting and receiving antenna. Concerning the definition of FSPL and the expression of FSPL is represented as follows:

$$FSPL = \left(\frac{4\pi d}{\lambda}\right)^2, \tag{11}$$

where d denotes a distance between Tx and Rx (Meters), Tx indicates transmitter, Rx indicates receiver, f indicates the frequency in Hertz, and λ indicates wavelength.

$$\lambda = \frac{c}{f}; c = 3 * 10^8 m/s. \tag{12}$$

4. Hardware Implementation

In this section, we will discuss the importance of customization of sensor node and gateway for real-time monitoring of the filling status of the bins. Here, the sensor node and gateway are developed based on LoRa modulation; additionally, the gateway is integrated with ESP 8266 Wi-Fi module for transmitting data over the internet. In the initial stage of designing the nodes, the power consumption parameter is the key as the nodes are installed in an outdoor environment where the electrical grid network should not be utilized for powering the nodes because it increases the infrastructure cost for installing a separate electric grid network. The power consumption parameter is the key in the initial stage of designing the node, as the nodes are meant to be installed in an outdoor environment where the electrical grid network cannot be used to power the nodes because it raises the cost of infrastructure for installing separate electric grid network. Concerning powering the sensor node, we embedded the power jack for supplying power to the node through batteries. The sensor node and gateway are integrating with the +5 V and +3.3 V voltage converter to supply the appropriate voltage.

As discussed earlier in recent studies, the selection of a communication module plays a critical role in preventing the power dissipation in the nodes during transmission of the data. So, communication module is chosen with low power consumption for transmitting the data to long-range. Another major component that is taken into account for designing the node is the computing unit. The computing unit is chosen based on computational power that is a criterion for processing the acquisition data. In our case, the computing power for processing the acquired data is of low complexity because the main objective of our study is to senses the filling data. Selecting the appropriate microcontroller/microprocessor is also a vital parameter for developing an energy-efficient node. After the completion of choosing suitable components for designing the node, the next stage is subjected to integrating all these components on the single board for executing a flexible, reliable, and compatible node.

Figure 2 illustrates the block diagram, PCB layout, and prototype of the sensor node. The node is embedding with an Atmega 328P controller and SX 1278 LoRa transceiver. In our study, the complexity of processing the data is low, we preferred the Atmega328P controller for computing the data as it consumes the least amount of current around milliamperes (mA). Here, we estimated the amount of battery required for the sensor node from Reference [34]. The sensor node sends data for every 30 min with airtime of 78.08 ms for 25-byte payload with SF7 and 125 BW. Therefore, to perform one transmission by the sensor node, it consumes 1.0962 mA of energy. If we had a battery of 2000 mAh of capacity, we could perform 1824 transmissions, i.e., performing 48 transmission per day (one transmission for every 30 min interval). With respect to battery capacity of 2000 mAh, the sensor node remains active for 38 days. If we place the battery capacity of 20,000 mAh, the sensor node performs the 18,240 transmission and the sensor node remains active for 380 days. The technical specifications of the Atmega328P microcontroller and SX1278 LoRa module are detailed in Tables 1 and 2.

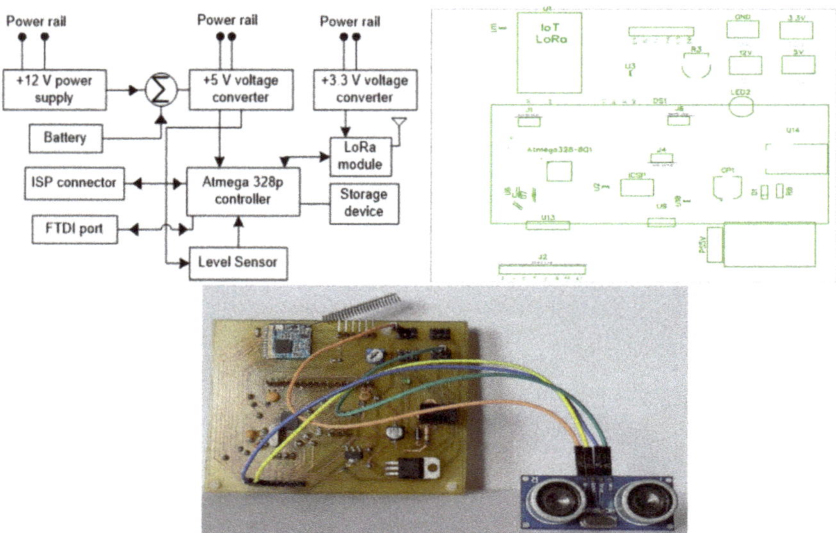

Figure 2. Sensor nodes.

Table 1. SX 1278 LoRa specifications [35].

Characteristic	Specification
Frequency	433 MHz
Network topology	Point-to-Multipoint, Point-to-Point, Mesh, and Peer-to-Peer
Modulation	FSK/GFSK/MSK/LoRa
Data rate	<300 kbps
Sensitivity	-136 dBm
Output power	+20 dB
Operating voltage	1.8 V to 3.6 V
Current	Tx: 120 mA, Rx: 10.8 mA
RSSI	127 dB
Link budget	168 dB

Table 2. Atmega328P specifications [36].

Characteristic	Specification
Controller	8-bit microcontroller
Architecture	RISC
Programming	In-system programming
Serial interface	Master/slave SPI
PWM channels	6
Pin 6 analog pins	14 digital pins
Operating voltage	2.7 V to 5.5 V
Current	Active state: 1.5 mA at 3 V–4 MHz, Power-down state: 1 µA at 3 V

Figure 3 illustrates the block diagram, schematic view, and prototype of the LoRa-based gateway. An Atmega 328P controller is built into the gateway, SX 1278 LoRa transceiver, and ESP 8266 Wi-Fi module. The objective of the gateway is for transmitting the data over multiple communication protocols, in this SX 1278 LoRa transceiver receives the data, the Atmega 328P controller triggers the ESP 8266 Wi-Fi for transmitting the data over internet protocol (IP). The technical specifications of ESP 8266 Wi-Fi are shown in Table 3.

Table 3. Technical specifications of ESP 8266 [37].

Parameter	Feature
Processor	Tensilica L106 32-bit processor
IEEE standard	802.11 b/g/n
Frequency	2.4 GHz
Data rate	72 Mbps
Network Protocols	Ipv4, TCP/UDP, HTTP
Tx power	20 dBm (802.11 b), 17 dBm (802.11 g) & 14 dBm (802.11 n)
Rx sensitivity	-91 dBm (802.11 b), -75 dBm (802.11 g) & -71 dBm (802.11 n)
Operating voltage	2.5 V to 3.6 V
Current	Average: 80 mA

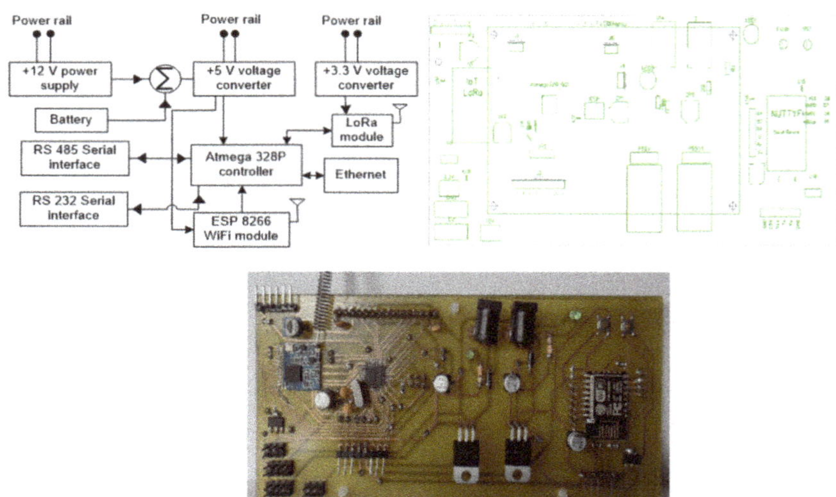

Figure 3. LoRa-based gateway.

5. LoRa Architecture for Waste Management

After the completion of the design and development of the sensor node and gateway, we now have to implement in the real-time set up for analyzing the performance of the nodes during the transmission of the data. In order of implementing the customized node for sensing and transmitting the filling level of the bins, we are proposing architecture for waste management that is based on basic LoRa architecture and as shown in Figure 4. Physical Layer, gateway, network server, and cloud server are the key layers that exist in this architecture. The sensor nodes are part of the physical layer, the function of the sensor node is to sense the filling level of the bins and transmits the sensory data to the gateway via LoRa (Long-range).

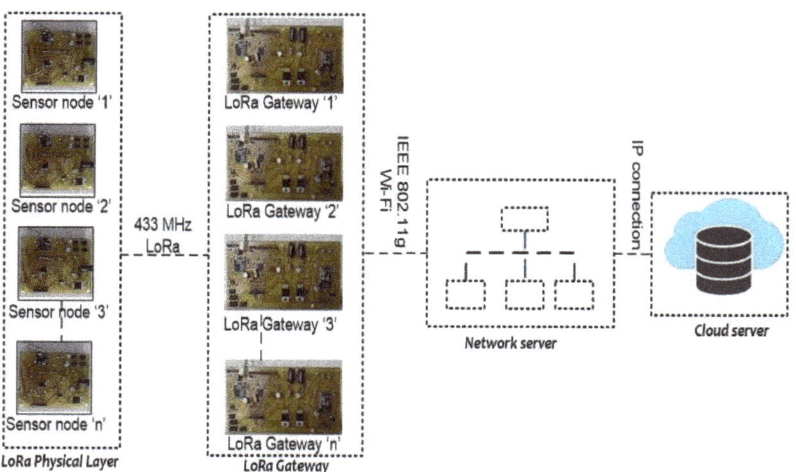

Figure 4. LoRa architecture for solid waste management.

The gateway in Figure 4 receives the messages from the sensor node to the network server. Since a transmission from a sensor node may be processed by more than single gateway, they establish a star-of-stars topology as standarised by the LoRaWAN architecture. Gateway utilizes the internet connection at their location to transmit data between the nodes and the network server. In proposed system, thhe gateways can communicate with one another via IEEE 802.11-based wireless mesh network in order of reaching the network server. This part of the system is independent of the LoRaWAN architecture. To optimize the system, we presume that the gateways have a good and direct connection to the network server.

6. Performance Analysis

In this section, we will be discussing the performance analysis of LoRa network. Initially, the section begins with the energy consumption of the LoRa nodes in different cases. Later on, the analysis of data rate/bit rate, LoRa sensitivity, time on air (ToA), link budget, and battery life of sensor node are elaborated.

6.1. Energy Consumption of Nodes

In previous studies, the simulation of LoRaWAN network is performed for studying the performance of the LoRa network by using the ns-3 module [38]. Before implementing the proposed model of waste management in real time environment, it is simulated on the virtual environment for examining the performance and capability of the model. In the virtual environment, we will analyze the LoRa network behavior in our model by varying certain parameters and aspects. FLoRa is a simulation framework, that utilizing the OMNeT++ discrete event simulation library. In spite of OMNeT++ framework, FLoRa is also based on the INET Framework, which is an open-source library for OMNeT++ and its function is to aid the procedure of experimentation for distinct network protocols [39]. In addition, the FLoRa structure ensures proper implementation of the LoRaWAN architecture [40] and a reliable LoRa radio physical layer model resulting from previous experimental findings [41]. The FLoRa architecture is convenient for simulating a complete LoRaWAN architecture that is based on star topology where it consists of four entities including LoRa nodes, gateways, a network server and application server [42].

Figure 5 also presents the working model of the FLoRa framework, where the LoRa nodes transmits the information to the gateway. The two nested submodules that handle the application layer (simpleLoRaApp) and the LoRa network service (LoRaNic) are shown in Figure 6. LoRaNode module includes the network module (LoRaNic) and the application module (simpleLoRaApp). LoRaNic module provides the LoRa network capability, radio module (LoRaRadio), and MAC module (LoRaMAC). In order of analyzing the behavior of LoRa network of waste management, LoRaNode is the main module in which parameters should to be modified. In this simulation we will generate the energy consumption of LoRa node by setting the parameters as CF = 433 MHz, BW = 125 KHz, SF = 7, CR = 4, No of packets = 2, network size = 480 m × 480 m, gateway distance = 320 m. number of nodes = 4, number of gateways = 1. The four nodes are represented as node '1', node '2', node '3', and node '4'. The deployment type of network is select as square shape. As the network size is 480 m * 480 m, the geometry position of nodes in FLoRa environment are represented as Node 1 (263.43 m, 284.56 m), Node 2 (411.81 m, 261.54 m), Node 3 (310 m, 184.5 m), Node 4 (130 m, 184.05 m), Node 5 (444.2 m, 401.31 m), Node 6 (311.12 m, 9.70 m), Node 7 (373.51 m, 67.36 m), Node 8 (227.3 m, 383.5 m), Node 9 (374.6 m, 325.8 m), Node 10 (68.80 m, 257.9 m), Node 11(41.8 m, 311.1 m), Node 12 (176.7 m, 399.6 m), Node 13 (373.5 m, 67.3 m), Node 14 (417.6 m, 469.7 m, Node 15 (383.5 m, 384.4 m, Node 16 (299.3 m, 374.6 m).

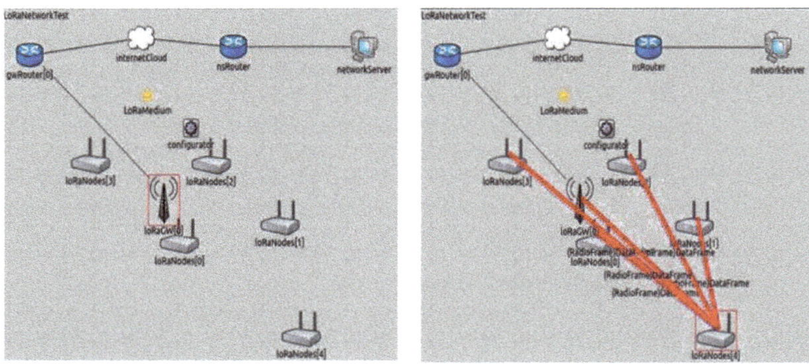

Figure 5. Framework for LoRa (FLoRa) graphical interface.

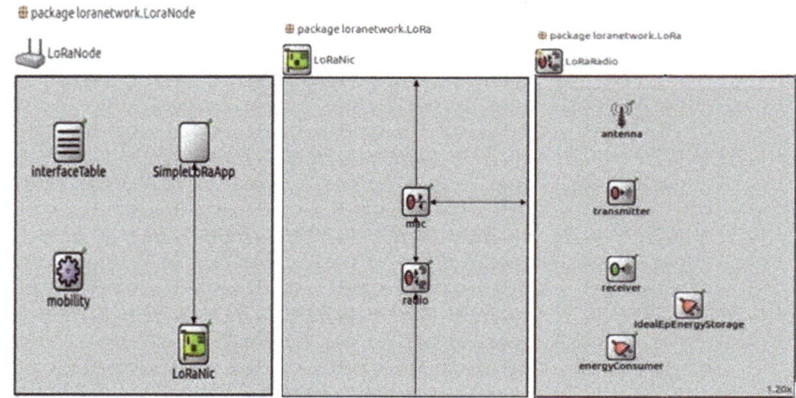

Figure 6. Sub-modules of LoRa Node.

In case '1', initially, we varied the code rate from 1 to 4, where no substantial changes are identified in the energy consumption of the LoRa node. Later on, we varied the transmission (Tx) power of the LoRa node from 2, 5, 8, 11, and 14 dBm and found the change in the energy consumption. We denoted the transmission power of the LoRa node as Tx1, Tx2, Tx3, Tx4, and Tx5. Here, remaining parameters, like CF, SF, B/W, CR, no of packets, network size, and gateway distance, are kept constant. Table 4 illustrates the energy consumption of the nodes '1'.

Table 4. Energy consumption of the nodes '1'.

Node	Tx Power = 2 dBm	Tx Power = 5 dBm	Tx Power = 8 dBm	Tx Power = 11 dBm	Tx Power = 14 dBm
1	2.45 mW	2.46 mW	2.46 mW	2.09 mW	2.18 mW
2	2.45 mW	2.46 mW	2.18 mW	1.9 mW	2.03 mW
3	1.89 mW	1.90 mW	2.74 mW	3.10 mW	3.24 mW
4	2.38 mW	2.39 mW	2.60 mW	2.31 mW	2.41 mW

Figure 7 illustrates the graphical representation of the energy consumption of nodes of four different cases. The simulation results show that node '1' and node '2' has the same amount of energy consumption at the transmission (Tx) power of 2 and 5 dBm. In this case of node '2', gradually, the energy consumption decreases at the Tx power of 11 dBm. Energy consumption of the node '3' is suddenly increased at the Tx power of 8 dBm, i.e., from 1.90 mW to 2.74 mW. Node '4' consumed the constant energy consumption for Tx power of 2 and 5 dBm; however, the energy consumption is triggered high at the Tx power of 8 dBm and suddenly the energy consumption is reduced to 2.31 mW. The part 'a' of Figure 7 reveals that the node '3' (red color line) is consuming high energy consumption of all nodes from the Tx power of 8 dBm. The part 'a' reveals that the node '3' (red color line) is consuming high energy consumption of all nodes from the Tx power of 8 dBm. In the case '2', we performed a simulation with 2 gateways by maintaining constant other parameters. The energy consumption of the 4 nodes are presented in Table 5. As the gateways have increased to two, the energy consumption of all nodes gradually decreased at the Tx power of 2 dBm.

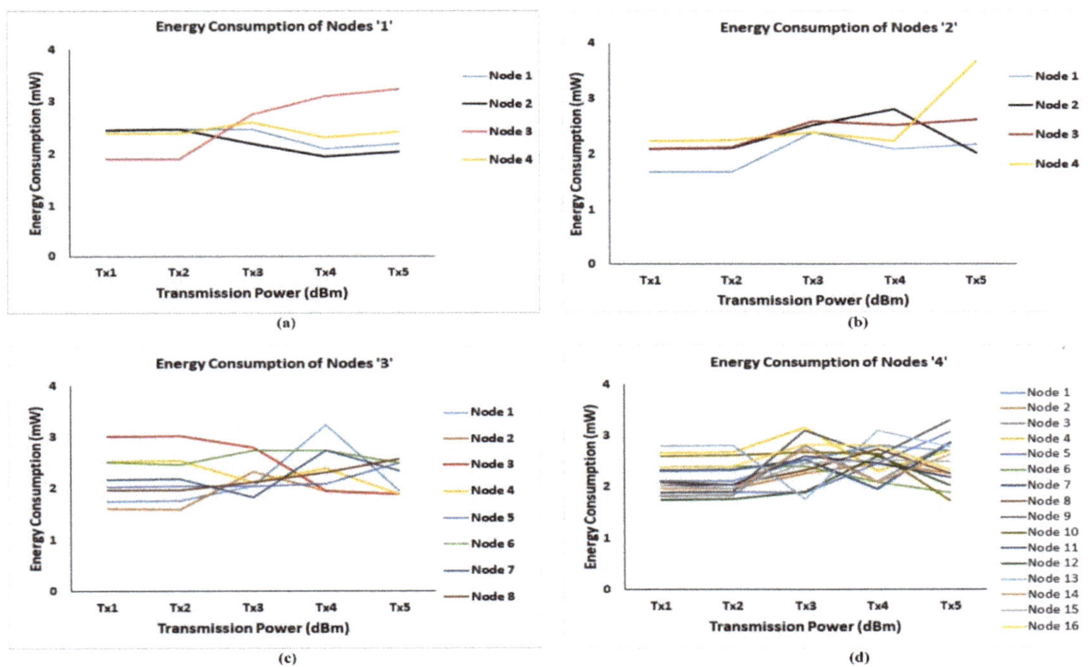

Figure 7. Energy consumption of nodes (a–d).

Table 5. Energy consumption of nodes '2'.

Node	Tx Power = 2 dBm	Tx Power = 5 dBm	Tx Power = 8 dBm	Tx Power = 11 dBm	Tx Power = 14 dBm
1	1.68 mW	1.69 mW	2.39 mW	2.10 mW	2.78 mW
2	2.10 mW	2.11 mW	2.53 mW	2.81 mW	2.03 mW
3	2.10 mW	2.12 mW	2.60 mW	2.52 mW	2.63 mW
4	2.24 mW	2.25 mW	2.39 mW	2.24 mW	3.69 mW

The energy consumption of Node '1 decreased for the Tx of 2 dBm, 5 dBm, and 8 dBm is decreased when compared with a single gateway. In Node '4', the energy consumption at all Tx power is decreased except for the Tx power of 14 dBm, where, at Tx power = 14 dBm, the energy consumption is the highest energy consumption among all nodes for all Tx power. Except for Tx power of 2 dBm and 5 dBm, the energy consumption of the Node '3' is decreased at the remaining Tx power. The graphical representation of the energy consumption is represented in Figure 7b. In case '3', we increased the number of nodes to 8 and the remaining parameters of the simulation are kept constant as assumed in the case '1'. As the number of nodes increased to the 8, the nodes are represented as node '1', node '2', node '3', node '4', node '5', node '6', node '7', and node '8'. The case '3' is represented as energy consumption of nodes '3', and values are presented in Table 6. Figure 7c illustrates the graphical representation of case '3'.

Table 6. Energy consumption of nodes '3'.

Node	Tx Power = 2 dBm	Tx Power = 5 dBm	Tx Power = 8 dBm	Tx Power = 11 dBm	Tx Power = 14 dBm
1	1.75 mW	1.76 mW	2.04 mW	3.25 mW	1.95 mW
2	1.61 mW	1.59 mW	2.32 mW	1.95 mW	1.88 mW
3	3.01 mW	3.02 mW	2.88 mW	1.95 mW	1.88 mW
4	2.52 mW	2.53 mW	2.11 mW	2.38 mW	1.88 mW
5	2.03 mW	2.04 mW	2.04 mW	2.09 mW	2.48 mW
6	2.51 mW	2.46 mW	2.74 mW	2.74 mW	2.48 mW
7	2.17 mW	2.18 mW	1.83 mW	2.74 mW	2.33 mW
8	1.96 mW	1.97 mW	2.11 mW	2.31 mW	2.56 mW

The amount of energy consumption of the first four nodes has varied concerned to the previous case. In this case, Node '1' has consumed a high amount of energy at the Tx power of 11 dBm, i.e., 3.25 dBm, and Node '2' has consumed a low amount of energy at the Tx power of 5 dBm, i.e., 1.61 dBm. At Tx power of 14 dBm, the energy consumption for the node '2', '3', and '4' is the same, i.e., 1.88 mW, and also for the Nodes '5' and '6', the energy consumption remains the same, i.e., 2.48 mW. For node '3', the energy consumption gradually decreases from 3.01 mW to 1.88 mW. The energy consumption of Node '5' gradually increased from 2.03 mW to 2.48 mW, and node '8' also increased from 1.96 mW to 2.56 mW. The graphical representation of the energy consumption is represented in Figure 7c. In case '4', we performed the simulation by increasing the number of nodes to 16, and the remaining parameters are kept constant. The case '4' is represented as energy consumption of nodes '4' and is presented in Table 7. The 16 nodes are represented as node '1', node '2', node '3', node '4', node '5', node '6', node '7', node '8', node '9', node '10', node '11', node '12', node '13', node '14', node '15', and node '16'. The high energy consumption of nodes is recorded at Tx of 11 dBm for the Node '13', i.e., 3.10 mW, and low energy consumption of nodes is recorded at Tx power of 14 dBm for the Node '10', i.e., 1.73 mW. Here, the energy consumption for Node '5' is increased, and, for Node '6', from the Tx power of 2 dBm to 14 dBm. The energy consumption for Node 1, 3, 4, 5, 6, 7, 9, 10, 11, 12, 14, and 16 are increased from Tx power of 2 dBm to 8 dBm and for node 2, 8 the energy consumption have increased from 2 dBm to 11 dBm. The graphical representation of the energy consumption is represented in Figure 7d.

6.2. Data Rate/Bit Rate

The data rate/bit rate is defined as the quantity of bits that are transferred during transmission between transmitter and receiver. We utilized Equation (3) for calculating the data rate of the LoRa. To calculate the data rate of LoRa, the input parameters, like CR, SF, and BW, are included in the equation. Figure 8 presents the data rate of LoRa from SF 7 to SF 12. The data rate is denoted in terms of bits per second (bps). It can be observed that in every SF, the data rate is in increasing at the point of BW 7 (62.5 kHz) and it exponentially raised after the BW 8 and reached maximum at the BW 10 (500 kHz). We can observe that increasing the SF7 is inversely affecting the data rate, as the data rate declines gradually from SF 7 to SF 12. In SF 7, the data rate of the LoRa touched 22,000 bps and in SF 12 the data rate is limited to 2000 bps. An increase in SF will lead to transmitting a low amount of data during transmission, so SF 7 is the optimal SF that needs to be considered for sending a large amount of the data.

Table 7. Energy consumption of nodes '4'.

Node	Tx Power = 2 dBm	Tx power =5 dBm	Tx Power = 8 dBm	Tx Power = 11 dBm	Tx Power = 14 dBm
1	2.11 mW	2.11 mW	2.53 mW	2.81 mW	2.71 mW
2	1.96 mW	1.97 mW	2.26 mW	2.46 mW	2.26 mW
3	1.82 mW	1.83 mW	2.81 mW	2.09 mW	2.86 mW
4	2.66 mW	2.67 mW	3.17 mW	2.31 mW	2.71 mW
5	1.89 mW	1.90 mW	1.89 mW	2.46 mW	3.08 mW
6	2.38 mW	2.39 mW	2.39 mW	2.09 mW	1.88 mW
7	2.31 mW	2.32 mW	2.53 mW	1.96 mW	2.86 mW
8	2.09 mW	2.04 mW	2.32 mW	2.74 mW	2.26 mW
9	1.89 mW	1.90 mW	3.10 mW	2.60 mW	3.31 mW
10	2.60 mW	2.61 mW	2.67 mW	2.64 mW	1.73 mW
11	2.03 mW	2.04 mW	2.60 mW	2.45 mW	2.18 mW
12	1.75 mW	1.76 mW	1.90 mW	2.60 mW	2.03 mW
13	2.80 mW	2.81 mW	1.76 mW	3.10 mW	2.78 mW
14	1.96 mW	1.97 mW	2.74 mW	2.09 mW	2.63 mW
15	2.03 mW	1.97 mW	2.46 mW	2.52 mW	2.48 mW
16	2.38 mW	2.39 mW	2.82 mW	2.81 mW	2.33 mW

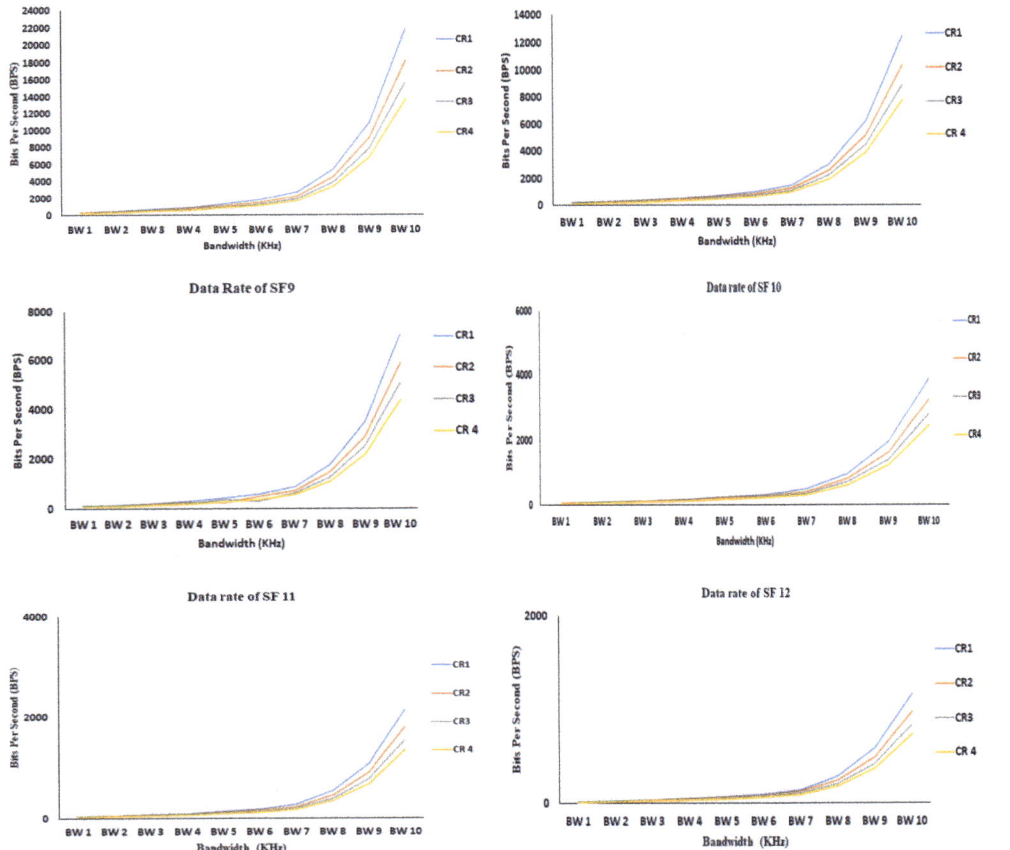

Figure 8. Data rate of LoRa from spreading factor (SF) 7 to SF 12.

6.3. LoRa Sensitivity

As discussed earlier in Section 3, the spreading factor, noise figure, and bandwidth are the input parameters for calculating the LoRa sensitivity of the receiver. Here, we utilized Equation (4) for calculating the sensitivity of the receiver. Noise fig. (SNR) value is different for distinct SF. SNR of the SF 7 is −7.5, SF 8 is −10, SF9 is −12.5, SF 10 is −15, SF 11 is 17.5, and SF 12 is −20. The above SNR value is considered for calculating the sensitivity of the receiver. As discussed in the data rate section, the same type of BW is utilized for sensitivity calculation. Generally, the sensitivity of power is in terms of a negative value; for example, −127 dBm, where the value going above this value indicates that the sensitivity is in decline. Figure 9 presents the sensitivity from the SF 7 to SF 12; the sensitivity power is highest for the SF 7 at the BW 10 (500 kHz), and the lowest sensitivity is observed for the SF 12 at the BW 1 (7.5 kHz). The sensitivity of every SF is gradually increasing after the BW 4 (20.8 kHz).

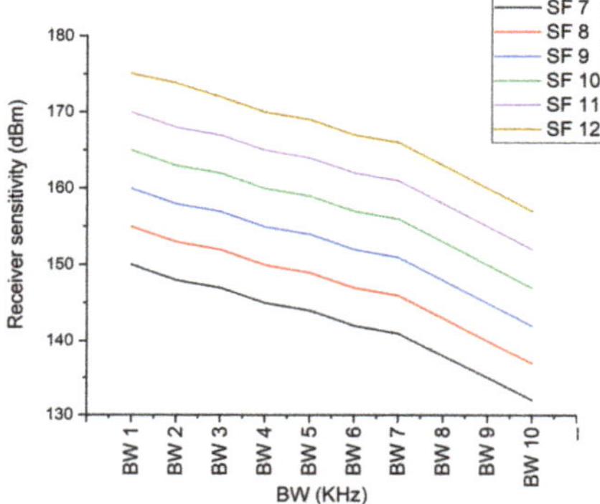

Figure 9. LoRa sensitivity from (SF 7–SF 12).

6.4. Time on Air (ToA)

In this study, to calculate the ToA, we set certain parameters, like payload = 25 bytes, preamble = 8 symbols, CF = 433 MHz, and Tx power = 14 dBm. The tool for calculating the ToA is considered from this [43]. Here, we considered three different BW, namely 125 kHz, 250 kHz, and 500 kHz, SF from 7 to 12, and CR from 1 to 4. The calculated results are recorded in the below tabular form. ToA is calculated at the distinct CR, namely 1, 2, 3, 4, and the ToA is illustrated in Tables 8–11. The table delivers that the ToA increases when the SF moves from 7 to 12 and ToA is reduced by half when the bandwidth increases to 125 kHz to 500 kHz. Figure 10 illustrates the ToA for the payload size of 25 bytes. The graph of the ToA at distinct CR reveals that the ToA starts increasing after the SF 9. Of all three BWs, the ToA is high for the 125 kHz BW and low for the 500 kHz BW. It is observed that the CR is directly proportional to the increase in the ToA; for example, at CR 1, the ToA of BW 1 is 56.58 ms, and, at the same BW 1, the ToA at CR 4 is 78.08 ms.

Table 8. Time on-air (ToA) (ms) at code rate (CR) 1.

SF	BW 1 = 125 KHz	BW 2 = 250 KHz	BW 3 = 500 KHz
SF 7	56.58	28.29	14.14
SF 8	102.91	51.46	25.73
SF 9	205.82	102.91	51.46
SF 10	370.69	185.34	92.67
SF 11	741.38	370.69	185.34
SF 12	1318.91	659.46	329.73

Table 9. ToA (ms) at CR 2.

SF	BW 1 = 125 KHz	BW 2 = 250 KHz	BW 3 = 500 KHz
SF 7	63.74	31.87	15.94
SF 8	115.2	57.6	28.8
SF 9	230.4	115.2	57.6
SF 10	411.65	205.82	102.91
SF 11	823.3	411.65	205.82
SF 12	1449.98	724.99	362.5

Table 10. ToA (ms) at CR 3.

SF	BW 1 = 125 KHz	BW 2 = 250 KHz	BW 3 = 500 KHz
SF 7	70.91	35.46	17.73
SF 8	127.49	63.74	31.87
SF 9	254.98	127.49	63.74
SF 10	452.61	226.3	113.15
SF 11	905.22	452.61	226.3
SF 12	1581.06	790.53	395.26

Table 11. ToA (ms) at CR 4.

SF	BW 1 = 125 KHz	BW 2 = 250 KHz	BW 3 = 500 KHz
SF 7	78.08	39.04	19.52
SF 8	139.78	69.89	34.94
SF 9	279.55	139.78	69.89
SF 10	493.57	246.78	123.39
SF 11	987.14	493.57	246.78
SF 12	1712.13	856.06	428.03

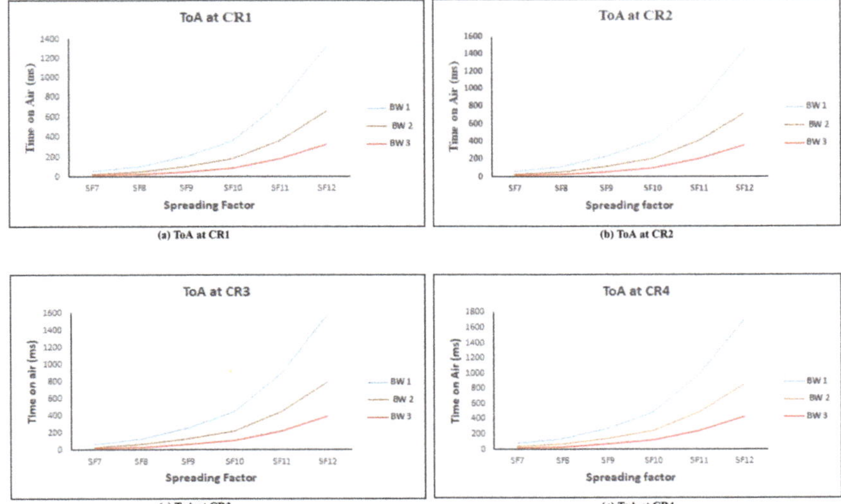

Figure 10. ToA for a payload of 25 bytes.

6.5. Link Budget

Link budget or received power is the summation of the losses and gains. The concept of the link budget is addressed in Section 3. In order to calculate the link budget, we considered Equation (11). We considered the three BW, namely 125 kHz, 250 kHz, and 500 KHz, and SF from 7 to 12.

As the link budget is the sum of the transmitted power (Tx) and receiver sensitivity, the increase in the Tx is leading to an increase in the link budget, and it can be observed in Figure 11. For every increment in the SF, there is also arises in the link budget. At 14 dBm, the link budget achieved maximum dBm for 125 kHz of SF 12. At 2 dBm, the link budget recorded low dBm for 500 kHz of SF7. It is concluded that the maximum link budget is achieved with high Tx power and low BW.

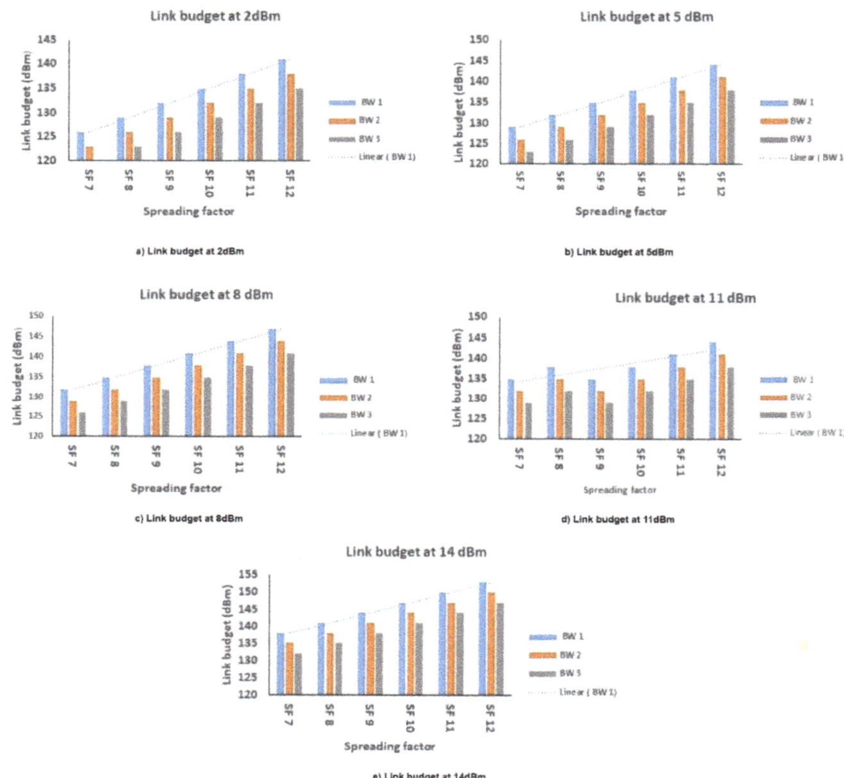

Figure 11. Link budget at distinct transmitter power (Tx).

6.6. Battery Life of Sensor Node

The calculating battery life of the sensor node is a necessary element in the hardware prototype. Here, we assumed the processing power of 15 mW for 5 ms during sensing value and power consumption at the sleep is 10 µW. SF 7 with 125 kHz BW and 2 dB Tx power are considered due to its optimal energy consumption during transmission. We subdivided the evaluation of battery life into three scenarios based on payload. Figure 12 presents the distinct battery life scenarios, where the battery life is represented as a time to live (TTL). In our study, we need to sense only the level measurement data, so, here, we considered a payload of 22 bytes, 25 bytes, and 28 bytes for analyzing the battery life at distinct periodicity. Here, periodicity is considered in terms of minutes likely, 15 min, 30 min, 45 min, and 60 min. The battery life of the sensor node is about months, and three distinct battery capacities are considered likely 260 mAh, 1000 mAh, and 2000 mAh Li-ion batteries. If the periodicity of sending the data from the sensor to the receiver end is 60 min, the battery life is long, and, in the case where the periodicity of sending the data of sensor is 15 min, the battery life is shorter. It is also observed that the battery life is the same for the (i) 260 mAh and 1000 mAh at a payload of 22 bytes and 25 bytes (ii) 1000 mAh at a payload of 22 bytes and 25 bytes.

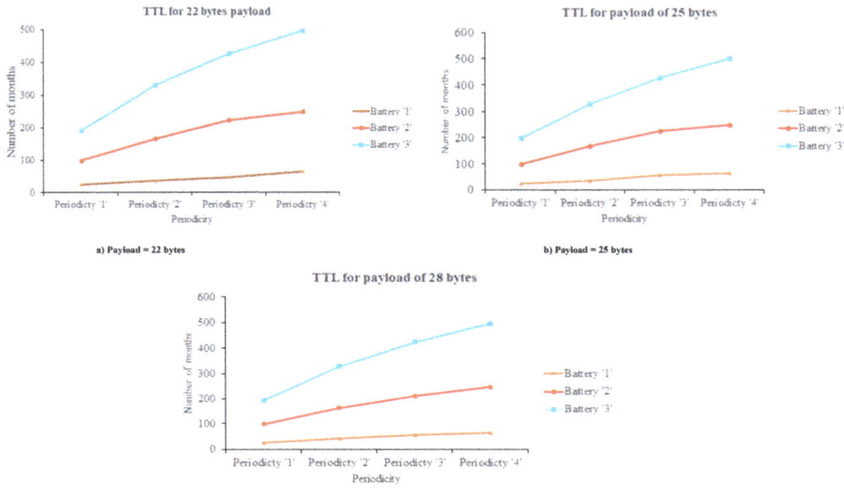

Figure 12. Time to live (TTL) of sensor node.

7. Result Analysis

This section covers the deployment of sensor node and gateway in a real-time environment, and we will also present the sensors data that recorded on the cloud server and also a comparison of the previous studies with proposed studies is mention in detail. To evaluate the coverage of the LoRa, we deployed our sensor nodes in the bins that are located in our university. The gateway is placed at 320 m from the sensor nodes. In this study, we designed the bins for evaluating the filling levels of the bins. LoRa-based sensor node is deployed in the four bins with HC-SR 04 ultrasonic sensor (https://www.sparkfun.com/products/15569 (accessed on 7 March 2021)) with the sensor that measures the level in the bins. LoRa network is deployed in the sensor nodes according to the architecture that is proposed in Figure 4. Sensor nodes are connected with a gateway for receiving the status of the bins via the 433 MHz LoRa module. The gateway is placed at a distance of 320 m from the bins. The gateway is effectively receiving the data from the sensor node and it is logging in to the cloud server through ESP 8266 Wi-Fi module. Figure 13 illustrates the deployment of the sensor node and gateway in the real-time environment. The status of the bins that are recorded in the cloud server with graphical representation is presented in Figure 14. Distinct color representation in the graph provides the % of waste-filled in the bins. The green color represents that the bin is filled 50% and the red color represents that the bin is filled 100%. Orange and blue color represents that bins are 75% filled. With this result, we conclude that the proposed architecture based on LoRa can provide real-time data of the bins on the cloud server. Figure 14 presents that bin '2' and bin '4' are having constant waste for the prescribed time format.

Figure 13. Sensor node deployed in the bins.

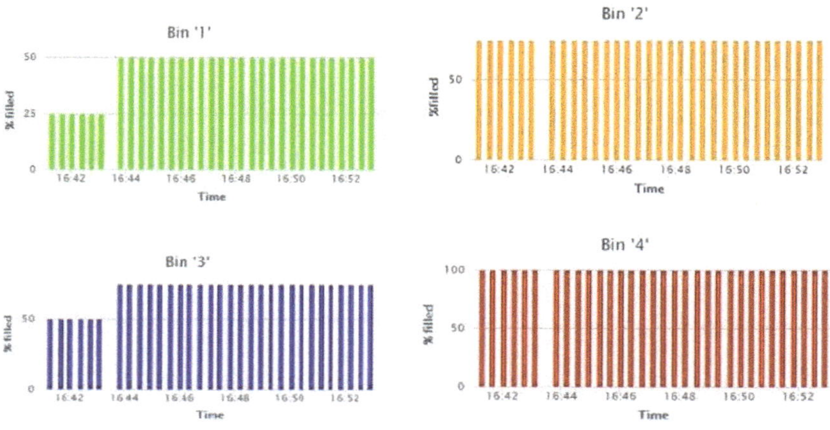

Figure 14. Bin status on the cloud server.

Table 12 illustrates the comparison of previous studies focused on LoRa-based waste management with our study. The comparison is done based on the certain parameters that are illustrated in Table 12. The proposed study is having designed customized sensor node and LoRa-based gateway for realizing the LoRa-based waste management. In our implementation, we performed the simulation on FLoRa where the energy consumption of the nodes is estimated by changing distinct network parameters, like BW, CR, CF, number of nodes, and gateways. Even the distinct evaluation metrics include data rate, ToA, receiver sensitivity, link budget, and also estimated the battery life of the sensor node. The simulation of LoRa network for the waste management on the FLoRa simulator is missing in the many articles related to the waste management and also the estimated lifetime of the customized sensor node are included in our implementation. We recorded the sensor data from the sensor node with our customized sensor node and gateway in the cloud server. The data in Table 12 concludes that the customized gateway with LoRa and ESP 8266 module are implemented in limited studies. Simulation-based analysis and plot of evaluation metrics of LoRa are not addressed in any studies. Finally, the realization of the sensor data on the cloud server is observed in the few studies. Table 13 illustrates the main achievements that are achieved in the results section. The real time implementation of the proposed architecture with the customized sensor node and gateway with cloud server is the significant achievement in the results.

Table 12. Comparison of our study with previous studies.

Research	MCU	Communication	Designed Sensor Node	Designed Gateway	Customized Node	Proof of Concept	Simulation-Based Analysis	Plot of Evaluation Metrics
[14]	Arduino Uno	SX 1272 LoRa & Waspmote	Yes	No	Yes	Yes	No	No
[27]	ATSAML21	SX 1276 LoRa	Yes	No	Yes	Yes	No	No
[28]	Atmega328P	SX 1278 LoRa	Yes	No	Yes	Yes	No	No
[29]	Atmega328P	SX 1272 LoRa	Yes	No	Yes	Yes	No	No
[44]	NA	NA	Yes	No	No	Yes	No	No
[45]	Arduino Uno	SX 1272 LoRa	No	No	No	No	No	No
[46]	Raspberry Pi3	IP67 LoRa gateway	No	Yes	Yes	Yes	No	No
[47]	Atmega328P	SX 1278 LoRa	Yes	Yes	Yes	Yes	No	No
Proposed study	Atmega328P	SX 1278 LoRa	Yes	Yes	Yes	Yes	Yes	Yes

Table 13. Achievements from the result section.

S.No	Parameters	Achievements
1	Energy consumption	Energy consumption of nodes are calculated in FLoRa simulation by changing distinct network parameters
2	Data rate	An increase in SF leads to transmission of a low data rate, so SF 7 is the optimal for sending a large amount of the data.
3	Sensitivity power	The sensitivity power is highest for the SF 7 at the 500 kHz.
4	ToA	ToA is good at 125 KHz and code rate 1, i.e., 56.58 ms.
5	Link budget	At 14 dBm, 125 kHz and SF 12, maximum link budget is achieved.
6	Battery life of sensor node	Periodicity of transmitting data = 60 min, the battery life is long, if Periodicity of transmitting data = 15 min, the battery life is short.
7	Real time experiment set up	Realized the objective of proposed architecture in Figure 4, where the sensor data is logging into the cloud server through the customized sensor node and gateway.

8. Conclusions

Wireless communication protocol plays key role for reliable transmission between the sensor nodes and cloud server in IoT. LoRa is one of the prominent and emergent wireless communication that meet the requirements of IoT in terms of minimum power consumption and long-range transmission. In this paper, we implemented the LoRaWAN architecture for the realizing of the real-time filling level of the bins with customized nodes. LoRa-based customized sensor nodes are deployed in the bins for establishing energy efficient nodes for transmitting the collected sensor data to the cloud server. FLoRa simulation based on LoRaWAN is performed for evaluating the energy consumption of the nodes by varying distinct parameters and also evaluated the metrics of LoRa network with plots. Finally, to show real application, an experimental set up is implemented for checking the performance of the proposed LoRa architecture for sensing the level of the bins in real-time scenario and logging the status of each of the bins in the cloud server.

Author Contributions: S.V.A. and R.S. made contributions to conception and manuscript writing; A.G. and M.A.A. examined and supervised this research and outcomes; O.S.F., M.R., W.E.-S., and D.P. revised and polished the manuscript. All authors have read and agreed to the published version of the manuscript.

Funding: This study was funded by the Deanship of Scientific Research, Taif University Researchers Supporting Project number (TURSP-2020/98), Taif University, Taif, Saudi Arabia.

Institutional Review Board Statement: Not applicable.

Informed Consent Statement: Not applicable.

Data Availability Statement: Not applicable.

Conflicts of Interest: The authors declare no conflict of interest.

Abbreviations

List of abbreviations used in study:

BW	Bandwidth
CF	Carrier Frequency
CR	Code Rate
CRC	Cyclic Redundancy Check
CSS	Chirp Spread Spectrum
FEC	Forward Error Correction
FLoRa	Framework for LoRa
FSPL	Free Space Path Loss
GSM	Global System for Mobile communication
GPRS	Global packet radio service
GPS	Global Positioning System
IEEE	Institute of Electrical and Electronics Engineers
IoT	Internet of Things
IP	Internet Protocol
LoRa	Long Range
LoRaWAN	LoRa Wide Area Network
MQTT	Message Queuing Telemetry Transport
PCB	Printed Circuit Board
RFID	Radio Frequency Identification
SF	Spreading Factor
SoC	System on a Chip
ToA	Time on Air
SNIR	Signal-to-Noise Ratio
VANET	Vehicular Ad Hoc Network
Wi-Fi	Wireless Fidelity
WSN	Wireless Sensor Network

References

1. THE 17 GOALS | Sustainable Development. Available online: https://sdgs.un.org/goals (accessed on 28 December 2020).
2. Manager. The Weight of Cities. 2018. Available online: https://www.resourcepanel.org/reports/weight-cities (accessed on 1 January 2021).
3. Kumar, S.; Smith, S.R.; Fowler, G.; Velis, C.; Kumar, S.J.; Arya, S.; Rena; Kumar, R.; Cheeseman, C. Challenges and opportunities associated with waste management in India. *R. Soc. Open Sci.* **2017**, *4*, 160764. [CrossRef] [PubMed]
4. Sinha, R.S.; Wei, Y.; Hwang, S.-H. A survey on LPWA technology: LoRa and NB-IoT. *ICT Express* **2017**, *3*, 14–21. [CrossRef]
5. Nižetić, S.; Šolić, P.; González-de-Artaza, D.L.; Patrono, L. Internet of Things (IoT): Opportunities, issues and challenges towards a smart and sustainable future. *J. Clean. Prod.* **2020**, *274*, 122877. [CrossRef] [PubMed]
6. Kumar, S.; Jasuja, A. Air quality monitoring system based on IoT using Raspberry Pi. In Proceedings of the IEEE International Conference on Computing, Communication and Automation, ICCCA 2017, Greater Noida, India, 5–6 May 2017; Institute of Electrical and Electronics Engineers Inc.: Greater Noida, India, 2017; pp. 1341–1346.
7. Gomez, C.; Chessa, S.; Fleury, A.; Roussos, G.; Preuveneers, D. Internet of Things for enabling smart environments: A technology-centric perspective. *J. Ambient. Intell. Smart Environ.* **2019**, *11*, 23–43. [CrossRef]

8. Hussain, A.; Draz, U.; Ali, T.; Tariq, S.; Irfan, M.; Glowacz, A.; Daviu, J.A.A.; Yasin, S.; Rahman, S. Waste Management and Prediction of Air Pollutants Using IoT and Machine Learning Approach. *Energies* **2020**, *13*, 3930. [CrossRef]
9. Jain, A.; Sharma, B.; Gupta, P. Internet of things: Architecture, security goals, and challenges—A survey. *Int. J. Innov. Res. Sci. Eng.* **2016**, *2*, 154–163.
10. Longhi, S.; Marzioni, D.; Alidori, E.; Buo, G.D.; Prist, M.; Grisostomi, M.; Pirro, M. Solid waste management architecture using wireless sensor network technology. In Proceedings of the 2012 5th International Conference on New Technologies, Mobility and Security, Istanbul, Turkey, 7–10 May 2012; pp. 1–5.
11. Dubey, S.; Singh, M.K.; Singh, P.; Aggarwal, S. Waste Management of Residential Society using Machine Learning and IoT Approach. In Proceedings of the International Conference Emerging Smart Computing and Informatics, ESCI 2020, Pune, India, 12–14 March 2020; pp. 293–297.
12. Madakam, S.; Ramaswamy, R.; Tripathi, S. Internet of Things (IoT): A Literature Review. *J. Comput. Commun.* **2015**, *3*, 164–173. [CrossRef]
13. Mekki, K.; Bajic, E.; Chaxel, F.; Meyer, F. A comparative study of LPWAN technologies for large-scale IoT deployment. *ICT Express.* **2019**, *5*, 1–7. [CrossRef]
14. Sheng, T.J.; Islam, M.S.; Misran, N.; Baharuddin, M.H.; Arshad, H.; Islam, M.R.; Chowdhury, M.E.H.; Rmili, H.; Islam, M.T. An internet of things based smart waste management system using LoRa and tensorflow deep learning model. *IEEE Access.* **2020**, *8*, 148793–148811. [CrossRef]
15. Lokhande, P.; Pawar, M.D. Garbage Collection Management System. *Int. J. Eng. Comput. Sci.* **2016**, *5*, 18800–18805. [CrossRef]
16. Kumar, G.N.; Swamy, C.; Nagadarshini, K.N. Efficient garbage disposal management in metropolitan cities using VANETs. *J. Clean Energy Technol.* **2014**, *2*, 258–262.
17. Lata, K.; Singh, S.S.K. IOT based smart waste management system using Wireless Sensor Network and Embedded Linux Board. *Int. J. Curr. Trends Eng. Res.* **2016**, *2*, 210–214.
18. Omara, A.; Gulen, D.; Kantarci, B.; Oktug, S.F. Trajectory-Assisted Municipal Agent Mobility: A Sensor-Driven Smart Waste Management System. *J. Sens. Actuator Netw.* **2018**, *7*, 29. [CrossRef]
19. Suryawanshi, S.; Bhuse, R.; Gite, M.; Hande, D. Waste Management System Based On IoT. *Int. Res. J. Eng. Technol.* **2018**, *5*, 1835–1837.
20. Cavdar, K.; Koroglu, M.; Akyildiz, B. Design and implementation of a smart solid waste collection system. *Int. J. Environ. Sci. Technol.* **2016**, *13*, 1553–1562. [CrossRef]
21. Karthikeyan, S.; Rani, G.S.; Sridevi, M.; Bhuvaneswari, P.T.V. IoT enabled waste management system using ZigBee network. In Proceedings of the 2nd IEEE International Conference Recent Trends ELECTRONICS Information and Communication Technology, Bangalore, India, 19–20 May 2017; pp. 2182–2187.
22. Reis, P.; Caetano, F.; Pitarma, R.; Gonçalves, C. IEcoSys—An intelligent waste management system. In *Advances in Intelligent Systems and Computing*; Springer: Berlin/Heidelberg, Germany, 2015; pp. 843–853.
23. Satria, D.; Hidayat, T. Implementation of wireless sensor network (WSN) on garbage transport warning information system using GSM module. *J. Phys. Conf. Ser.* **2019**, *1175*, 12054. [CrossRef]
24. Zavare, S.; Parashare, R.; Patil, S.; Rathod, P.; Babanne, V. Smart City waste management system using GSM. *Int. J. Comput. Sci. Trends Technol.* **2017**, *5*, 74–78.
25. Augustin, A.; Yi, J.; Clausen, T.; Townsley, W.M. A study of Lora: Long range & low power networks for the internet of things. *Sensors* **2016**, *16*, 1466.
26. Saari, M.; bin Baharudin, A.M.; Sillberg, P.; Hyrynsalmi, S.; Yan, W. LoRa—A survey of recent research trends. In Proceedings of the 41st International Convention on Information and Communication Technology, Electronics and Microelectronics (MIPRO), Opatija, Croatia, 21–25 May 2018; pp. 872–877.
27. Lozano, Á.; Caridad, J.; Paz, J.F.D.; González, G.V.; Bajo, J. Smart waste collection system with low consumption LoRaWAN nodes and route optimization. *Sensors* **2018**, *18*, 1465. [CrossRef]
28. Cerchecci, M.; Luti, F.; Mecocci, A.; Parrino, S.; Peruzzi, G.; Pozzebon, A. A low power IoT sensor node architecture for waste management within smart cities context. *Sensors* **2018**, *18*, 1282. [CrossRef]
29. Addabbo, T.; Fort, A.; Mecocci, A.; Mugnaini, M.; Parrino, S.; Pozzebon, A.; Vignoli, V. A LoRa-based IoT Sensor Node for Waste Management Based on a Customized Ultrasonic Transceiver. In Proceedings of the IEEE Sensors Applications Symposium, Sophia Antipolis, France, 11–13 March 2019; pp. 1–6.
30. Lavric, A.; Petrariu, A.I.; Coca, E.; Popa, V. Lora traffic generator based on software defined radio technology for lora modulation orthogonality analysis: Empirical and experimental evaluation. *Sensors* **2020**, *20*, 4123. [CrossRef] [PubMed]
31. Semtech Corporation. SX1272/3/6/7/8 LoRa Modem Design Guide, AN1200.13. 2013. p. 9. Available online: https://www.rs-online.com/ (accessed on 7 March 2021).
32. Ortín, J.; Cesana, M.; Redondi, A. How do ALOHA and listen before talk coexist in LoRaWAN? In Proceedings of the IEEE 29th Annual International Symposium on Personal, Indoor and Mobile Radio Communications, Bologna, Italy, 9–12 September 2018; pp. 1–7.
33. Balanis, C.A. *Antenna Theroy: Analysis and Design*; John Wiley & Sons: Hoboken, NJ, USA, 2012; p. 109.
34. Sendra, S.; García, L.; Lloret, J.; Bosch, I.; Vega-Rodríguez, R. LoRaWAN network for fire monitoring in rural environments. *Electronics* **2020**, *9*, 531. [CrossRef]

35. Semtech Corporation. SX1276/77/78/79—137 MHz to 1020 MHz Low Power Long Range Transceiver. 2016. Available online: https://www.scribd.com/document/399165229/SX1276-1278 (accessed on 7 March 2021).
36. Atmel Corporation. Data Sheet ATmega328P. 2015; pp. 1–294. Available online: http://ww1.microchip.com/downloads/en/DeviceDoc/Atmel-7810-Automotive-Microcontrollers-ATmega328P_Datasheet.pdf (accessed on 9 March 2021).
37. ESP8266 Wi-Fi MCU I Espressif Systems. 2020. Available online: https://www.espressif.com/en/products/socs/esp8266 (accessed on 9 March 2021)
38. Magrin, D.; Centenaro, M.; Vangelista, L. Performance evaluation of LoRa networks in a smart city scenario. In Proceedings of the IEEE International Conference on Communications, Paris, France, 21–25 May 2017; pp. 1–7.
39. OMNeT++ Discrete Event Simulator. Available online: https://omnetpp.org/ (accessed on 18 February 2021).
40. Home I FLoRa—A Framework for LoRa Simulations. Available online: https://flora.aalto.fi/ (accessed on 18 February 2021).
41. Slabicki, M.; Premsankar, G.; Francesco, M.D. Adaptive configuration of LoRa networks for dense IoT deployments. In Proceedings of the NOMS 2018–2018 IEEE/IFIP Network Operations and Management Symposium, Taipei, Taiwan, 23–27 April 2018; pp. 1–9.
42. Petajajarvi, J.; Mikhaylov, K.; Roivainen, A.; Hanninen, T.; Pettissalo, M. On the coverage of LPWANs: Range evaluation and channel attenuation model for LoRa technology. In Proceedings of the 14th International Conference on ITS Telecommunications, Copenhagen, Denmark, 2–4 December 2015; pp. 55–59.
43. LoRaTool. Available online: https://www.loratools.nl (accessed on 1 April 2021).
44. Lundin, A.C.; Ozkil, A.G.; Schuldt-Jensen, J. Smart cities A case study in waste monitoring and management. In Proceedings of the 50th Hawaii International Conference on System Sciences, Waikoloa, HI, USA, 4–7 January 2017; pp. 1392–1401.
45. Ziouzios, D.; Dasygenis, M. A smart bin implementantion using LoRa. In Proceedings of the 2019 4th South-East Europe Design Automation, Computer Engineering, Computer Networks and Social Media Conference (SEEDA-CECNSM 2019), Piraeus, Greece, 20–22 September 2019; Institute of Electrical and Electronics Engineers Inc.: Piraeus, Greece, 2019.
46. Beliatis, M.J.; Mansour, H.; Nagy, S.; Aagaard, A.; Presser, M. Digital waste management using LoRa network a business case from lab to fab. In Proceedings of the Global Internet Things Summit, GIoTS 2018, Bilbao, Spain, 4–7 June 2018; pp. 1–6.
47. Khoa, T.A.; Phuc, C.H.; Lam, P.D.; Nhu, L.M.B.; Trong, N.M.; Phuong, N.T.H.; Dung, N.V.; Nguyen, N.; Ta, H.N.; Duc, D.N.M. Waste Management System Using IoT-Based Machine Learning in University. *Wirel. Commun. Mob. Comput.* **2020**, *2020*, 6138637.

Article

Temporal Changes in Air Quality According to Land-Use Using Real Time Big Data from Smart Sensors in Korea

Sung Su Jo, Sang Ho Lee and Yountaik Leem *

Department of Urban Engineering, Hanbat National University, Daejeon 34158, Korea;
gr181203@hanbat.ac.kr (S.S.J.); lshsw@hanbat.ac.kr (S.H.L.)
* Correspondence: ytleem@hanbat.ac.kr; Tel.: +82-42-821-1189

Received: 29 September 2020; Accepted: 4 November 2020; Published: 9 November 2020

Abstract: This study analyzed the changes in particulate matter concentrations according to land-use over time and the spatial characteristics of the distribution of particulate matter concentrations using big data of particulate matter in Daejeon, Korea, measured by Private Air Quality Monitoring Smart Sensors (PAQMSSs). Land-uses were classified into residential, commercial, industrial, and green groups according to the primary land-use around the 650-m sensor radius. Data on particulate matter with an aerodynamic diameter <10 μm (PM10) and <2.5 μm (PM2.5) were captured by PAQMSSs from September-October (i.e., fall) in 2019. Differences and variation characteristics of particulate matter concentrations between time periods and land-uses were analyzed and spatial mobility characteristics of the particulate matter concentrations over time were analyzed. The results indicate that the particulate matter concentrations in Daejeon decreased in the order of industrial, housing, commercial and green groups overall; however, the concentrations of the commercial group were higher than those of the residential group during 21:00–23:00, which reflected the vital nighttime lifestyle in the commercial group in Korea. Second, the green group showed the lowest particulate matter concentration and the industrial group showed the highest concentration. Third, the highest particulate matter concentrations were in urban areas where commercial and business functions were centered and in the vicinity of industrial complexes. Finally, over time, the PM10 concentrations were clearly high at noon and low at night, whereas the PM2.5 concentrations were similar at certain areas.

Keywords: smart sensor; real time big data; land-use; air quality; particulate matter (PM10 PM2.5)

---

1. Introduction

There has been an increase in interest in air quality owing to its effects on the health and quality of life of communities in urban areas [1]. Particularly, the effect of particulate matter influxes to cities from pollutants originating outside the cities [2] and the effect of pollutants from China, such as yellow smog [3], are factors that may amplify particulate matter concentrations in South Korea [4]. Previous studies have reported that particulate matter can have fatal impacts on vulnerable groups, including elderly people, pregnant women, and children, and that it has a close relationship with mortality rates; for instance, in the case of particulate matter with an aerodynamic diameter < 10 μm (PM10), mortality rates from disease increase by 0.3% as the concentration increases by 10 μg/m^3 [5–10].

In 2013, the International Agency for Research on Cancer under the World Health Organization (WHO) classified particulate matter as a first-class carcinogen. Accordingly, communities began to pay attention to information on the atmospheric environment (such as particulate matter generated in urban areas), and hence, relevant data were required. Recently, air quality has emerged as the most serious urban and social problem in Korea [11]. As a result, the demand for home appliances, such as air purifiers, has increased rapidly [12]. Smart city plans are being promoted by Korean local governments to address urban problems, such as air quality [13]. In the smart city plan that was presented after 2018, a number of services were proposed to solve the problem of particulate matter [13]. National Air Quality Monitoring Sensors (NAQMSs) were implemented at 502 locations nationwide (as of September 2020) and have been continuously recording atmospheric environmental data, including concentrations of PM10, particulate matter with an aerodynamic diameter < 2.5 μm (PM2.5), O_3, NO_2, CO, and SO_2. The collected data are then provided to the general public through an internet portal in Korea.

The popularization of smart sensors led by the advancement of information and communications technology (ICT) has enabled private companies to promptly provide urban environment data, such as PM10 and PM2.5 concentrations, to communities. An application called 'Air Map Korea' is one example. It collects atmospheric environmental data (including of particulate matter) through Private Air Quality Monitoring Smart Sensors (PAQMSSs) from 2400 locations across the country and provides them to the public. PAQMSSs were installed by a Korean private telecommunications company at a location where particulate matter pollution is growing seriously. The purpose of the PAQMSSs project is to measure and provide data for particulate matter at the height of citizen's breathing [14]. The majority of the nationally operated NAQMSs are located on the roofs of buildings. Considering the spread of particulate matter, it is important to measure the fine dust at the height at which citizens breathe [15]. There are 4.7 times more PAQMSSs than NAQMSs, which comprise 504 sensors across the nation that are managed by the national government.

NAQMSs guarantee reliable atmospheric environmental data; however, their high cost (USD 20,300/sensor) in addition to the difficulty in implementation at multiple locations are limitations of NAQMSs that constrain their range of coverage in urban areas. In contrast, PAQMSSs collect big data on the atmospheric environment across a wider range through affordable smart sensors and provide the data to the public free of charge.

A total of 12 NAQMSs are located in the city of Daejeon, implying that each sensor covers approximately 45 km^2 in the entire city area; in the case of urbanized areas, each sensor covers approximately 8 km^2. Given the results of previous studies that found that dust concentrations varied by land-use [16–19], NAQMSs do not provide accurate information regarding the air quality of spaces where people live and work. Affordable PAQMSSs (134 sensors) have been implemented by a private company throughout Daejeon and provide more accurate particulate matter information to the public. For instance, each PAQMSS in the entire city of Daejeon covers approximately 4 km^2, and in urbanized areas, each sensor covers approximately 0.7 km^2. In the case of urbanized areas in Daejeon, the area covered by PAQMSSs is approximately 11.4 times larger than that covered by NAQMSs.

Particulate matter research has been conducted from both humanitarian and environmental aspects. Studies in the humanities involve the relationships between, and the implications of, the number of vehicle registrations, industrial locations, traffic facilities, and particulate matter effects [20–23], the implications of particulate matter according to land-use and seasons [24–26], relationships between particulate matter, population density, and traffic volume [27–30], characteristics of particulate matter concentrations according to transportation, green areas, and building distribution [31–33], and changes in particulate matter concentrations on urban heat islands [34,35].

Several studies have been conducted regarding environmental aspects, such as the relationships between particulate matter and weather conditions (such as temperature, wind direction, wind speed, and precipitation) [36–38], characteristics of particulate matter concentrations reflecting green area structures and vegetation indices [39,40], and the effects of plants and vegetation in reducing particulate matter [41,42].

The majority of previous studies used statistical methods to analyze particulate matter based on relationships between humanitarian and environmental factors and were conducted using data collected from a limited number of NAQMSs. Although studies have been conducted on spatial aspects, as well as the implications of relationships between particulate matter risks to health, sources of occurrence, humanities, and the environment [18], there have been insufficient studies related to particulate matter distribution using spatial information. Therefore, this study analyzed changes in particulate matter concentrations according to time and land-use and the spatial characteristics of the distribution of particulate matter concentrations according to real-time using big data of PM10 and PM2.5 in Daejeon measured by PAQMSSs.

The study was conducted from September-October, (i.e., fall) 2019 in the city of Daejeon, South Korea. Data from September-October were used for the following reason: particle matter concentrations are relatively lower in Korea from September to October than in other seasons [26,43]. This means that there is little effect of influx of yellow dust from other countries (e.g., China) and variable control was done naturally [43]. Accordingly, it is possible to accurately identify which land-use has the highest impact on particulate matter concentrations. First, five time periods were classified with consideration of human behavior: AM1 (03:00–05:00), AM2 (07:00–09:00), Noon (11:00–13:00), PM1 (17:00–19:00), and PM2 (21:00–23:00). Second, the study determined the mean distance (650-m buffer) with the intention of considering PAQMSS locations and appropriately including areas based on land-use by utilizing a nearest neighbor analysis (NNA). Third, land-uses at locations where PAQMSSs were implemented were classified into four groups: residential, commercial, industrial, and green, according to the land-use ratio based on the 650-m buffer, and k-means clustering was conducted. Next, the differences and variation characteristics of the particulate matter concentrations between time and land-use groups were analyzed using nonparametric test methods, i.e., Kruskal–Wallis test and Mann–Whitney U test. Finally, the inverse distance-weighted method (IDWM) was used to determine the spatial mobility characteristics of particulate matter concentrations over time.

2. Literature Review

Types of particulate matter are determined by their aerodynamic diameter as either PM10 (<10 μm) or PM2.5 (<2.5 μm). The size of PM10 is approximately one-fifth to one-seventh of the diameter of a human hair, whereas PM2.5 is about one-twentieth to one-thirtieth [44]. There are natural and artificial sources of particulate matter, which is defined as invisible dust, including not only solid particles in the air but also smoke emitted from fossil fuels [44]. Examples of natural sources are soil and pollen, and artificial sources are generated from industries and human activities, such as exhaust fumes from cars, tire dust, and crematory fumes [44]. PM2.5 contains SO_2, NO_2, CO, and heavy metals and is a secondary pollutant generated when air pollutants, such as sulfur oxides and nitrogen oxides, combine and undergo chemical reactions [44].

Studies of air quality related to PM10 and PM2.5 that may have a critical impact on humans have been undertaken. Hwang et al. [16] assessed the status of particulate matter pollution using PM10 data obtained from 11 NAQMSs in the city of Daegu, South Korea from 2006-2008 and weather data, including wind direction and wind speed. Additionally, in this study, NAQMSs were divided into residence, commerce, industry, and green groups according to the location characteristics, and the implications of weather factors on particulate matter were analyzed depending on the land-use. The results showed that PM10 concentrations in fall and winter were higher than those in spring or summer and that the particulate matter concentrations in industrial areas were twice as high as those in residential areas. In addition, it was reported that particulate matter concentrations would be higher during days without wind and with fog.

Jeong [22] conducted a spatial distribution analysis using IDWM on the average annual PM10 concentration data collected via NAQMSs from 2000–2005 in Seoul, Korea. The results showed that the particulate matter concentration decreased in the order of winter, spring, fall, and summer and that considerable amounts of PM10 were generated in areas with traffic, dense populations, and large-scale construction sites. In other words, it concluded that particulate matter concentrations were not high across the entire city of Seoul but rather tended to be higher in certain areas.

Jeong and Lee [29] analyzed the particulate matter distribution in Seoul over time, focusing on PM10 and PM2.5 data captured by NAQMSs on the 17th and 18th January 2018. The study used IDWM to identify the relationships between land-use, traffic volume, and particulate matter. Results showed that the distribution of particulate matter concentrations exhibited different spatial and temporal patterns and that commercial areas and traffic increased the particulate matter concentrations, whereas green areas reduced the particulate matter concentrations.

Jeon et al. [18] conducted an analysis to determine whether there were local differences in the influence of variables on PM10 concentrations, based on the Seoul metropolitan area, using geographically weighted ridge regression and ordinary least squares as research methods. The independent variable was PM10 and the selected dependent variables were natural factors (temperature, precipitation, atmospheric congestion, date, etc.) and human factors (transportation, industrial, residential, commercial, livestock facilities, etc.). The results showed that the lower the precipitation and air movement, the higher the particulate matter concentration. In addition, particulate matter concentrations in livestock or industrial facilities were higher than those in residential or commercial facilities. Overall, the study showed that different factors affected particulate matter concentrations.

Choi et al. [4] investigated differences in particulate matter concentrations depending on land-use and seasons using PM10 and PM2.5 data collected from NAQMSs in Seoul in 2016. The ratio of the urbanized areas/forest areas located within a 3-km radius of the NAQMSs were divided into three groups; in all cases, the highest PM10 and PM2.5 concentrations occurred in spring and the lowest occurred in summer. Additionally, among the three groups, when the ratio of the forest areas was higher than that of the urbanized areas, particulate matter concentrations were reduced, and this effect was more pronounced in summer than in winter [4].

Choi et al. [26] analyzed the land-use type with the greatest impact on particulate matter using PM10 and PM2.5 data in Seoul in 2016. Based on correlation and regressions, the study reported that particulate matter had a negative correlation with forest areas and a positive correlation with urbanized areas. Moreover, the results showed that broad-leaved forests are more effective in reducing particulate matter than coniferous forests [26].

The preceding studies had the following limitations. First, although it has been shown that particulate matter concentrations differ depending on land-use, the focus has been on interpreting figures, such as statistics, and there remains a lack of studies on temporal and spatial distributions. Second, although NAQMSs enable accurate identification of the widespread generation of particulate matter, they are not densely located, and, hence, further studies using PAQMSSs are required. Given these limitations, this study analyzed changes in particulate matter concentrations according to time and land-use and determined the spatial mobility characteristics of the distribution of particulate matter concentrations using PM10 and PM2.5 big data of particulate matter in Daejeon measured by PAQMSSs.

3. Data and Method

This study utilized PM10 and PM2.5 concentration data measured by PAQMSSs that collect and manage data from 134 locations in Daejeon, from September-October 2019. Among them, data collected by 123 PAQMSSs were used; 11 PAQMSSs were excluded because missing values were identified due to data transmission errors, etc. The data did not satisfy the normality test, and the total number of data points was 108,072.

The results of basic statistical analysis, including the maximum, minimum, and mean values, are summarized in Table 1. The PAQMSSs (134 locations) operated in Daejeon secured approximately 12 times more branches than the NAQMSs (10 locations). This indicated that PAQMSSs should be used to analyze changes in the PM10 and PM2.5 concentrations in more detail across the entire city of Daejeon. Existing studies show that the data generated by PAQMSSs are as reliable as the nationally-managed NAQMSs [45]. In this study, it was verified that there was no difference between NAQMSs and PAQMSSs data using paired samples t-test. Therefore, this study secured the reliability of the data.

Table 1. Descriptive statistics of PM10 and PM2.5 (unit: $\mu g/m^3$).

Type of Particulate Matter	No. of PAQMSSs	n	Concentration				
			Min.	Max.	Mean	SD	Variance
PM10	123	180,072	14.75	62.32	31.89	5.40	54.71
PM2.5	123	180,072	8.71	32.68	16.72	3.85	27.75

Figure 1 shows the mean particulate matter concentration over time using data from 123 PAQMSSs to identify the trends of the particulate matter concentrations during fall (September-October). The PM10 concentrations exhibited a pattern of being low at dawn, increasing during the afternoon, and then decreasing in the evening. Particularly, concentrations were highest during Noon (11:00–13:00) and slightly increased after 21:00. PM2.5 showed a similar pattern to PM10 but with less deviation.

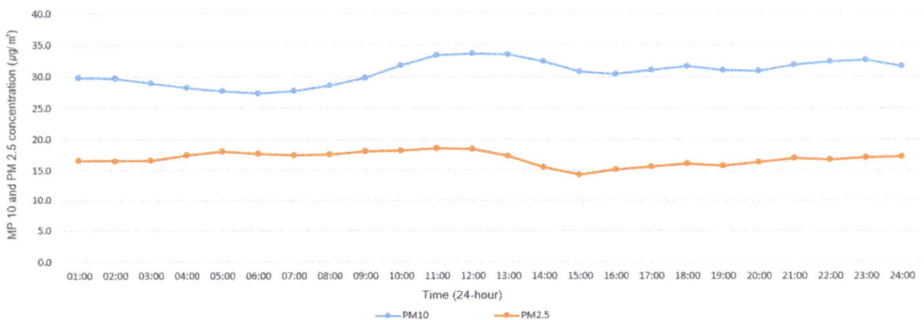

Figure 1. Mean concentrations of particulate matter (PM10, PM2.5) over time obtained from Private Air Quality Monitoring Smart Sensors (PAQMSSs) in Daejeon.

The analysis methods used in this study were NNA, k-means clustering, Kruskal–Wallis test, Mann–Whitney U test, and IDWM. NNA was undertaken to consider the distances between each PAQMSS and set a buffer for calculation of the optimum land-use ratio focusing on PAQMSSs. The k-means clustering method was introduced to classify PAQMSSs into four groups according to characteristics: residence, commerce, industry, and green. Cluster analysis using big data can be classified into supervised learning-based K-Nearest Neighbor (KNN) and unsupervised learning-based k-means clustering. This study used k-means clustering based on unsupervised learning because it was determined to be more suitable in this study to classify clusters based on the characteristics of each datum (unsupervised basis). This method involves dividing land-use ratios resulting from each PAQMSS into k groups, with the limitation of estimating the optimal number of k. In this study, k was divided into four groups based on land-use.

Nonparametric test methods (Kruskal–Wallis test) were used for statistical verification of concentration differences between the five time periods (AM1, AM2, Noon, PM1, and PM2) and land-use groups (residential, commercial, industrial, and green). After the differences between groups and time periods were statistically verified, the Mann–Whitney U test was used to verify differences between groups within the same time period. This is a nonparametric test method that can test PM10 and PM2.5 concentration differences between detailed groups. The significance of the Mann–Whitney U test was determined by the significance level of correction by Bonferroni correction and Kruskal–Wallis tests, and the Mann–Whitney U test was used when data did not satisfy normality.

PAQMSS data points expressing PM10 and PM2.5 concentrations were plotted on a map using IDWM. IDWM is a method of inversely weighting distances from observation points, wherein a lower weight indicates a larger distance [46]. This study used IDWM to identify regional differences in particulate matter concentrations. Spatial interpolation methods such as kriging and spline using statistical methods exist; however, this study used IDWM due to the lack of normality of the data [47,48].

4. Results and Discussion

4.1. Classification of Land-Use Group around PAQMSSs

The land-use ratio of the area surrounding PAQMSSs depends on the buffer range. In this study, the mean distance between the PAQMSSs was calculated using NNA; therefore, land-use area ratios were appropriately included while considering each PAQMSS' location and the corresponding distances. The calculated distance among PAQMSSs was derived as the 650-m-radius buffer, taking into account the minimum and maximum distances of PAQMSSs (Figure 2).

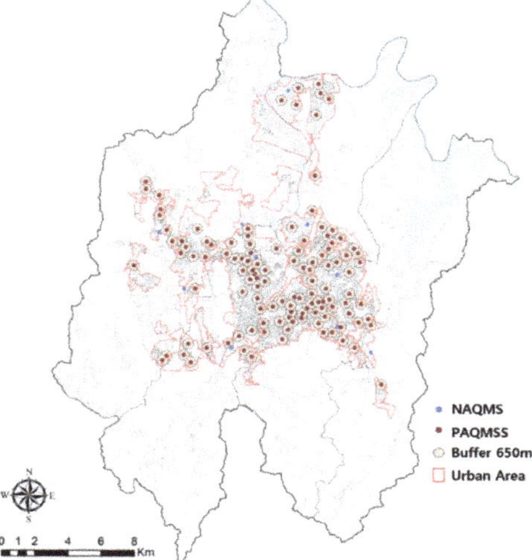

Figure 2. Air Quality Monitoring Sensors (AQMSs) map with 650-m buffer.

A 650-m-diameter buffer centered on PAQMSSs covered > 30.2% of the urbanized areas in Daejeon, i.e., the area covered is 12.6 times larger than that covered by National Air Quality Monitoring Sensors (NAQMSs). Land-use types within the range of the 650-m-diameter buffer were simplified into residence, commerce, industry, green, and roads, and their ratios were determined as 30.5%, 26.3%, 2.7%, 19.8%, and 24.7%, respectively (Table 2). Residential areas accounted for the largest proportion, followed by commercial areas, roads, green areas, and industrial areas. K-means clustering analysis was used to analyze the land-use characteristics of 123 PAQMSSs based on the land-use area ratio and classified the 123 PAQMSSs into four groups: Group 1, Group 2, Group 3, and Group 4 (Figures 3 and 4; Table 3).

Table 2. Land-use ratio and area in 650-m buffer (unit: m^2).

	Residential	Commercial	Industrial	Green	Transport
Min.	0.0 (0.0%)	4653.5 (1.2%)	0.0 (0.0%)	0.0 (0.0%)	9866.1 (3.4%)
Max.	274,022.4 (83.1%)	252,028.1 (76.5%)	254,441.7 (77.1%)	263,251.6 (79.6%)	178,822.2 (54.2%)
Mean	98,948.3 (30.5%)	84,881.1 (26.3%)	6114.2 (2.7%)	63,022.8 (19.8%)	78,729.0 (24.7%)
SD	63,844.4 (19.4%)	54,864.4 (17.5%)	29,954.9 (9.2%)	57,243.1 (17.7%)	32,028.7 (10.5%)

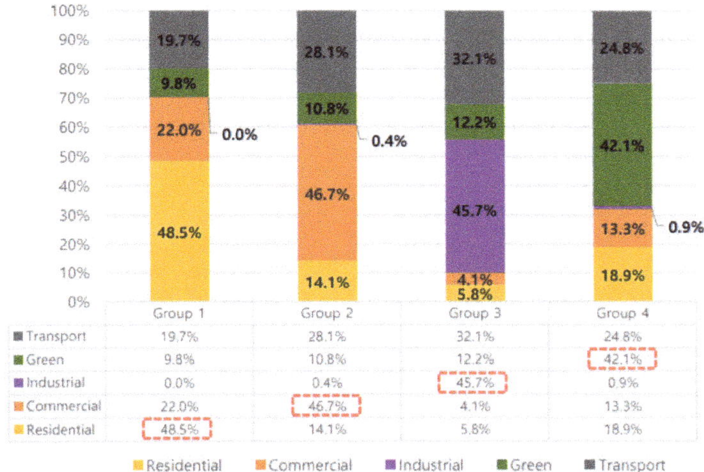

	Group 1	Group 2	Group 3	Group 4
Transport	19.7%	28.1%	32.1%	24.8%
Green	9.8%	10.8%	12.2%	42.1%
Industrial	0.0%	0.4%	45.7%	0.9%
Commercial	22.0%	46.7%	4.1%	13.3%
Residential	48.5%	14.1%	5.8%	18.9%

Figure 3. Residential, commercial, industrial, and green area ratio by groups.

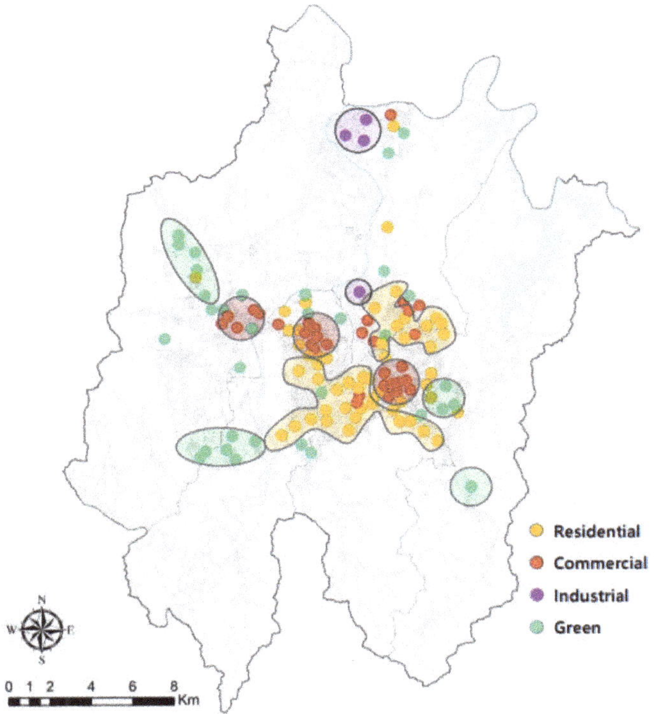

Figure 4. Map of clustered Private Air Quality Monitoring Smart Sensors (PAQMSSs).

Table 3. Results of k-means clustering analysis.

	Classification of Clustering				F	*p*-Value
	Group 1 (*n* = 52)	Group 2 (*n* = 33)	Group 3 (*n* = 4)	Group 4 (*n* = 34)		
Residential	0.485	0.141	0.058	0.189	104.155	0.000
Commercial	0.220	0.467	0.041	0.133	79.290	0.000
Industrial	0.000	0.004	0.457	0.009	155.736	0.000
Green	0.098	0.108	0.122	0.421	86.732	0.000
Transport	0.197	0.281	0.321	0.248	7.422	0.000
Group Characteristics	Residential Group	Commercial Group	Industrial Group	Green Group	-	

The characteristics of PAQMSSs were defined by selecting the largest land-use from the areas within a 650-m diameter from the PAQMSSs. For example, PAQMSSs with a residential area of 100 m^2, commercial area of 30 m^2, industrial area of 35 m^2, green area of 8 m^2, and transport area of 15 m^2 within a 650-m diameter belonged to the residential group because the residential area was larger than that of other areas. Each group was defined in this way as a residential, commercial, industrial and green group. Group 1 included 52 PAQMSSs with the largest proportion of residences (48.5%); therefore, it had residential characteristics. Group 2 included 33 PAQMSSs with the largest proportion of commerce (46.7%) and, therefore, was classified as having commercial characteristics. Group 3 included four PAQMSSs and its industrial ratio was 45.7%. This group had industrial characteristics and a higher road ratio than other groups. In Group 4, green areas accounted for the largest percentage (42.1%) and 34 PAQMSSs were included; accordingly, this group was classified as having green characteristics. In this manner, the highest land-use ratio was defined as the characteristic of the group. In terms of characteristics per group, Groups 1, 2, and 4 had the lowest industrial ratios and Group 3 included the lowest residential and commercial ratios. For the green ratio, all groups except Group 4 had low ratios. Group 3 the highest ratio for roads and Group 1 had the lowest.

The spatial distribution of PAQMSSs included in the four groups is shown in Figure 3. PAQMSSs classified as residential groups were widely distributed across Daejeon's urbanized areas (yellow dots). Commercial groups of PAQMSSs were concentrated in old and new urban areas that may be considered as core areas in Daejeon (red dots). Industrial groups were located around industrial complexes and were classified as representative industrial areas within the urbanized areas (green dots), and green groups were located on the outskirts of the urbanized areas in Daejeon (Figure 4). The classification of land-use groups through the k-means clustering method and the result of the PAQMSSs distribution chart well-reflected the group characteristics when compared to the current land-use in Daejeon.

4.2. Changes in Particulate Matter Concentrations according to Land-Use and Time Period

Tables 4 and 5 present the differences in PM10 and PM2.5 concentrations over time and between land-use groups and the differences in concentrations between groups within the same time frame. In the cases when the PM10 and PM2.5 concentrations differed between land-use groups over time in the nonparametric test, an additional analysis of the differences between groups within the same time frame was undertaken using the Mann–Whitney U test. In addition, this study used IDWM to visualize and analyze the spatial distribution characteristics of PM10 and PM2.5 concentrations within the regions.

Table 4. Differences in PM10 concentration by land-use over time.

Period	Group	PM10 (μg/m^3) Mean $\pm$ SD	Kruskal–Wallis	Paired Comparison					
				Residential–Commercial	Residential–Industrial	Residential–Green	Commercial–Industrial	Commercial–Green	Industrial–Green
AM1 (03:00-05:00)	Residential	28.4 $\pm$ 14.4	$\chi^2 = 31.104$ $p = 0.001$ *	$Z = -3.620$ $p = 0.000$ **	$Z = -1.417$ $p = 0.000$ **	$Z = -5.314$ $p = 0.000$ **	$Z = -0.367$ $p = 0.002$ **	$Z = -1.327$ $p = 0.005$ **	$Z = -1.175$ $p = 0.000$ **
	Commercial	28.2 $\pm$ 11.9							
	Industrial	29.7 $\pm$ 10.0							
	Green	26.4 $\pm$ 12.3							
AM2 (07:00-09:00)	Residential	30.8 $\pm$ 14.9	$\chi^2 = 20.490$ $p = 0.001$ *	$Z = -3.278$ $p = 0.001$ **	$Z = -1.560$ $p = 0.001$ **	$Z = -4.086$ $p = 0.000$ **	$Z = -0.016$ $p = 0.001$ **	$Z = -0.641$ $p = 0.001$ **	$Z = -0.362$ $p = 0.001$ **
	Commercial	29.9 $\pm$ 12.7							
	Industrial	32.2 $\pm$ 9.7							
	Green	27.2 $\pm$ 13.4							
Noon (11:00-13:00)	Residential	34.6 $\pm$ 17.0	$\chi^2 = 29.267$ $p = 0.001$ *	$Z = -1.923$ $p = 0.004$ **	$Z = -5.118$ $p = 0.001$ **	$Z = -0.323$ $p = 0.004$ **	$Z = -4.093$ $p = 0.000$ **	$Z = -1.967$ $p = 0.001$ **	$Z = -4.945$ $p = 0.000$ **
	Commercial	34.1 $\pm$ 16.8							
	Industrial	35.8 $\pm$ 13.5							
	Green	28.3 $\pm$ 27.3							
PM1 (17:00-19:00)	Residential	32.3 $\pm$ 18.2	$\chi^2 = 26.899$ $p = 0.001$*	$Z = -3.975$ $p = 0.000$ **	$Z = -3.950$ $p = 0.000$ **	$Z = -2.649$ $p = 0.001$ **	$Z = -1.894$ $p = 0.000$ **	$Z = -1.218$ $p = 0.003$ **	$Z = -2.541$ $p = 0.000$ **
	Commercial	32.2 $\pm$ 17.2							
	Industrial	34.4 $\pm$ 14.5							
	Green	27.8 $\pm$ 18.5							
PM2 (21:00-23:00)	Residential	31.4 $\pm$ 14.7	$\chi^2 = 23.205$ $p = 0.001$ *	$Z = -3.667$ $p = 0.000$ **	$Z = 1.580$ $p = 0.000$ **	$Z = -4.246$ $p = 0.000$ **	$Z = -0.156$ $p = 0.001$ **	$Z = -0.395$ $p = 0.001$ **	$Z = -0.362$ $p = 0.000$ **
	Commercial	32.6 $\pm$ 12.6							
	Industrial	32.8 $\pm$ 11.1							
	Green	27.6 $\pm$ 12.4							

* $p < 0.05$, ** $p < 0.0083$ (Adjusted by Bonferroni correction method = 0.05/6).

Table 5. Differences in PM 2.5 concentration by land-use over time.

Period	Group	PM2.5 (μg/m^3) Mean ± SD	Kruskal–Wallis	Paired Comparison					
				Residential – Commercial	Residential – Industrial	Residential – Green	Commercial – Industrial	Commercial – Green	Industrial – Green
AM1 (03:00-05:00)	Residential Commercial Industrial Green	17.9 ± 8.6 17.1 ± 10.2 18.6 ± 7.7 16.2 ± 8.6	χ^2 = 32.139 p = 0.000 *	z = −3.451 p = 0.001 **	Z = −0.251 p = 0.002 **	Z = −5.425 p = 0.000 **	Z = −1.318 p = 0.001 **	Z = −−1.576 p = 0.001 **	Z = −2.185 p = 0.000 **
AM2 (07:00-09:00)	Residential Commercial Industrial Green	18.3 ± 8.3 17.7 ± 9.5 19.7 ± 7.7 16.5 ± 9.2	χ^2 = 29.273 p = 0.000 *	Z = −3.282 p = 0.001 **	Z = −2.692 p = 0.007 **	Z = −4.943 p = 0.000 **	Z = −1.125 p = 0.001 **	Z = −1.344 p = 0.000 **	Z = −0.0447 p = 0.000 **
Noon (11:00-13:00)	Residential Commercial Industrial Green	18.6 ± 7.1 18.1 ± 8.2 20.2 ± 6.5 16.6 ± 7.9	χ^2 = 5.041 p = 0.000 *	Z = −1.603 p = 0.001 **	Z = −1.656 p = 0.000 **	Z = −0.047 p = 0.000 **	Z = −0.831 p = 0.000 **	Z = −−1.427 p = 0.001 **	Z = −1.502 p = 0.000 **
PM1 (17:00-19:00)	Residential Commercial Industrial Green	17.5 ± 7.2 17.4 ± 8.6 21.8 ± 6.4 15.2 ± 7.9	χ^2 = 39.001 p = 0.000 *	Z = −3.253 p = 0.001 **	Z = −1.556 p = 0.000 **	Z = −5.510 p = 0.000 **	Z = −2.942 p = 0.003 **	Z = −1.880 p = 0.000 **	Z = −3.994 p = 0.000 **
PM2 (21:00-23:00)	Residential Commercial Industrial Green	17.4 ± 6.7 17.9 ± 7.9 19.8 ± 6.7 14.5 ± 6.6	χ^2 = 23.195 p = 0.000 *	Z = −2.718 p = 0.007 **	Z = −2.207 p = 0.001 **	Z = −4.550 p = 0.000 **	Z = −0.947 p = 0.001 **	Z = −1.436 p = 0.000 **	Z = −0.203 p = 0.000 **

* p < 0.05, ** p < 0.0083 (adjusted by Bonferroni correction method = 0.05/6).

The mean PM10 concentrations from September-October in Daejeon were moderate (26.4 ± 12.3 µg/m^3~35.8 ± 13.5 µg/m^3) according to the WHO (Table 4). There were significant differences in the PM10 concentrations between all land-use groups within the same time period ($p < 0.0083$).

The PM10 concentrations between land-use groups (residence, commerce, industry, and green) showed differences in all time periods ($p < 0.05$). The PM10 concentrations were the lowest in the green group, and the concentrations were high in the order of industrial, residential, and commercial. However, the PM10 concentration during PM2 was 1.2 µg/m^3 higher in the commercial group than in the residential group (Table 4).

The industrial and green groups showed the largest differences in PM10 concentration, with differences of 12.5% (AM1), 18.4% (AM2), 26.5% (Noon), 23.7% (PM1), and 18.8% (PM2). PM10 concentrations showed the biggest difference during Noon and the smallest difference during AM1. The PM10 concentrations of the green group (with high forest ratios) were low and the concentrations of the industrial group (with PM10 emission sources, e.g., industry and roads) were high, indicating that the particulate matter concentrations varied depending on the land-use ratio [4]. The land-use groups presenting the smallest differences were the residential and commercial groups; differences between these groups were 0.7% (AM1), 3.0% (AM2), 1.5% (Noon), 0.3% (PM1), and −3.7% (PM2). Unlike the differences between industrial and green groups, the PM10 concentrations between residential and commercial groups had the biggest difference during PM2 and the smallest difference during PM1.

In particular, it is believed that the commercial group had higher PM10 concentrations than the residential group during PM2 owing to the increased human activities in commercial areas. Industrial groups had higher PM10 concentrations than other groups due to the greater amount of fuel used in industrial areas [43]. The PM10 concentration patterns showed that the concentrations in the residential, commercial, and industrial groups gradually decreased after reaching the peak during Noon (Figure 5). This was understood to be because most of the activities in a city (such as vehicle operation and movement of people) are carried out during the day.

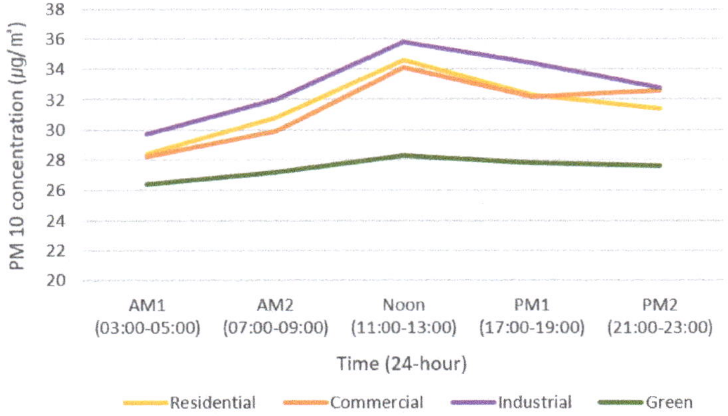

Figure 5. Differences in PM10 concentration by land-use over time.

Except for the industrial group, the land-use groups with large green area ratios showed lower PM10 concentrations. Moreover, as residential area ratios were high, the particulate matter concentrations were characterized to be high, and the PM10 concentrations decreased as green area ratios increased.

The spatial distribution changes in the PM10 and PM2.5 concentrations were analyzed using IDWM (Figure 6). The PM10 concentrations in Daejeon were high in the central area where commercial and business functions were concentrated and in old and new urban areas. Furthermore, the PM10 concentrations were high in the industrial areas where industrial complexes were located. The PM10 concentration gradually began to increase from the northeast over time and spread throughout Daejeon during Noon. Subsequently, it showed a gradually decreasing distribution of the concentrations from the southwest (Figure 6).

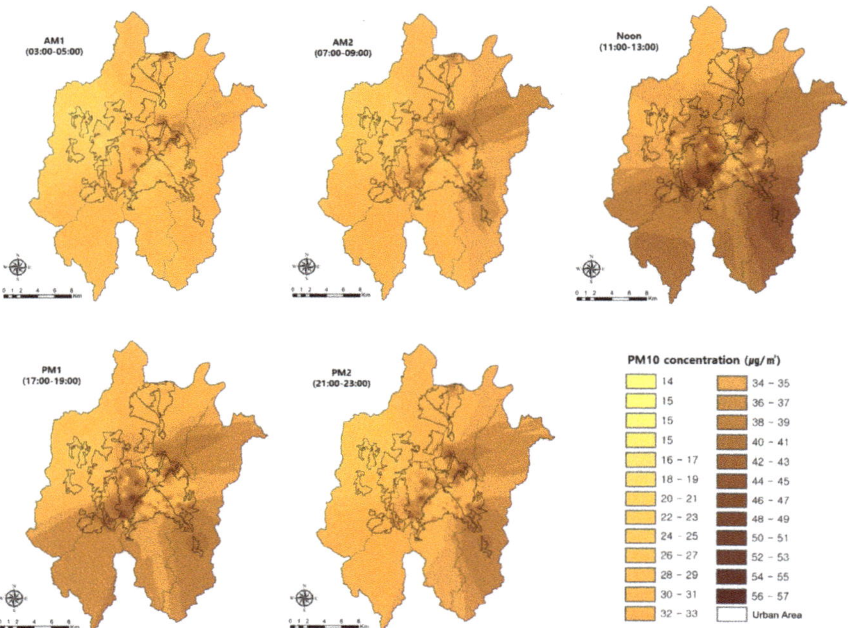

Figure 6. Changes in spatial distribution characteristics of PM10 concentration over time. AM1, AM2, Noon, PM1, and PM2.

High PM10 concentrations were maintained in the central area where commercial and business functions were concentrated and in the industrial area where industrial complexes were located. The mean PM2.5 concentration was moderate (14.5 ± 6.6 μg/m³~21.8 ± 6.4 μg/m³), similar to that of PM10.

Analyses of the PM2.5 concentration changes provided the following results: there were differences in the PM2.5 concentrations between the land-use groups at all times and land-use groups at the same time ($p < 0.05$, $p < 0.0083$). The PM2.5 concentrations were the lowest in the green group for all time periods, and similarly to PM10, they were high in the order of industrial, residential, and commercial. However, the PM10 concentrations during PM2 were 0.5 μg/m³ higher in the commercial group than in the residential group.

The industrial and green groups showed the biggest differences in PM2.5 concentration. The differences between the two groups were 14.8% (AM1), 19.4% (AM2), 21.7% (Noon), 43.4% (PM1), and 36.6% (PM2). The PM2.5 concentration showed the largest differences during PM1 and the smallest differences during AM2, which differed from the results of PM10. This is determined to be a phenomenon in which pollutants (PM10) generated in industrial areas combine with surrounding O_3 water vapor, resulting in higher PM2.5 concentrations. This was influenced by the fact that all industrial areas in Daejeon are located near rivers [49].

PM2.5 concentration patterns were similar to those of PM10; however, a constant PM2.5 concentration was characteristically maintained in the residential and commercial groups (Figure 7). In addition, the industrial group showed a phenomenon of peaking during PM1, and the green group showed a steeply declining pattern after Noon. Characteristically, a phenomenon was observed whereby the PM2.5 concentration of the commercial group was higher than that of the residential group during PM2, which was the same pattern as PM10. This is due to the greater movement and energy consumption of vehicles and people in the commercial group than in the residential group during PM2 [29].

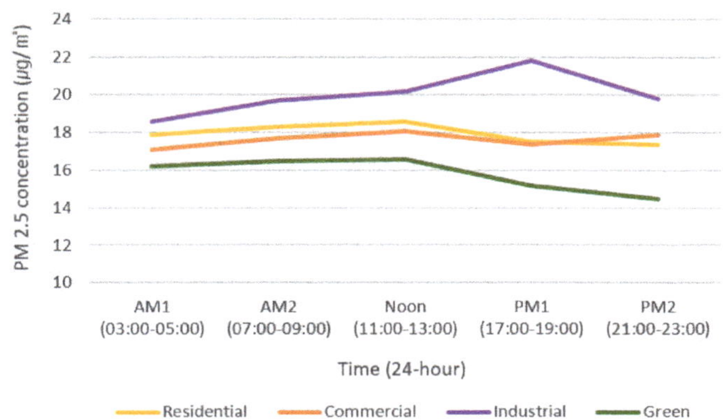

Figure 7. Changes in PM2.5 concentration by land-use group over time.

The residential and commercial groups had the smallest differences in PM 2.5 concentrations, which were 4.7% (AM1), 3.4% (AM2), 2.8% (Noon), 0.6% (PM1), and −2.8% (PM2). The difference in the PM2.5 concentrations between residential and commercial groups was the largest during AM1 and the smallest during PM1. Different from the PM10 results, PM2.5 concentrations were higher at dawn, which indicates the PM2.5 is not easily resolved overnight.

The spatial distribution of PM2.5 was similar to that of PM10, whereby the highest concentrations occurred where commercial and business functions were concentrated (Figure 8). However, a constant PM2.5 concentration was maintained at certain locations. Its features were evident in the commercial and industrial groups, i.e., the PM2.5 concentrations were the highest during Noon. Moreover, PM2.5 concentrations were maintained at specific locations, rather than being widely distributed overall. In the central and industrial areas, PM2.5 concentrations were high regardless of time, and the distribution of concentrations had similar characteristics to those of PM10.

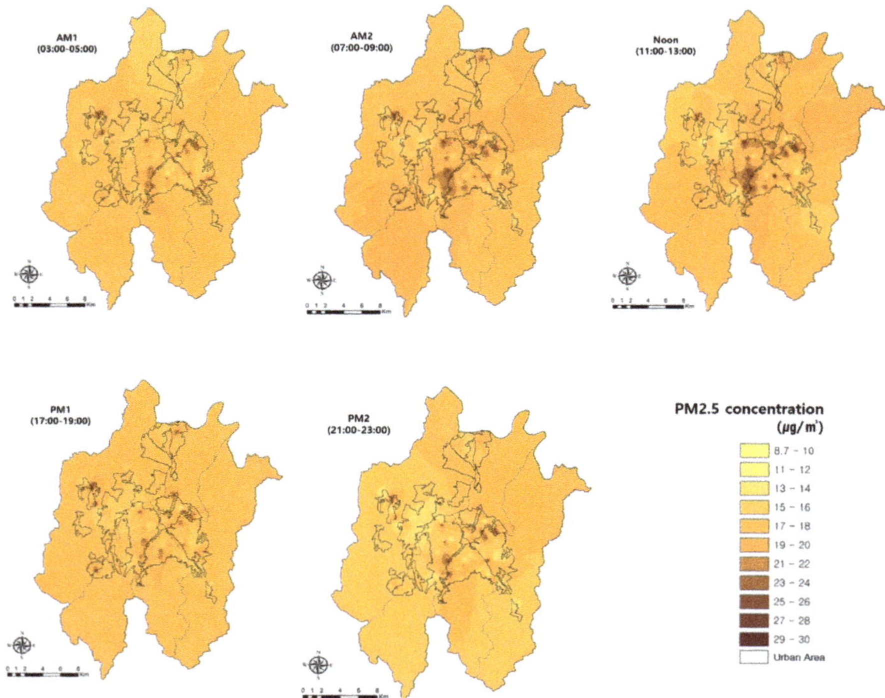

Figure 8. Changes in spatial distribution characteristics of PM2.5 concentrations over time. AM1, AM2, Noon, PM1, and PM2.

5. Conclusions

This study analyzed changes in particulate matter concentrations and the spatial characteristics of the distribution of those concentrations according to time and land-use, using PM10 and PM2.5 big data measured in Daejeon by a private company's PAQMSSs from September-October 2019. The results are summarized as follows: first, the land-use types within the range of 650-m-diameter buffers based on 123 PAQMSSs were simplified to residences, commerce, industry, green, and roads, with ratios of 30.5%, 26.3%, 2.7%, 19.8%, and 24.7%, respectively. According to the grouping based on the ratios, four groups (residence, commerce, industry, and green) were classified. Analyses of the highest land-use ratio in each group identified residence (48.5%) in Group 1, commerce (46.7%) in Group 2, industry (45.7%) in Group 3, and green (42.1%) in Group 4. Then, the highest land-use ratio in each group was defined as being characteristic of the group.

Second, the PM10 and PM2.5 data showed moderate levels of particulate matter concentrations, and there were significant differences in the concentrations between groups over time and between groups at the same time (with the exception of during PM2). Particulate matter concentrations were high during all time periods in the order of the industry, residence, commerce, and green (with the exception of during PM2); however, concentrations in the industrial group were higher than those in the residential group. This may be a result of the increasing mixed land-use due to intensive zoning control. The weakening of zoning control has resulted in a large supply of residential areas in commercial areas, which has caused an increase in nighttime activities.

Third, the groups presenting the biggest concentration differences were both PM10 and PM2.5 in the residential and green groups. In addition, the particulate matter concentration in the green group (with high forest ratios) was low, and the concentration in the industrial group (with high industrial and road ratios) was high. This indicates that the concentrations varied depending on the land-use ratio, which is in agreement with previous studies. Industrial areas use more fuel and have higher emissions of pollutants from combustion facilities and production processes than commercial and residential areas [49]. Moreover, PM10 showed the biggest differences during Noon, whereas PM2.5 showed differences during PM1. The reason for this difference is believed to be that PM10 combined with O_3 and water vapor and was transformed to PM2.5 by chemical reactions.

Fourth, PM10 and PM2.5 concentrations tended to be high in old and new urban areas (where commercial and business functions were concentrated), and where the industrial complex was located. Moreover, overall, the particulate matter concentrations were low in the morning (AM1 and AM2), highest in the afternoon (Noon), and gradually increased in the evening (PM1 and PM2). PM10 concentrations clearly showed variations over time, whereas the PM2.5 concentrations had distribution characteristics that remained stable in certain areas. The results of this study show that the PM10 concentration can be resolved naturally over time, however PM2.5 showed a stagnation phenomenon, whereby it was not easily diluted naturally from concentrated areas [50,51]. Addressing this problem involves promoting a long-term policy to reduce the occurrence of particulate matter pollution and, simultaneously, providing physical measures, such as parks and green areas, to minimize the effects of particulate matter on the human body. In other words, firstly, in order to prevent this phenomenon, a policy that minimizes air pollutants is needed. Secondly, green space is important for particulate matter management, as revealed in previous studies [26]. Sufficient parks and green areas should be provided to absorb fine dust generated in industrial, commercial, and residential areas.

This study examined whether particulate matter concentrations changed depending on time and land-use and analyzed the characteristics of the spatial distribution of particulate matter. The residential, commercial, and industrial areas representing urbanized areas were found to increase the particulate matter concentration and the green area was identified as a factor in decreasing the concentration [29,52]. This study provides guidelines for establishing particulate matter reduction policies since environmental policies are significant for pollution reduction. The limitation of the study was that the analysis only considered changes in the particulate matter concentrations in the city of Daejeon. However, the locations of sources of particulate matter vary and the various causes are complex and interwoven. Therefore, studies on a range of aspects are required, including the degree of influence of particulate matter between areas and the causes of occurrence, by expanding the current research scope.

Author Contributions: Data curation, S.S.J.; Formal analysis, S.S.J.; Writing—original draft, S.S.J.; Writing—review and editing, S.H.L. and Y.L.; Supervision, Y.L. All authors have read and agreed to the published version of the manuscript.

Funding: This work was supported by the National Research Foundationof Korea (NRF) grant funded by the Korea government (MSIT) (No. 2019R1F1A1062708).

Acknowledgments: The authors express their gratitude to the City of Daejeon for providing the material.

Conflicts of Interest: The authors declare no conflict of interest.

References

1. Dominici, F.; Greenstone, M.; Sunstein, C.R. Particulate matter matters. *Science* **2014**, *344*, 257–259. [CrossRef]
2. Hatakeyama, S.; Takami, A.; Sakamaki, F.; Mukai, H.; Sugimoto, N.; Shimizu, A.; Bandow, H. Aerial measurement of air pollutants and aerosols during 20–22 March 2001 over the East China Sea. *J. Geophys. Res. Atmos.* **2004**, *109*. [CrossRef]
3. Park, S.-Y.; Kim, Y.-J.; Kim, C.-H. Characteristics of Long-Range Transport of Air Pollutants due to Different Transport Patterns over Northeast Asia. *J. Korean Soc. Atmos. Environ.* **2012**, *28*, 142–158. [CrossRef]
4. Choi, T.-Y.; Moon, H.-G.; Kang, D.-I.; Cha, J.-G. Analysis of the seasonal concentration differences of particulate matter according to land cover of Seoul—Focusing on forest and urbanized area. *J. Environ. Impact Assess.* **2018**, *27*, 635–646. [CrossRef]

5. Dockery, D.W.; Pope, C.A.; Xu, X.; Spengler, J.D.; Ware, J.H.; Fay, M.E.; Ferris, B.G.; Speizer, F.E. An Association between Air Pollution and Mortality in Six U.S. Cities. *N. Engl. J. Med.* **1993**, *329*, 1753–1759. [CrossRef]

6. Styer, P.; McMillan, N.; Gao, F.; Davis, J.; Sacks, J. Effect of outdoor airborne particulate matter on daily death counts. *Environ. Health Perspect.* **1995**, *103*, 490–497. [CrossRef]

7. Schwartz, J.; Dockery, D.W.; Neas, L.M. Is Daily Mortality Associated Specifically with Fine Particles? *J. Air Waste Manag. Assoc.* **1996**, *46*, 927–939. [CrossRef]

8. Bae, H.-J. Effects of Short-term Exposure to PM_{10} and $PM_{2.5}$ on Mortality in Seoul. *Korean J. Environ. Health Sci.* **2014**, *40*, 346–354. [CrossRef]

9. Park, J.-K.; Choi, Y.-J.; Jung, W.-S. An analysis on the distribution characteristics of PM10 concentration and its relation to the death from Asthma in Seoul, Korea. *J. Environ. Sci. Int.* **2015**, *24*, 961–968. [CrossRef]

10. Xing, Y.-F.; Xu, Y.-H.; Shi, M.-H.; Lian, Y.-X. The impact of PM2.5 on the human respiratory system. *J. Thorac. Dis.* **2016**, *8*, E69–E74.

11. Lee, S.H.; Park, H.M. Analysis of Characteristics of Fine Particulate Matter Concentrations Using Statistical Methods. *Int. J. Highw. Eng.* **2019**, *21*, 109–120. [CrossRef]

12. Kim, S.-H.; Kim, N.-H.; Lee, M.-J.; Han, N.-Y.; Lee, H.-C. Analysis of influence of humidity on fine dust sensor through 2k factorial design. *Asia Pac. J. Multimed. Serv. Converg. Art Humanit. Sociol.* **2019**, *9*, 765–773. [CrossRef]

13. Bae, S.H. Smart City Policy Considering Spatial Characteristics. *J. Korean Reg. Sci. Assoc.* **2019**, *35*, 25–32. [CrossRef]

14. Available online: https://iot.airmapkorea.kt.com/info/ (accessed on 23 October 2020).

15. Eom, Y.-S.; Park, B.-R.; Kim, S.-G.; Kang, D.-H. A Case Study on Placement of Portable Air Cleaner Considering Outdoor Particle Infiltration into an Elementary School Classroom. *J. Korean Inst. Arch. Sustain. Environ. Build. Syst.* **2020**, *14*, 158–170. [CrossRef]

16. Hwang, Y.-J.; Lee, S.-J.; Do, H.-S.; Lee, Y.-K.; Son, T.-J.; Kwon, T.-G.; Han, J.-W.; Kang, D.-H.; Kim, J.-W. The Analysis of PM10Concentration and the Evaluation of Influences by Meteorological Factors in Ambient Air of Daegu Area. *J. Korean Soc. Atmos. Environ.* **2009**, *25*, 459–471. [CrossRef]

17. Do, H.-S.; Choi, S.-J.; Park, M.-S.; Lim, J.-K.; Kwon, J.-D.; Kim, E.-K.; Song, H.-B. Distribution Characteristics of the Concentration of Ambient PM-10 and PM-2.5 in Daegu Area. *J. Korean Soc. Environ. Eng.* **2014**, *36*, 20–28. [CrossRef]

18. Jeon, C.-W.; Cho, D.; Zhu, L. Exploring the Spatial Heterogeneity of Particulate Matter (PM10) using Geographically Weighted Ridge Regression (GWRR). *J. Korean Cartogr. Assoc.* **2018**, *18*, 91–104. [CrossRef]

19. Hong, S.-H.; Architecture, P.N.U.D.O.L.; Kang, R.-Y.; An, M.-Y.; Kim, J.-S.; Jung, E.-S. Study on the Impact of Roadside Forests on Particulate Matter between Road and Public Openspace in front of Building Site. *Korean J. Environ. Ecol.* **2018**, *32*, 323–331. [CrossRef]

20. Oh, K.-S.; Chung, H.-B. The influence of urban development density on air pollution. *J. Korea Plan. Assoc.* **2007**, *42*, 197–210.

21. Kim, H.-J. Analysis on Relationship between Urban Development Characteristics and Air Pollution level—A Case of Seoul Metropolitan Region. *J. Korea Plan. Assoc.* **2014**, *49*, 151. [CrossRef]

22. Jeong, J.-C. A Spatial Distribution Analysis and Time Series Change of PM10 in Seoul City. *J. Korean Assoc. Geogr. Inf. Stud.* **2014**, *17*, 61–69. [CrossRef]

23. Park, J.-K.; Choi, Y.-J.; Jung, W.-S. Understanding on Regional Characteristics of Particular Matter in Seoul. *J. Environ. Sci. Int.* **2017**, *26*, 55–65. [CrossRef]

24. Duan, J.; Tan, J.; Wang, S.; Hao, J.; Chai, F. Size distributions and sources of elements in particulate matter at curbside, urban and rural sites in Beijing. *J. Environ. Sci.* **2012**, *24*, 87–94. [CrossRef]

25. Wang, S.; Zhou, C.; Wang, Z.; Feng, K.; Hubacek, K. The characteristics and drivers of fine particulate matter (PM2.5) distribution in China. *J. Clean. Prod.* **2017**, *142*, 1800–1809. [CrossRef]

26. Choi, T.-Y.; Kang, D.-I.; Cha, J.-G. An analysis of the correlation between Seoul's monthly particulate matter concentrations and surrounding land cover categories. *Korean Soc. Environ. Impact Assess.* **2019**, *28*, 568–579. [CrossRef]

27. Weng, Q.; Yang, S. Urban Air Pollution Patterns, Land Use, and Thermal Landscape: An Examination of the Linkage Using GIS. *Environ. Monit. Assess.* **2006**, *117*, 463–489. [CrossRef]

28. Kim, Y.; Guldmann, J.-M. Impact of traffic flows and wind directions on air pollution concentrations in Seoul, Korea. *Atmos. Environ.* **2011**, *45*, 2803–2810. [CrossRef]

29. Jeong, J.-C.; Lee, P.S.-H. Spatial distribution of particulate matters in comparison with land-use and traffic volume in Seoul, Republic of Korea. *J. Cadastre Land Inf.* **2018**, *48*, 123–138. [CrossRef]

30. Hwang, S.-Y.; Moon, J.-Y.; Kim, J.-J. Relationship analysis between fine dust and traffic in Seoul using R. *Inst. Int. Broadcast. Commun.* **2019**, *19*, 139–149. [CrossRef]

31. Abhijith, K.V.; Kumar, P.; Gallagher, J.; McNabola, A.; Baldauf, R.; Pilla, F.; Pulvirenti, B. Air pollution abatement performances of green infrastructure in open road and built-up street canyon environments— A review. *Atmos. Environ.* **2017**, *162*, 71–86. [CrossRef]

32. Fan, S.; Li, X.; Han, J.; Cao, Y.; Dong, L. Field assessment of the impacts of landscape structure on different-sized airborne particles in residential areas of Beijing, China. *Atmos. Environ.* **2017**, *166*, 192–203. [CrossRef]

33. Chen, J.; Yu, X.; Sun, F.; Lun, X.; Fu, Y.; Jia, G.; Zhang, Z.; Liu, X.; Mo, L.; Bi, H. The Concentrations and Reduction of Airborne Particulate Matter (PM10, PM2.5, PM1) at Shelterbelt Site in Beijing. *Atmosphere* **2015**, *6*, 650–676. [CrossRef]

34. Jonsson, P.; Bennet, C.; Eliasson, I.; Lindgren, E.S. Suspended particulate matter and its relations to the urban climate in Dar es Salaam, Tanzania. *Atmos. Environ.* **2004**, *38*, 4175–4181. [CrossRef]

35. He, G.-X.; Yu, C.W.F.; Lu, C.; Deng, Q. The Influence of Synoptic Pattern and Atmospheric Boundary Layer on PM10 and Urban Heat Island. *Indoor Built Environ.* **2013**, *22*, 796–807. [CrossRef]

36. Tai, A.P.K.; Mickley, L.J.; Jacob, D.J. Correlations between fine particulate matter (PM2.5) and meteorological variables in the United States: Implications for the sensitivity of PM2.5 to climate change. *Atmos. Environ.* **2010**, *44*, 3976–3984. [CrossRef]

37. Park, C.-S. Variations of PM10 concentration in Seoul during 2015 and relationships to weather condition. *J. Assoc. Korean Photo Geogr.* **2017**, *27*, 47–64. [CrossRef]

38. Li, X.; Ma, Y.; Wang, Y.; Liu, N.; Hong, Y. Temporal and spatial analyses of particulate matter (PM_{10} and $PM_{2.5}$) and its relationship with meteorological parameters over an urban city in northeast China. *Atmos. Res.* **2017**, *198*, 185–193. [CrossRef]

39. Jin, S.; Guo, J.; Wheeler, S.; Kan, L.; Che, S. Evaluation of impacts of trees on PM2.5 dispersion in urban streets. *Atmos. Environ.* **2014**, *99*, 277–287. [CrossRef]

40. Nguyen, T.; Yu, X.; Zhang, Z.; Liu, M.; Liu, X. Relationship between types of urban forest and PM2.5 capture at three growth stages of leaves. *J. Environ. Sci.* **2015**, *27*, 33–41. [CrossRef] [PubMed]

41. Nowak, D.J.; McHale, P.J.; Ibarra, M.; Crane, D.; Stevens, J.C.; Luley, C.J. Modeling the Effects of Urban Vegetation on Air Pollution. In *Air Pollution Modeling and Its Application XII*; Springer Science and Business Media LLC: Boston, MA, USA, 1998; pp. 399–407.

42. Nowak, D.J.; Crane, D.E.; Stevens, J.C. Air pollution removal by urban trees and shrubs in the United States. *Urban For. Urban Green.* **2006**, *4*, 115–123. [CrossRef]

43. Park, S.; Shin, H. Analysis of the Factors Influencing PM2.5 in Korea: Focusing on Seasonal Factors. *J. Environ. Policy Adm.* **2017**, *25*, 227–248. [CrossRef]

44. Ministry of Environment. *Particulate Matter, What Is It*; Ministry of Environment Press: Sejong, Korea, 2016; pp. 1–72.

45. Je, M.-H.; Jung, S.-H. Urban Heat Island Intensity Analysis by Landuse Types. *J. Korea Cont. Assoc.* **2018**, *18*, 1–12. [CrossRef]

46. Nam Goung, S.J.; Choi, K.Y.; Hong, H.J.; Yoon, D.K.; Kim, Y.S.; Park, S.H.; Kim, Y.K.; Lee, C.M. Study on the selection and application of a spatial analysis model appropriate for selecting the radon priority management target area. *J. Environ. Health Sci.* **2019**, *45*, 82–96. [CrossRef]

47. Lee, J.-H.; Ryu, J.-E.; Choi, Y.-Y.; Chung, H.-I.; Jeon, S.-W.; Lim, J.-H.; Choi, H.-S. Spatial estimation of forest species diversity index by applying spatial interpolation method—Based on 1st Forest Health Management data. *J. Korean Soc. Environ. Restor. Technol.* **2019**, *22*, 1–14. [CrossRef]

48. Yim, J.S.; Lee, G.H. Estimating Urban Temperature by Combining Remote Sensing Data and Terrain Based Spatial Interpolation Method. *J. Korean Cartogr. Assoc.* **2017**, *17*, 75–88. [CrossRef]

49. Lee, K.S.; Kim, T.H.; Lee, J.O.; Lee, Y.L. Repeated Measures ANOVA for Fine Dust of Industrial Complex 3 and 4 in Daejeon, South Korea. *J. Korean Soc. Hazard Mitig.* **2019**, *19*, 235–245. [CrossRef]

50. Gouriou, F.; Morin, J.-P.; Weill, M.-E. On-road measurements of particle number concentrations and size distributions in urban and tunnel environments. *Atmos. Environ.* **2004**, *38*, 2831–2840. [CrossRef]

51. Oberdörster, G.; Oberdörster, E.; Oberdörster, J. Nanotoxicology: An Emerging Discipline Evolving from Studies of Ultrafine Particles. *Environ. Health Perspect.* **2005**, *113*, 823–839. [CrossRef]
52. Yoon, T.D. A Study on the Relationship between Urban Characteristics and the Concentration of the Fine-Particle (PM$_{10}$) for Creating Pleasant Air-Environment. Master Thesis, Korea University, Seoul, Korea, August 2019.

Publisher's Note: MDPI stays neutral with regard to jurisdictional claims in published maps and institutional affiliations.

Article

Distributed Artificial Intelligence-as-a-Service (DAIaaS) for Smarter IoE and 6G Environments

Nourah Janbi [1], Iyad Katib [1], Aiiad Albeshri [1] and Rashid Mehmood [2,*

[1]   Department of Computer Science, Faculty of Computing and Information Technology, King Abdulaziz University, Jeddah 21589, Saudi Arabia; Njanbi0006@stu.kau.edu.sa (N.J.); IAKatib@kau.edu.sa (I.K.); AAAlbeshri@kau.edu.sa (A.A.)

[2]   High Performance Computing Center, King Abdulaziz University, Jeddah 21589, Saudi Arabia

*   Correspondence: RMehmood@kau.edu.sa

Received: 19 August 2020; Accepted: 9 October 2020; Published: 13 October 2020

Abstract: Artificial intelligence (AI) has taken us by storm, helping us to make decisions in everything we do, even in finding our "true love" and the "significant other". While 5G promises us high-speed mobile internet, 6G pledges to support ubiquitous AI services through next-generation softwarization, heterogeneity, and configurability of networks. The work on 6G is in its infancy and requires the community to conceptualize and develop its design, implementation, deployment, and use cases. Towards this end, this paper proposes a framework for Distributed AI as a Service (DAIaaS) provisioning for Internet of Everything (IoE) and 6G environments. The AI service is "distributed" because the actual training and inference computations are divided into smaller, concurrent, computations suited to the level and capacity of resources available with cloud, fog, and edge layers. Multiple DAIaaS provisioning configurations for distributed training and inference are proposed to investigate the design choices and performance bottlenecks of DAIaaS. Specifically, we have developed three case studies (e.g., smart airport) with eight scenarios (e.g., federated learning) comprising nine applications and AI delivery models (smart surveillance, etc.) and 50 distinct sensor and software modules (e.g., object tracker). The evaluation of the case studies and the DAIaaS framework is reported in terms of end-to-end delay, network usage, energy consumption, and financial savings with recommendations to achieve higher performance. DAIaaS will facilitate standardization of distributed AI provisioning, allow developers to focus on the domain-specific details without worrying about distributed training and inference, and help systemize the mass-production of technologies for smarter environments.

Keywords: internet of everything (IoE); 6th generation (6G) networks; artificial intelligence; Distributed AI as a Service (DAIaaS); fog computing; edge computing; cloud computing; smart airport; smart districts

---

1. Introduction

We are living in unprecedented times. Artificial intelligence (AI) has taken us by storm, helping us to make decisions in everything we do, even in finding the "true love" of our life and selecting the "significant other" [1]. Siri, Cortana, Google Assistant, Bixby, Alexa, Uber, Databot, Socratic, and Fyle are among the many apps that we use on an hourly basis, if not non-stop. The number of industries benefitting from AI is growing, such as recommender systems, autonomous vehicles, renewable energy, agriculture, healthcare, transportation, security, finance, smart cities and societies, and the list goes on [2–4]. The global market for AI is estimated to reach 126 billion U.S. dollars in 2025, from $10.1 billion in 2018 [5].

AI allows us to embed "smartness" in our environments by intelligently monitoring and acting on it [6]. Internet of Everything (IoE) extends the Internet of Things (IoT) paradigm and integrates various entities in this ecosystem including sensors, things, services, people, and data [7]. The grand challenge for such IoE enabled smart environments is related to the 4Vs of big data analytics [8]—volume, velocity, variety, and veracity—that is, to devise optimal strategies for migration and placement of data and analytics in these ubiquitous environments. The networking infrastructure would have to be smart to support these services and address the challenges.

Various deployments of the Fifth Generation (5G) of wireless systems have begun to appear across the globe, promising mobile internet at unseen speeds. However, a radical change is needed to support extreme-scale ubiquitous AI services [9–12]. the Sixth Generation networks (6G) pledge this through next-generation softwarization, heterogeneity, and configurability of networks [13,14]. 6G will provide much higher speeds, reliability, capacity, and efficiency at lower latencies [15] through various enabling technologies such as higher spectrum and satellite communications [9,10,16,17], the use of AI to optimize network operations, and the use of fog and edge computing [18,19]. Figure 1 (see Section 2 for elaboration) depicts an envisioned view of smart societies that are enhanced with 6G and IoE technologies, showing also the distinguishing characteristics of 6G.

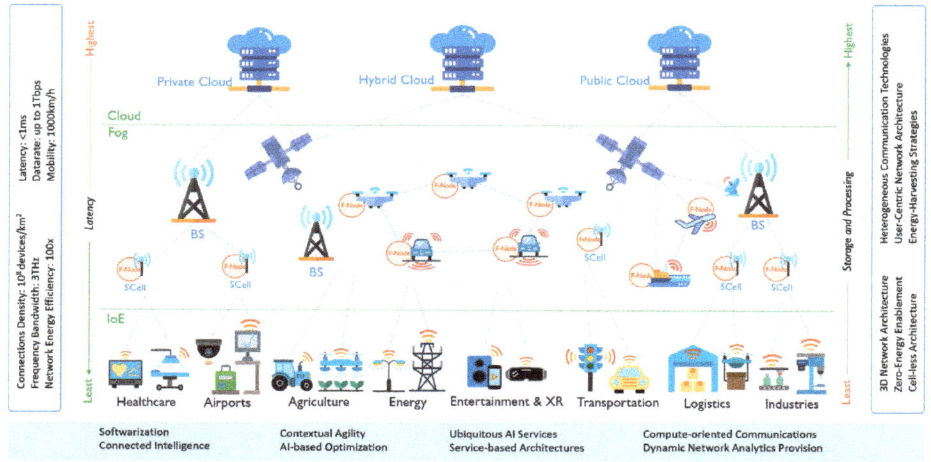

Figure 1. Sixth generation (6G)-internet of everything (IoE) enhanced smart societies.

The work on 6G is in its infancy and requires the community to conceptualize and develop its design, implementation, deployment, and use cases. Towards this end, this paper proposes a framework for Distributed AI as a Service (DAIaaS) provisioning for IoE and 6G environments. The AI service is "distributed" because the actual training and inference computations are divided into smaller, concurrent, computations suited to the large, medium, and smaller resources available with cloud, fog, and edge layers (see Figure 1). The AI service could be delivered by the Internet Service Providers (ISP) or other players. Multiple DAIaaS provisioning configurations for distributed training and inference are proposed to investigate the design choices and performance bottlenecks of DAIaaS. Specifically, we have developed three case studies (smart airport, smart district, and distributed AI delivery models) with eight scenarios (various usage configurations of cloud, fog, and edge layers) comprising nine applications and AI delivery models (smart surveillance, passport and passenger control, federated learning, etc.) and 50 distinct sensor and software modules (camera, ultrasonic sensor, electric power sensor, smart bins, data pre-processing, AI model building, data fusion, motion detector, object tracker, etc.).

The smart airport case study models the recently inaugurated King Abdulaziz International Airport, KAIA, Jeddah, and the smart district case study models the King Abdullah Economic City (KAEC), a smart city, both in Saudi Arabia. These two case studies model real-life physical environments and provide the details about actual sensors and computing operations and are used to understand and develop the DAIaaS framework and various AI service provisioning strategies. Using the knowledge gained from the first two case studies, the third case study investigates various distributed AI delivery models as a service (DAIaaS) without regard to the specific high-level applications in the underlying environments.

The evaluation of the DAIaaS framework using the three case studies is reported in terms of the end-to-end delay, network usage, energy consumption, and financial savings with recommendations to achieve higher performance. The results show a range of scenarios and configurations and how these affect the performance metrics and the related costs. Moreover, we demonstrate through these investigations and results that the challenging task for designing and deploying DAIaaS is the service placement because, for example, the edge devices might fail due to the limitations in their computation capabilities when the computing resource demands are high (as is the case for AI applications). Similarly, moving data too often to the cloud may lead to an inability to provision the required latencies for the edge devices.

The benefit of the DAIaaS framework is to standardize distributed AI provisioning at all the layers of digital infrastructure. It will allow application, sensor, and IoE developers to focus on the various domain-specific details, relieve them from worries related to the how-to of distributed training and inference and help systemize and mass-produce technologies for smarter environments. Moreover, to address the challenges noted by Viswanathan and Mogensen [19] and others, DAIaaS will provide unified interfaces that will facilitate the process of joint software development across different application domains. Therefore, we believe this work will have a far-reaching impact on developing next-generation digital infrastructure for smarter societies. To the best of our knowledge, this is the first work where distributed AI as a service has been proposed, modeled, and investigated.

The rest of this paper is organized as follows. Section 2 provides the background and reviews the related works. Section 3 explains our methodology and design, detailing how the various scenarios, applications, modules, and networks are modeled. Sections 4–6 give details that are specific to each of the three case studies and provide the performance analysis. Finally, conclusions are presented with future directions in Section 7.

2. Background and Related Works

We revisit Figure 1 that depicts a potential view of smart societies enhanced with 6G and IoE technologies. We consider the digital infrastructure of smart societies to be organized in three layers: IoE, Fog, and Cloud Layers. IoE Layer at the bottom comprises devices, sensors, and actuators from various application domains, transportation, energy, etc. The devices and sensors generate big data [20–22] that must be continuously processed and analyzed to make smart decisions and communicated to the IoE devices for actuation and other purposes. A sensor's data may also be aggregated with other sensors for context-awareness, enhanced decision making, exploratory analyses, cross-sectoral and global optimizations, and other reasons (e.g., see [23]). The Fog Layer consists of fog nodes placed in various 6G connection providers (e.g., base stations) closer to the edge devices in the IoE Layer. Fog nodes will provide storage and computation power at the proximity of edges and with the 6G capabilities, they will achieve ultra-low latency. The Cloud Layer at the top consists of various types of private, public, and hybrid data centers (clouds) that will provide high computation and storage resources but with higher latency to the edges. Various distinguishing characteristics of 6G are mentioned in the boxes around the figure.

In the rest of the section, we explain and review the works related to the five core technologies used in our work. These are AI, IoE, edge-fog-cloud computing, smart societies, and 6G, discussed in Sections 2.1–2.5, respectively.

2.1. Artificial Intelligence (AI)

Artificial intelligence is a field of study that focuses on the creation of intelligent machines that can learn, work, and react intelligently like humans. Deep learning (DL), machine learning (ML), neural network (NN), pattern recognition, computer vision, natural language processing, clustering, etc. are tools that can be used to train computers to accomplish specific tasks such as computer vision and natural language processing (NLP) [23–25]. AI models usually rely on data to build their knowledge therefore big data and data collected from a huge number of devices and sensors, as in IoE, has provided the fuel for the AI models [4,20–22]. The increasing volume of data generated from many connected, heterogeneous, and distributed objects (IoT/IoE) and the continuous development and evolution of networks and communication technologies have motivated the emergence of Distributed Artificial Intelligence (DAI) [26]. In DAI, AI models are distributed into multi-agents (or multiple processes) that are cooperatively sharing knowledge to solve or act either separately or to build a global knowledge for the whole system. Agents or sub-models can be residing either inside a single machine or across multiple machines to perform AI training or inference in a distributed or parallel way. One approach is to partition the AI model into sub-models or sub-tasks that can be run concurrently using parallel processing techniques such as pipelining. Wang et al. [27] have developed a framework that pipelines the processing of partitioned NN layers across heterogeneous cores for faster inference. An alternative is data partitioning, where the dataset is split across concurrently-running models, and results are aggregated later. These techniques are useful with massive AI models and data. A discussion of model and data parallelism is available in [28,29].

Another approach is Edge Intelligence (EdgeAI) where the AI model is distributed across network edges. Several works have discussed the convergence of edge and AI [30–35]. AI model can be pre-trained then modified and optimized to be appropriate to run in the resource-constrained edges. A discussion of DL optimizations at both software and hardware levels for edge AI is covered in [36]. The collaboration between edge and cloud is also a possible model where some of the pre-processing and less-intensive computations are placed in the edge and global analysis located in the cloud. Parra et al. [37] have developed a distributed attack detection system for IoT where different AI models are used in both edge and cloud to provide local and global detection systems. Federated Learning (FL) is a DAI model where multi-agents collaboratively share their local knowledge for faster convergence and to make better decisions. FL concept, applications, challenges, and methods have been reviewed in [38,39]. Smith and Hollinger [40] developed a distributed robotic system that collaboratively shares their knowledge for a single goal (environment exploration). On the other hand, in [23] autonomous vehicles share knowledge to improve their own decisions.

Artificial Intelligence-as-a-Service (AIaaS) has also been considered in [41–43] as a natural extension of the usual "as-a-service" ("aaS") service delivery models available with cloud providers. This allows developers to focus on their domain-specific details and conveniently add AI capability to their software. Several works have shown the benefits of AIaaS in developing and supporting applications that require AI capabilities. Casati et al. [42] proposed a framework and architecture that facilitate the deployment of cloud-based AIaaS for smarter enterprise management solutions. Milton et al. [43] conducted real experiments utilizing Google's Dialogflow API (a simple AIaaS provided by Google) [44] to develop a chatbot. The AI services that are provided by the top cloud providers such as Google, Amazon, Microsoft, and IBM, are discussed in [41].

2.2. Internet of Everything (IoE)

Internet of Everything (IoE) has emerged as a concept that extends the Internet of Things (IoT) [45] to include processes, people, data, and things [7]. In the core of IoE, sensors are usually embedded with "everything" to monitor, identify the status and act intelligently to generate new opportunities for the society. There are a variety of sensors designed for different purposes such as temperature, pressure, biosensors, light, position, velocity, etc. which are discussed in [46]. A massive number of sensors are expected to be deployed everywhere to support such applications and many others in different areas including industrial, traffic, smart cities, and healthcare systems [47,48].

There are some IoE works that have looked at challenges in sensors connection and data collection and processing. AlSuwaidan [49] adopted Cloud as a Service (CaaS) and the Fog-to-Cloud concept to overcome the challenge of integrating, storing, and migrating distributed data. Lv and Kumar [50] proposed the software definition to sensors in the 6G/IoE network, along with Software Defined Network (SDN) technology to provide better control. Aiello et al. [51] have developed a self-contextualizing service for IoE that separates the logical part from the physical contexts. Others such as Badr et al. [52] focused on energy harvesting and Ryoo et al. [53] covered security and privacy concerns in IoE.

2.3. Edge, Fog, and Cloud Computing

The continuous increase in the number of IoE sensors and edge devices joining the network required a shift in the paradigm that pushes data processing closer to the data sources. Edge and fog computing are two architectures that aim to bring processing closer to users at the network edges. While some in the literature do not differentiate between fog and edge [54], we and many others differentiate between them [55,56], depending on where the computation is performed. In edge computing, processes are localized in the edge devices to produce instant results. On the other hand, fog computing is an intermediate layer extending the cloud layer that brings the functions of cloud computing closer to the users [57]. Fog nodes are devices that can provide resources for services, and they might be resource-limited devices such as access points, routers, switches, and base stations, or resource-rich machines such as Cloudlet and IOx [58].

Discussions on fog computing and other edge paradigms are provided in [50,59] and their role in IoT is covered in [60]. Nath et al. [61] proposed an optimization algorithm to manage communication between the IoE cluster and the cloud. Wang et al. [62] adopted imitation learning for online task scheduling in vehicular edge computing. Badii at al. [63] have developed a platform for managing smart mobility and transport in network edges. Tammemäe et al. [64] proposed service architecture to support the self-awareness in fog and IoE.

2.4. Smart Cities, Societies, and Ecosystems

Smart cities and smart ecosystems employ different information and communication technologies (ICT) to intelligently monitor, collect, analyze, respond to environmental changes [2,65–75]. The population growth in the urban area and the advancement in technologies have increased the demand for more sustainable cities that adopt smarter, effective, and efficient ways to manage the urban area and integrate various aspects of the ecosystem [76]. This includes introducing smartness to the infrastructure, operations, services, industries, education, security, and many more. In this context, IoE will be the base that will enable and integrate city services, people, things, and data. Deploying sensors all over the city including the one that is attached to the people, such as smartwatches, or at their mobile devices will provide great services and unlimited innovation opportunities.

Several works have looked at the design of the applications for smart societies. Ahad et al. [77] have developed a smart educational environment based on IoE to produce a learning analytics system that evaluates the learning process and achievements. Al-dhubhani et al. [78] have proposed a smart border security system where sensors and different sources of data are used to make decisions and take actions. Queralta et al. [79] proposed an IoE-based architecture that employs a heterogeneous group of vehicles to improve traveling quality. Alam et al. [80] have developed an object recognition method for autonomous driving to improve the accuracy of vehicle recognition. Many other proposals on smart societies [81] exist, such as in transportation [25,65,69,71,82], healthcare [6], disaster management [83], logistics [66,84], and more.

2.5. Sixth Generation Networks (6G)

6G is the next generation of cellular networks that is expected to overcome the limitation of current fifth-generation (5G) deployment and fulfill requirements of the future fully connected digital society [10]. There are some publications [10,13–19] that have discussed the future vision of 6G cellular networks, requirements, enabling technologies, and challenges. The main challenges are coming from the expected continuous increase in the number of sensors joining the network and the popularity of IoE-based smart services [9]. Extensive improvements in the speed and capacity of communication can be achieved by adopting a higher spectrum and utilize various communication technologies [15]. 6G networks are expected to be an ultra-dense heterogeneous network [85]. Both terrestrial (cellular network) and non-terrestrial (e.g., satellites, drones, and planes) infrastructures will be employed to provide continuous and reliable network services [10]. Though the heterogeneity of network architecture, communication links, devices, and applications will increase network control and operation complexity [17]. Therefore, 6G is expected to take the 5G softwarization and virtualization to their next level by empowering the network with AI approaches to optimize the network operation [13,14]. Moreover, service-oriented operations should offer higher flexibility and looser integration with various network components, which will facilitate deployment, configuration, and management of new applications and services [16]. Distributed artificial intelligence with user-centric network architectures will be a fundamental component of the 6G networks to reduce communication overhead, and provide autonomous and real-time decisions [10]. Energy efficiency is also one of the critical requirements for 6G networks that must be taken into account from antenna design to the zero-energy nodes for low rate sensing applications [9]. It is expected from 6G to have 10–100 times higher energy efficiency than 5G to accommodate joining devices and applications with lowest-cost and eco-friendly deployment [11].

To summarize, digital infrastructures that will support smart societies require rich and flexible AI capabilities. 6G pledges to support ubiquitous AI services, however, the work on 6G is in its infancy and requires the community to contribute to its realization, such as in designing AI models, data management, service placements, job scheduling, and communication management and optimization for both application developers and service providers. Solutions are required to reduce the complexity of the systems and to allow application developers to focus on the various domain-specific details rather than worrying about the how-to of distributed training and inference. Table 1 summarizes some of the reviewed literature and compares it with our work. Note in the table that none of the published proposals have incorporated all the key technologies for next-generation digital infrastructure. The particular differentiating factor of our work is the DAIaaS framework and its detailed evaluation.

Table 1. Summary of relevant research.

Research	IoE	Edge/Fog	Smart Societies	6G	Distributed AI (DAI)	as a Service (aaS)	AIaaS	DAIaaS
Letaief et al. [13]		x		x	x	x		
Smith and Hollinger [40]		x	x		x			
AlSuwaidan [49]	x		x			x		
Lv and Kumar [50]	x	x		x				
Aiello et al. [51]	x		x			x		
Nath et al. [61]	x	x	x		x			
Wang et al. [62]	x	x	x		x			
Badii at al. [63]	x	x	x					
Ahad et al. [77]	x		x					
Casati et al. [42]			x			x	x	
Milton et al. [43]	x		x			x	x	
This work	x	x	x	x	x	x	x	x

3. Methodology and Design

In Section 1, we have already given an overview of our methodology in terms of the motivation for the three case studies, and the comprising scenarios, applications and AI delivery models, and sensor and software modules. This section discusses our methodology and design, and the main components of our simulations in detail. Section 3.1 describes the devices used in the edge, fog, and cloud layers. Section 3.2 introduces the applications and delivery models used in our work and how these are modeled in the simulations. Section 3.3 explains how the network infrastructure is modeled. Finally, Section 3.4 defines the performance metrics used for performance evaluation.

Software and Hardware: We have used the iFogSim [86] simulation software to model and evaluate DAIaaS. We selected it because it allows simulating a range of applications, modules, placements, data streams, sensors, edges, fogs, cloud datacenters, and communication links. All experiments are executed on the Aziz supercomputer (Jeddah, Saudi Arabia), which comprises 492 nodes with 24 cores each. The supercomputer allowed us to run many large simulations with different configurations concurrently on different nodes.

3.1. Cloud, Fog, and IoE Layers

Figure 1 shows a high-level view of smarter environments, supported by 6G and IoE, comprising three main layers: IoE, Fog, and Cloud, and this has been explained in some detail in Section 2. Each layer contains devices with distinct resource capabilities that are represented in the simulations using various parameters. We have determined the values for these parameters considering the specification of the devices that are available today. Table 2 list the three types of devices (edge, fog, and cloud devices) and the associated parameters. The computational capabilities of these devices are represented in the simulations with certain values of MIPS (Million Instructions Per Second) and RAM (Random Access Memory). The communication capabilities are simulated using uplink and downlink bandwidth. Each device is also characterized by specific power consumption in the idle and busy states. For example, The MIPS parameter for cloud for one virtual machine (VM) has the highest MIPS and RAM (220,000 and 40,000) values compared to the fog and edge devices.

Table 2. Device configurations.

Device Parameter	Cloud (VM)	Fog Device	Edge Device
MIPS	220,000	50,000	5000
RAM (MB)	40,000	4000	1000
Uplink Bandwidth (Mbps)	100	10,000	10,000
Downlink Bandwidth (Mbps)	10,000	10,000	10,000
Busy Power (W)	16×103	107.339	87.53
Idle Power (W)	16×83.25	83.4333	82.44

3.2. Distributed Applications and AI Delivery Models

A smart city would have various applications running on it simultaneously. Each application (see Sections 4 and 5) and AI delivery models (see Section 6) has a set of modules (m) that performs some computations on the data they receive. The modules are organized in a directed graph (DG) with edges between them to represent data or workload (w) passing between the modules. This is depicted in Figure 2 using the Smart Surveillance application, which we use in this section as an example. For each workload received by the module, it will be processed and a new workload will be generated depending on the configured mapping between workloads and the selectivity rate in case more than one exists. A workload (w) can be characterized by its CPU requirements (w_c) in terms of million instructions (MI) required by the module to process the workload as well as its network requirements (w_n) in terms of bytes to be transferred between the two modules over the network.

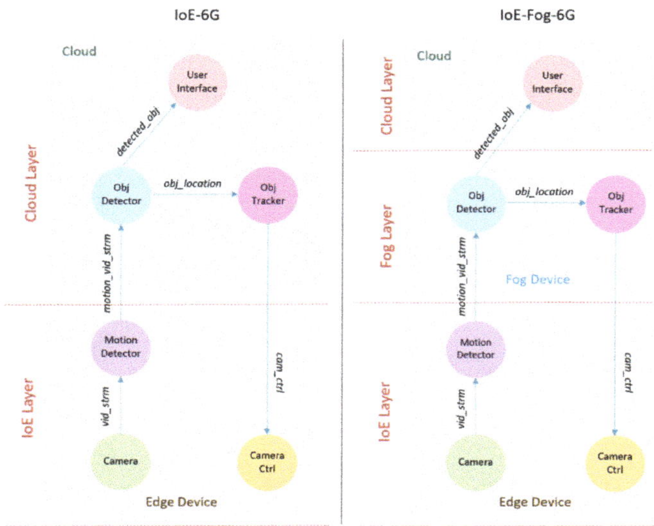

Figure 2. Smart surveillance application module.

Table 3 lists the various workloads used in the Smart Surveillance application. We will come back to it later after explaining the application in Figure 2. The figure shows that the modules can be placed in different layers (i.e., edge, fog, or cloud devices) depending on the scenario (IoE-6G versus IoE-Fog-6G). The Smart Surveillance application uses CCTV (Closed-Circuit Television) cameras to detect and track objects in a specific area, such as in [86]. The CCTV cameras generate live video streams. Therefore, this application has a high computation requirement especially in a crowded environment such as airports or pedestrian areas, where many people and objects must be tracked, identified, and analyzed carefully for security reasons. The application consists of six modules: Camera, Motion Detector, Object Detector (Obj Detector), Object Tracker (Obj Tracker), Camera Control (Camera Ctrl), and User Interface.

Table 3. Smart surveillance: workload configuration.

Workload Type	Source Module	Destination Module	CPU Requirement (w_c) (MI)	Network Requirement (w_n) (Bytes)
vid_strm	Camera	Motion Detector	1000	20K
motion_vid_strm	Motion Detector	Obj Detector	2000	2000
detected_obj	Obj Detector	User Interface	500	2000
obj_location	Obj Detector	Obj Tracker	1000	100
cam_ctrl	Obj Tracker	Camera Ctrl	50	100

The Camera contains the sensor and Camera Ctrl contains the pan–tilt–zoom (PTZ), which is the actuator in the camera that adjusts the camera zoom depending on the PTZ parameters. The Motion Detector is always located in the smart cameras and it receives live video streams (*vid_strm*) from the Camera and when motion is detected it transfers the motion video stream (*motion_vid_strm*) to the Obj Detector module. The Obj Detector module is located in the cloud in the IoE-6G scenario, and in the fog node in the IoE-Fog-6G scenario. It receives video streams (*motion_vid_strm*) from the Motion Detector and intelligently detects objects and activates Obj Tracker if it hasn't been activated before for the same object. The Obj Detector module sends two workloads: the detected object (*detected_obj*) to the User Interface and the object location (*obj_location*) to the Obj Tracker. The Obj Tracker module is located in the cloud in the IoE-6G scenario, and in the fog node in the IoE-Fog-6G scenario. It receives coordinates of the tracked objects (*obj_location*) and calculates the PTZ configuration, which is sent to the Camera Ctrl using the workload, camera control (*cam_ctrl*). The User Interface is always located in the cloud and it receives a video stream of the tracked objects (*detected_obj*) from the Obj Detector. Each application contains one or more *application loop*, which is defined as a series of modules (a tuple) to measure the end-to-end delay between the start and the end of the loop. The Smart Surveillance application contains one control loop represented by the tuple of modules (Camera, Motion Detector, Obj Detector, Obj Tracker, Camera Ctrl). The end-to-end delay of each *application loop* defined in Section 3.4 and is computed as part of the application and network performance.

Table 3 lists the configuration of each workload, specified with its source and destination modules and resource requirements. For example, Row 1 in the table shows that the workload *vid_strm* requires 1000 million instructions (MI) and 20000 bytes to be transferred from the Camera module to the Motion Detector module.

3.3. Network Infrastructure

We have defined five categories of devices, Cloud, Gateway, Fog, Edge, and Sensor/Actuator. These devices operate in different layers and accordingly the expected link latency between them varies. Table 4 lists the various types of links (L) and their defined latency (t_l) in ms. These latencies are set based on the expected 6G link latencies between the layers. For instance, the configured link latency between Cloud and Gateway is set to 100 ms while links between Gateway, Fog, and Edge are set to 2 ms because they are closer to each other. The link latency between Edges and their Sensor/Actuator is set to 1ms because they are expected to be part of the edge devices. We have deliberately used modest values for latencies compared to the expected 6G latencies to keep some levels of performance margins.

Table 4. Links latency configurations.

Link (L)	Latency (ms)
Cloud-Gateway	100
Gateway-Fog	2
Fog-Edge	2
Edge-Sensor/actuator	1

3.4. Performance Metrics

For evaluation purposes, three performance metrics are monitored: network usage, application loops end-to-end delay, and network energy consumption.

The network usage (U) is the average load on the network in bytes per second. U is computed by Equation (1) where t_l represents the latency of a link l and t_L is a set of all links latency. W_n is the network requirement of workload w and W_n is a set of all workload's network requirements. T is the total simulation time.

$$Network\ usage\ U = \frac{\sum_{t_l \in t_L, w_n \in W_n} w_n * t_l}{T} \tag{1}$$

The application's loop end-to-end delay allows us to evaluate the response time of the applications in different scenarios. For every application loop type (a), we calculate the average end-to-end delay (D_a) from the first module to the last module in a specific loop using Equation (2). $T_{s(i)}$ is the start time and $t_{e(i)}$ is the end time of loop number (i) of type a, and I is the total number of loops of type a.

$$Loop\ delay\ D_{a \in A} = \frac{\sum t_{e(i)} - t_{s(i)}}{I}, \ 0 < i < I \tag{2}$$

The network energy consumption is calculated per hour (E_h) using Equation (3) where ε is the estimated energy and U is the network usage. To estimate the network energy consumption (E_h), we used the energy estimation of a gigabyte transfer on the network from [87] Table 5 shows their forecasted energy consumption rate for the wireless access network (WAN) for 2010, 2020, and 2030. The average energy consumption of 2020 = 0.54 kWh/GB used as ε value.

$$Energy\ consumption\ E_h = 3600 * \varepsilon * U \tag{3}$$

Table 5. Estimated network energy consumption [87].

Estimated Energy Consumption	2010 (kWh/GB)	2020 (kWh/GB)	2030 (kWh/GB)
Best	5.65	0.05	0.002
Worst	14.78	1.04	0.048
Average	10.22	0.54	0.025

In addition to the network energy consumption, the estimated daily Cost of energy is also calculated based on the electricity price in Dollar per kWh for Saudi Arabia from [88] using Equation (4). β is the electricity price in dollar per kWh and E_h is the energy consumption per hour.

$$Cost\ C_d = 24 * \beta * E_h \tag{4}$$

4. Case Study 1: Smart Airport

In this section, we present and discuss our first case study (Smart Airport) including the use of IoE in smart airports and their applications, the experiment design, configuration, and results.

4.1. IoE in Smart Airports

According to the International Air Transport Association (IATA), it predicted that the number of passengers will double to 8.2 billion by 2037 [89]. This expected increase in the number of passengers will put huge pressure on the aviation industry, especially in the current infrastructure [89]. IoE will also play an important role in enhancing passenger experience and offering a great opportunity for both airlines and airports [90]. Many devices and sensors can be deployed to support the smartness in the airport such as surveillance cameras, Radio-frequency identification (RFID), various sensors (e.g., air quality sensor), wearable devices (e.g., watches), avionics devices (e.g., flight recorders), biometric devices and/or, digital regulators (e.g., electricity). Using data collected from these devices, many smart airport applications might be adopted such as Baggage Tracking, security applications, indoor navigation systems, and airport operation and administrations.

4.2. Smart Airport: Architectural Overview

In the first case study, we selected the new King Abdulaziz International Airport (KAIA), Jeddah, Saudi Arabia, for evaluation. Figure 3 shows the layout of the simulated airport including the main components of the system. The whole airport landscape is divided into small areas, where each area is covered by a gateway router that works as a fog device. This router provides a connection for all edge devices in that area. Three types of edge devices are simulated: smart camera, barcode readers (at gates), counter devices, and each of them is connected to a specific type of sensor or actuator. Figure 4 shows the detailed architectural design of the IoE-6G (a) and IoE-Fog-6G (b) scenarios. Three applications are shown Smart Surveillance, Smart Gate Control, and Smart Counter. Although both scenarios have the same physical infrastructures, the application modules placement differ in them. In the following sections, we will discuss Smart Gate Control and Smart Counter, while third application, Smart Surveillance, we explained already in the previous section.

Figure 3. King AbdulAziz International Airport (Smart Airport Layout).

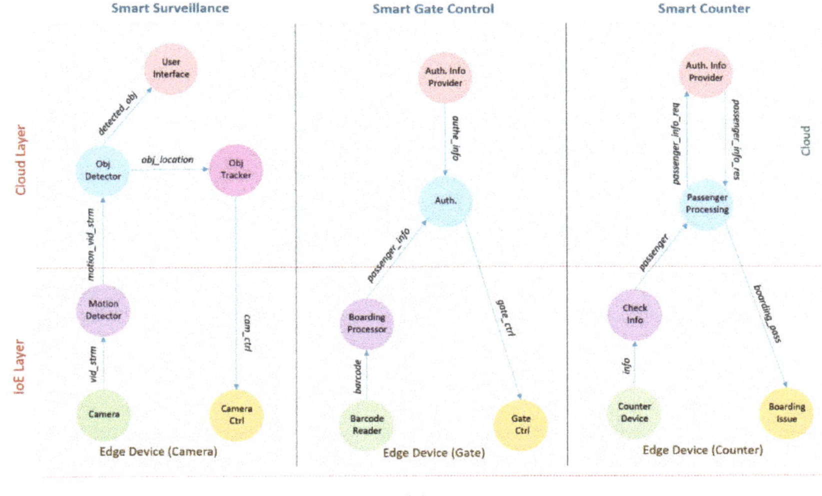

(**a**)

Figure 4. *Cont.*

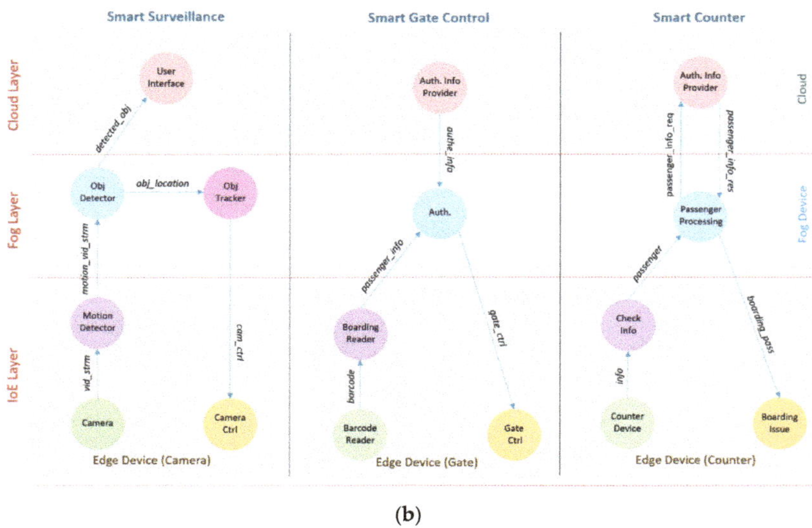

(b)

Figure 4. Smart airport: (**a**) IoE-6G scenario and (**b**) IoE-Fog-6G Ssenario.

4.3. Application: Smart Counter

The Smart Counter application is responsible for counter operation where passengers finish their check-in procedures. The application consists of five modules: Barcode Reader, Check Information (Check Info), Passenger Processing, Authentication Information Provider (Auth. Info Provider), and Boarding Issue. The Barcode Reader uses light sensors to read passports or ID cards. The Check Info module receives passenger information (*info*) from the counter and passes it to the Passenger Processing module. Passenger Processing is located in the cloud at the IoE-6G scenario, and in fog at the IoE-Fog-6G scenario. It receives passenger (*passenger*) orders from the counter and requests passenger information (*passenger_info_req*) from the Auth. Info Provider to perform the check-in process. The Auth. Info Provider will send the result back (*passenger_info_res*) to Passenger Processing module. After authentication, the boarding pass information (*boarding_pass*) will be sent to the Boarding Issue actuator. Auth. Info Provider is always located in the cloud. In the case of the IoE-Fog-6G scenario, it has an extra role, that is designed to increase the data locality and provide faster service at the smart gates. When a passenger arrives at the counter and after the check-in, the passenger authentication information (*authe_info*) will be sent to the fog node where the passenger boarding gate is located. In this way, when the passenger arrives at the gate, his information will be available at the fog which will enhance the response time of the gate. Auth. Info Provider will send periodically the passenger's data to the proper fog node, specifically to the Authenticator (Auth.) module of the Gate Control application, which will be discussed next.

4.4. Application: Smart Gate Control

The Smart Gate Control application is responsible for processing passengers boarding passes at boarding gates. The application consists of five modules: Counter Device, Boarding Processor, Authenticator (Auth.), Authentication Information Provider (Auth. Info Provider), and Gate Control (Gate Ctrl). The Barcode Reader uses light sensors to read boarding pass code. The Boarding Processor module is always located in the smart gate. It receives barcode information (*barcode*) from the Barcode Reader, and it sends the code to the Auth. for authentication. The Auth. is located in the cloud on the IoE-6G scenario, and fog node on the IoE-Fog-6G scenario. It receives passenger info (*passenger_info*) from the Boarding Processor and authenticates the passenger. After that, a decision will be sent as a control signal (*gate_ctrl*) to the Gate Ctrl, so it acts depending on that. The Auth. Info Provider is placed in the cloud and is responsible for communication with the Auth. Info Provider in the Smart Counter application.

4.5. Experiment Configurations

The main configuration parameters of the airport simulation are the number of areas, cameras, gates, and counters. As mentioned in the previous section, the whole airport is divided into areas, each with a fog device and a set of cameras, gates, and counters. KAIA main building area is around 1.2 km^2, therefore we assumed that each simulated area is around 100 m^2 and ranged our areas parameter from 100 to 1000 to cover the 1 km^2, having 10 configurations in total. Each area has two cameras (ranging from 200 to 2000 cameras), and in total 40 gates and 120 counters distributed across different areas. These configurations aim to show the architecture performance when the whole environment scales up from 360 to 2160 devices.

Table 6 lists the configurations of the Smart Airport sensors including the type of workload they generate, and the distribution of inter-arrival time. Camera has deterministic distribution as it generates workloads regularly every 5 ms while gate and counters have uniform distribution (5 to 20 ms) as they depend on the passenger arrival. Table 7 lists the various workloads alternating between modules for the Smart Gate Control and Smart Counter applications, while the Smart Surveillance workloads is listed in Table 3.

Table 6. Smart Airport: Sensors Configuration.

	Camera	Barcode Reader	Counter Device
Workload type	*vid_strm*	*barcode*	*info*
Distribution (ms)	Deterministic Distribution (5)	Uniform Distribution (5,20)	Uniform Distribution (5,20)

Table 7. Smart airport: workloads configuration.

Workload Type	Source Module	Destination Module	CPU Req. (MI)	Network Req. (Byte)
barcode	Barcode Reader	Boarding Processor	100	1000
passenger_info	Boarding Processor	Authenticator (Auth.)	2000	1000
gate_ctrl	Authenticator	Gate Ctrl	100	100
auth_info	Auth. Info Provider	Authenticator	100	100
info	Counter Device	Check Info	100	1000
passenger	Check Info	Passenger Processing	500	1000
passenger_info_req	Passenger Processing	Auth. Info Provider	1000	1000
passenger_info_res	Auth. Info Provider	Passenger Processing	1000	100
counter control	Passenger Processing	Counter Ctrl	100	500

4.6. Results and Analysis

This section will present our results of IoE-6G and IoE-Fog-6G scenarios for the 10 configurations in terms of network usage, application loop end-to-end delay, and energy consumption. Figure 5a shows the network usage in GB/s. Deploying modules on fog devices in the IoE-Fog-6G scenario decrease the volume of data sent to the cloud by 36% for 2160 devices. The difference between the two scenarios network usage increased from 13% to 36% with the increase in the number of devices which approves that IoE-Fog-6G architecture will have a greater impact when the network scale-up and will alleviate traffic jams around the datacenter.

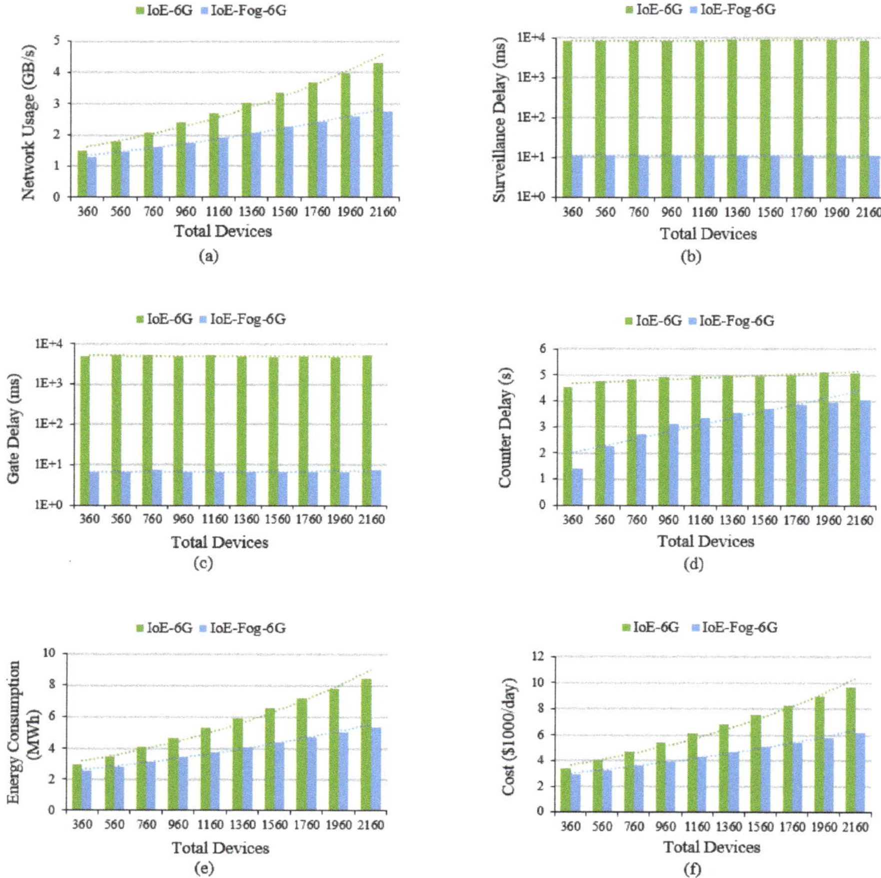

Figure 5. Smart airport case study results: (**a**) Total network usage, (**b**) Smart surveillance application average loop end-to-end delay on a log scale, (**c**) Smart gate application average loop end-to-end Delay on a log scale, (**d**) Smart counter application average loop end-to-end delay, (**e**) Network energy consumption, and (**f**) Estimated energy cost.

Figure 5b shows the average application loop end-to-end delay of the two scenarios for the Smart Surveillance control loop. The main control loop for Smart Surveillance is the tuple of modules (Camera, Motion Detector, Obj Detector, Obj Tracker, Camera Ctrl). There is a huge difference in the delay of the applications between the two scenarios. The IoE-Fog-6G provided a faster response than the IoE-6G as most of the processing is done on the edge and fog devices. In addition, because the data here is a stream of video, a huge amount of time is reduced when the unnecessary transformation is avoided. In Figure 5c, the average application loop end-to-end delay for Smart Gate is shown. The tuple (Barcode Reader, Boarding Processor, Auth., Gate Ctrl) is the main control loop for the Smart Gate. Similar to Surveillance, the delay is significantly less in the IoE-Fog-6G scenario. In addition, here we can see the benefit of proactive caching that the Smart Counter performs when it processes a passenger as the passenger authentication information is transformed into the gate fog. This guarantees the availability of authentication information before the passenger arrival which improved the loop latency. Finally, the average application loop end-to-end delay for a Smart Counter is shown Figure 5d. The tuple (Counter Device, Check Info, Passenger Processing, Auth. Info Provider, Passenger Processing,

Counter Ctrl) is the main control loop for the Smart Counter application. The result here differs from the results of Surveillance and Gate applications because of the way modules are placed in this application. As shown in Figure 4, the Auth. Info Provider module is always located in the cloud in both scenarios, so when the number of devices increases the pressure on the datacenter increase which means higher delay.

Figure 5e shows the average energy consumption for network data transfer. The energy consumption is reduced in the IoE-Fog-6G scenario by around 3 MWh with 2160 devices, and this difference is also expected to be larger with more devices joining. Similarly, Figure 5f shows the cost of energy for 2160 devices reduced by $3500 per day with fog deployment which is around $1,260,000 per year saving.

5. Case Study 2: Smart District

In this section, the second case study is presented which evaluates the effectiveness of deploying edge/fog layer to the IoE/6G network in a smart district system. This case study differs from the smart airport case study as it represents an outdoor area and uses the 6G base stations as fog devices. In the following subsections, we will discuss the use of IoE in the smart district and its applications. Then, our experiment design, configuration, and results will be presented.

5.1. IoE in Smart District

The smart district is the building block for smart cities and various applications can be involved to provide the smartness that will improve the quality of life. Challenges such as energy and utility provision, healthcare, education, transport, waste management, environment, and many others must be solved efficiently and effectively in a smart ecosystem [91]. Smart sensors and IoE devices provide real-time monitoring of the district and act inelegantly. Applications such as parking guidance systems, parking lots monitoring, parking lot reservation, parking entrance, and security management, use sensors such as infrared sensors, ultrasonic sensors, inductive loop detectors, cameras, RFID, magnetometer, and/or microwave radar [92]. For energy provisioning and management systems, smart grids might be utilized to provide a real-time monitoring and control using sensors that can read parameters such as voltage, current, power flow, and temperature [93].

5.2. Smart District: Architectural Overview

King Abdullah Economic City (KAEC) is one of the new cities in Saudi Arabia that aims to provide a new way of living, working, and playing. KAEC is around 173 km^2, so we choose one of its districts that is the Bayla Sun district which is around 4 km^2. Bayla Sun is one of the active areas at KAEC and has different facilities including, a residential area, college, parks, hotels, and a resort, a fire station, restaurants, and many others. Figure 6 shows a screenshot of Bayla Sun district from google map. The active area is divided into small areas of a size around 100 m^2. Each is supported by one fog device (6G station). Three applications of the smart district are simulated Smart Surveillance, Smart Meters, and Smart Bins. Three types of edge devices: Smart Camera, Smart Meter, and Smart Bin are considered, and each of them is connected to its specific sensor or actuator depending on the system. The IoE-6G or IoE-Fog-6G scenarios have the same physical infrastructures but different modules placemen. Figure 7 shows the detailed architectural design of both scenarios and their application modules placement. In the following subsections, we will discuss the Smart Meter and Smart Bin applications. Smart Surveillance application were already explained in the previous sections.

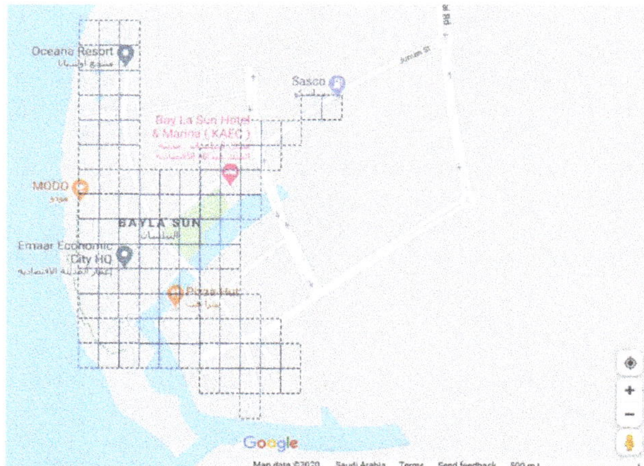

Figure 6. Simulated King Abdullah Economic City's (KAEC's) Bayla Sun District Layout.

5.3. Application: Smart Meter

The Smart Meter application is the energy management and analysis system in the district where voltage and current sensors monitor electricity usage and detect power cut in a real-time manner. This application consists of six modules: Meter, Meter Monitor, Electricity Controller (Elect Controller), Outage Notifier, User Interface, and Meter Control (Meter Ctrl). The Meter Monitor module is always located in the smart meter device. It receives reading (*meter_ reading*) from the Meter sensors and it sends the reading to the Elect Controller. It also detects an outage and sends an outage signal (*outage_ status*) to the Outage Notifier. The Electricity Controller module is the main processing module and it is located in the cloud on the IoE-6G scenario and fog node on the IoE-Fog-6G scenario. It receives status (*meter_ status*) from Meter Monitor to be evaluated and will send electricity analysis (*elect_analysis*) to the User Interface and controls (*ctrl_params*) to Meter Ctrl. The Outage Notifier module is also located in the cloud on the IoE-6G scenario and in meter (edge) on the IoE-Fog-6G scenario and it receives an outage signal (*outage_ status*) from the Meter Monitor and sends an outage signal to the local operator. The User Interface is always located in the cloud and it receives electricity analysis (*elect_analysis*) and presented to the user.

5.4. Application: Smart Bin

The Smart Bin application is the smart waste management system that optimize and monitor waste collection and recycling in a real-time manner. Smart Bins have sensors that use ultrasonic beams to sense fill-levels and type of waste, such as mixed waste, paper, glass, or metal. This application consists of six modules: Bin, Bin Monitor, Bins Coordinator (Bins Coord), Full Notifier, User Interface, and Bin Control (Bin Ctrl). The Bin Monitor module is always located in the smart bin device. It reads the bin fill-levels from the sensors and it sends the reading (*bin_ reading*) to the Bins Coord. It also detects a full bins state and sends a full signal (*full_status*) to the Full Notifier for real-time responses. The Bins Coord module is the main waste management module and it is located in the cloud on the IoE-6G scenario and fog node on the IoE-Fog-6G scenario. It receives bin status (*bin_ status*) from Bin Monitor to be analyzed and will send waste conditions (*waste_cond*) to the User Interface and controls (*ctrl_params*) to Bin Ctrl. The Full Notifier module is located in the cloud on the IoE-6G scenario and bin on the IoE-Fog-6G scenario. It receives a full signal (*full_status*) from the Bin Monitor and sends a full signal to the local operator for collection. The User interface module similar to other applications is always located in the cloud. It receives waste conditions (*waste_cond*) from the Bin Monitor and presents it to the user.

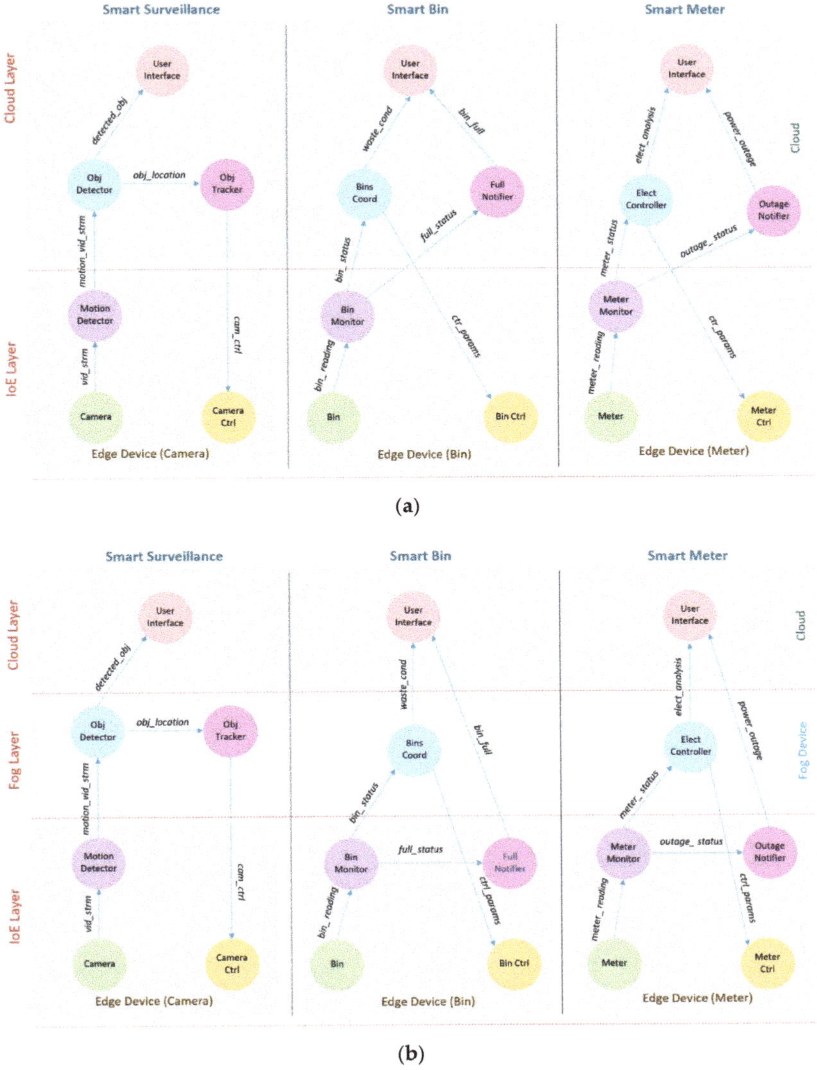

(a)

(b)

Figure 7. Smart District: (a) IoE-6G Scenario and (b) IoE-Fog-6G Scenario.

5.5. Experiment Configurations

The main configuration parameters of the district simulation are the number of areas, cameras, meters, and bins. The number of areas is fixed to 100 to represent the active areas of KAEC's Bayla Sun district as shown in Figure 6. In each area, we specify the number of cameras, meters, and bins on it. In this study, 10 configurations were also simulated. The aim here is to show the architecture performance when the granularity of IoE devices increases. Therefore, the number of cameras, meters, and bins is increased from 1 to 10 per area (fog device) which means that the total number of end devices ranges from 300 to 2100 devices. The sensors in this case study are configured to periodically generate workloads following a deterministic distribution of 5 ms. Table 8 list the properties of all workloads alternating between application modules for the two applications, while the Smart Surveillance is listed in Table 3.

Table 8. Smart District: Workloads Configuration.

Workload Type	Source Module	Destination Module	CPU Req. (MI)	Network Req. (Byte)
meter_ reading	Meter	Meter Monitor	100	500
outage_ status	Meter Monitor	Outage Notifier	500	2000
meter_ status	Meter Monitor	Elect Controller	1000	2000
elect_analysis	Elect Controller	User Interface	1000	500
ctrl_params	Elect Controller	Meter Ctrl	500	50
bin_ reading	Bin	Bin Monitor	100	500
full_status	Bin Monitor	Full Notifier	200	2000
bin_ status	Bin Monitor	Bins Coord	800	2000
waste_cond	Bins Coord	User Interface	1000	500
ctrl_params	Bins Coord	Bin Ctrl	500	50

5.6. Results and Analysis

This section discusses the Smart District results of both IoE-6G and IoE-Fog-6G scenarios for the 10 configurations simulated in terms of network usage, application loop end-to-end delay, and energy consumption. Figure 8a shows that the difference in network usage between the two scenarios remains study with the growth in the granularity of IoE devices, at an average of 38% reduction in network usage. This clearly shows the role of edge/fog deployment in reducing the pressure on the 6G network, even when the number of IoE devices grows, by performing computation near the users and avoid transferring data to data centers as much as possible.

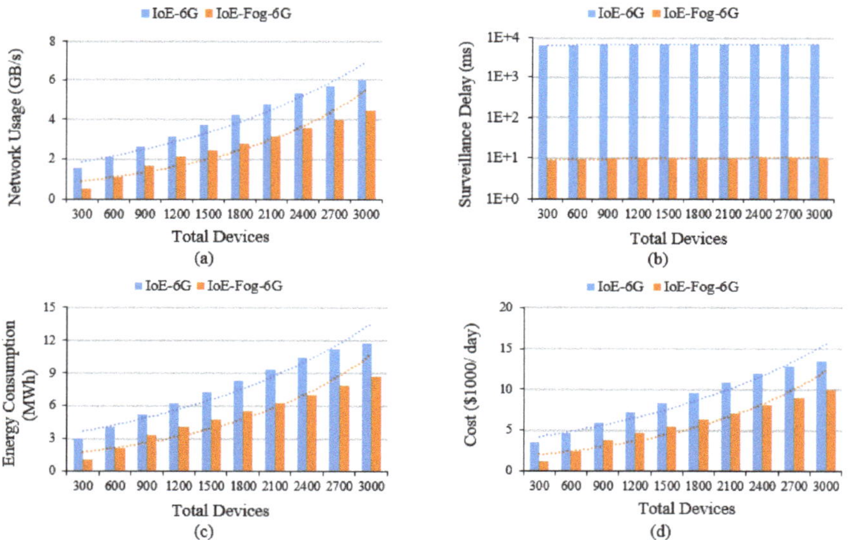

Figure 8. Smart district case study results: (**a**) Total network usage, (**b**) Smart surveillance application average loop end-to-end delay on a log scale, (**c**) Network energy consumption, and (**d**) Estimated energy cost.

We evaluated the end-to-end delay of the three applications, Smart Surveillance, Smart Meter, and Smart Bin Figure 8b shows the Smart Surveillance application result, while the other applications results are not presented here as they have a similar pattern which is due to the similarity in the module placement. All applications showed a significant reduction in the end-to-end delay of the application's main control loop due to the local analysis on the fog. Smart Surveillance delay ranged from 8 to 10 ms for IoE-Fog-6G and from 6 to 7 s for IoE-6G scenario. Smart Meter and Smart Bin delay ranged from 10 to 12 ms for IoE-Fog-6G and from 4 to 5 s for IoE-6G scenario. The slight increase in the delay is due to the growth in the granularity of IoE devices which increases the pressure on fog devices with more workloads arriving at them.

Figure 8c shows the average energy consumption of network data transfer decreased by about 3 MWh for 3000 devices when the fog is deployed in the network at the IoE-Fog-6G scenario. The cost of energy, as shown in Figure 8d, also decreased at the same rate with a total saving of around $1,280,000 per year for 3000 devices.

6. Distributed Artificial Intelligence-as-a-Service (DAIaaS)

AI is critical in embedding smartness into smart cities and societies. Due to the exponential increase in the number IoE devices, a pressing need to reduce latencies for real-time sensing and control, privacy constraints, and other challenges, the existing cloud-based AIaaS model, even with fog and edge computing support, is not sustainable. Distributed Artificial Intelligence-as-a-Service (DAIaaS) will facilitate standardization of distributed AI provisioning in smart environments, which in turn will allow developers of applications, networks, systems, etc., to focus on the domain-specific details without worrying about distributed training and inference. Eventually, it will help systemize the mass-production of technologies for smarter environments. We describe in this section DAIaaS and investigate it using four different scenarios.

6.1. DAIaaS: Architectural Overview

Figure 9 shows four DAIaaS provisioning scenarios (Scenarios A, B, C, and D) for distributed training and inference to investigate various design choices and their performance. DAIaaS comprises several modules that represent typical core operations in AI workflow, including Data Collection, Data Aggregation (Data Agg.), Data Fusion, Data Pre-Processing (Data Prep.), Model Building, and Analytics. In an AI application, data (*data*) generated from various sensors connected to the edge devices are sent to the Data Collection module. The collected data ($data_c$) is then sent by Data Collection to the Data Aggregation module that will combine them in a unified form and structure. Aggregated data ($data_{ag}$) then will be passed to the Data Fusion module where the data from various sources will be fused to reduce uncertainty and produce enhanced forms of the data. Then, fused data ($data_f$) will be pre-processed by the Data Pre-Processing module where any missing data, noises, and drift are treated, and data are reduced and transformed as necessary. Finally, Preprocessed data ($data_p$) are passed to the Model Building to train or retrain the model. When the model (*model*) is ready, the Analytics module will represent the inference step where the model is used to produce a decision or prediction as a result (*results*). We discuss next the four scenarios in detail.

6.2. Scenario A: Training/Retraining and Inference at Cloud

In the first DAIaaS scenario (Figure 9a), all the data are sent to the datacenter (clouds) to be processed. Therefore, all the AI computation and processing modules are located in the datacenter including Data Agg., Data Fusion, Data Prep., Model Building, and Analytics, except the Data Collection module that will be located on the edges to receive data from sensors. Figure 9a shows these modules, their arrangements in the two network layers, and the workloads between them as a directed graph. Table 9 lists the different workloads passing between the modules and the required resources in terms of the network and CPU computations. We will see in Section 6.5 that this scenario will facilitate a high-level of computing and storage resources, allowing the applications to run higher accuracy models on large volumes of data at the expense of higher delays.

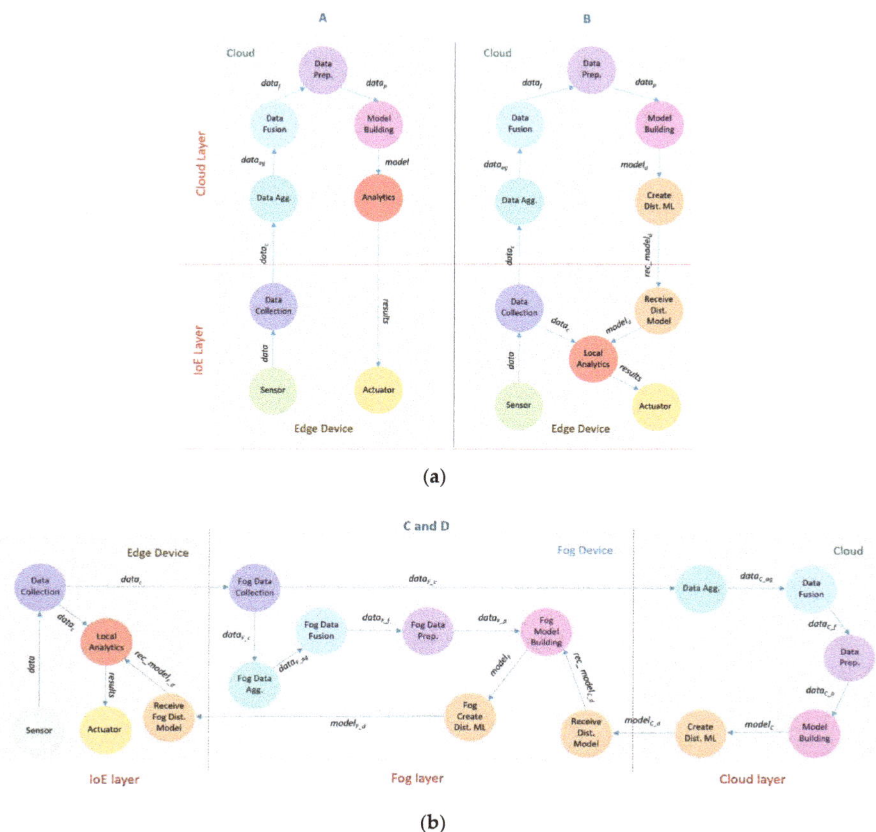

Figure 9. DAIaaS: (**a**) Scenarios A and B and (**b**) Scenarios C and D.

6.3. Scenario B: Training/Retraining at Cloud & Inference at Edge

The model is built, trained, and retrained on the cloud in the second DAIaaS scenario (see depiction in Figure 9a, and the workload configuration in Table 9), but a smaller version of the model is built and sent to the edge devices. Edges in this scenario are responsible for inference, and passing the inference results to the actuators. The cloud layer contains the modules: Data Agg., Data Fusion, Data Prep., Model Building, and Create Distributed Model (Create Dist. ML). The Create Dist. ML module is an extra module that generates a smaller version of the model called Dist. model ($model_d$) that is sent it to the edges. Techniques such as distillation, pruning, and quantization can be used to reduce the model size with minimum effect on model accuracy. The Data Collection, Local Analytics, and Receive Distributed Model (Receive Dist. Model) modules are placed at the edge. The Local Analytics module will use the model received from the cloud to generate results in a real-time manner. The Receive Dist. Model module is required to receive the distributed model (rec_model_d) from the cloud and passed to the Local Analytics.

The Data Collection module will use the Local Analytics module 90% of the time but because the edge-local model will be outdated after a while, it will offload the data to the cloud 10% of the time to retrain the model and receive a new model. We will see in Section 6.5 that this scenario will facilitate a high-level of computing and storage resources, however, the model accuracy will be affected due to smaller, somewhat outdated models running locally on edges with the advantage of faster response times due to analytics at the edge.

Table 9. DAIaaS: Workloads configuration.

Workload Type	Source Module	Destination Module	CPU Req. (MI)	Network Req. (Byte)
Scenario A				
data	Sensor	Data Collection	100	RD = 20 K
Collected data ($data_c$)	Data Collect	Data Aggregation	200	RD
Aggregated data ($data_{ag}$)	Data Aggregation	Data Fusion	100 K	DA = RD × E
Fused data ($data_f$)	Data Fusion	Data Prep.	150 K	DF = DA × 0.80
Preprocessed data ($data_p$)	Data Prep.	Model Build	150 K	DP = DF × 0.50
model	Model Build	Analytics	200 K	1 MG
results	Analytics	Actuator	100 K	1000
Scenario B				
data	Sensor	Data Collection	100	RD = 20 K
Collected data ($data_c$)	Data Collection	Data Aggregation	200	RD
Aggregated data ($data_{ag}$)	Data Aggregation	Data Fusion	100 K	DA = RD × E
Fused data ($data_f$)	Data Fusion	Data Pre-Processing	150 K	DF = DA × 0.80
Preprocessed data ($data_p$)	Data Pre-Processing	Model Building	150 K	DP = DF × 0.50
model	Model Building	Create Dist. ML	200 K	M = 1 MB
Dist. model ($model_d$)	Create Dist. ML	Receive Dist. Model	200 K	DM = M/E 5K < DM < 50K
Rec dist. model (rec_model_d)	Receive Dist. Model	Local Analytics	1000	DM
Collected data ($data_c$)	Data Collection	Local Analytics	200	RD
Results	Local Analytics	Actuator	3000	1000
Scenarios C and D				
data	Sensor	Data Collection	100	RD = 20000
Cloud collected data ($data_{C\_c}$)	Fog Data Collection	Data Aggregation	200	FDC = RD × E
Cloud aggregated data ($data_{C\_ag}$)	Data Aggregation	Data Fusion	100 K	DA = FDC × F
Cloud fused data ($data_{C\_f}$)	Data Fusion	Data Pre-Processing	150 K	DF = DA × 0.80
Cloud preprocessed data ($data_{C\_p}$)	Data Pre-Processing	Model Building	150 K	DP = DF × 0.50
Cloud model ($model_C$)	Model Building	Create Dist. ML	200 K	M = 1 MB
Cloud dist. model ($model_{C\_d}$)	Create Dist. ML	Rec. Dist. Model	200 K	DM = M/F 20K < DM < 200 K
Rec cloud model ($rec\_model_{C\_d}$)	Rec. Dist. Model	Fog Model Building	10 K	DM
Collected data ($data_c$)	Data Collection	Fog Data Collection	200	RD
Fog collected data ($data_{F\_c}$)	Fog Data Collection	Fog Data Aggregation	200	FDC = RD × E
Fog aggregated data ($data_{F\_ag}$)	Fog Data Aggregation	Fog Data Fusion	20 K	FDA = FDC × E
Fog fused data ($data_{F\_f}$)	Fog Data Fusion	Fog Data Pre-Processing	30 K	FDF = FDA × 0.80
Fog preprocessed data ($data_{F\_p}$)	Fog Data Pre-Processing	Fog Model Building	30 K	FDP = FDF × 0.50
Fog model ($model_F$)	Fog Model Building	Fog Create Dist. ML	40 K	FM = M/F 20 K < FM < 200 K
Fog dist. model ($model_{F\_d}$)	Fog Create Dist. ML	Rec. Fog Dist. Model	40 K	DFM = FM/E 5 K < DFM < 50 K
Rec fog model($rec\_model_{F\_d}$)	Rec. Fog Dist. Model	Local Analytics	1000	DFM
Collected data ($data_c$)	Data Collection	Local Analytics	200	RD
results	Local Analytics	Actuator	3000	1000

6.4. Scenarios C and D: Training/Retraining at Cloud and Fog and Inference at Edge

The scenarios C and D contain the extra (fog) layer (see Figure 9b and Table 9). The model building and retraining happens in both cloud and fog layers but the model retraining at the cloud is done less often to reduce latency. The main AI modules (Data Agg., Data Fusion, Data Prep., Model Building, and Create Dist. ML) are located in both the datacenter and the fog. There is no direct communication between the edge and the cloud. The Create Dist. ML module is required in both the cloud and fog layers to create smaller models namely Cloud dist. model ($model_{C\_d}$) and Fog dist. model ($model_{F\_d}$). These smaller models are able to fit within the available resources in their parent layers. The Receive Dist. Model modules are needed in the fog to receive the distributed model from the cloud ($rec\_model_{C\_d}$). Similarly, it is needed on the edge to receive the distributed model from the fog ($rec\_model_{F\_d}$). The edges will have the Receive Dist. Model, Data Collection, and Local Analytics modules, which will work similar to Scenario B to generate results in a real-time manner.

Figure 9b shows that both scenarios C and D have the same modules and arrangements, however, they differ in the time they are retrained on the cloud. In Scenarios C and D, the Data Collection module will use the Local Analytics module 90% of the time, as was the case in Scenario B, however, in this case, it will offload the data to the fog layer 10% of the time to retrain the model and receive a new model. In Scenarios C, the fog will retrain the model using the new data received from its edge devices 90% of the time locally and 10% of the time on cloud. In Scenario D, this split is 99% at fog and 1% at cloud. These scenarios provide relatively lower accuracy than Scenarios A and B but with the benefits of lower latencies.

6.5. Results and Analysis

We now discuss the results for the four scenarios. All four scenarios are investigated using the same number of edge devices (varying between 50, 100, up to 500). Scenarios A and B have no fog devices, while in Scenarios C and D, the number of fogs is fixed at 50. Figure 10a shows the network usage of the DAIaaS model for the four scenarios. Note that the network usage of scenario A is exponentially increasing compared to the other scenarios with the number of edges increases. This clearly shows that offering AI as services on different levels (edge and fog) will reduce the pressure on the 6G network compared to using merely Cloud AIaaS (as in scenario A). In the case of 500 edges, the usage is reduced from 6 GB/s in scenario A to 2 GB/s in scenarios B, C, and D which is a three-fold improvement.

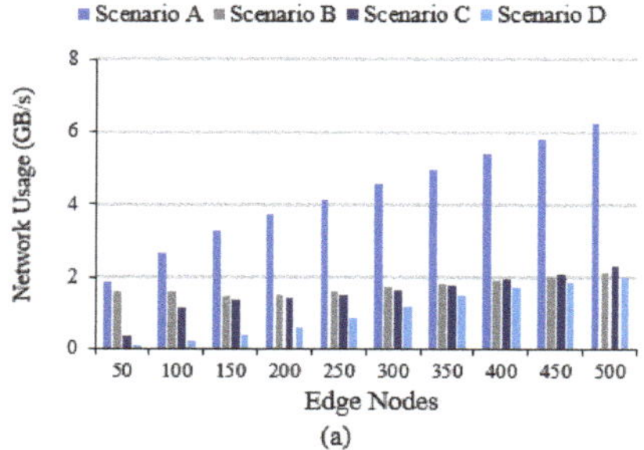

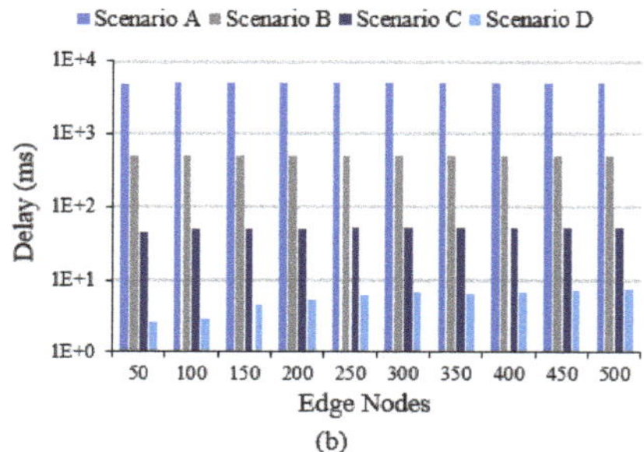

Figure 10. DAIaaS results: (**a**) Total network usage and (**b**) Average loop end-to-end delay for all requests on a log scale.

Figure 10b shows the average end-to-end delay for all AI applications in the four scenarios. The number of loops differs in each scenario, and also the frequency of each loop depends on the configuration. Scenario A has one loop, which is the tuple (Data Collection, Data Agg., Data Fusion, Data Prep., Model Building, Analytics, Actuator). Scenario B has two loops one to the cloud for model creation (Data Collection, Data Agg., Data Fusion, Data Prep., Model Building, Create Dist. ML, Receive Dist. Model) and the second in the edge to get the results (Data Collection, Local Analytics, Actuator). Scenario C and D have three loops, one to the cloud for model creation (Fog Data Collection, Data Agg., Data Fusion, Data Prep., Model Building, Create Dist. ML, Receive Dist. Model), the second to the fog, also, for model creation (Data Collection, Fog Data Collection, Fog Data Agg., Fog Data Fusion, Fog Data Prep., Fog Model Building, Fog Create Dist. ML, Receive Fog Dist. Model) and the third in the edge to get the results (Data Collection, Local Analytics, Actuator). Note in Figure 10b that the delay has been reduced significantly from Scenario A (all AI at the cloud) to other scenarios because in the other scenarios data are processed less often in the cloud. Scenario D has the lowest delay of around 8 ms (for 500 edges) as only 1% of the time data travels to the cloud while in scenario A the highest delay of around 5 s (for 500 edges) is because 100% of the time data travels to the cloud.

Figure 11a shows the average energy consumption of the network data transfer for the four scenarios. The energy consumption decreases by more than 7 MWh (for 500 edges) from scenario A at cloud (12 MWh) to scenarios B, C, and D (4 MWh). The cost of energy also decreases at the same rate, as shown in Figure 11b, with a total saving of around $3.1 million per year for 500 edges.

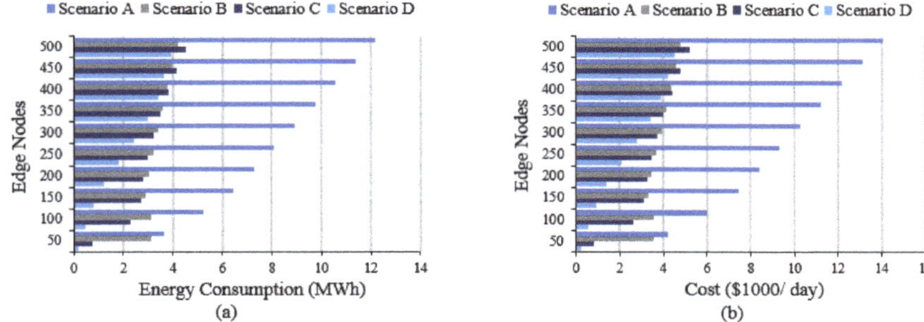

Figure 11. DAIaaS results: (**a**) Network energy consumption and (**b**) Network energy cost.

7. Conclusions and Future Work

In this paper, we proposed a framework for DAIaaS provisioning for IoE and 6G environments and evaluated it using three case studies comprising eight scenarios, nine applications and delivery models, and 50 distinct sensor and software modules. These case studies have allowed us to investigate various design choices for DAIaaS and helped to identify the performance bottlenecks. The first two case studies modelled real-life physical environments. We were able to see the benefits of various computation placement policies, allowing us to reduce the end-to-end delay, network usage, energy consumption, and annual energy cost by 99.8%, 33%, 3 MW, and 36%, on average, respectively. The third case study investigated various AI delivery models, without regard to the underlying applications. Again, we were able to identify various design choices that allowed us to reduce the end-to-end delay, network usage, energy consumption, and annual energy cost by 99%, 66%, 8 MW, and 66%, on average, respectively. We also showed that certain design choices may lead to lower performance (e.g., higher latencies) at the cost of higher AI accuracies and vice versa.

To the best of our knowledge, this is the first work where distributed AI as a service has been proposed, modeled, and investigated. This work will have a far-reaching impact on developing next-generation digital infrastructure for smarter societies by facilitating standardization of distributed AI provisioning, allowing developers to focus on the domain-specific details without worrying about distributed training and inference, and by helping systemize the mass-production of technologies for smarter environments. The future work will focus on improving the depth and breadth of the DAIaaS framework in terms of the case studies, applications, sensors, and software modules, and AI delivery models, and thereby, on developing new strategies, models, and paradigms for the provision of distributed AI services.

Author Contributions: Conceptualization, N.J. and R.M.; methodology, N.J. and R.M.; software, N.J.; validation, N.J. and R.M.; formal analysis, N.J. and R.M.; investigation, N.J. and R.M.; resources, R.M., I.K., and A.A.; data curation, N.J.; writing—original draft preparation, N.J. and R.M.; writing—review and editing, R.M., A.A., and I.K.; visualization, N.J.; supervision, R.M.; project administration, R.M., I.K., and A.A.; funding acquisition, R.M., A.A., and I.K. All authors have read and agreed to the published version of the manuscript.

Funding: This project was funded by the Deanship of Scientific Research (DSR) at King Abdulaziz University, Jeddah, under grant number RG-10-611-38. The authors, therefore, acknowledge with thanks the DSR for their technical and financial support.

Acknowledgments: The experiments reported in this paper were performed on the Aziz supercomputer at KAU.

Conflicts of Interest: The authors declare no conflict of interest.

References

1. Jespersen, L. Is AI the Answer to True Love? 2021.AI. 2018. Available online: https://2021.ai/ai-answer-true-love/ (accessed on 21 September 2020).
2. Yigitcanlar, T.; Butler, L.; Windle, E.; DeSouza, K.C.; Mehmood, R.; Corchado, J.M. Can Building "Artificially Intelligent Cities" Safeguard Humanity from Natural Disasters, Pandemics, and Other Catastrophes? An Urban Scholar's Perspective. *Sensors* **2020**, *20*, 2988. [CrossRef] [PubMed]
3. Mehmood, R.; See, S.; Katib, I.; Chlamtac, I. Smart Infrastructure and Applications: Foundations for Smarter Cities and Societies. In *EAI/Springer Innovations in Communication and Computing*; Springer International Publishing: New York, NY, USA; Springer Nature Switzerland AG: Cham, Switzerland, 2020; p. 692.
4. Bibri, S.E.; Krogstie, J. The core enabling technologies of big data analytics and context-aware computing for smart sustainable cities: A review and synthesis. *J. Big Data* **2017**, *4*, 1–50. [CrossRef]
5. Statista. Global AI Software Market Size 2018–2025. Tractica. 2020. Available online: https://www.statista.com/statistics/607716/worldwide-artificial-intelligence-market-revenues/ (accessed on 21 September 2020).
6. Alotaibi, S.; Mehmood, R.; Katib, I.; Rana, O.; Albeshri, A. Sehaa: A Big Data Analytics Tool for Healthcare Symptoms and Diseases Detection Using Twitter, Apache Spark, and Machine Learning. *Appl. Sci.* **2020**, *10*, 1398. [CrossRef]
7. Vaya, D.; Hadpawat, T. Internet of Everything (IoE): A New Era of IoT. In *Lecture Notes in Electrical Engineering*; Springer Verlag: Berlin/Heidelberg, Germany, 2020; pp. 1–6.
8. Usman, S.; Mehmood, R.; Katib, I. Big Data and HPC Convergence for Smart Infrastructures: A Review and Proposed Architecture. In *Smart Infrastructure and Applications Foundations for Smarter Cities and Societies*; Springer: Cham, Switzerland, 2020; pp. 561–586.
9. Latva-Aho, M.; Leppänen, K. Key Drivers and Research Challenges for 6G Ubiquitous Wireless Intelligence. In *6G Research Visions 1*; University of Oulu: Oulu, Finland, 2019.
10. Giordani, M.; Polese, M.; Mezzavilla, M.; Rangan, S.; Zorzi, M. Toward 6G Networks: Use Cases and Technologies. *IEEE Commun. Mag.* **2020**, *58*, 55–61. [CrossRef]
11. Khan, L.U.; Yaqoob, I.; Imran, M.; Han, Z.; Hong, C.S. 6G Wireless Systems: A Vision, Architectural Elements, and Future Directions. *IEEE Access* **2020**, *1*. [CrossRef]
12. Muhammed, T.; Albeshri, A.; Katib, I.; Mehmood, R. UbiPriSEQ: Deep Reinforcement Learning to Manage Privacy, Security, Energy, and QoS in 5G IoT HetNets. *Appl. Sci.* **2020**, *10*, 7120. [CrossRef]
13. Letaief, K.B.; Chen, W.; Shi, Y.; Zhang, J.; Zhang, Y.J.A. The Roadmap to 6G: AI Empowered Wireless Networks. *IEEE Commun. Mag.* **2019**, *57*, 84–90. [CrossRef]
14. Gui, G.; Liu, M.; Tang, F.; Kato, N.; Adachi, F. 6G: Opening New Horizons for Integration of Comfort, Security and Intelligence. *IEEE Wirel. Commun.* **2020**. [CrossRef]
15. NTT Docomo, Inc. *White Paper—5G Evolution and 6G*; NTT Docomo, Inc.: Tokyo, Japan, 2020.
16. Taleb, T.; Aguiar, R.; Yahia, I.G.B.; Christensen, G.; Chunduri, U.; Clemm, A.; Costa, X.; Dong, L.; Elmirghani, J.; Yosuf, B.; et al. *White Paper on 6G Networking*; University of Oulu: Oulu, Finland, 2020.
17. Saad, W.; Bennis, M.; Chen, M. A Vision of 6G Wireless Systems: Applications, Trends, Technologies, and Open Research Problems. *IEEE Netw.* **2020**, *34*, 134–142. [CrossRef]
18. Lovén, L.; Leppänen, T.; Peltonen, E.; Partala, J.; Harjula, E.; Porambage, P.; Ylianttila, M.; Riekki, J. EdgeAI: A vision for distributed, edge-native artificial intelligence in future 6G networks. In Proceedings of the 1st 6G Wireless Summit, Levi, Finland, 24–26 March 2019; pp. 1–2.
19. Chen, S.; Liang, Y.-C.; Sun, S.; Kang, S.; Cheng, W.; Peng, M. Vision, Requirements, and Technology Trend of 6G: How to Tackle the Challenges of System Coverage, Capacity, User Data-Rate and Movement Speed. *IEEE Wirel. Commun.* **2020**. [CrossRef]
20. Arfat, Y.; Usman, S.; Mehmood, R.; Katib, I. Big Data Tools, Technologies, and Applications: A Survey. In *Smart Infrastructure and Applications*; Springer: Cham, Switzerland, 2020; pp. 453–490.
21. Arfat, Y.; Usman, S.; Mehmood, R.; Katib, I. Big Data for Smart Infrastructure Design: Opportunities and Challenges. In *Smart Infrastructure and Applications*; Springer: Cham, Switzerland, 2020; pp. 491–518.
22. Alam, F.; Mehmood, R.; Katib, I.; Albogami, N.N.; Albeshri, A. Data Fusion and IoT for Smart Ubiquitous Environments: A Survey. *IEEE Access* **2017**, *5*, 9533–9554. [CrossRef]

23. Alam, F.; Mehmood, R.; Katib, I.; Altowaijri, S.M.; Albeshri, A. TAAWUN: A Decision Fusion and Feature Specific Road Detection Approach for Connected Autonomous Vehicles. *Mob. Networks Appl.* **2019**. [CrossRef]

24. Alomari, E.; Mehmood, R.; Katib, I. Road Traffic Event Detection Using Twitter Data, Machine Learning, and Apache Spark. In Proceedings of the 2019 IEEE SmartWorld, Ubiquitous Intelligence & Computing, Advanced & Trusted Computing, Scalable Computing & Communications, Cloud & Big Data Computing, Internet of People and Smart City Innovation (SmartWorld/SCALCOM/UIC/ATC/CBDCom/IOP/SCI), Leicester, UK, 19–23 August 2019; pp. 1888–1895.

25. Alomari, E.; Katib, I.; Mehmood, R. Iktishaf: A Big Data Road-Traffic Event Detection Tool Using Twitter and Spark Machine Learning. *Mob. Networks Appl.* **2020**. [CrossRef]

26. Shi, Z. *Advanced Artificial Intelligence*; World Scientific: Singapore, 2019.

27. Wang, S.; Ananthanarayanan, G.; Zeng, Y.; Goel, N.; Pathania, A.; Mitra, T. High-Throughput CNN Inference on Embedded ARM big.LITTLE Multi-Core Processors. *IEEE Trans. Comput. Des. Integr. Circuits Syst.* **2019**, 1. [CrossRef]

28. Mayer, R.; Jacobsen, H.A. Scalable Deep Learning on Distributed Infrastructures: Challenges, techniques, and tools. *ACM Comput. Surv.* **2020**. [CrossRef]

29. Tang, Z.; Shi, S.; Chu, X.; Wang, W.; Li, B. Communication-Efficient Distributed Deep Learning: A Comprehensive Survey. Available online: https://arxiv.org/abs/200306307 (accessed on 24 September 2020).

30. Wang, X.; Han, Y.; Leung, V.C.; Niyato, D.; Yan, X.; Chen, X. Convergence of Edge Computing and Deep Learning: A Comprehensive Survey. *IEEE Commun. Surv. Tutor.* **2020**. [CrossRef]

31. Zhou, Z.; Chen, X.; Li, E.; Zeng, L.; Luo, K.; Zhang, J. Edge Intelligence: Paving the Last Mile of Artificial Intelligence with Edge Computing. *Proc. IEEE* **2019**, *107*, 1738–1762. [CrossRef]

32. Park, J.; Samarakoon, S.; Bennis, M.; Debbah, M. Wireless Network Intelligence at the Edge. *Proc. IEEE* **2019**, *107*, 2204–2239. [CrossRef]

33. Chen, J.; Ran, X. Deep Learning with Edge Computing: A Review. *Proc. IEEE* **2019**. [CrossRef]

34. Isakov, M.; Gadepally, V.; Gettings, K.M.; Kinsy, M.A. Survey of Attacks and Defenses on Edge-Deployed Neural Networks. In Proceedings of the 2019 IEEE High Performance Extreme Computing Conference (HPEC), Boston, MA, USA, 24–26 September 2019; pp. 1–8.

35. Rausch, T.; Dustdar, S. Edge Intelligence: The Convergence of Humans, Things, and AI. In Proceedings of the 2019 IEEE International Conference on Cloud Engineering (IC2E), Prague, Czech Republic, 24–27 June 2019; pp. 86–96.

36. Marchisio, A.; Hanif, M.A.; Khalid, F.; Plastiras, G.; Kyrkou, C.; Theocharides, T.; Shafique, M. Deep Learning for Edge Computing: Current Trends, Cross-Layer Optimizations, and Open Research Challenges. In Proceedings of the 2019 IEEE Computer Society Annual Symposium on VLSI (ISVLSI), Miami, FL, USA, 15–17 July 2019; pp. 553–559.

37. Parra, G.D.L.T.; Rad, P.; Choo, K.K.R.; Beebe, N. Detecting Internet of Things attacks using distributed deep learning. *J. Netw. Comput. Appl.* **2020**, *163*, 102662. [CrossRef]

38. Li, T.; Sahu, A.K.; Talwalkar, A.; Smith, V. Federated Learning: Challenges, Methods, and Future Directions. *IEEE Signal Process. Mag.* **2020**, *37*, 50–60. [CrossRef]

39. Yang, Q.; Liu, Y.; Chen, T.; Tong, Y. Federated Machine Learning: Concept and applications. *ACM Trans. Intell. Syst. Technol.* **2019**. [CrossRef]

40. Smith, A.J.; Hollinger, G.A. Distributed inference-based multi-robot exploration. *Auton. Robot.* **2018**, *42*, 1651–1668. [CrossRef]

41. Pathak, N.; Bhandari, A.; Pathak, N.; Bhandari, A. The Artificial Intelligence 2.0 Revolution. In *IoT, AI, and Blockchain for .NET*; Apress: Berkeley, CA, USA, 2018; pp. 1–24.

42. Casati, F.; Govindarajan, K.; Jayaraman, B.; Thakur, A.; Palapudi, S.; Karakusoglu, F.; Chatterjee, D. Operating Enterprise AI as a Service. In *Lecture Notes in Computer Science (Including Subseries Lecture Notes in Artificial Intelligence and Lecture Notes in Bioinformatics)*; Springer: Berlin/Heidelberg, Germany, 2019; pp. 331–344.

43. Milton, R.; Hay, D.; Gray, S.; Buyuklieva, B.; Hudson-Smith, A. Smart IoT and Soft AI. In *IET Conference Publications*; Institution of Engineering and Technology (IET): London, UK, 2018.

44. Dialogflow. 2020. Available online: https://cloud.google.com/dialogflow (accessed on 24 September 2020).

45. Yu, J.; Zhang, P.; Chen, L.; Liu, J.; Zhang, R.; Wang, K.; An, J. Stabilizing Frame Slotted Aloha Based IoT Systems: A Geometric Ergodicity Perspective. *IEEE J. Sel. Areas Commun.* **2020**, *8716*, 1. [CrossRef]

46. Shilpa, A.; Muneeswaran, V.; Rathinam, D.D.K.; Santhiya, G.A.; Sherin, J. Exploring the Benefits of Sensors in Internet of Everything (IoE). In Proceedings of the 2019 5th International Conference on Advanced Computing & Communication Systems (ICACCS), Coimbatore, India, 15–16 March 2019; pp. 510–514.

47. Markets and Markets Blog. Smart Sensor Market. 2020. Available online: http://www.marketsandmarketsblog.com/smart-sensor-market.html (accessed on 24 September 2020).

48. Sirma, M.; Kavak, A.; Inner, B. Cloud Based IoE Connectivity Engines for The Next Generation Networks: Challenges and Architectural Overview. In Proceedings of the 2019 1st International Informatics and Software Engineering Conference (UBMYK), Ankara, Turkey, 6–7 November 2019; pp. 1–6.

49. Alsuwaidan, L. Data Management Model for Internet of Everything. In *Lecture Notes in Computer Science (including subseries Lecture Notes in Artificial Intelligence and Lecture Notes in Bioinformatics)*; Springer: Berlin/Heidelberg, Germany, 2019; pp. 331–341.

50. Lv, Z.; Kumar, N. Software defined solutions for sensors in 6G/IoE. *Comput. Commun.* **2020**, *153*, 42–47. [CrossRef]

51. Aiello, G.; Camillo, A.; Del Coco, M.; Giangreco, E.; Pinnella, M.; Pino, S.; Storelli, D. A context agnostic air quality service to exploit data in the IoE era. In Proceedings of the 2019 4th International Conference on Smart and Sustainable Technologies (SpliTech), Split, Croatia, 18–21 June 2019.

52. Badr, M.; Aboudina, M.M.; Hussien, F.A.; Mohieldin, A.N. Simultaneous Multi-Source Integrated Energy Harvesting System for IoE Applications. In Proceedings of the 2019 IEEE 62nd International Midwest Symposium on Circuits and Systems (MWSCAS), Dallas, TX, USA, 4–7 August 2019; pp. 271–274.

53. Ryoo, J.; Kim, S.; Cho, J.; Kim, H.; Tjoa, S.; DeRobertis, C. IoE Security Threats and You. In Proceedings of the 2017 International Conference on Software Security and Assurance (ICSSA), Altoona, PA, USA, 24–25 July 2017; pp. 13–19.

54. Sunyaev, A.; Sunyaev, A. Fog and Edge Computing. In *Internet Computing*; Springer International Publishing: New York, NY, USA, 2020; pp. 237–264.

55. Muhammed, T.; Mehmood, R.; Albeshri, A.; Katib, I. UbeHealth: A Personalized Ubiquitous Cloud and Edge-Enabled Networked Healthcare System for Smart Cities. *IEEE Access* **2018**, *6*, 32258–32285. [CrossRef]

56. Khan, L.U.; Yaqoob, I.; Tran, N.H.; Kazmi, S.M.A.; Dang, T.N.; Hong, C.S. Edge Computing Enabled Smart Cities: A Comprehensive Survey. *IEEE Internet Things J.* **2020**, 1. [CrossRef]

57. Negash, B.; Rahmani, A.M.; Liljeberg, P.; Jantsch, A. Fog Computing Fundamentals in the Internet-of-Things. In *Fog Computing in the Internet of Things*; Springer: Berlin/Heidelberg, Germany, 2018; pp. 3–13. [CrossRef]

58. Yi, S.; Li, C.; Li, Q. A Survey of Fog Computing. In Proceedings of the 2015 Workshop on Mobile Big Data, Association for Computing Machinery (ACM), New York, NY, USA, 22–25 June 2015; pp. 37–42.

59. Yousefpour, A.; Fung, C.; Nguyen, T.; Kadiyala, K.; Jalali, F.; Niakanlahiji, A.; Kong, J.; Jue, J.P. All one needs to know about fog computing and related edge computing paradigms: A complete survey. *J. Syst. Arch.* **2019**, *98*, 289–330. [CrossRef]

60. Bonomi, F.; Milito, R.; Zhu, J.; Addepalli, S. Fog computing and its role in the internet of things. In Proceedings of the First Edition of the MCC Workshop on Mobile Cloud Computing, Association for Computing Machinery (ACM), New York, NY, USA, 13–17 August 2012; pp. 13–16.

61. Nath, S.; Seal, A.; Banerjee, T.; Sarkar, S.K. Optimization Using Swarm Intelligence and Dynamic Graph Partitioning in IoE Infrastructure: Fog Computing and Cloud Computing. In *Communications in Computer and Information Science*; Springer: Berlin, Germany, 2017; pp. 440–452.

62. Wang, X.; Ning, Z.; Guo, S.; Wang, L. Imitation Learning Enabled Task Scheduling for Online Vehicular Edge Computing. *IEEE Trans. Mob. Comput.* **2020**, 1. [CrossRef]

63. Badii, C.; Bellini, P.; DiFino, A.; Nesi, P. Sii-Mobility: An IoT/IoE Architecture to Enhance Smart City Mobility and Transportation Services. *Sensors* **2019**, *19*, 1. [CrossRef] [PubMed]

64. Tammemäe, K.; Jantsch, A.; Kuusik, A.; Preden, J.S.; Õunapuu, E. Self-Aware Fog Computing in Private and Secure Spheres. In *Fog Computing in the Internet of Things: Inteliligence at the Edge*; Springer International Publishing: New York, NY, USA, 2017; pp. 71–99.

65. Aqib, M.; Mehmood, R.; Alzahrani, A.; Katib, I.; Albeshri, A.; Altowaijri, S.M. Rapid Transit Systems: Smarter Urban Planning Using Big Data, In-Memory Computing, Deep Learning, and GPUs. *Sustainability* **2019**, *11*, 2736. [CrossRef]

66. Mehmood, R.; Meriton, R.; Graham, G.; Hennelly, P.; Kumar, M. Exploring the influence of big data on city transport operations: A Markovian approach. *Int. J. Oper. Prod. Manag.* **2017**, *37*, 75–104. [CrossRef]

67. Aqib, M.; Mehmood, R.; Alzahrani, A.; Katib, I.; Albeshri, A.; Altowaijri, S.M. Smarter Traffic Prediction Using Big Data, In-Memory Computing, Deep Learning and GPUs. *Sensors* **2019**, *19*, 2206. [CrossRef]

68. Mehmood, R.; Alam, F.; Albogami, N.N.; Katib, I.; Albeshri, A.; Altowaijri, S.M. UTiLearn: A Personalised Ubiquitous Teaching and Learning System for Smart Societies. *IEEE Access* **2017**, *5*, 2615–2635. [CrossRef]

69. Suma, S.; Mehmood, R.; Albeshri, A. Automatic Detection and Validation of Smart City Events Using HPC and Apache Spark Platforms. In *Smart Infrastructure and Applications: Foundations for Smarter Cities and Societies*; Springer: Berlin/Heidelberg, Germany, 2020; pp. 55–78.

70. Alomari, E.; Mehmood, R. Analysis of Tweets in Arabic Language for Detection of Road Traffic Conditions. In *Lecture Notes of the Institute for Computer Sciences, Social Informatics and Telecommunications Engineering, LNICST*; Springer: Cham, Switzerland, 2018; pp. 98–110.

71. Arfat, Y.; Suma, S.; Mehmood, R.; Albeshri, A. Parallel Shortest Path Big Data Graph Computations of US Road Network Using Apache Spark: Survey, Architecture, and Evaluation. In *Smart Infrastructure and Applications Foundations for Smarter Cities and Societies*; Springer: Cham, Switzerland, 2020; pp. 185–214.

72. Bosaeed, S.; Katib, I.; Mehmood, R. *A Fog-Augmented Machine Learning based SMS Spam Detection and Classification System*; Institute of Electrical and Electronics Engineers (IEEE): Piscataway, NJ, USA, 2020; pp. 325–330.

73. Usman, S.; Mehmood, R.; Katib, I.; Albeshri, A. ZAKI+: A Machine Learning Based Process Mapping Tool for SpMV Computations on Distributed Memory Architectures. *IEEE Access* **2019**, *7*, 81279–81296. [CrossRef]

74. Usman, S.; Mehmood, R.; Katib, I.; Albeshri, A.; Altowaijri, S.M. ZAKI: A Smart Method and Tool for Automatic Performance Optimization of Parallel SpMV Computations on Distributed Memory Machines. *Mob. Networks Appl.* **2019**. [CrossRef]

75. Ahmad, N.; Mehmood, R. Enterprise Systems for Networked Smart Cities. In *Smart Infrastructure and Applications: Foundations for Smarter Cities and Societies*; Springer: Cham, Switzerland; pp. 1–33.

76. Kuchta, R.N.R.; Novotný, R.; Kuchta, R.; Kadlec, J. Smart City Concept, Applications and Services. *J. Telecommun. Syst. Manag.* **2014**, *3*, 1–8. [CrossRef]

77. Ahad, M.A.; Tripathi, G.; Agarwal, P. Learning analytics for IoE based educational model using deep learning techniques: Architecture, challenges and applications. *Smart Learn. Environ.* **2018**, *5*, 1–16. [CrossRef]

78. Al-dhubhani, R.; Al Shehri, W.; Mehmood, R.; Katib, I.; Algarni, A.; Altowaijri, S. Smarter Border Security: A Technology Perspective. In Proceedings of the 1st International Symposium on Land and Maritime Border Security and Safety, Jeddah, Saudi Arabia, 15–19 October 2017; pp. 131–143.

79. Queralta, J.P.; Gia, T.N.; Tenhunen, H.; Westerlund, T. Collaborative Mapping with IoE-based Heterogeneous Vehicles for Enhanced Situational Awareness. In *SAS 2019 IEEE Sensors Applications Symposium Conference Proceedings, LNCST*; Institute of Electrical and Electronics Engineers (IEEE): Piscataway, NJ, USA, 2019; pp. 1–6. [CrossRef]

80. Alam, F.; Mehmood, R.; Katib, I. D2TFRS: An Object Recognition Method for Autonomous Vehicles Based on RGB and Spatial Values of Pixels. In *Lecture Notes of the Institute for Computer Sciences, Social Informatics and Telecommunications Engineering, LNICST*; Springer: Cham, Switzerland, 2018; Volume 224, pp. 155–168.

81. Mehmood, R.; Bhaduri, B.; Katib, I.; Chlamtac, I. *Smart Societies, Infrastructure, Technologies and Applications. Lecture Notes of the Institute for Computer Sciences, Social Informatics and Telecommunications Engineering (LNICST)*; Springer: Berlin, Germany, 2018; p. 367.

82. Alomari, E.; Mehmood, R.; Katib, I. Sentiment Analysis of Arabic Tweets for Road Traffic Congestion and Event Detection. In *Smart Infrastructure and Applications: Foundations for Smarter Cities and Societies*; Springer International Publishing: Cham, Switzerland, 2020; pp. 37–54.

83. Aqib, M.; Mehmood, R.; Alzahrani, A.; Katib, I. A Smart Disaster Management System for Future Cities Using Deep Learning, GPUs, and In-Memory Computing. In *Smart Infrastructure and Applications. EAI/Springer Innovations in Communication and Computing*; Springer: Cham, Switzerland, 2020; pp. 159–184.

84. Mehmood, R.; Graham, G. Big Data Logistics: A health-care Transport Capacity Sharing Model. *Procedia Comput. Sci.* **2015**, *64*, 1107–1114. [CrossRef]

85. Zhang, Z.; Xiao, Y.; Ma, Z.; Xiao, M.; Ding, Z.; Lei, X.; Karagiannidis, G.K.; Fan, P. 6G Wireless Networks: Vision, Requirements, Architecture, and Key Technologies. *IEEE Veh. Technol. Mag.* **2019**, *14*, 28–41. [CrossRef]

86. Gupta, H.; Dastjerdi, A.V.; Ghosh, S.K.; Buyya, R. iFogSim: A toolkit for modeling and simulation of resource management techniques in the Internet of Things, Edge and Fog computing environments. *Software Pr. Exp.* **2017**, *47*, 1275–1296. [CrossRef]

87. Andrae, A.S.G.; Edler, T. On Global Electricity Usage of Communication Technology: Trends to 2030. *Challenges* **2015**, *6*, 117–157. [CrossRef]

88. Global Petrol Prices. 2019. Saudi Arabia Electricity Price. Available online: http://www.efficiency-from-germany.info/ENEFF/Redaktion/DE/Downloads/Publikationen/Zielmarktanalysen/marktanalyse_saudi_arabien_2011_gebaeude.pdf?__blob=publicationFile&v=4 (accessed on 24 September 2020).

89. International Air Transport Association (IATA). IATA Forecast Predicts 8.2 Billion Air Travelers in 2037. IATA Press Release No. 62. 2018. Available online: https://www.iata.org/pressroom/pr/Pages/2018-10-24-02.aspx (accessed on 24 September 2020).

90. Karakus, G.; Karşıgil, E.; Polat, L. The Role of IoT on Production of Services: A Research on Aviation Industry. In Proceedings of the International Symposium for Production Research, Vienna, Austria, 28–30 August 2018; pp. 503–511.

91. Lazaroiu, C.; Roscia, M. *Smart District through IoT and Blockchain*; Institute of Electrical and Electronics Engineers (IEEE): Piscataway, NJ, USA, 2017; pp. 454–461.

92. Paidi, V.; Fleyeh, H.; Håkansson, J.; Nyberg, R.G. Smart parking sensors, technologies and applications for open parking lots: A review. *IET Intell. Transp. Syst.* **2018**, *12*, 735–741. [CrossRef]

93. Song, E.Y.; Fitzpatrick, G.J.; Lee, K.B. Smart Sensors and Standard-Based Interoperability in Smart Grids. *IEEE Sens. J.* **2017**, *17*, 7723–7730. [CrossRef] [PubMed]

MDPI

St. Alban-Anlage 66

4052 Basel

Switzerland

Tel. +41 61 683 77 34

Fax +41 61 302 89 18

www.mdpi.com

Sensors Editorial Office

E-mail: sensors@mdpi.com

www.mdpi.com/journal/sensors